Environmental Science and Engineering

Environmental Engineering

Series editors

Ulrich Förstner, Hamburg, Germany
Robert J. Murphy, Tampa, USA
W. H. Rulkens, Wageningen, The Netherlands

For further volumes:
http://www.springer.com/series/3172

Leonardo Di G. Sigalotti · Jaime Klapp
Eloy Sira
Editors

Computational and Experimental Fluid Mechanics with Applications to Physics, Engineering and the Environment

Springer

Editors
Leonardo Di G. Sigalotti
Eloy Sira
Centro de Física
Instituto Venezolano de Investigaciones Científicas
San Antonio de Los Altos, Estado Miranda
Venezuela

Jaime Klapp
Instituto Nacional de Investigaciones Nucleares and Cinvestav-Abacus
La Marquesa, Estado de México
Mexico

ISSN 1431-2492
ISBN 978-3-319-00190-6 ISBN 978-3-319-00191-3 (eBook)
DOI 10.1007/978-3-319-00191-3
Springer Cham Heidelberg New York Dordrecht London

Library of Congress Control Number: 2013955901

Printed on acid-free paper

Springer is part of Springer Science+Business Media (www.springer.com)

Preface

Fluid flows occur everywhere in nature and occupy a relevant place in our technological world as well as in the running of a vast number of industrial processes. They are not only essential to life, but also to understand fundamental physical processes at all measurable scales, from the nanometric world to the cosmological scales. The principles of fluid mechanics are used in almost every form of mechanical and chemical engineering, with far-reaching effects on the technological advances that lead to the multitude of products which determine the high standard of living that nowadays we take for granted. Fluid flows are also known to be at the heart of health, biological, and environmental sciences, including the flows in the human body and its energy supply, the multitude of flows in the entire fauna and flora, and the atmospheric flow processes, which influence the weather and the climate. Thus, fluid flows are vital and their understanding is an essential part of the general education of humans.

This book presents a collection of papers dealing with recent advances in computational and experimental fluid mechanics with applications to physics and engineering. Among these papers, a few ones are reviews outlining the impact of fluid mechanics on important active research areas such as weather prediction and climate change, cancer research, and cosmology. The present collection includes research work presented at the I Workshop of the Venezuelan Society of Fluid Mechanics, held in the *Margarita Island,* Venezuela, on November 5–9, 2012 under the auspices of the Instituto Venezolano de Investigaciones Científicas, IVIC, and the Fondo Nacional de Ciencia, Innovación y Tecnología, FONACIT, of Venezuela. The book begins with invited lectures held during the Workshop by renowned national and international scientists and engineers, covering a wide range of topics, followed by a number of invited seminars presented by young researchers and graduate students working actively in the field of fluid mechanics and related areas.

The I Workshop of the Venezuelan Society of Fluid Mechanics represented a unique opportunity to provide a forum for the presentation of state-of-the-art research in theoretical, experimental, and applied fluid mechanics oriented to engineering technology, where scientists, coming from different universities and research institutions of the country, together with mechanical, chemical, and petroleum engineers from public and private enterprises, with a huge experience in applied industrial problems, have participated in fruitful discussions on fundamental and technical aspects, paving the way for future collaborations.

The Workshop will be organized every 2 years. The 5 days of oral sessions accommodated 45 talks and had close to 60 attendees with 10 international and 20 national researchers, and more than 30 graduate and undergraduate students. The wide variety of topics presented included free-surface and interface flows, such as drops and bubbles, turbulent flows, multiphase flows with applications to biological and oil extraction systems, shock structure and acoustic waves, opto-fluids, granular fluids, astrophysical and cosmological flows, and computational fluid dynamics. Among the renowned researchers, Joseph J. Niemela, from The Abdus Salam International Centre for Theoretical Physics, ICTP, Trieste, Italy, showed the results of controlled laboratory experiments of turbulent diffusion of heat at high Rayleigh numbers; Dominique Legendre, from the Institut de Mécanique des Fluides de Toulouse, IMFT, Toulouse, France, presented numerical simulations of sliding drops on an inclined solid surface; Catalina Stern-Forgach, from the Department of Physics of the Universidad Nacional Autónoma de México, UNAM, Mexico, described the results of experimental measurements of shock structure and acoustic waves inside a supersonic jet; and José R. Castrejón-Pita, coming from the Department of Engineering of the University of Cambridge, Cambridge, United Kingdom, spoke of the relevance of the breakup of liquid surfaces to industry and discussed current issues faced by researchers working in the field of droplet dynamics. Interesting lectures on bubble growth in viscous liquids were given by Abraham Medina and Abel López-Villa, both from the Instituto Politécnico Nacional (I.P.N.) of Mexico, while Julián Chela-Flores, from The Abdus Salam International Centre for Theoretical Physics, ICTP, Trieste, Italy, gave a magisterial conference on how fluid mechanics is playing a major role in space exploration for understanding the cosmic distribution of life. The theoretical physics of granular fluids and an introductory view of the jamming transition problem were given by Leonardo Trujillo, from the IVIC's Centre of Physics. Other interesting talks were presented by Humberto Cabrera, from the IVIC's Department of Applied Physics, on the Soret effect in binary fluid mixtures; by Luis R. Rojas-Solórzano, from the Department of Energy Conversion and Transport of the Universidad Simón Bolívar, USB, Caracas, Venezuela, who described a multiphase approach to model blood flow in micro-tubes; and Miguel R. Paiva-Rojas, from the Refining and Industrialization Department of the Instituto Tecnológico Venezolano del Petróleo, PDVSA-Intevep, Los Teques, Venezuela, who spoke on the estimation of the gas–liquid–solid phase distribution in a cold slurry bubble column system for hydro-conversion processes. Other local speakers gave short oral presentations on computational and experimental drop dynamics, compositional flows applied to the oil industry, granular and porous media flows, and astrophysical flows.

The short oral presentations were organized by themes: Drops, Particles, and Waves; Multiphase and Multicomponent Flow, Granular and Porous Media Flow; and Astrophysical and Relativistic Flow. The book is aimed to undergraduate and graduate students, as well as to physicists, chemists, and engineers dealing with fluid mechanics from both the experimental and theoretical point of view. The material is also adequate for both teaching and research. The invited lectures and the other selected contributions are introductory and use a minimum of mathematics.

The editors are deeply indebted to the several institutions that made possible the realization of the I Workshop of the Venezuelan Society of Fluid Mechanics. In particular, we thank the Instituto Venezolano de Investigaciones Científicas, IVIC, and the Fondo Nacional de Ciencia, Innovación y Tecnología, FONACIT, of Venezuela for providing financial support. We are also grateful to the Instituto Tecnológico Venezolano del Petróleo (PDVSA-Intevep), the Centro de Investigaciones de Astronomía Francisco José Duarte, CIDA, the FUNDACITE-Miranda, and the Mexican institutions: Consejo Nacional de Ciencia y Tecnología, CONACYT, Consejo Mexiquense de Ciencia y Tecnología, COMECYT, Instituto Nacional de Investigaciones Nucleares, ININ, and Cinvestav-Abacus of the Instituto Politécnico Nacional, I.P.N.

We also acknowledge the Organizing Committee of the I Workshop of the Venezuelan Society of Fluid Mechanics: Armando Blanco, Leonardo Trujillo, Jorge Troconis, Eric Plaza, Franklin Peña-Polo, and Joselyn Sequera for their invaluable contribution to the final manuscript.

Caracas, June 2013

Leonardo Di G. Sigalotti
Jaime Klapp
Eloy Sira

Acknowledgments

The production of this book has been sponsored by the Instituto Venezolano de Investigaciones Científicas (IVIC), the Fondo Nacional de Ciencia, Innovación y Tecnología (FONACIT) of Venezuela, the Sociedad Venezolana de Mecánica de Fluidos (SVMF), the Instituto Nacional de Investigaciones Nucleares (ININ), and Cinvestav-Abacus of the Instituto Politécnico Nacional (I.P.N.).

Contents

Contributors

Alejandro Acevedo-Malavé Centro Multidisciplinario de Ciencias, Instituto Venezolano de Investigaciones Científicas, Mérida 5101, Estado Mérida, Venezuela, e-mail: alaceved@ivic.gob.ve

José M. Alvarado-Reyes Departamento de Física, Facultad de Ciencias, Universidad Nacional Autónoma de México, México D.F., ; Circuito Exterior S/N, Ciudad Universitaria, CO 04510 México D.F., Mexico, e-mail: mar_ney2003@yahoo.com.mx

Carlos E. Alvarado-Rodríguez Departamento de Física, Instituto Nacional de Investigaciones Nucleares, Km 36.5, Carretera México-Toluca, 52750 La Marquesa, Estado de México, Mexico; División de Ciencias Naturales y Exactas, Universidad de Guanajuato, Campus Guanajuato, Noria Alta S/N, Guanajuato, Mexico, e-mail: q_l_o@hotmail.com

L. Álvarez Escuela de Ingeniería y Ciencias Aplicadas, Universidad de Oriente, Núcleo de Anzoátegui, Avenida Argimiro Gabaldón, Municipio Bolívar, Barcelona, Estado Anzoátegui, Venezuela, e-mail: lalvarezm57@gmail.com

Jorge Ancheyta Laboratorio de Síntesis e Hidrotratamiento, Instituto Mexicano del Petróleo, Eje Central Lázaro Cárdenas Norte 152, Colonia San Bartolo Atepehuacan, 07730 México D.F, Mexico, e-mail: jancheyt@imp.mx

Efrén D. J. Andrades UBIP Project, Sección de Biotecnología, Instituto de Investigaciones, Escuela de Farmacia, Facultad de Farmacia y Bioanálisis, Universidad de Los Andes, Mérida, Estado Mérida, Venezuela, e-mail: andradesefren@cantv.net

Silvia E. Andrades-Grassi UBIP Project, Sección de Biotecnología, Instituto de Investigaciones, Escuela de Farmacia, Facultad de Farmacia y Bioanálisis, Universidad de Los Andes, Mérida, Estado Mérida, Venezuela, e-mail: silvigrassi@gmail.com

Victor J. Andrades-Grassi UBIP Project, Sección de Biotecnología, Instituto de Investigaciones, Escuela de Farmacia, Facultad de Farmacia y Bioanálisis, Universidad de Los Andes, Mérida, Estado Mérida, Venezuela, e-mail: vicandrades@gmail.com

Carlos Araujo Facultad de Ingeniería, Universidad del Zulia, Avenida Goajira, Núcleo Técnico, Maracaibo, Estado Zulia, Venezuela, e-mail: cdaraujo@fing.luz.edu.ve

Solange C. Araujo Refining and Industrialization Department, Instituto Tecnológico Venezolano del Petróleo (PDVSA-Intevep), Urbanización El Tambor, Los Teques 1201, Estado Miranda, Venezuela, e-mail: araujosc@pdvsa.com

Yessica Arellano Instituto Tecnológico Venezolano del Petróleo (PDVSA-Intevep), Urbanización El Tambor, Los Teques 1201, Estado Miranda, Venezuela, e-mail: arellanoy@pdvsa.com

Guillermo Arreaga-García Departamento de Investigación de la Universidad de Sonora, Apartado Postal 14740, C.P. 83000 Hermosillo, Sonora, Mexico, e-mail: garreaga@cifus.uson.mx

Dimas C. Belisario Dipartimento di Oncologia, Università degli Studi di Torino, Via Santena 5/BIS, Torino 10126, Italy, e-mail: caro.belisario@gmail.com

Armando Blanco Departamento de Mecánica, Universidad Simón Bolívar, Caracas 1080, Venezuela, e-mail: ajblanco@usb.ve

William Blanco Instituto Tecnológico Venezolano del Petróleo (PDVSA-Intevep), Urbanización El Tambor, Los Teques 1201, Estado Miranda, Venezuela, e-mail: blancowa@pdvsa.com

Adriana Brito Instituto Tecnológico Venezolano del Petróleo (PDVSA-Intevep), Urbanización El Tambor, Los Teques 1201, Estado Miranda, Venezuela, e-mail: britoah@pdvsa.com

Joaquín L. Brito Laboratorio de Fisicoquímica de Superficies, Centro de Química, Instituto Venezolano de Investigaciones Científicas, Apartado Postal 20632, Caracas 1020-A, Venezuela, e-mail: joabrito@ivic.gob.ve

Ramón Cabello Instituto Tecnológico Venezolano del Petróleo (PDVSA-Intevep), Urbanización El Tambor, Los Teques 1201, Estado Miranda, Venezuela, e-mail: cabellor@pdvsa.com

Humberto Cabrera Departamento de Física Aplicada, Laboratorio de Óptica Aplicada, Instituto Venezolano de Investigaciones Científicas, Loma de Los Guamos, Vía Jají, Mérida, Estado Mérida, Venezuela; The Abdus Salam International Centre for Theoretical Physics, Strada Costiera 11, 34014 Trieste, Italy, e-mail: cabrera_25@yahoo.com

José R. Castrejón-Pita Department of Engineering, University of Cambridge, 17 Charles Babbage Road, Cambridge CB3 0FS, UK, e-mail: jrc64@cam.ac.uk

Jorge L. Cervantes-Cota Departamento de Física, Instituto Nacional de Investigaciones Nucleares, Km 36.5, Carretera México-Toluca, 52750 La Marquesa, Estado de México, Mexico, e-mail: jorge.cervantes@inin.gob.mx

Julián Chela-Flores The Abdus Salam International Centre for Theoretical Physics, Strada Costiera 11, 34014 Trieste, Italy; Fundación Instituto de Estudios Avanzados, Carretera Nacional Hoyo de la Puerta, Valle de Sartenejas, Baruta, Caracas 1015-A, Estado Miranda, Venezuela, e-mail: chelaf@ictp.it

José Colmenares Instituto Tecnológico Venezolano del Petróleo (PDVSA-Intevep), Urbanización El Tambor, Los Teques 1201, Estado Miranda, Venezuela, e-mail: jcolmenares@cantv.net

Virginia Contreras-Andrade Escuela de Ingeniería Química, Universidad de Los Andes, Mérida 5101, Estado Mérida, Venezuela, e-mail: virginia.coa@gmail.com

Eduardo de la Cruz-Sánchez Departamento de Física, Instituto Nacional de Investigaciones Nucleares, Km 36.5, Carretera México-Toluca, 52750 La Marquesa, Estado de México, Mexico, e-mail: eduardo.delacruz@inin.gob.mx

Alberto A. Díaz Centro de Física, Instituto Venezolano de Investigaciones Científicas, Apartado Postal 20632, Caracas 1020-A, Venezuela, e-mail: aadv777@gmail.com

Nelson Fernández Escuela de Ingeniería Mecánica, Instituto de Mecánica de Fluidos, Universidad Central de Venezuela, Caracas 1051, Venezuela, e-mail: nelson.fernandez@ucv.ve

Francisco García Escuela de Ingeniería Mecánica, Instituto de Mecánica de Fluidos, Universidad Central de Venezuela, Caracas 1051, Venezuela

Juan J. González Facultad de Ingeniería, Universidad del Zulia, Avenida Goajira, Núcleo Técnico, Maracaibo, Estado Zulia, Venezuela, e-mail: gonzalezlopezjj@gmail.com

Hilda C. Grassi UBIP Project, Sección de Biotecnología, Instituto de Investigaciones, Escuela de Farmacia, Facultad de Farmacia y Bioanálisis, Universidad de Los Andes, Mérida, Estado Mérida, Venezuela, e-mail: cristinagrassi@gmail.com

Jordan A. Guerra Physics Department, School of Arts and Science, The Catholic University of America, 620 Michigan Avenue, Nebraska, Washington, D.C. 20064, USA, e-mail: jordan.guerra@gmail.com

Ian M. Hutchings Department of Engineering, University of Cambridge, 17 Charles Babbage Road, Cambridge CB3 0FS, UK, e-mail: imh2@cam.ac.uk

Jaime Klapp Departamento de Física, Instituto Nacional de Investigaciones Nucleares, Km 36.5, Carretera México-Toluca, 52750 La Marquesa, Estado de México, Mexico; Departamento de Matemáticas, Cinvestav del Instituto Politécnico Nacional, 07360 México D.F., Mexico, e-mail: jaime.klapp@inin.gob.mx

Mauricio A. Labarca Facultad de Ingeniería, Universidad del Zulia, Avenida Goajira, Núcleo Técnico, Maracaibo, Estado Zulia, Venezuela, e-mail: mauriciolabarca4tw@gmail.com

Dominique Legendre Institut de Mécanique des Fluides de Toulouse, Université de Toulouse, CNRS-INPT-UPS, 2 Allée du Professeur Camille Soula, 31400 Toulouse, France, e-mail: dominique.legendre@imft.fr

Abel López-Villa SEPI ESIME Azcapotzalco, Instituto Politécnico Nacional, Avenida de las Granjas 682, Colonia Santa Catarina, Azcapotzalco, México D.F. 02250, Mexico, e-mail: abelvilla77@hotmail.com

Nelson MacQuhae Escuela de Ingeniería Mecánica, Instituto de Mecánica de Fluidos, Universidad Central de Venezuela, Caracas 1051, Venezuela, e-mail: nelsonm@microsoft.com

Marco Maglio Institut de Mécanique des Fluides de Toulouse, Université de Toulouse, CNRS-INPT-UPS, 2 Allée du Professeur Camille Soula, 31400 Toulouse, France, e-mail: marco.maglio@imft.fr

Juan F. Marín Centro de Física, Instituto Venezolano de Investigaciones Científicas, Apartado Postal 20632, Caracas 1020-A, Venezuela; Escuela de Física, Facultad de Ciencias, Universidad Central de Venezuela, Apartado Postal 20513, Avenida Los Ilustres, Los Chaguaramos, Caracas 1020-A, Venezuela, e-mail: juanfmarinm@gmail.com

Tonatiuh Matos Departamento de Física, Centro de Investigación y de Estudios Avanzados (Cinvestav) del Instituto Politécnico Nacional, Apartado Postal 14-740, 07000 México D.F., Mexico, e-mail: tmatos@fis.cinvestav.mx

Estela Mayoral Departamento de Física, Instituto Nacional de Investigaciones Nucleares, Km 36.5, Carretera México-Toluca, 52750 La Marquesa, Estado de México, Mexico, e-mail: estela.mayoral@inin.gob.mx

Abraham Medina SEPI ESIME Azcapotzalco, Instituto Politécnico Nacional, Avenida de las Granjas 682, Colonia Santa Catarina, 02250 Azcapotzalco, México D.F., Mexico, e-mail: abraham_medina_ovando@yahoo.com

Gerardo E. Medina-Ramírez UBIP Project, Sección de Biotecnología, Instituto de Investigaciones, Escuela de Farmacia, Facultad de Farmacia y Bioanálisis, Universidad de Los Andes, Mérida, Estado Mérida, Venezuela, e-mail: medinag@ula.ve

L. Mendoza Instituto Tecnológico Venezolano del Petróleo (PDVSA-Intevep), Urbanización El Tambor, Los Teques 1201, Estado Miranda, Venezuela, e-mail: mendozale@pdvsa.com

J. B. Moreno Department of Thermodynamics and Transport Phenomena, Universidad Simón Bolívar, Apartado Postal 89000, Caracas, Venezuela, e-mail: morenoj@usb.ve

T. M. Mubita Department of Thermodynamics and Transport Phenomena, Universidad Simón Bolívar, Apartado Postal 89000, Caracas, Venezuela, e-mail: 08-86884@usb.ve

Eduardo Nahmad-Achar Instituto de Ciencias Nucleares, Universidad Nacional Autónoma de México, Apartado Postal 70-543, 04510 México D.F., Mexico, e-mail: nahmad@nucleares.unam.mx

Joseph J. Niemela The Abdus Salam International Centre for Theoretical Physics, Strada Costiera 11, 34014 Trieste, Italy, e-mail: niemela@ictp.it

Erick A. Pacheco UBIP Project, Sección de Biotecnología, Instituto de Investigaciones, Escuela de Farmacia, Facultad de Farmacia y Bioanálisis, Universidad de Los Andes, Mérida, Estado Mérida, Venezuela, e-mail: alejandropc@gmail.com

Miguel V. Paiva-Rojas Refining and Industrialization Department, Instituto Tecnológico Venezolano del Petróleo (PDVSA-Intevep), Urbanización El Tambor, Los Teques 1201, Estado Miranda, Venezuela, e-mail: paivams@pdvsa.com

Franklin Peña-Polo Centro de Física, Instituto Venezolano de Investigaciones Científicas, Apartado Postal 20632, Caracas 1020-A, Venezuela, e-mail: franklin.pena@gmail.com

Luis D. Pérez-Guerra Instituto Tecnológico Venezolano del Petróleo (PDVSA-Intevep), Urbanización El Tambor, Los Teques 1201, Estado Miranda, Venezuela, e-mail: perezlu@pdvsa.com

Luis F. Pérez Centro de Física, Instituto Venezolano de Investigaciones Científicas, Apartado Postal 20632, Caracas 1020-A, Venezuela, e-mail: alfa_1100@hotmail.com

Juan C. Petit Centro de Física, Instituto Venezolano de Investigaciones Científicas, Apartado Postal 20632, Caracas 1020-A, Venezuela, e-mail: jcpetit71@gmail.com

Eric Plaza Instituto Zuliano de Investigaciones Tecnológicas, La Cañada de Urdaneta, Km. 15, Vía Palmarejo Viejo, Estado Zulia, Venezuela, e-mail: eplaza@inzit.gob.ve

José M. Ramírez Centro de Física, Instituto Venezolano de Investigaciones Científicas, Apartado Postal 20632, Caracas 1020-A, Venezuela, e-mail: josem@ivic.gob.ve

S. Ramírez Laboratorio de Síntesis e Hidrotratamiento, Instituto Mexicano del Petróleo, Eje Central Lázaro Cárdenas Norte 152, Colonia San Bartolo Atepehuacan, 07730 México D.F., Mexico, e-mail: sramire@imp.mx

Mario A. Rodríguez-Meza Departamento de Física, Instituto Nacional de Investigaciones Nucleares, Km 36.5, Carretera México-Toluca, 52750 La Marquesa, Estado de México, Mexico, e-mail: marioalberto.rodriguez@inin.gob.mx

Snell Rojas Centro de Física, Instituto Venezolano de Investigaciones Científicas, Apartado Postal 20632, Caracas 1020-A, Venezuela, e-mail: snell0808@hotmal.com

Luis R. Rojas-Solórzano Department of Energy Conversion and Transport, Universidad Simón Bolívar, Apartado Postal 89000, Caracas, Venezuela; School of Engineering, Nazarbayev University, Astana 010000, Republic of Kazakhstan, e-mail: rrojas@usb.ve

Elizabeth T. Romero-Guzmán Departamento de Física, Instituto Nacional de Investigaciones Nucleares, Km 36.5, Carretera México-Toluca, 52750 La Marquesa, Estado de México, Mexico, e-mail: elizabeth.romero@inin.gob.mx

César Ruiz-Ferrel Carretera del Departamento del D.F., Universidad Tecnológica del Valle de Toluca, Km 7.5, Santa María Atarasquillo, Lerma, Estado de México, Mexico, e-mail: cesarunder90@gmail.com

H. Salazar Escuela de Ingeniería y Ciencias Aplicadas, Núcleo de Anzoátegui, Universidad de Oriente, Avenida Argimiro Gabaldón, Municipio Bolívar, Barcelona, Estado Anzoátegui, Venezuela, e-mail: salazar.heider@gmail.com

Iván Sánchez Centro de Física, Instituto Venezolano de Investigaciones Científicas, Apartado Postal 20632, Caracas 1020-A, Venezuela, e-mail: ivanjo@gmail.com

Ledys Y. Sánchez Laboratorio de Cinética y Catálisis, Departamento de Química, Facultad de Ciencias, Universidad de Los Andes, Mérida 5101, Estado Mérida, Venezuela, e-mail: ledys@yahoo.com

Leonardo Di G. Sigalotti Centro de Física, Instituto Venezolano de Investigaciones Científicas, Apartado Postal 20632, Caracas 1020-A, Venezuela, e-mail: leonardo.sigalotti@gmail.com

Eloy Sira Centro de Física, Instituto Venezolano de Investigaciones Científicas, Apartado Postal 20632, Caracas 1020-A, Venezuela, e-mail: eloy.sira@gmail.com

Francisco Solórzano-Araujo Universidad Tecnológica del Valle de Toluca, Carretera del Departamento del D.F., Km 7.5, Santa María Atarasquillo, Lerma, Estado de México, Mexico, e-mail: frank_trov17@hotmail.com

Catalina Stern-Forgach Departamento de Física, Facultad de Ciencias, Universidad Nacional Autónoma de México, México D.F., Mexico; Circuito Exterior S/N, Ciudad Universitaria, CO 04510 México D.F., Mexico, e-mail: catalinastern@gmail.com

Jorge Troconis Centro de Física, Instituto Venezolano de Investigaciones Científicas, Apartado Postal 20632, Caracas 1020-A, Venezuela, e-mail: jtroconi@ivic.gob.ve

Orlenys Troconis Escuela de Física, Universidad Central de Venezuela, Avenida Los Ilustres, Los Chaguaramos, Apartado Postal 20513, Caracas 1020-A, Venezuela, e-mail: otroconis@fisica.ciens.ucv.ve

Jorge Trujillo Multiphase Solutions Inc. Kenny Australia, Level 3, Wood Group House, 432 Murray Street, Perth, WA 6000, Australia, e-mail: jorgenicolas71@gmail.com

Leonardo Trujillo Centro de Física, Instituto Venezolano de Investigaciones Científicas, Apartado Postal 20632, Caracas 1020-A, Venezuela; The Abdus Salam International Centre for Theoretical Physics, Strada Costiera 11, 34014 Trieste, Italy, e-mail: leonardo.trujillo@gmail.com

Hailleen Varela Centro de Física, Instituto Venezolano de Investigaciones Científicas, Apartado Postal 20632, Caracas 1020-A, Venezuela, e-mail: hailleeng3@gmail.com

Carlos R. Vera-Lagos Marketing Department, Laboratorios Servier S.A., Avenida Sanatorio del Ávila, Centro Empresarial Ciudad Center, Torre F, Piso 3, Boleíta Norte, Caracas 1071, Venezuela, e-mail: carlos.vera@ve.netgrs.com

Yanet Villasana Laboratorio de Fisicoquímica de Superficies, Centro de Química, Instituto Venezolano de Investigaciones Científicas, Apartado Postal 20632, Caracas 1020-A, Venezuela, e-mail: ytvillas@gmail.com

Miguel Zavala Departamento de Física, Instituto Nacional de Investigaciones Nucleares, Km 36.5, Carretera México-Toluca, 52750 La Marquesa, Estado de México, Mexico, e-mail: zvl_2006@yahoo.com.mx

Part I
Invited Lectures

Environmental Fluid Mechanics: Applications to Weather Forecast and Climate Change

Leonardo Di G. Sigalotti, Eloy Sira, Jaime Klapp and Leonardo Trujillo

Abstract Virtually all economic sectors as well as many public and private activities are affected in some measure by changes in weather and climate. Uncertainties in the scope and severity of these changes pose financial and social risks for individuals, businesses, and government agencies, with direct influence on food security and production, transport, health, electricity generation, and water resources. The vulnerability of human settlement to extreme weather and climate episodes is a further aspect that must be emphasized. Hence, achieving accurate weather and climate forecasts has important implications to modern society. In this chapter, we present an overview of the basic fluid-mechanical principles that govern the behaviour of weather and climate. We shall mainly focus on the numerical modelling of weather prediction and climate projections, spanning the range from the very first attempts, based on simple barotropic models, to the development of general circulation models of the atmosphere and ocean to the most recent multi-model ensemble forecasting systems.

L. Di G. Sigalotti (✉) · E. Sira · L. Trujillo
Centro de Física, Instituto Venezolano de Investigaciones Científicas, IVIC, Apartado Postal 20632, Caracas 1020-A, Venezuela
e-mail: leonardo.sigalotti@gmail.com

E. Sira
e-mail: eloy.sira@gmail.com

J. Klapp
Departamento de Física, Instituto Nacional de Investigaciones Nucleares, ININ, Km. 36.5, Carretera México-Toluca, 52750 La Marquesa, Estado de México, Mexico
e-mail: jaime.klapp@inin.gob.mx

J. Klapp
Departamento de Matemáticas, Cinvestav del Instituto Politécnico Nacional (I. P. N.), 07360 México, D. F., Mexico
e-mail: jaime.klapp@hotmail.com

L. Trujillo
The Abdus Salam International Centre for Theoretical Physics, ICTP, Strada Costiera 11, 34014 Trieste, Italy
e-mail: leonardo.trujillo@gmail.com

L. Di G. Sigalotti et al. (eds.), *Computational and Experimental Fluid Mechanics with Applications to Physics, Engineering and the Environment*, Environmental Science and Engineering,
DOI:10.1007/978-3-319-00191-3_1,

1 Introduction

There are many different kinds of naturally occurring fluid flows in the environment. Natural fluid motions are vital, and there is a general strong incentive to study them, particularly those of air in the atmosphere and of water from underground aquifers to surface flows in rivers, lakes, and oceans. Environmental concerns have encouraged interdisciplinarity to a degree that has been increasing in proportion to the acuity of the problems, giving rise to a body of knowledge that comprises several disciplines, including hydrology, meteorology, climatology, and oceanography among others. Whereas the particular objectives of each of these disciplines, such as weather forecasting in meteorology and climate change projections in climatology, encourage disciplinary segregation, environmental concerns compel experts in those disciplines to base their models on the solution of the equations of fluid dynamics.

The threat of climate change is one of the greatest challenges currently facing society. Because of the increased threats imposed by global warming and the increasing severity and occurrence of storms and natural disasters, improving our understanding of the climate system has become an international priority. In simple words, climate refers to the average of weather conditions. Descriptions of the climate generally encompass statistical information concerning the mean and variability of relevant quantities, as temperature, precipitation, and wind, over a multi-year time period. Fluctuations in the Earth system result naturally from interactions between the ocean, the atmosphere, the land, the frozen portion of the Earth's surface (or cryosphere), and the changes in the Earth's energy balance arising from volcanic eruptions and variations in the Sun's intensity. Although global warming has been accepted as incontrovertible, humans continue to alter the composition of the atmosphere, primarily through the burning of fossil fuels. The build up of greenhouse gases and trace constituents is another factor that contributes to changes in the Earth's heat energy balance. Its impact on the planet has been detected and is projected to become increasingly more important in the coming decades and centuries.

Today, a fundamental tool used for predicting weather and climate changes is the use of numerical models, i.e., mathematical models run as computer simulations. However, the basic ideas of weather forecasting and climate modelling were developed about more than a century ago, long before the construction of the first electronic computers (Phillips 1970; Lynch 2008). At these early times, observations were rather sparse and irregular, especially for the upper air and over the oceans, making weather forecasting very imprecise and unreliable. The basic laws of physics, fluid motion, and chemistry played no role and were replaced by the forecaster with crude techniques of extrapolation, knowledge of local climatology, and guesswork based on mere intuition. It was not until the beginning of the last century that meteorologists started to recognize that fluid mechanics and thermodynamics represent the set of fundamental physical principles that govern the flow of the atmosphere (Abbe 1901; Bjerknes 1904; Willis and Hooke 2006). In particular, Abbe (1901) proposed the first mathematical approach to forecasting, and shortly after Bjerknes (1904) introduced the idea that rational forecasting should consist of a diagnostic step, in which the initial state of the atmosphere is determined observationally and represented in charts giving the distribution of the variables at different levels, and a prognostic step, in which the laws of fluid motion are used to calculate the changes of this state over time. Non-linear advection—the transport of fluid properties and characteristics by the motion of the fluid itself—was identified as the primary physical process. However, he employed a graphical approach, rather than numerical methods, for solving the fluid dynamics equations and building up new charts describing the atmosphere some hours later, with the process being repeated iteratively until the desired forecast length was achieved.

The beginning of modern numerical weather prediction (NWP) was pioneered by Richardson (1922), who first attempted a direct solution of the equations of motion using finite difference methods (Lynch 2006). His work impelled profound developments in the theory of meteorology and is the foundation upon which modern forecasting is built. Since then, the advances in numerical analysis, which enabled the design of stable algorithms, the development of the digital computer technology, and the invention of the radiosonde, and its introduction in a global network, providing timely observations of the atmosphere in three-space dimensions (i.e., in latitude, longitude, and height), have completed the task. A definite impulse to modern meteorology was given later on by Charney (1947, 1948, 1950), who developed a set of equations known as the *quasi-geostrophic vorticity* system for calculating the large-scale motions of planetary-scale waves (Charney 1948), giving the first convincing physical explanation for the development of mid-latitude cyclones–his baroclinic instability theory. This theory was capable of producing a quantitatively accurate prediction of the atmospheric flow (Charney et al. 1950; Platzman 1979). In 1979 he leaded an ad hoc study group on carbon dioxide and climate for the United States National Research Council, with their final written report being one of the earliest modern scientific assessments about global warming (Charney et al. 1979). They estimated that doubling of CO_2 emissions will produce a global warming near 3 °C with a probable error of ± 1.5 °C, which is quite close to the best estimate value of about 3 °C for the global temperature increase given by the Intergovernmental Panel on Climate Change (IPCC) Fourth Assessment Report published in 2007.

With the advances in computer technology, numerical weather predictions have achieved breakthrough improvements in many aspects. In the 1960s, operational forecasts started to use models based on numerical solutions of the *primitive* equations—a set of non-linear differential equations, consisting of a form of the familiar Navier-Stokes equations, a continuity equation, and a thermal energy equation (Charney 1955; Hinkelmann 1959; Phillips 1960; Smagorinsky 1963). A six-level primitive equation model was introduced into operations at the National Meteorological Center in Washington in June, 1966, running on a CDC 6600 computer (Shuman and Hovermale 1968). Manipulating the vast datasets and performing the complex calculations necessary to modern weather prediction require some of the most powerful supercomputers in the world. Even with the increasing power of supercomputers, the forecast skill of NWP models extends to about only 6 days. The density and quality of observations used as input to the forecasts and the deficiencies in the models themselves are important factors affecting the accuracy of the predictions.

A more fundamental problem lies in the chaotic nature of the fluid-dynamics equations used to simulate the atmosphere. In addition, these equations need to be supplemented with parameterizations that attempt to capture the phenomenology of small-scale processes, including solar and terrestrial radiation, moisture content (cloudiness and relative humidity), surface hydrology (precipitation, evaporation, snow melt and run-off), heat exchange, soil, vegetation, surface water, and the effects of terrain. On the other hand, the development of regional (limited area) models has facilitated accurate forecasting of the tracks of tropical cyclones and hurricanes as well as of air quality (Shuman 1989; van Dop and Steyn 1991). The inclusion of the interactions of land and vegetation with the atmosphere has led to more realistic forecasts (Xue et al. 1996).

The chaotic nature of the atmospheric flow imposes a limit on predictability, as inherent errors in the initial state grow rapidly and render the forecast useless after some days. A numerical prediction method, known as ensemble forecasting, which is a form of Monte Carlo analysis has been introduced in which multiple numerical predictions, each starting from slightly different initial conditions, are run and the combined outputs are used to deduce probabilistic information about future changes in the atmosphere (Molteni et al. 1996; Toth and Kalnay 1997; Buizza et al. 1999). With this approach, probability forecasts for a wide range of weather events are currently generated and disseminated for use in the operational centres. For instance, seasonal forecasts, with a range of 6 months, are prepared at the European Centre for Medium-Range Weather Forecasts (ECMWF) and at the National Center for Environmental Prediction (NCEP) in Washington. They are made using a coupled atmosphere/ocean model, and a large number of forecasts are combined in an ensemble each month. In particular, these forecast ensembles have demonstrable skills for tropical regions with recent impressive predictions for the onset of El Niño and La Niña events. However, in middle latitudes, as in Europe, no significant skill has yet been achieved by these models. In fact, seasonal forecasting for middle latitudes remains one of the great problems facing us today.

Weather and climate are different in the sense that climate predictions do not need knowledge of weather in detail. A good analogy of the difference between weather and climate is to consider a swimming pool. Suppose that the pool is being slowly filled. If someone dives into it, this will certainly generate waves on the water surface. The waves represent the weather, while the average water level is the climate. A new diver jumping into the pool next day will produce more waves, but the water level will be higher as more water has flowed into the pool. In the atmosphere the ‘water hose’ is increasing the amount of greenhouse gases, which will cause the climate to warm even though we still have a changing weather (waves). Thus, climate scientists use models to forecast the average water level in the pool and not the waves. However, climate modelling derives from efforts first formulated to numerically predict the weather. The first successful long-range simulation of the general circulation of the atmosphere was developed in 1956 (Phillips 1956), which realistically depicted monthly and seasonal patterns in the troposphere (Cox 2002). This work had a galvanizing effect on the meteorological community and thereafter several general circulation models (GCMs) were developed. One early model of particular interest has been that developed at the National Center for Atmospheric Research (NCAR) (Kasahara and Washington 1967). By the early 1980s, NCAR has developed the Community Climate Model (CCM), which has been continuously refined into the next 20 years (Williamson 1983; Williamson et al. 1987; Williamson and Olson 1994), with the Community Atmosphere Model (CAM 3.0) being the latest version (Collins et al. 2004). On the other hand, coupled atmosphere/ocean climate models such as HadCM3 and HadGEM are used at the Hadley Centre for Climate Prediction and Research in the United Kingdom for a wide range of climate studies (Lynch 2006). Advanced models, such as the atmospheric GCM ECHAM5 developed at the Max Planck Institute for Meteorology (Roeckner et al. 2003), are under continuing refinements and extensions, and are increasing in sophistication and comprehensiveness. Most of them simulate not only the atmosphere and oceans but also a wide range of geophysical, chemical, and biological processes and feedbacks. In particular, these models, now called Earth System Models, are applied to the practical problem of weather prediction and also to the study of climate variability and mankind's impact on it.

2 Weather Modelling and Prediction

The atmosphere is a fluid (composed mostly of air) that covers the entire Earth surface. Most of the phenomena which we associate with day-to-day weather occur in its lowest layer, called the troposphere, which ranges in thickness from about 8 km at the poles to 16–20 km over the equator. The troposphere is denser than the layers of the atmosphere above it and contains up to 75 % of the mass of the atmosphere, with approximate composition of 78 % nitrogen, 21 % oxygen, and 1 % small concentrations of other trace gases. Nearly all atmospheric water vapour (or moisture) and aerosols are found in the troposphere. Since temperature decreases with altitude, warm air near the surface of the Earth can readily rise, being less dense than the colder air above it. This induces a vertical movement, or convection, of air which generates clouds and ultimately rain from the moisture within the air, giving rise to much of the weather we experience in our daily lifes.

The troposphere is capped by the tropopause, a boundary region of stable temperature, separating the troposphere from the stratosphere, where the air temperature begins to rise. Such a temperature increase prevents much of the air convection beyond the troposphere, and consequently most weather phenomena, including towering cumulonimbus thunderclouds, are confined to the troposphere. For instance, most commercial aircrafts fly in the lower stratosphere, just above the tropopause where clouds are usually absent, as also are significant weather perturbations (Petty 2008). However, vigorous thunderstorms as, for example, those of tropical origin may overshoot into the lower stratosphere and undergo low-frequency vertical oscillations of an hour-order duration, or less (Shenk 1974), which in turn may induce low-frequency atmospheric gravity waves capable of affecting both atmospheric and oceanic currents in the region (Bromirski et al. 2010). Sometimes the temperature does not decrease with height in the troposphere, but rather increases, which is known as a temperature inversion. In general, temperature inversions limit or prevent the vertical mixing of air, causing a state of atmospheric stability. This can lead to episodes of air pollution, where air becomes stagnant and pollutants emitted at ground level remain trapped underneath the temperature inversion zone (Phalen and Phalen 2012).

Among the most significant scientific advances of the past century is our ability to simulate complex physical systems using numerical methods and predict their evolution. One outstanding example is the development of GCMs of the atmosphere and ocean, which can be used to predict the weather for several days in advance with a high degree of confidence and gain insight into the factors that cause changes in the climate as well as into their likely timing and severity. Here we shall review the most important numerical weather prediction models, which were the precursors to climate prediction systems, viewed as a problem in non-linear fluid mechanics.

2.1 Barotropic Models

Barotropic models are short-range prediction models that include only the reversible part of atmospheric physics. That is, the atmosphere is treated as a one-component gas consisting of dry air so that irreversible processes, such as non-adiabatic heating and cloud formation, are not taken into account. The barotropic model was the first kind of NWP model ever successfully implemented (Charney 1948; Charney et al. 1950). It is probably the simplest model that can realistically model atmospheric flow around the Earth. Meteorologists use the word barotropic to describe an atmosphere where *isosteric* surfaces—surfaces of constant specific volume—and *isobaric* surfaces—surfaces of constant pressure—coincide. In other words, the gradient of the specific volume (or density) and the gradient of pressure are parallel and proportional to each other so that the density is a function of pressure (adiabatic atmosphere).

Typical barotropic models are based on a set of equations known as the *quasi-geostrophic* system (Charney et al. 1950). These equations are derived from the Euler equations of motion by assuming that the Coriolis force resulting from horizontal air currents exactly balances the horizontal pressure gradients (geostrophic balance), while in the vertical direction hydrostatic equilibrium is assumed. If the atmosphere is divergence-free, the curl of the Euler equations of motion reduces to the barotropic vorticity equation (Bennett et al. 1993):

$$\frac{D\zeta}{Dt} = 0, \tag{1}$$

where D/Dt is the substantial time derivative and ζ is the absolute vorticity defined by

$$\zeta = m^2 \left[\frac{\partial}{\partial x}\left(\frac{v}{m}\right) - \frac{\partial}{\partial y}\left(\frac{u}{m}\right) \right] + f, \tag{2}$$

where v and u are the horizontal geostrophic wind components in the direction of the map coordinates x and y, respectively, m is the map factor, and $f = 2\Omega \sin\phi$ is the Coriolis frequency. Here Ω is the angular velocity of planetary rotation and ϕ is the latitude. Since the model has non-divergent flow, a streamfunction Ψ can be defined by

$$v = m\frac{\partial \Psi}{\partial x}, \qquad u = -m\frac{\partial \Psi}{\partial y}, \tag{3}$$

so that

$$\zeta = m^2\nabla^2\Psi + f. \tag{4}$$

In low-pressure systems, where the Rossby number (Ro) is small, the effects of planetary rotation are large compared to the net wind acceleration, allowing the use of the geostrophic approximation given by Eqs. (1–4) (Marshall and Plumb 2008). Typical barotropic models for operational weather prediction were based on an extended version of Eqs. (1–4) to account for small deviations from strict geostrophic balance—the so-called semi or quasi-geostrophic equations (Phillips 1970; Chynoweth and Sewell 1991). Since the pioneering work of Charney (1948) and (Charney et al. 1950), the quasi-geostrophic equations have become an accepted system of approximate equations for the study of mid-latitude motions of the atmosphere on a sypnotic scale, while allowing for the presence of mesoscale phenomena such as the atmospheric fronts.

A barotropic instability is a wave instability associated with shear in a jet-like current and this appears to be of central importance in the tropics. Early attempts of forecasting in the tropics with a barotropic atmospheric model were addressed to predict upper-air flow patterns in the tropical Pacific areas of both the Northern and Southern hemispheres (Jordan 1956; Vederman et al. 1966). A similar model was applied to forecasts of flow patterns at 500 mb level in the Indian region (Shukla and Saha 1970). Barotropic prediction models have also provided the basis for a significant advance of the state of the art of tropical cyclone motion and hurricane track forecasting in the range from one to several days (Bennett et al. 1993; Sanders and Burpee 1968; Sanders et al. 1980; DeMaria 1985). Although there are some situations where tropical cyclone motion can only be modelled using a more general form of the basic equations as, for example, in the case when a vortex interacts with a vertically-sheared basic current, there has been evidence that some aspects of tropical cyclone motion can be described with simple barotropic models. For instance, the SANBAR model (Burpee 2008)—a barotropic tropical cyclone track prediction model designed for the North Atlantic tropical cyclone basin and used operationally during 1973–1984 and 1985–1989, was recognized to be superior to other forecast methods for medium range track forecasts of low-latitude Atlantic tropical cyclones (Neumann and Pelissier 1981). It has also been shown that for the Australian/Southwest Pacific region many aspects of tropical cyclone motion can be explained using a theory based on a barotropic vorticity equation (Holland 1983, 1984). In fact, calculations of the terms in the full form of the vorticity equation, using aircraft and rawinsonde composite data, have shown that the dominant contribution to the local vorticity change in the regions near the tropical cyclone centre comes from the horizontal advection term (Chan 1984).

Barotropic NWP models have also been used to demonstrate the close coupling existing between the westwards propagating African waves and the broad scale African monsoons on the time scale of 3–5 days (Krishnamurti et al. 1980). It is well-known today that about 80 % of all tropical cyclones on the globe forms near or within the intertropical convergence zone (ITCZ) (Gray 1979). In satellite images, the ITCZ is sometimes observed to undulate, forming cloud patterns. At times, such an undulating ITCZ breaks down into several tropical disturbances within which tropical cyclones may form (Gray 1979; Zehr 1993). The resulting tropical cyclones and typhoons then move into higher latitudes, allowing the ITCZ to reform and perhaps start the cycle over again (Guinn and Schubert 1993). These undulations are a clear signature of easterly waves in the tropical troposphere. Easterly waves have early been recognized to play an important role in tropical cyclogenesis (Riehl 1945). These have since been observed in the Atlantic Ocean and West Africa (Reed et al. 1977; Chen and Ogura 1982), in the Pacific Ocean (Nitta et al. 1985; Nitta and Takayabu 1985; Tai and Ogura 1987; Heta 1991), and in the South China Sea and India (Saha et al. 1981). All these studies concluded that easterly waves occur in the lower tropical troposphere and have typical wavelengths and speeds in the ranges from 2,000 to 4,000 km and 5–8 ms^{-1}, respectively. While nearly 60 % of all Atlantic tropical cyclones originates from African easterly waves (Avila and Clark 1989), observational and numerical studies indicate that they result from a convectively modified form of combined barotropic and baroclinic instability of the African easterly jet, which has maximum winds of 10–15 ms^{-1} near 700 mb and 15 °N (Norquist et al. 1977; Thorncroft and Hoskins 1994a,b). Barotropic model simulations based on the shallow-water equations have suggested that the ITCZ break-down may play a role in producing the observed tendencies for tropical storms to cluster in time and form polewards of the central latitude of the ITCZ and to the east of existing tropical storms (Nieto Ferreira and Schubert 1997). More recently, barotropic instability calculations have also been employed to investigate the possible importance of barotropic shear variations for explaining the effect of the Madden-Julian oscillation on hurricane formation over the eastern and western North Pacific (Hartmann and Maloney 2001).

In spite of its numerous applications during more than 40 years, the quasi-geostrophic modelling was abandoned because of the development of more efficient ways of integrating the primitive equations (Bengtsson 1999). On the other hand, the incorporation of physical processes, radiation, clouds, precipitation processes, etc. was by far more complicated to implement in the quasi-geostrophic models, and this was an additional reason not to use them any longer in NWP.

2.2 Baroclinic Models

The occurrence of large vertical temperature gradients in the troposphere can lead to the formation of convective air currents, which transport the excess energy away from the surface to higher altitudes where the air is significantly cooler. When this happens we say that the atmosphere is statically unstable. In analogous manner, when the latitudinal temperature distribution is such that a large equator-to-pole temperature gradient exists, the atmosphere will break down into wind flows to move the excess energy from the regions of excess (warm tropics) to regions of deficit (cool poles). In this case, the atmosphere is said to be baroclinically unstable. This imbalance of energy is essentially due to an excess of radiational heating in the tropical latitudes. In a stratified fluid, a source term of the form $\nabla\rho \times \nabla p/\rho^2$ appears in the vorticity equation whenever *isopycnic* (constant density) surfaces and *isobaric* surfaces are not aligned, which is responsible for the baroclinic contribution to the local vorticity (Marshall and Plumb 2008). In meteorology, a baroclinic atmosphere is one in which the density depends on both the temperature and the pressure.

The most important application of the baroclinic instability is the cyclogenesis process at mid-latitudes, which represents the development of sypnotic scale weather disturbances. In other words, it is the leading mechanism shaping the cyclones and anticyclones that influence weather at mid-latitudes. For instance, in the ocean the baroclinic instability is responsible for the generation of mesoscale eddies that play a role in the transport of tracers, which are used in oceanography to deduce flow patterns in the ocean (Davis 1991). In general, vorticity is the curl of the velocity field and its evolution can be broken into contributions from advection (as vortex tubes move with the flow), stretching and twisting (as vortex tubes are pulled or twisted by the flow), and baroclinic vorticity generation (Nadiga and Aurnou 2008). Therefore, the study of the evolution of these baroclinic instabilities is a crucial part of developing theories of mid-latitude weather. The birth of baroclinic NWP models started with the classical work of Charney (1947) and Eady (1949). The energy source for baroclinic instability is the potential energy associated with the environmental flow, and since then meteorologists have become aware that baroclinic instability can develop even in situations of rapid rotation (small Ro) and strong stable stratification (large Richardson number, Ri) as is typically observed in the atmosphere, where Ri is a dimensionless number that serves to quantify the ratio of potential to kinetic energy.

Since a tropical cyclone is a huge tropospheric convection cell and the axis of the horizontal wind circulation remains almost vertical during the movement, there was a need to develop baroclinic prediction models capable of simulating the three-dimensional atmospheric motion more closely than single-level barotropic models. After the Electronic Numerical Integrator and Computer (ENIAC) forecast chaired by Charney in the 1950s in Aberdeen, Maryland (Platzman 1979), several baroclinic models were developed in the next few years, which were all based on the quasi-geostrophic system of equations (Phillips 1951, 1954; Charney and Phillips 1953; Matsumoto 1956; Wiin-Nielsen 1959; Kasahara 1960). Most of these models were employed to evaluate the instantaneous movement velocity of tropical cyclones from multi-level data, i.e., the atmosphere is divided into two, or more, levels where prognostic and diagnostic variables are evaluated from known data at these levels. However, it was soon argued that early experiments with baroclinic models capable of generating additional kinetic energy from the store of available potential energy failed (Ellsaesser 1968), and that the multi-level models were worse than the single-level barotropic forecasts (Bengtsson 1964; Shuman 1989). One major cause of the failure was due to a net accumulation of kinetic energy in the models owing to the presence of the baroclinicity source and the absence of a dissipative sink of kinetic energy. Therefore, the single-level model was preferred when regular operational weather forecasting commenced in 1958.

2.3 Primitive Equation Models

As numerical weather prediction passed its infancy, the quasi-geostrophic approximation was replaced by the primitive equations. On the basis that these equations would simulate the atmospheric dynamics and energetics more realistically than the filtered equations, Hinkelmann (1951) first tackled the issue of suitable initial conditions for integration of the primitive equations, followed by other important studies of initialization (Charney 1955; Phillips 1960). The first applications of the primitive equations were a success (Hinkelmann 1959; Smagorinsky 1963) and soon thereafter, they started to be used in operational settings in 1966 at the Deutscher Wetterdienst in West Germany (Reiser 1986) and at the National Meteorological Center in Washington (Shuman and Hovermale 1968), followed by the United Kingdom Meteorological Office in 1972 and the Australian Bureau of Meteorology in 1977 (Leslie and Dietachmayer 1992; Lynch 2008).

The primitive equations are a set of non-linear differential equations that form the basis for any NWP scheme. Their precise form depends on the coordinate system used to represent the vertical structure of the atmosphere, which may be either the pressure (p), the geometrical height (z), or the potential temperature (θ) (Kasahara 1974). In particular, models based on pressure as a vertical coordinate must be distinguished into three types: pressure, log pressure, and the so-called σ-system, where $\sigma = p/p_0$ and p_0 is the Earth's surface pressure (Phillips 1957). The use of pressure as a vertical coordinate became very popular during the 1950s and 1960s (Hinkelmann 1959; Eliassen 1949; Leith 1965). However, this scheme has certain computational disadvantages in the vicinity of mountains because the lower limit of the atmosphere is not a coordinate surface. In fact, there have been very few attempts to incorporate details of the Earth's orography in these models. To overcome this difficulty, the σ-system was proposed in which the Earth's surface is always a coordinate surface (Phillips 1957; Smagorinsky et al. 1965; Sela and Bostelman 1973). Moreover, the use of the potential temperature (defined as $\theta = T(p_0/p)^{\kappa}$, where $\kappa = R_g/c_p$, R_g is the specific gas constant, c_p the specific heat at constant pressure, and T the temperature) as a vertical coordinate in primitive-equation models commenced in the 1970s (Eliassen and Raustein 1968, 1970; Shapiro 1973). Although the approach is particularly suitable for resolving details of frontal structure, it still faces the same degree of complexity in handling the lower boundary conditions as in the isobaric coordinate system. While the representation of bottom topography has historically been crude, the choice of vertical coordinates is perhaps the single most important feature that differentiates between models and is still an active area of research.

In the pressure as well as in the height and potential temperature coordinate systems, special procedures were implemented to take into account the effects of the Earth's orography, consisting of examining the height of the mountains and shaping them as lateral boundary conditions at the grid points. Although the σ-system is not free of shortcomings, the idea of transforming the Earth's surface to a coordinate surface has also been applied to the height and potential temperature coordinate systems as well. A comprehensive overview of models using all three vertical coordinates as well as a concise review of the equations of oceanic motion, sub-grid-scale parameterizations, and numerical approximation techniques can be found in Haidvogel and Beckmann (1999). A convenient way to introduce a general system that utilizes any well defined variable as a vertical coordinate has been discussed by Kasahara (1974). For example, in the z-system any fluid quantity will be a function of the Cartesian coordinates (x, y, z) and time t, while in the generalized coordinate system (the s-system), the independent variables would be (x, y, s, t) such that $s = s(x, y, z, t)$. When x, y, and t are held fixed, this equation gives a single-valued monotonic relation between s and z. The basic primitive equations for large-scale atmospheric flows written in the s-system are as follows: the horizontal equation of motion

$$\frac{D\mathbf{v}}{Dt} = -\frac{1}{\rho}\nabla_s p - g\nabla_s z - f\mathbf{k}\times\mathbf{v} + \mathbf{F}, \tag{5}$$

where $\mathbf{v} = u\mathbf{i} + v\mathbf{j}$ is the horizontal velocity, with u and v being its x-and y-components, $\nabla_s = \nabla_z + (\partial s/\partial z)\nabla_s z(\partial/\partial s)$ is the gradient operator in the s-system, $\mathbf{k}$ is the unit vector along the s-coordinate, f is the Coriolis frequency $(= 2\Omega \sin\phi)$, Ω is the angular velocity of the Earth's rotation, ϕ the geographical latitude, ρ the atmosphere density, g the Earth's gravitational acceleration, p the pressure, $\mathbf{F}$ the frictional force per unit area, and

$$\frac{D}{Dt} = \left(\frac{\partial}{\partial t}\right)_s + \mathbf{v}\cdot\nabla_s + \dot{s}\frac{\partial}{\partial s}, \tag{6}$$

is the total time derivative in the s-system, where $\dot{s}$ is the generalized vertical velocity; the continuity equation

$$\frac{D}{Dt}\ln\left(\rho\frac{\partial z}{\partial s}\right) + \nabla_s\cdot\mathbf{v} + \frac{\partial\dot{s}}{\partial s} = 0; \tag{7}$$

the hydrostatic equation

$$\rho\frac{\partial z}{\partial s} = -\frac{1}{g}\frac{\partial p}{\partial s}; \tag{8}$$

the ideal gas law

$$p = \rho R_g T; \tag{9}$$

and the first law of thermodynamics

$$\frac{D}{Dt} \ln \theta = \frac{Q}{c_p T}, \tag{10}$$

where θ is the potential temperature as defined in the text above and Q is the rate of heating/cooling per unit mass per unit time. Equations (5–10) constitute the basic set of dynamical principles for NWP. In predicting the atmospheric flow, we must define appropriate boundary conditions as required by any solution of the problem. In general, it is convenient to choice the upper boundary condition as a vertical coordinate surface, $s = s_T =$ const., so that there is no mass transport through it ($\dot{s} = 0$). As a lower boundary condition of the atmosphere, it is usually assumed that there is no mass flow through the Earth's surface, which is located at fixed altitude H above the mean sea level $z = 0$. In the s-system, the Earth's surface is expressed by

$$s = s_H = s(x, y, H, t), \tag{11}$$

where the value of s at $z = H$ may vary with time and space. Since the air at the Earth's surface may move only along the Earth's surface itself, the lower boundary condition must read

$$\dot{s} = \frac{\partial s_H}{\partial t} + \mathbf{v}_H \cdot \nabla s_H, \tag{12}$$

at $s = s_H$. If the Earth's surface coincides with a constant s-surface, then Eq. (12) becomes $\dot{s} = 0$ at $s = s_H$. It is worth noticing that many worldwide groups were also examining how to used "hybrid" coordinates, where the vertical coordinate may be a function of height in the mixed layer, a function of isentropes in the interior, and some function of the terrain in the bottom boundary layer (Spall and Robinson 1989; Arakawa and Konor 1996; Rõõm et al. 2007).

For prediction of large-scale weather phenomena, it is important to add to the above set of equations the prediction of the water vapour field. Water vapour is a dynamically active constituent of the tropical atmosphere which, though to a significant extent locally controlled by vertical advection, precipitation, and surface evaporation, is also affected by horizontal advection. Water vapour affects the flow in turn, because a humid atmosphere supports deep, precipitating convection more readily than a dry atmosphere. For instance, precipitation heats the atmosphere, and this heating drives the flow. The differential equation for the specific humidity q, defined as the mass of water vapour per unit mass of air, in the s-system has the form (Sobel 2002):

$$\frac{Dq}{Dt} = Q_q, \tag{13}$$

where Q_q represents sources and sinks of moisture due to unresolved processes, such as transport of water vapour as well as loss by condensation (Yanai et al. 1973). Similarly, Q in Eq. (10) represents sources and sinks of heat, such as radiative transfer of electromagnetic energy. For example, $Q = Q_c + Q_R + Q_d$, where Q_c is the apparent source of heat associated with buoyant moist convection (i.e., release of latent heat by condensation of water vapour or freezing of liquid water as well as transport of heat), Q_R represents radiative heating or cooling, and Q_d represents diffusive or turbulent transport by motions that are not directly associated with deep convection (Yanai et al. 1973). In order to obtain a closed dynamical system, we need to parameterize these sources and sinks as functions of the large-scale state variables $\mathbf{v}, q$, and T. In particular, Q_c and Q_q are determined by a convective parameterization (Arakawa 1993). A detailed discussion of the parameterization problem is precluded here and the reader is referred to a few useful textbooks for a detailed account (Emanuel 1994; Smith 1997).

Weather models that have grid-boxes with sides between 5 and 25 km can explicitly represent convective clouds, although they need to parameterize the cloud microphysics which occur at much smaller scales (Narita and Ohmori 2007). For example, the formation of large-scale clouds (stratus-type) is more physically based and form when the relative humidity reaches some prescribed value. On the other hand, the amount of solar radiation reaching the ground and the formation of cloud droplets, which occur on the molecular scale, must be parameterized before they can be included in any model. Atmospheric drag produced by mountains must also be parameterized because limitations in the resolution of elevation contours may produce significant underestimates of the actual drag (Stensrud 2009). A parameterization of the surface flux of energy between the ocean and the atmosphere is also required in order to determine realistic sea surface temperatures and type of sea ice found near the ocean's surface (McGuffie and Henderson-Sellers 2005). In addition, the impact of multiple cloud layers as well as soil type, vegetation type, and soil moisture are factors that must be taken into account in NWP models (Melnikova and Vasilyev 2005; Stensrud 2009). Within air quality models, parameterizations are required to take into account atmospheric emissions from multiple relatively tiny sources, as roads, urban areas, fields, and factories, within specific grid-boxes (Baklanov et al. 2009).

In the last three decades a myriad of primitive-equation models has been reported in the literature, most of which have found applications in ocean dynamics and tropical cyclone forecasting (Arakawa and Suarez 1983; Beckers 1991; Song and Haidvogel 1994; Ezer and Mellor 1997; Barnier et al. 1998; Fraedrich and Frisius 2001). For testing and operational models, the process of entering observational data to generate initial conditions is called initialization. On land, terrain maps that are available at resolutions down to 1 km are employed to facilitate atmospheric circulation models within regions of rugged topography. This permits depict features such as downslope winds, lee waves—atmospheric standing waves due to wind flows towards a mountain, —and related cloudiness that affects the incoming solar radiation (Stensrud 2009). In country-based weather services, the main input data is produced by observations from radiosondes (placed in weather balloons that measure relevant atmospheric parameters and transmit them to a fixed receiver) and from weather satellites. Permanent weather observation stations either report hourly in METAR reports—the most popular format in the world for the transmission of weather data—or every 6 h in SYNOP (surface synoptic observations) reports. In general, these observations are irregularly spaced and so they must be processed by data assimilation and objective analysis methods, which perform quality control and obtain values at locations usable by NWP models (Krishnamurti 1995). Many of these models are global, primitive-equation models based on finite-difference techniques, where the world is represented as discrete points on a spherical grid in latitude and longitude (Chaudhari et al. 2007), while a few other models are based on spectral methods that solve for a range of wavelengths. Today, information from weather satellites is used where traditional data sources are not available. Research projects use reconnaissance aircrafts to fly in and around weather systems of interest, such as tropical cyclones. In particular, reconnaissance aircrafts are also used over the open oceans during the cold season into systems which cause significant uncertainty in forecast guidance, or which are expected to be of high impact from 3 to 7 days into the future over the downstream continent.

The horizontal domain of a NWP model can be either global, covering the entire globe, or regional—also known as limited-area models,—covering only part of the Earth. The latter models allow for the use of finer grid spacing than global models because the available computational resourses are focused on a specific area, thereby allowing explicit resolution of small-scale meteorological phenomena that cannot be represented on the coarser grid of a large-scale, or global, model. In general, regional models use information from global models to specify boundary conditions at the edge of their domain and eventually allow systems from outside the limited area to move into it. For instance, high-resolution models (also called mesoscale models), such as the Weather Research and Forecasting (WRF) model, which was created through a partnership including the National Oceanic and Atmospheric Administration (NOAA), NCAR, and more than 150 other organizations and universities in the United States and other countries, and the Nonhydrostatic Mesoscale Model (NMM), which was designed for forecasting operations at various National Weather Service offices in the United States, are primitive-equation codes based either on hybrid or σ vertical coordinates that are employed to explore ways of improving the accuracy of hurricane track, intensity, and rainfall forecasts, among other meteorological questions.

3 Climate Modelling

Climate is a complex, large-scale phenomenon that emerges from complicated interactions among small-scale physical systems. As mentioned by Schmidt (2007) in his Physics Today's article on the physics of climate modelling: *the task climate modellers have set for themselves is to take their knowledge of the local interactions of air masses, water, energy, and momentum and from that knowledge to explain the climate system's large-scale features, variability, and response to external pressures, or "forcings". That is a formidable task, and though far from complete, the results so far have been surprisingly successful.*

Computer models of the coupled atmosphere-land surface-ocean-sea ice system are essential scientific tools for understanding and predicting natural and human-caused changes in the Earth's climate. Recently, these models have added more components such as interactive atmospheric aerosols, atmospheric chemistry, and representations of the carbon cycle. There is no doubt that the study of climate change and its impacts are of enormous importance for our future and that global climate models are perhaps the best means we have of anticipating the likely changes. In general, climate models are used for a variety of purposes, which range from the study of the dynamics of the climate system to projections of future climate. In recent years, the most talked-about use of climate models has been to project temperature changes resulting from increases in atmospheric concentrations of greenhouse gases.

3.1 Phenomena of Interest in Climate Modelling

A number of well-known phenomena may contribute to climate change over short and long periods, which include the global carbon cycle, El Niño-Southern Oscillation (ENSO) climate pattern and its counterpart La Niña, the greenhouse warming, the atmospheric chemistry, the ocean circulation, and extreme events such as mesoscale storms and volcanic eruptions.

3.1.1 The Global Carbon Cycle

In the geological history of the Earth, carbon has been cycling among large reservoirs in the land (including plants and fossil fuels), oceans, and the atmosphere. This natural cycling of CO_2 usually takes millions of years to move large amounts of carbon from one system to another. However, atmospheric carbon dioxide comes increasingly from human activities, which together with other trace (greenhouse) gases in the atmosphere absorb radiation emitted from the Earth, thereby trapping heat in the atmosphere and contributing to its warming. For instance, since the Industrial Revolution in the nineteenth century, the amount of CO_2 in the atmosphere has risen by 30 % as a result of the sustained increase in burning of fossil fuels (oil and natural gas) and other carbon based fuels, principally wood and coal, due to the rise of industry and transportation emissions.

There are two large reservoirs of carbon that are capable of taking significant amounts of CO_2 out of the atmosphere at comparable rates: the oceans and the land plants. A comprehensive study of the ocean storage of CO_2 derived from human activity based on a decade-long survey of carbon distributions in the Atlantic, Pacific, and Indian oceans indicate that the oceans have taken up to 118 billion metric tons of CO_2 from human sources (anthropogenic CO_2) between the period from 1800 to 1994, implying that the oceanic sink accounts for $\sim$48 % of the total fossil-fuel and cement-manufacturing emissions (Sabine et al. 2004).

3.1.2 Greenhouse Gases and Aerosols

Trace (or greenhouse) gases in the atmosphere, such as water vapour, carbon dioxide, ozone, methane, nitrous oxide, and carbon monoxide, are present in the atmosphere in a tiny percentage ($\sim$1 %) compared to its total composition, mostly nitrogen and oxygen. However, such a small amount contributes significantly to long-term changes in the Earth's climate. They absorb and re-emit some of the outgoing energy radiated from the Earth's surface, retaining the excess heat in the lower atmosphere and affecting the surface energy balance of the planet. Some greenhouse gases remain in the atmosphere for decades or even centuries, warming the atmosphere and resulting in long-term changes to global climate. The factors that influence the Earth's energy balance are quantified in terms of radiative forcing. While some greenhouse gases, like carbon dioxide, have always been present in the atmosphere, some others may be new compounds, introduced into the air by man-made mechanisms such as manufacturing processes. This human-induced (anthropogenic) warming has had a discernible influence on many physical and biological systems, and future warming is projected to have important impacts on the sea level rise, increased frequency and severity of extreme weather events, loss of biodiversity, and agricultural productivity.

Cumulative anthropogenic emissions of CO_2 are recognized to be a major cause of global warming (Botzen et al. 2008), with the developed countries contributing to more than 80 % of industrial CO_2 emissions (Höhne et al. 2010). A recent analysis estimates that water vapour accounts for about 50 % of the Earth's greenhouse effect, with clouds—formed by suspended water droplets and ice crystals (Kiehl and Trenberth 1997)—contributing 25 %, carbon dioxide 20 %, and other minor trace gases and aerosols accounting for the remaining 5 % (Schmidt et al. 2010). Though to a relatively minor extent, aerosols—fine solid particles of various types and concentrations suspended in the atmosphere such as smoke, dust, smog, ashes, pollen, and other sources (Hinds 1999)—can also affect the behaviour of the Earth system. For example, aerosols can absorb and scatter radiation, which can cause either warming or cooling of the atmosphere. Therefore, they are important in the formation and behaviour of clouds, and can influence the water cycle and shift the Earth's radiative balance.

3.1.3 El Niño and La Niña

El Niño is a natural fluctuation of the global climate system. Originally it was the name given to the periodic warming of ocean waters along the tropical South American coasts and out along the Equator to the dateline. Today, the name is used to describe the whole El Niño-Southern Oscillation (ENSO) phenomenon. During El Niño events, warmer than average sea surface temperatures occur in the central and eastern equatorial Pacific accompanied by high air surface pressure in the western Pacific, while during La Niña—the opposite extreme of the ENSO cycle,—cooler than average see surface temperatures predominate in the equatorial central and eastern Pacific accompanied by low air surface pressure in the western Pacific (Trenberth et al. 2007). ENSO is an important component of the climate system since El Niño/La Niña phases impact weather on a global scale.

Under normal conditions, i.e., when neither El Niño nor La Niña are present, the Walker circulation—parcels of air following a closed circulation in the zonal and vertical directions in the lower tropical atmosphere—is seen at the sea surface in the form of easterly trade winds that move air and water warmed by the Sun towards the west (Briggs and Smithson 1986). During El Niño events, the trade winds weaken, leading to a rise in sea surface temperature in the eastern equatorial Pacific and a reduction of up-welling off South America. Heavy rainfall and flooding occur over Peru, and drought over Australia and Indonesia. The supplies of nutrient-rich water off the South American coasts are cut off due to the reduced up-welling, adversely affecting fisheries in that region. In the tropical South Pacific the pattern of occurrence of tropical cyclones shifts eastwards, so there are more cyclones than normal in areas such as the Cook Islands and French Polynesia. Conversely, during La Niña events, the trade winds strengthen and the pattern is a more intense version of the normal conditions, with an even colder tongue of sea surface temperatures in the eastern equatorial Pacific. Typically, this anomaly happens at irregular intervals of 2–12 years, and lasts 9 months–2 years, with an average period length of 5 years (Philander 1990).

The strong El Niño event of 1982–1983 has inspired innovative climate research, which has resulted in greater predictability of ENSO. In particular, the NOAA's research laboratories have taken a leadership role in furthering ENSO observations and research to improve understanding, predictions, and impacts. This not only serves society's need for information about weather and climate, but also helps plan and respond to weather and climate impacts. For example, ENSO has widespread impacts on a global scale such as drought, wildfires, crop failure, starvation, increased tropical storm/hurricane activity, damage to ecosystems, flooding, and increased spreading of infectious diseases. Understanding and predicting ENSO has resulted in more accurate climate predictions and, hence, in a reduction of its impacts through better planning. For example, scientists are now taking their understanding of ENSO a step further by comparing comprehensive descriptions of these events from the observed record with those simulated by numerical prediction models (Emile-Geay et al. 2013a,b).

3.1.4 Atmospheric Chemistry

The composition and chemistry of the atmosphere is important primarily because of the interactions between the atmosphere and living organisms. As a matter of fact, the composition of the atmosphere changes as a result of natural events such as volcano emissions, lightning—massive electrostatic discharges between electrically charged regions within clouds, or between a cloud and the Earth's surface,—and bombardment by solar wind particles, and also as a result of air pollution derived from human activities. Well-known examples of problems currently addressed by atmospheric chemistry include ozone depletion, acid rains, photochemical smog, greenhouse gases, and global warming (Seinfeld and Pandis 2006).

Progress in atmospheric chemistry is often driven by the interplay between observations, laboratory measurements, and numerical modelling. One common trade-off in numerical modelling is between the number of chemical compounds and reactions that are modelled and the representation of chemical transport and mixing in the atmosphere. Typical box models might include hundreds, or even thousands, of chemical reactions but will only have a rather crude representation of mixing in the atmosphere. In contrast, existing 3D models based on primitive equations represent many of the physical processes of the atmosphere but due to constraints on computational resources will have far fewer chemical reactions and compounds. A trend today is to incorporate atmospheric chemistry as modules in existing climate models.

3.1.5 The Ozone Layer

The ozone layer is a deep layer located in the stratosphere between 30 and 90 km above the ground, encircling the Earth and where most of the atmospheric ozone (O_3) is concentrated. It is well-known that though ozone represents only a small fraction of the gas present in the atmosphere, it plays a protective role by shielding humans and other types of life from the harmful ultraviolet (UV) radiation that comes from the Sun (Seinfeld and Pandis 2006). Ozone on the Earth's stratosphere is a bluish gas created by UV light striking oxygen molecules containing two oxygen atoms (O_2) and separating them into individual oxygen atoms, which can then recombine with other O_2 molecules to form O_3.

Over the last two or three decades, the ozone layer has become more widely appreciated by the public as it was realized that certain industrial processes and consumer products result in the atmospheric emission of chemicals, such as chlorofluorocarbons and hydrochlorofluorocarbons, which have contributed to the depletion of the ozone layer through a complex series of chemical reactions (Steger and Bowermaster 1990). There is also evidence that natural sources of bromides and chlorides from ocean spray and volcanos can contribute to depletion of the ozone (Steger and Bowermaster 1990). As a consequence of these discoveries, an international treaty was signed in 1973, called the Montreal Protocol, and since then other international agreements were also put in place to limit the emissions of human-made, ozone-depleting substances. As a result of these efforts, it is expected that the ozone layer will progressively recover in the coming decades.

Since ozone is also a greenhouse gas in the upper atmosphere, it will have an impact on Earth's climate. For instance, the increase of primary greenhouse gases may affect how the ozone layer will recover in the coming years. Therefore, understanding precisely how ozone abundances will change in the future with diminished chlorofluorocarbon emissions and increased emission of greenhouse gases remains an important challenge for atmospheric scientists. On the other hand, satellite data after the volcanic eruptions of El Chichón (Mexico) in 1982 and Mount Pinatubo (the Philippines) in 1991 showed a 15–20 % ozone loss at high latitudes, and a greater than 50 % loss over the Antarctic, suggesting that volcanic eruptions can play a significant role in reducing ozone levels. Eruption-generated particles, or aerosols, appear to provide surfaces upon which chemical reactions with chlorine- and bromine-bearing compounds from human-made chlorofluorocarbons take place. Thus, although volcanic aerosols provide a catalyst for ozone depletion, the real culprits in destroying ozone are human-generated chlorofluorocarbons (Solomon 1990; Newman et al. 2007).

Ozone depletion in the Earth's ozone layer is seen to occur most severely in the polar regions. The discovery of the Antarctic ozone hole announced in 1985 (Farman et al. 1985) came as a shock to the scientific community, because the observed decline in polar ozone was far larger than anyone had anticipated (Zehr 1994). A review of the status of the ozone hole based on continued total-ozone measurements at Halley, Antarctica, reported in 2002 (Jones and Shanklin 1995), indicated that the ozone hole continued to deepen and that ozone loss extends into the months of January and February with a significant increase in UV-B radiation over the Antarctica in summer. The evolution of the ozone hole in the Antarctic stratosphere is continually monitored and improved measurements of ozone depletion are currently being reported (Huck et al. 2007). Significant depletion also occurs in the Arctic ozone layer during the late winter and spring period (between January and April). However, the maximum depletion was generally less severe than that observed in the Antarctic, with no large and recurrent ozone hole having taken place in the Arctic. However, an unprecedented large Arctic ozone hole was detected in 2011 (Manney et al. 2011). The hole possibly formed because the Arctic stratosphere remained cold longer than usual between December 2010 and March 2011. This way, cold air allowed water vapour and nitric acid to condense into polar stratospheric clouds, which catalyzed the conversion of chlorine into chemically active forms that destroyed ozone.

3.1.6 Paleoclimatology

A credibility test for existing climate models is their ability to simulate past climatic periods as the Cretaceous and the Last Glacial Maximum, which represent abnormally warm and cold climates, respectively. However, paleoclimatology also studies the climate prior to the widespread availability of records of temperature, precipitation, and other instrumental data. Unfortunately, records of past climate changes from satellites and human measurements generally cover less than 150 years, which are too short to examine the full range of climate variability. Therefore, it is crucial to examine climate changes going back to hundreds and thousands of years using paleoclimatic records from tree rings, corals, sediments, microfossils, glaciers, and other natural proxy sources (Cronin 2010).

Understanding how climate has changed on interannual to interdecadal time scales in the past can help scientists understand how climate may change in the future. For example, since the paleoclimate record shows that the Earth's climate system is capable of undergoing abrupt changes, drastic changes in the frequency and intensity of extreme events may be a symptom of this process. The study of past climate change also helps us understanding how humans influence the Earth's climate. For instance, the climate record over the last 1,000 years clearly shows that temperatures have increased significantly in the twentieth century, and that this warming was likely to have been unprecedented during all this period. The paleoclimatic record may also help unravel how much of this warming can be explained by natural causes and how much by human influences.

3.1.7 Global Ocean Circulation

The ocean is the major driver of global climate. It redistrubutes large amounts of heat around the planet via global ocean currents through regional scale up-welling and down-welling, and via a process called *thermohaline circulation* (Di Lorenzo et al. 2008; D'Orgeville and Peltier 2009), which refers to large-scale currents that are driven by fluxes of heat and freshwater across the sea surface and subsequent interior mixing of heat and salt (Rahmstorf 2003). Although winds and tides are important in creating turbulence, this driving mechanism is clearly distinct from wind-driven circulation: thermohaline circulation requires thermohaline surface forcing caused by differences in temperature and salinity of the water, while wind-driven circulation does not. Marine and coastal ecosystems, as we know them today, have adapted over time to the ocean circulation patterns. In addition, global climate change alters the factors that impact ocean circulation, such as wind, precipitation, temperature, and salinity patterns. These changes in forcing mechanisms may also lead to an increase in storm activity, thereby affecting local weather.

On the other hand, thermohaline circulation, which behaves as a conveyor belt, originates in the northern Atlantic Ocean where cold, dense waters sink to the deep ocean. These waters travel across ocean basins to the tropics where they warm and up-well to the surface, which are then drawn to polar latitudes to replace the cold sinking waters. During this process, heat is transferred to the atmosphere, causing the water to become cold and dense, and thus renewing the conveyor cycle. On the other hand, the salinity and the density of polar waters could be reduced by the melting of polar ice, which, in turn could weaken the rate at which the water sinks and alter the movement of heat around the Earth. Moreover, changes in global air temperatures over land and the ocean, as well as increased temperature variations, will alter atmospheric pressure gradients that drive the strength of winds over the ocean. Stronger winds are expected to induce a rapid, intense up-welling that provides a large influx of nutrients in a short amount of time, which can also increase the frequency and distribution of hypoxic events—low oxygen zones (Grantham et al. 2004; Chan et al. 2008). On the other hand, increased variability of winds due to global climate change may cause stronger and longer ENSO regimes (Yeh et al. 2009).

3.1.8 Extreme Weather Events

When a meteorological event comes as a surprise, such as a very hot summer, a unexpectedly mild winter, a flood, a drought, or a tornado, climate change is usually mentioned as one possible underlying cause. Yet, climate scientists warn us about the intrinsic erratic nature of weather, as well as on the difficulties of disentangling the climate change contribution to weather variability. However, changes in some types of extreme events have already been observed as, for example, increases in the frequency and intensity of heat waves and heavy precipitation events. Since 1950, the number of heat waves has increased and widespread increases have occurred in warm nights (Trenberth et al. 2007). In addition, the extent of regions affected by droughts has also increased as precipitation over land has marginally decreased, while evaporation has increased due to warmer conditions. In general, the number of heavy daily precipitations that lead to flooding has also risen, but not everywhere. On the other hand, it is well-known that tropical storm and hurricane frequencies vary from year to year, but evidence suggests substantial increases in intensity and duration since the 1970s (Trenberth et al. 2007). In the extra-tropics, variations in tracks and intensity of storms are a reflection of variations in major features of the atmospheric circulation, such as the North Atlantic oscillation.

In a warmer future climate, there will be an increased risk of more intense, more frequent, and longer-lasting heat waves. The European heat wave of 2003 was a clear example of the type of extreme heat event lasting from several days to over a week that is likely to become more common in a warmer future climate (Meehl et al. 2007). Most atmosphere/ocean GCMs predict increased dryness during summer and increased wetness during winter in most parts of northern middle and high latitudes (Meehl et al. 2007). Therefore, along with the risk of droughts, there will be an increased chance of intense precipitation and flooding due to the greater water-holding capacity of a warmer atmosphere so that intense and heavy downpours will be interspersed with longer relatively dry periods.

There is evidence from modelling studies that future tropical cyclones could become more severe, with greater wind speeds and more intense rainfalls (Bender et al. 2010). While it has been suggested that such changes may already be underway, there are clear indications that the average number of Category 4 and 5 hurricanes per year has increased over the past 30 years (McQuaid 2012). However, the overall frequency of Atlantic hurricanes is not expected to increase dramatically as the climate warms (Knutson et al. 2008; Zhao et al. 2009). In fact, the signal forced by greenhouse gases is a long-term trend, and a period of 30 years is too short to be able to distinguish a long-term trend from the multi-decadal fluctuations that are known to exist in the Atlantic (Landsea 2007). While the effects of global warming on hurricanes is still a matter of debate, climatic changes are responsible for the rise of the global sea levels at a rate of about 1.7 mm per year between 1950 and 2009, and at an accelerated pace of 3.3 mm from 1993 on, due to the expansion of warmer waters, ice melting in the poles, and shift of rainfall patterns. The northeast Atlantic coast is one region where this phenomenon is underway. A recent study has shown that sea levels from North Carolina to Canada have been rising at three to four times the global average since 1950 (Sallenger et al. 2012). By definition, higher seas mean higher storm surges, and hence huge storms. Whether amplified by global warming or not, they can go from destructive to catastrophic, implying that danger is compounded by the fact that most coastal fortifications were built when sea levels were lower, on the assumption that conditions would not change.

3.2 General Circulation Models

A general circulation model, often shortened as GCM, uses essentially the same partial differential equations of motion as a NWP model. The abbreviation GCM also refers to a global climate model, which is almost the same as a general circulation model, except that the former is used when the model is dealing specifically with global climate change. Although the main purpose of GCMs is to numerically predict changes in climate as a result of slow changes in some boundary conditions or physical parameters, such as, for example, the greenhouse gas concentration, they can also be used for weather forecasting as well as for understanding climate. In general, NWP models are employed to predict the weather in the short (from 1 to 3 days) and medium (from 4 to 10 days) range in the future, while GCMs are run much longer in time (from years to decades and decades to centuries) to learn about the climate in a statistical sense. A good NWP model can accurately predict the movement and evolution of atmospheric disturbances such as frontal systems and tropical cyclones. Although GCMs are capable to do this as well, most of them err so much after about 2 weeks or so, becoming useless for a perspective of weather forecasting in the long term. For example, an error in the sea surface temperature of a few centigrades, or even a small but systematic bias in cloudiness throughout the model, matter little to a NWP model, but for a GCM these factors are of great importance because they are relevant over a long-term evolution.

State-of-the-art GCMs use models capable of simulating surface and deep ocean circulations coupled to atmospheric GCMs. These models can be further coupled to dynamic models of sea ice and conditions on land. Coupled atmosphere/ocean GCMs are the models most often used to make predictions of future climate. These are very data intensive and require the most powerful supercomputers in the world to run. A recent trend in GCMs is to apply them as components of Earth System Models, which consist of coupling GCMs to ice sheet models for the dynamics of the Greenland and Antarctic ice sheets as well as to one or more chemical transport models (CTMs) for species relevant to climate. For example, a carbon CTM may allow a GCM to better predict changes in CO_2 concentrations resulting from changes in anthropogenic emissions. In addition, this approach allows accounting for inter-system feedbacks as may be the effects of climate change on the recovery of the ozone hole (Allen 2004). Uncertainties in climate prediction depend on uncertainties in chemical, physical, and social models (Kerr 2001). In other words, even though progress has been made in incorporating more realistic chemistry and physics in the models, significant uncertainties and unknowns remain, especially regarding the future course of human population, industry, and technology.

The first long-range simulation of the general circulation of the atmosphere was carried out by Phillips (1956), using a two-level, quasi-geostrophic model on a β-plane channel with rudimentary physics. Following Phillips' seminal work, several GCMs were developed. One early model of particular interest is the Kasahara-Washington model (Kasahara and Washington 1967), which was developed at NCAR. After several attempts to create a basic representation of large-scale atmospheric flow, scientists at Princeton University's Geophysical Fluid Dynamics Laboratory (GFDL) produced a model that incorporated large eddies, making the simulation much more representative of the atmosphere (Smagorinsky 1963; Smagorinsky et al. 1965). This experiment was deemed a major success and the model was considered to be the first true GCM. With this success research groups at UCLA's Lawrence Livermore National Laboratory (LLNL) and NCAR began to develop their own models (Ghan et al. 1982; Williamson 1983; Cess et al. 1985; Cess and Potter 1987; Williamson et al. 1987; Williamson and Olson 1994; Collins et al. 2004). Even with drastic advances in technology and scientific knowledge, climatologists still have to make many compromises in terms of realistically modelling the Earth. For example, until recently most models focused only on atmospheric circulation—ECHAM5 is an example of a relatively recent atmospheric GCM code developed by the Max Planck Institute for Meteorology (Roeckner et al. 2003)—and it was only during the 1990s that the first atmosphere/ocean coupled models began to appear. One important drawback of these models was an extremely coarse resolution so that many processes had to be parameterized, small-scale disturbances like thunderstorms and cyclones were not accounted for, and peninsulas, islands, and great lakes did not exist. While fine resolution may be ideal, a balance must always be struck between model resolution and computer power available.

As climate models evolved through the 1990s, scientists began to shift from reproducing general circulation to experimenting with the feedbacks of climatic processes due to increasing greenhouse gases, changing ocean currents, and the way the model responds to forced perturbations such as ENSO. As the next generation of models comes out, improvements and sophistications make them more reliable for global predictions and more capable of regional analyses. Examples of such models are the latest version of the Community Climate System Model CCSM3 (Gent 2006; Collins et al. 2006), the HadCM3—a well established coupled climate model that is cheap to run in current computers (Gordon et al. 2000; Pope et al. 2000)—and HadGEM1—a state-of-the-art global environment model (Johns et al. 2006; Martin et al. 2006; Stott et al. 2006)—both used at the Hadley Centre for climate modelling in the United Kingdom, the GISS GCM ModelE—the current incarnation of the GISS series of coupled atmosphere/ocean models developed by the National Aeronautics and Space Administration (NASA) (Schmidt et al. 2006),—and the European Centre for Medium-Range Weather Forecasts (ECMWF) coupled global model (Bechtold et al. 2008a,b). In particular, these models provide the ability to simulate many different configurations of Earth System Models, including interactive atmospheric chemistry, aerosols, carbon cycle and other tracers, as well as the standard atmosphere, ocean, sea ice, and land surface components.

Today, citizens and policy-makers want to know what heat waves, droughts, or floods are likely to occur in their particular region. Since the attention of the community turned to making predictions in ever more detail, only models that incorporate a much more realistic ocean and clouds would be able to calculate that. A scheme for representing clouds was developed in the 2000s at the Max Planck Institute for Meteorology (Gramelsberger 2010). This code uses 79 equations to describe the formation of stratiform clouds, incorporating a variety of constants, some known precisely from experiments or observations, and some others that had to be adjusted. The computation for each grid cell was a challenge even for the fastest supercomputers. Looking farther afield, the future climate system could not be determined very accurately until ocean/atmosphere GCMs are linked interactively with models for changes in vegetation. Dark forests and bright deserts do not only respond to climate, but also influence it. Since the early 1990s, and particularly during the last decade, the more advanced GCMs had incorporated dynamic global vegetation models suitable for use in NWP models and coupled GCMs, allowing for the simulation of vegetation-atmosphere interactions, photosynthesis and respiration processes as well as the representation of regional properties of vegetation (Quillet et al. 2010).

4 Predictability and Ensemble Forecasting

The accuracy of a given forecast depends on the internal error growth of the model, the model accuracy, and the errors in the initial state. When solving the equations at a global scale, the boundaries are periodic and the problem is an *initial value* problem. The initial conditions are integrated forward in time to obtain future states of the system. However, due to the intrinsic non-linear nature of the equations, the information of the initial conditions is lost within a few days, and the exact state of the system (weather) becomes unpredictable. In other words, given the chaotic nature of the atmosphere, we can never create a perfect forecast system because it is impossible to observe every detail of the atmosphere's initial state. Therefore, tiny errors in the initial conditions will be amplified, always imposing a limit to how far ahead we can predict any detail. If, on the other hand, we are interested in climate predictions, statistics of the system can still be obtained. In this case, the initial conditions become unimportant and the problem reduces to a *boundary value* problem. Only the response of the climate to external forcings such as changes in solar radiation, concentration of greenhouse gases, etc., is of interest.

The first attempt to address this problem in NWPs has been to calculate how errors of the initial state are likely to grow in particular meteorological situations. For instance, Epstein (1969) first proposed using an ensemble of stochastic Monte Carlo simulations to produce means and variances for the state of the atmosphere, and successively it was demonstrated that these simulations produced adequate forecasts only when the ensemble probability distribution was a representative sample of the probability distribution of the atmosphere (Leith 1974). Accepting the findings from chaos theory about the sensitivity of the prediction to uncertainties in the initial conditions, it has become a common practice to undertake a set of forecasts, or ensemble, with the same model but starting the runs from slightly different initial conditions. Small perturbations are added to the reference model, with amplitudes selected to be within the accuracy of the initial state.

Starting in 1992, ensemble forecasts have been used operationally by the ECMWF and the NCEP to account for the stochastic nature of weather processes. In particular, the ECMWF has made major contributions to this technique and has over the last years also developed and improved an operational system for ensemble prediction (Molteni et al. 1996; Mullen and Buizza 2002; Buizza et al. 1999, 2003, 2007). The ECMWF weather prediction model is run 51 times from slightly different initial conditions. One forecast, called the EPS control forecast, is run from the operational ECMWF analysis, followed by 50 additional integrations, called the perturbed members, which are designed to represent the uncertainties inherent in the operational analysis. The initial perturbations are generated using the singular vector technique to simulate the initial probability density (Barkmeijer et al. 1999). In contrast, the NCEP ensemble—the Global Ensemble Forecasting System—uses bred vectors, which are related to Lyapunov vectors and created by adding initially random perturbations to the model (Toth and Kalnay 1997; Kalnay 2003).

Ensemble forecastings are being used for many proposed problems, including global weather, hurricane track, intensity forecasts, and seasonal climate simulations. Seasonal forecasts, with a range of 6 months, are currently made using coupled GCMs by combining large numbers of forecasts in an ensemble each month, with impressive predictions for tropical regions and for the onset of El Niño and La Niña events. In the same way that many forecasts from a single model can be used to form an ensemble, multiple models can also be combined to produce an ensemble forecast. This approach is called *multi-model ensemble forecasting*, and it has been shown to improve forecasts when compared to a single model-based approach (Krishnamurti et al. 2000; Weigel et al. 2008, 2009; Zhou and Du 2010). One recent multi-model concept to medium-range weather forecasts is the THORPEX Interactive Grand Global Ensemble (TIGGE) (Bougeault et al. 2010). However, a recent comparison of the TIGGE multi-model forecasts with reforecast-calibrated ECMWF ensemble forecasts in extra-tropical regions has shown that the latter were of comparable or superior quality to the multi-model predictions (Hagedorn et al. 2012). The reforecast calibration procedure is particularly helpful at locations with clearly detectable systematic errors such as areas with complex orography or coastal grid points, while the multi-model approach might be advantageous in situations where it is able to suggest alternative solutions not predicted by the single-model of choice. Therefore, it would be desirable in the not so distant future to explore the relative merits of multi-model versus reforecast-calibrated predictions for other user-relevant variables such as precipitation and wind speed. Moreover, models within a multi-model ensemble can be adjusted for their various biases, which is a process known as *superensemble forecasting*. This type of forecast significantly reduces errors in the model output (Cane and Milelli 2010).

5 Future Perspectives and Challenges

Decision-makers from 155 nations agreed in 2009 to establish the world's first framework for climate services, an effort that will supply on-demand climate predictions to governments, businesses, and individuals. By providing tailored information on how climate change will affect certain regions and sectors, the Global Framework for Climate Services will help the world better adapt to the challenges of climate variability and change. This vision marks a new era in climate science, one in which seasonal weather forecasting and long-term climate projections will merge seamlessly, giving rise to decadal climate predictions that have the skill and reliability of weather forecasts. Provision of these data to local planners and policy-makers will really be a service to society.

Evidence that climate predictions can provide precise and accurate guidance about how the long-term future may evolve is basically lacking. In this sense, scientists and decision-makers alike should think of climate models as just one of a range of tools to explore future possibilities. Unfortunately, predictive skill is unknown for climate at the decade-to-century timescale. Unlike weather forecasts, whose value in informing decision-making can routinely be tested over time by comparison with observed weather patterns, there is currently no such empirical evidence with which to test the skill of climate predictions. Certainly, as knowledge of the climate system and how it responds to greenhouse gases improves, model predictions will change, as will their probability distributions.

The sophistication of prediction models is closely linked to the available computer power. The advances in digital computer technology as well as the developments in atmospheric dynamics, instrumentation, and observing practice have all pointed towards increasing forecast accuracy apace over the half-century of NWP activities, and progress continues on several fronts. However, some formidable challenges remain. The effective computational coupling between the dynamical processes and physical parameterizations is one of these big challenges. On the other hand, nowcasting is the process of predicting changes over periods of a few hours. Current numerical methods provide guidance which occasionally falls short of what is required to take effective action and avert disasters. Although greatest value is obtained by a systematic combination of NWP products with conventional observations as well as radar and satellite imageries, much remains to be done to develop optimal nowcasting systems. On the other side, the chaotic nature of the atmosphere imposes limitations to the validity of deterministic forecasting. Ensemble forecasts provide probabilistic guidance, but so far their use has proved to be quite difficult in many cases. While reasonably good progress in seasonal forecasting for the tropics has been made, long-range forecasts for temperate regions remain a further challenge. Accompanied to this is the modelling and prediction of climate change, a matter of increasing importance and concern. As technology continues advancing at a faster rate than ever, we may be optimistic that future developments will lead to notable improvements in both weather and climate prediction.

Acknowledgments This work has been partially supported by the Fondo Nacional de Ciencia, Tecnología e Innovación of Venezuela (FONACIT) under grant PC 201204710 (contract 112-1077) and the Consejo Nacional de Ciencia y Tecnología of Mexico (CONACyT) under the project CONACyT-EDOMEX-2011-C01-165873.

References

Abbe C (1901) The physical basis of long-range weather forecasts. Mon Weather Rev 29:551–561

Allen J (2004) Tango in the atmosphere: ozone and climate change. NASA Earth Observatory (http://www.theozonehole.com/climate.html)

Arakawa A, Suarez MJ (1983) Vertical differencing of the primitive equations in sigma coordinates. Mon Weather Rev 111:34–45

Arakawa A (1993) Closure assumptions in the cumulus parameterization problem. In: Emanuel KA, Raymond DJ (eds) The representation of cumulus convection in numerical models. American Meteorological Society, Boston, pp 1–16

Arakawa A, Konor CS (1996) Vertical differencing of the primitive equations based on the Charney-Phillips grid in hybrid σ-p vertical coordinates. Mon Weather Rev 124:511–528

Avila LA, Clark GB (1989) Atlantic tropical systems of 1988. Mon Weather Rev 117:2260–2265

Baklanov A, Grimmond S, Mahura A, Athanassiadou M (2009) Meteorological and air quality models for urban areas. Springer, Heidelberg

Barkmeijer J, Buizza R, Palmer TN (1999) 3D-Var Hessian singular vectors and their potential use in the ECMWF ensemble prediction system. Quart J Royal Meteorol Soc 125:2333–2351

Barnier B, Marchesiello P, Pimenta de Miranda A, Molines J-M, Coulibaly M (1998) A sigma-coordinate primitive equation model for studying the circulation in the South Atlantic. Part I: Model configuration with error estimates. Deep Sea Res Ocean Res Pap 45:543–572

Bechtold P, Köhler M, Jung T, Leutbecher M, Rodwell M, Vitart F (2008a) Advances in simulating atmospheric variability with IFS cycle 32r3. ECMWF Newslett 114:29–38

Bechtold P, Köhler M, Jung T, Doblas-Reyes F, Leutbecher M et al (2008b) Advances in simulating atmospheric variability with the ECMWF model: from sypnotic to decadal time-scales. Quart J Royal Meteorol Soc 134:1337–1351

Beckers JM (1991) Application of a GHER 3D general circulation model to the Western Mediterranean. J Marine Syst 1:315–332

Bender MA, Knutson TR, Tuleya RE, Sirutis JJ, Vecchi GA et al (2010) Modeled impact of anthropogenic warming on the frequency of intense Atlantic hurricanes. Science 327:454–458

Bengtsson L (1964) Some numerical experiments on the effect of the variation of static stability in two-layer quasi-geostrophic models. Tellus 16:328–348

Bengtsson L (1999) From short-range barotropic modelling to extended-range global weather prediction: A 40-year perspective. Tellus B 51:13–32

Bennett AF, Leslie LM, Hagelberg CR, Powers PE (1993) Tropical cyclone prediction using a barotropic model initialized by a generalized inverse method. Mon Weather Rev 121:1714–1729

Bjerknes V (1904) Das Problem der Wettervorhersage, betrachtet vom Standpunkte der Mechanik und der Physik. Meteorologische Zeitschrift 21:1–7

Botzen WJW, Gowdy JM, van den Bergh JCJM (2008) Cumulative CO_2 emissions: shifting international responsabilities for climate debt. Clim Pol 8:569–576

Bougeault P, Toth Z, Bishop C, Brown B, Burridge D et al (2010) The THORPEX interactive grand global ensemble. Bull Am Meteorol Soc 91:1059–1072

Briggs D, Smithson P (1986) The fundamentals of physical geography. Rowman & Littlefield Publishers, New Jersey

Bromirski PD, Sergienko OV, MacAyeal DR (2010) Transoceanic infragravity waves impacting Antarctic ice shelves. Geophys Res Lett 37:L02502

Buizza R, Miller M, Palmer TN (1999) Stochastic representation of model uncertainties in the ECMWF ensemble prediction system. Q J Royal Meteorol Soc 125:2887–2908
Buizza R, Richardson DS, Palmer TN (2003) Benefits of increased resolution in the ECMWF ensemble system and comparison with poor-man's ensembles. Quart J Royal Meteorol Soc 129:1269–1288
Buizza R, Bidlot J-R, Wedi N, Fuentes M, Hamrud M et al (2007) The new ECMWF VAREPS (Variable Resolution Ensemble Prediction System). Quart J Royal Meteorol Soc 133:681–695
Burpee RW (2008) The Sanders barotropic tropical cyclone track prediction model (SANBAR). Meteorol Monogr 33:233–240
Cane D, Milelli M (2010) Multimodel superensemble technique for quantitative precipitation forecasts in Piemonte region. Nat Hazards Earth Syst Sci 10:265–273
Cess RD, Potter GL, Ghan SJ, Gates WL (1985) The climatic effects of large injections of atmospheric smoke and dust: a study of climate feedback mechanisms with one- and three-dimensional climate models. J Geophys Res Atmos 90:12937–12950
Cess RD, Potter GL (1987) Exploratory studies of cloud radiative forcing with a general circulation model. Tellus A 39:460–473
Chan JC-L (1984) An observational study of the physical processes responsible for tropical cyclone motion. J Atmos Sci 41:1036–1048
Chan F, Barth JA, Lubchenco J, Kirincich A, Weeks H et al (2008) Emergence of anoxia in the California current large marine ecosystem. Science 319:920–920
Charney JG (1947) The dynamics of long waves in a baroclinic westerly current. J Meteorol 4:136–162
Charney JG (1948) On the scale of atmospheric motions. Geofysiske Publikasjoner 17:3–17
Charney JG, Fjörtoft R, von Neumann J (1950) Numerical integration of the barotropic vorticity equation. Tellus 2:237–254
Charney JG, Phillips NA (1953) Numerical integration of the quasi-geostrophic equations for barotropic and simple baroclinic flows. J Meteorol 10:71–99
Charney JG (1955) The use of the primitive equations of motion in numerical prediction. Tellus 7:22–26
Charney JG, Arakawa A, Baker DJ, Bolin B, Dickinson RE et al (1979) Carbon dioxide and climate: a scientific assessment. Report to the Climate Research Board. Assembly of Mathematical and Physical Sciences,National Research Council, Washington, 22 pp
Chaudhari HS, Lee KM, Oh JH (2007) Weather prediction and computational aspects of icosahedral-hexagonal gridpoint model GME. In: Kwon JH, Ecer A, Periaux J, Satofuka N, Fox P (eds) Parallel computational fluid mechanics: parallel computings and its applications. Elsevier, Amsterdam, pp 223–230
Chen Y-L, Ogura Y (1982) Modulation of convective activity by large-scale flow patterns observed in GATE. J Atmos Sci 39:1260–1279
Chynoweth S, Sewell MJ (1991) A concise derivation of the semi-geostrophic equations. Quart J Roy Meteorol Soc 117:1109–1128
Collins WD, Rasch PJ, Boville BA, Hack JJ, McCaa JR et al (2004) Description of the NCAR Community Atmosphere Model (CAM 3.0). NCAR Technical note NCAR/TN−464+STR. National Center of Atmospheric Research, Boulder, Colorado, 214 pp
Collins WD, Bitz CM, Blackmon ML, Bonan GB, Bretherton CS et al (2006) The community climate system model version 3 (CCSM3). J Clim 19:2122–2143
Cox JD (2002) Storm watchers: the turbulent history of weather prediction from Franklin's Kite to El Niño. Wiley, New Jersey
Cronin TN (2010) Paleoclimates: understanding climate change past and present. Columbia University Press, New York
Davis RE (1991) Lagrangian ocean studies. Ann Rev Fluid Mech 23:43–64
DeMaria M (1985) Tropical cyclone motion in a nondivergent barotropic model. Mon Weather Rev 113:1199–1210

D'Orgeville M, Peltier WR (2009) Implications of both statistical equilibrium and global warming simulations with CCSM3. Part I: On the decadal variability in the North Pacific basin. J Clim 22:5277–5297

Eady ET (1949) Long waves and cyclone waves. Tellus 1:33–52

Eliassen A (1949) The quasi-static equations of motion with pressure as independent variable. Geofysiske Publikasjoner 17(3):5–44

Eliassen A, Raustein E (1968) A numerical integration experiment with a model atmosphere based on isentropic surfaces. Meteorologiske Annaler 5:45–63

Eliassen A, Raustein E (1970) A numerical integration experiment with a six-level atmospheric model with isentropic information surface. Meteorologiske Annaler 5:429–449

Ellsaesser HW (1968) Diagnosis of early baroclinic NWP models. J Appl Meteorol 7:153–159

Emanuel KA (1994) Atmospheric convection. Oxford University Press, Oxford

Emile-Geay J, Cobb KM, Mann ME, Wittenberg AT (2013a) Estimating central equatorial Pacific SST variability over the past millennium. Part I: Methodology and validation. J Clim 26: 2302–2328

Emile-Geay J, Cobb KM, Mann ME, Wittenberg AT (2013b) Estimating central equatorial Pacific SST variability over the past millennium. Part II: Reconstructions and implications. J Clim 26:2329–2352

Epstein ES (1969) Stochastic dynamic prediction. Tellus 21:739–759

Ezer T, Mellor GL (1997) Simulations of the Atlantic Ocean with a free surface sigma coordinate ocean model. J Geophys Res Oceans 102:15647–15657

Farman JC, Gardiner BG, Shanklin JD (1985) Large losses of total ozone in Antarctica reveal seasonal ClO_x/NO_x interaction. Nature 315:207–210

Fraedrich K, Frisius T (2001) Two-level primitive equation baroclinic instability on an f-plane. Q J Roy Meteorol Soc 127:2053–2068

Gent PR (2006) Preface to special issue on community climate system model (CCSM). J Clim 19:2121–2121

Ghan SJ, Lingaas JW, Schlesinger ME, Mobley RL, Gates WL (1982) A documentation of the OSU two-level atmospheric general circulation model. Climate Research Institute Report No. 35. Oregon State University, Corvallis, Oregon

Gordon C, Cooper C, Senior CA, Banks H, Gregory JM et al (2000) The simulation of SST, sea ice extents and ocean heat transports in a version of the Hadley Centre coupled model without flux adjustments. Clim Dyn 16:147–168

Gramelsberger G (2010) Conceiving processes in atmospheric models-General equations, subscale parameterizations, and superparameterizations. Stud History Philos Sci B 41:233–241

Grantham BA, Chan F, Nielsen KJ, Fox DS, Barth JA et al (2004) Upwelling-driven nearshore hypoxia signals ecosystem and oceanographic changes in the northeast Pacific. Nature 429:749–754

Gray WM (1979) Hurricanes: their formation, structure, and likely role in the tropical circulation. In: Shaw DB (ed) Meteorology over the tropical oceans. Royal Meteorol Society, London, pp 155–218

Guinn TA, Schubert WH (1993) Hurricane spiral bands. J Atmos Sci 50:3380–3403

Hagedorn R, Buizza R, Hamill TM, Leutbecher M, Palmer TN (2012) Comparing TIGGE multimodel forecasts with reforecast-calibrated ECMWF ensemble forecasts. Quart J Royal Meteorol Soc 138:1814–1827

Haidvogel DB, Beckmann A (1999) Numerical ocean circulation modeling. World Scientific Publishing, London

Hartmann DL, Maloney ED (2001) The Madden-Julian oscillation, barotropic dynamics, and North Pacific tropical cyclone formation. Part II: stochastic barotropic modeling. J Atmos Sci 58:2559–2570

Heta Y (1991) On the origin of tropical disturbances in the equatorial Pacific. J Meteorol Soc Jpn 69:337–351

Hinds WC (1999) Aerosol technology: properties, behavior, and measurement of airborne particles. Wiley, New York
Hinkelmann K (1951) Der Mechanismus des meteorologischen Lärmes. Tellus 3:285–296
Hinkelmann K (1959) Ein numerisches experiment mit den primitiven Gleinchungen. In: Bolin B, Eriksson E (eds) The atmosphere and the sea in motion-scientific contributions to the Rossby Memorial volume. Rockefeller Institute Press, New York, pp 486–500
Höhne N, Blum H, Fuglestvedt J, Skeie RB, Kurosawa A, et al. (2010) Contributions of individual countries' emissions to climate change and their uncertainty. In: Climate change. Springer Science+Business, Media B.V., Dordrecht.10.1007/s10584-010-9930-6, p 33.
Holland GJ (1983) Tropical cyclone motion: environmental interaction plus a beta effect. J Atmos Sci 40:328–342
Holland GJ (1984) Tropical cyclone motion: a comparison of theory and observation. J Atmos Sci 41:68–75
Huck PE, Tilmes S, Bodeker GE, Randel WJ, McDonald AJ, Nakajima H (2007) An improved measure of ozone depletion in the Antarctic stratosphere. J Geophys Res Atmos 112:D11104
Johns TC, Durman CF, Banks HT, Roberts MJ, McLaren AJ et al (2006) The new Hadley Centre climate model (HadGEM1): evaluation of coupled simulations. J Clim 19:1327–1353
Jones AE, Shanklin JD (1995) Continued decline of total ozone over Halley, Antarctica, since 1985. Nature 376:409–411
Jordan CL (1956) An experiment in low-latitude numerical prediction with the barotropic model. J Meteorol 13:223–235
Kalnay E (2003) Atmospheric modeling, data assimilation and predictability. Cambridge University Press, Cambridge
Kasahara A (1960) The numerical prediction of hurricane movement with a two-level baroclinic model. J Meteorol 17:357–370
Kasahara A, Washington WM (1967) NCAR global general circulation model of the atmosphere. Mon Weather Rev 95:389–402
Kasahara A (1974) Various vertical coordinate systems used for numerical weather prediction. Mon Weather Rev 102:509–522
Kerr RA (2001) Global warming: Rising global temperature, rising uncertainty. Science 292:192–194
Kiehl JT, Trenberth KE (1997) Earth's annual global mean energy budget. Bull Am Meteorol Soc 78:197–208
Knutson TR, Sirutis JJ, Garner ST, Vecchi GA, Held IM (2008) Simulated reduction in Atlantic hurricane frequency under twenty-first century warming conditions. Nat Geosci 1:359–364
Krishnamurti TN, Pasch RJ, Ardanuy P (1980) Prediction of African waves and specification of squall lines. Tellus 32:215–231
Krishnamurti TN (1995) Numerical weather prediction. Ann Rev Fluid Mech 27:195–224
Krishnamurti TN, Kishtawal CM, Zhang Z, LaRow T, Bachiochi D, Williford E (2000) Multimodel ensemble forecasts for weather and seasonal climate. J Clim 13:4196–4216
Landsea CW (2007) Counting Atlantic tropical cyclones back to 1900. Eos Trans Am Geophys Union 88:197–202
Leith CE (1965) Numerical simulation of the Earth's atmosphere. In: Alder B, Fernbach S, Rotenberg M (eds) Methods in computational physics, vol 4. Academic Press, New York, pp 1–28
Leith CE (1974) Theoretical skill of Monte Carlo forecasts. Mon Weather Rev 102:409–418
Leslie LM, Dietachmayer GS (1992) Real-time limited area numerical weather prediction in Australia: a historical perspective. Aust Meteorol Mag 41:61–77
Di Lorenzo E, Schneider N, Cobb KM, Franks PJS, Chhak K et al (2008) North Pacific Gyre Oscillation links ocean climate and ecosystem change. Geophys Res Lett 35:L08607
Lynch P (2006) The emergence of numerical weather prediction: Richardson's dream. Cambridge University Press, Cambridge
Lynch P (2008) The origins of computer weather prediction and climate modeling. J Comput Phys 227:3431–3444

Manney GL, Santee ML, Rex M, Livesey NJ, Pitts MC et al (2011) Unprecedented Arctic ozone loss in 2011. Nature 478:469–475

Marshall J, Plumb RA (2008) Atmosphere, ocean, and climate dynamics: an introductory text. Elsevier Academic Press, Amsterdam

Martin GM, Ringer MA, Pope VD, Jones A, Dearden C, Hinton TJ (2006) The physical properties of the atmosphere in the new Hadley Centre global environmental model (HadGEM1). Part I: model description and global climatology. J Clim 19:1274–1301

Matsumoto S (1956) A kinematic method to compute the steering velocity of a typhoon. J Meteorol Soc Jpn 34:1–10

McGuffie K, Henderson-Sellers A (2005) A climate modelling primer. Wiley, Chichester

McQuaid J (2012) Hurricanes and climate change. (www.pbs.org/wgbh/nova/earth/hurricanes-climate.html).

Meehl GA, Stocker TF, Collins WD, Friedlingstein P, Gaye AT, et al (2007) Global climate projections. In: Solomon S, Qin D, Manning M, Chen Z, Marquis M, Averyt KB, Tignor M, Miller HL (eds) Climate change 2007: The physical science basis. Contribution of working group I to the 4th assessment report of the intergovernmental panel on climate change. Cambridge University Press, Cambridge pp 747–845.

Melnikova IN, Vasilyev AV (2005) Short-wave solar radiation in the earth's atmosphere: calculation, observation, interpretation. Springer, Heidelberg

Molteni F, Buizza R, Palmer TN, Petroliagis T (1996) The ECMWF ensemble prediction system: methodology and validation. Q J Royal Meteorol Soc 122:73–119

Mullen SL, Buizza R (2002) The impact of horizontal resolution and ensemble size on probabilistic forecasts of precipitation by the ECMWF ensemble prediction system. Weather Forecast 17:173–191

Nadiga BT, Aurnou JM (2008) Hands-on oceanography. A tabletop demonstration of atmospheric dynamics: Baroclinic instability. Oceanography 21:196–201

Narita M, Ohmori S (2007) Improving precipitation forecasts by the operational nonhydrostatic mesoscale model with the Kain-Fritsch convective parameterization and cloud microphysics. In: 12th Conference on Mesoscale Processes. American Meteorological Society, Boston. http://ams.confex.com/ams/pdfpapers/126017.pdf. p 6

Neumann CJ, Pelissier JM (1981) Models for the prediction of tropical cyclone motion over the North Atlantic: an operational evaluation. Mon Weather Rev 109:522–538

Newman PA, Daniel JS, Waugh DW, Nash ER (2007) A new formulation of equivalent effective stratospheric chlorine (EESC). Atmos Chem Phys 7:4537–4552

Nieto Ferreira R, Schubert WH (1997) Barotropic aspects of ITCZ breakdown. J Atmos Sci 54:261–285

Nitta T, Nagakomi Y, Suzuki Y, Hasegawa N, Kadokura A (1985) Global analysis of the lower-tropospheric disturbances in the tropics during the northern summer of the FGGE year. Part I: global features of the disturbances. J Meteorol Soc Jpn 63:1–19

Nitta T, Takayabu Y (1985) Global analysis of the lower tropospheric disturbances in the tropics during the northern summer of the FGGE year. Part II: Regional characteristics of the disturbances. Pure Appl Geophys 123:272–292

Norquist DC, Recker EE, Reed RJ (1977) The energetics of African wave disturbances as observed during phase III of GATE. Mon Weather Rev 105:334–342

Petty GW (2008) A first course in atmospheric thermodynamics. Sundog Publishing, Madison

Phalen RF, Phalen RN (2012) Introduction to air pollution science. A public health perspective, Jones and Bartlett Learning, Burlington

Philander SG (1990) El Niño, La Niña, and the Southern Oscillation. Academic Press, San Diego

Phillips NA (1951) A simple three-dimensional model for the study of large-scale extratropical flow patterns. J Meteorol 8:381–394

Phillips NA (1954) Energy transformations and meridional circulations associated with simple baroclinic waves in a two-level, quasi-geostrophic model. Tellus 6:273–286

Phillips NA (1956) The general circulation of the atmosphere: a numerical experiment. Q J Royal Meteorol Soc 82:123–164
Phillips NA (1957) A coordinate system having some special advantages for numerical forecasting. J Meteorol 14:184–185
Phillips NA (1960) On the problem of initial data for the primitive equations. Tellus 12:121–126
Phillips NA (1970) Models for weather prediction. Ann Rev Fluid Mech 2:251–292
Platzman GW (1979) The ENIAC computations of 1950-gateway to numerical weather prediction. Bull Am Meteorol Soc 60:302–312
Pope VD, Gallani ML, Rowntree PR, Stratton RA (2000) The impact of new physical parametrizations in the Hadley Centre climate model: HadAM3. Clim Dyn 16:123–146
Quillet A, Peng C, Garneau M (2010) Toward dynamic global vegetation models for simulating vegetation-climate interactions and feedbacks: recent developments, limitations, and future challenges. Envtl Rev 18:333–353
Rahmstorf S (2003) The current climate: thermohaline circulation. Nature 421:699–699
Reed RJ, Norquist DC, Recker EE (1977) The structure and properties of African wave disturbances as observed during phase III of GATE. Mon Weather Rev 105:317–333
Reiser H (1986) Obituary of Karl Heinz Hinkelmann. World Meteorol Org Bull 35:393–394
Richardson LF (1922) Weather prediction by numerical process. Cambridge University Press, Cambridge
Riehl H (1945) Waves in the easterlies and the polar front in the tropics. Issue 17 of miscellaneous reports. University of Chicago, Department of Meteorology, p 79
Roeckner E, Bäuml G, Bonaventura L, Brokopf R, Esch M, et al (2003) The atmospheric general circulation model ECHAM5. Part I. Model description. Report No. 349, Max Planck Institute for Meteorology, Hamburg, Germany, ISSN 0937–1060, p 127.
Rõõm R, Männik A, Luhamaa A (2007) Non-hydrostatic semi-elastic hybrid-coordinate SISL extension of HIRLAM. Part I: Numerical scheme. Tellus A 59:650–660
Sabine CL, Feely RA, Gruber N, Key RM, Lee K et al (2004) The oceanic sink for anthropogenic CO_2. Science 305:367–371
Saha K, Sanders F, Shukla J (1981) Westward propagating predecessors of monsoon depressions. Mon Weather Rev 109:330–343
Sallenger AH Jr, Doran KS, Howd PA (2012) Hotspot of accelerated sea-level rise on the Atlantic coast of North America. Nat Clim Chang 2:884–888
Sanders F, Burpee RW (1968) Experiments in barotropic hurricane track forecasting. J Appl Meteorol 7:313–323
Sanders F, Adams AL, Gordon NJB, Jensen WD (1980) Further development of a barotropic operational model for predicting paths of tropical storms. Mon Weather Rev 108:642–654
Schmidt GA, Ruedy R, Hansen JE, Aleinov I, Bell N et al (2006) Present-day atmospheric simulations using GISS ModelE: comparison to in situ, satellite, and reanalysis data. J Clim 19:153–192
Schmidt GA (2007) The physics of climate modeling. Phys Today 60:72–73
Schmidt GA, Ruedy RA, Miller RL, Lacis AA (2010) Attribution of the present-day total greenhouse effect. J Geophys Res Atmos 115:D20106
Seinfeld JH, Pandis SN (2006) Atmospheric chemistry and physics: from air pollution to climate change. Wiley, New Jersey
Sela J, Bostelman WJ (1973) A vertically integrated primitive-equation model. Mon Weather Rev 101:871–876
Shapiro MA (1973) Numerical weather prediction and analysis in isentropic coordinates. Atmos Technol 3:67–70
Shenk WE (1974) Cloud top height variability of strong convective cells. J Appl Meteorol 13:917–922
Shukla J, Saha KR (1970) Application of non-divergent barotropic model to predict flow patterns in the Indian region. J Meteorol Soc Japan 48:405–409
Shuman FG, Hovermale JB (1968) An operational six-layer primitive equation model. J Appl Meteorol 7:525–547

Shuman FG (1989) History of numerical weather prediction at the National Meteorological Center. Weather Forecast 4:286–296
Smagorinsky J (1963) General circulation experiments with the primitive equations. I. The basic experiment. Mon Weather Rev 91:99–164
Smagorinsky J, Manabe S, Holloway JL Jr (1965) Numerical results from a nine-level general circulation model of the atmosphere. Mon Weather Rev 93:727–768
Smith RK (1997) The physics and parameterization of moist atmospheric convection. Kluwer, Dordrecht
Sobel AH (2002) Water vapor as an active scalar in tropical atmospheric dynamics. Chaos 12: 451–459
Solomon S (1990) Progress towards a quantitative understanding of Antarctic ozone depletion. Nature 347:347–354
Song Y, Haidvogel D (1994) A semi-implicit ocean circulation model using a generalized topography-following coordinate system. J Comput Phys 115:228–244
Spall MA, Robinson AR (1989) A new open ocean, hybrid coordinate primitive equation model. Math Comput Simul 31:241–269
Steger W, Bowermaster J (1990) Saving the earth: a citizen's guide to environmental action. Knopf, New York
Stensrud DJ (2009) Parameterization schemes: keys to understanding numerical weather prediction models. Cambridge University Press, Cambridge
Stott PA, Jones GS, Lowe JA, Thorne P, Durman C et al (2006) Transient climate simulations with the HadGEM1 climate model: causes of past warming and future climate change. J Clim 19:2763–2782
Tai K-S, Ogura Y (1987) An observational study of easterly waves over the eastern Pacific in the northern summer using FGGE data. J Atmos Sci 44:339–361
Thorncroft CD, Hoskins BJ (1994) An idealized study of African easterly waves. I: a linear view. Q J Roy Meteorol Soc 120:953–982
Thorncroft CD, Hoskins BJ (1994) An idealized study of African easterly waves. II: a nonlinear view. Q J Roy Meteorol Soc 120:983–1015
Toth Z, Kalnay E (1997) Ensemble forecasting at NCEP and the breeding method. Mon Weather Rev 125:3297–3319
Trenberth KE, Jones PD, Ambenje P, Bojariu R, Easterling D, et al (2007) Observations: Surface and atmospheric climate change. In: Solomon S, Qin D, Manning M, Chen Z, Marquis M, Averyt KB, Tignor M, Miller HL (eds) Climate change 2007: the physical science basis. Contribution of working group I to the 4th assessment report of the intergovernmental panel on climate change. Cambridge, Cambridge University Press, pp 235–336.
van Dop H, Steyn DG (1991) Air pollution modeling and its applications VIII, vol 15. Springer Science + Business Media, New York.
Vederman J, Hirata GH, Manning EJ (1966) Forecasting in the tropics with a barotropic atmospheric model. Mon Weather Rev 94:337–344
Weigel AP, Liniger MA, Appenzeller C (2008) Can multi-model combination really enhance the prediction skill of probabilistic ensemble forecasts? Quart J Royal Meteorol Soc 134:241–260
Weigel AP, Bowler NE (2009) Comment on Can multi-model combination really enhance the prediction skill of probabilistic ensemble forecasts? Quart J Royal Meteorol Soc 135:535–539
Wiin-Nielsen A (1959) On barotropic and baroclinic models, with special emphasis on ultra-long waves. Mon Weather Rev 87:171–183
Williamson DL (1983) Description of NCAR Community Climate Model (CCM0B). NCAR Technical Note NCAR/TN−210+STR, National Center of Atmospheric Research, Boulder, Colorado, NTIS No. PB83 23106888, p 88.
Williamson DL, Kiehl JT, Ramanathan V, Dickinson RE, Hack JJ (1987) Description of NCAR Community Climate Model (CCM1). NCAR Technical note NCAR/TN−285+STR. National Center of Atmospheric Research, Boulder, Colorado, p 112

Williamson DL, Olson JG (1994) Climate simulations with a semi-Lagrangian version of the NCAR Community Climate Model. Mon Weather Rev 122:1594–1610
Willis EP, Hooke WH (2006) Cleveland Abbe and American meteorology, 1871–1901. Bull Am Meteorol Soc 87:315–326
Xue Y, Fennessy MJ, Sellers PJ (1996) Impact of vegetation properties on U.S. summer weather prediction. J Geophys Res Atmos 101:7419–7430
Yanai M, Esbensen S, Chu J-H (1973) Determination of bulk properties of tropical cloud clusters from large-scale heat and moisture budgets. J Atmos Sci 30:611–627
Yeh S-W, Kug J-S, Dewitte B, Kwon M-H, Kirtman BP, Jin F-F (2009) El Niño in a changing climate. Nature 461:511–514
Zehr RM (1993) Recognition of mesoscale vortex initiation as stage one of tropical cyclogenesis. Paper presented at the 20th conference on hurricanes and tropical meteorology, 10–14 May, San Antonio. American Meteorological Society, Boston, pp 405–408.
Zehr SC (1994) Accounting for the ozone hole: scientific representations of an anomaly and prior incorrect claims in public settings. Sociol Quart 35:603–619
Zhao M, Held IM, Lin S-J, Vecchi GA (2009) Simulations of global hurricane climatology, inter-annual variability, and response to global warming using a 50-km resolution GCM. J Clim 22: 6653–6678
Zhou B, Du J (2010) Fog prediction from a multimodel mesoscale ensemble prediction system. Weather Forecast 25:303–322

Turbulent Diffusion of Heat at High Rayleigh Numbers

Joseph J. Niemela

Abstract Thermal convection is observed in controlled laboratory experiments at very high Rayleigh numbers using a relatively large apparatus filled with low temperature helium gas. The low temperature environment offers two advantages toward the study of turbulent convection; namely the favorable properties of the working fluid in achieving very high Rayleigh numbers and the low thermal mass of the heated metallic surfaces at cryogenic temperatures. The latter property is exploited in order to provide a means of measuring an effective thermal diffusion coefficient of the buoyancy-driven turbulence by propagating thermal waves into the bulk and observing the damping of their amplitude with distance. The diffusivity measured directly in this way compares well with values inferred from the time-independent measurements of the global turbulent heat transfer at Rayleigh numbers of order 10^9 but are significantly different at Rayleigh numbers of order 10^{13} which can be interpreted as a consequence of the formation of well developed bulk turbulence decoupled from the thermal boundary layers at the heated horizontal surfaces.

1 Introduction

Thermal convection is common to many natural and engineering systems, and turbulence in these flows is more the rule than the exception, especially for large scale natural phenomena. Familiar examples are convection in stars, in the outer core of the earth, and in the atmosphere. Considering stellar convection alone, we could argue that turbulent convection itself is the most ubiquitous type of flow that we know of. Unfortunately, the values of the principal control parameter for stellar convection are quite high and pose problems for laboratory experiments, notwithstanding the

J. J. Niemela (✉)
The Abdus Salam International Centre for Theoretical Physics, ICTP, Strada Costiera 11, 34014 Trieste, Italy
e-mail: niemela@ictp.it

L. Di G. Sigalotti et al. (eds.), *Computational and Experimental Fluid Mechanics with Applications to Physics, Engineering and the Environment*, Environmental Science and Engineering,
DOI:10.1007/978-3-319-00191-3_2,

obvious difficulties in matching fluid properties and boundary conditions. If we concentrate on the first problem, namely the principal control parameter, and simplify the rest of the problem by selecting an artificial system with well defined boundary conditions, we have the possibility to make some progress. This assumes, of course, that in doing this we retain the essential physics of the problem. It is important to note that even a relatively simple system becomes highly nonlinear at high values of the control parameter, and therefore poses a challenging problem in itself. The system referred to here is Rayleigh-Bénard convection (RBC), in which a layer of fluid is contained between two heated horizontal surfaces. The upper (lower) surface is cooled (heated) so that a mechanically unstable density gradient is formed across the fluid layer, which is assumed to be thin, in the sense that we can neglect compressibility effects. In actuality, compressibility cannot be entirely neglected, and a correction for adiabatic temperature gradients is necessary. We also confine our attention to fluids for which the coefficient of thermal expansion has a reversed sign, requiring the opposite heating arrangement (e.g. water below 4 °C). Any parcel of, say, hot fluid near the lower boundary is subject to buoyancy forces which promote its vertical rise, leading to convective currents, with the same consideration applying to cold fluid near the top of the layer. The degree to which buoyancy overcomes dissipative processes and can lead to convection is given by the dimensionless Rayleigh number Ra given by

$$\mathrm{Ra} = \frac{g\alpha \Delta T H^3}{\nu\kappa}, \tag{1}$$

where α, ν and κ are, respectively, the isobaric thermal expansion coefficient, kinematic viscosity and thermal diffusivity of the fluid, ΔT the temperature difference across the fluid layer of height H and g the acceleration due to gravity. The interplay between heat and momentum diffusion is important, especially for turbulent convection, and that is characterized by the Prandtl number

$$\mathrm{Pr} = \frac{\nu}{\kappa}. \tag{2}$$

2 Apparatus

The apparatus and methodology for the present work has been described in detail in Niemela and Sreenivasan (2003, 2006, 2008). In brief, the fluid—in this case helium gas near 5 K—is held between two OHFC copper plates separated vertically by thin cylindrical stainless steel sidewalls with a fixed diameter of 50 cm. The height of the fluid layer in these experiments was either 50 or 12.5 cm, so that the diameter-to-height aspect ratio was $\Gamma = 1$ or 4, respectively. The OHFC copper used for the heated horizontal surfaces was annealed to have a conductivity near $1\,\mathrm{kW\,m^{-1}K^{-1}}$ at helium temperatures. More importantly, it is nearly five orders of magnitude larger than the molecular conductivity of the working fluid so that the corresponding Biot number is extremely small. This means that constant temperature conditions are assured even when the fluid is turbulent. Heating is provided by a serpentine metallic film encased in mylar and sandwiched to the top and bottom plates by additional copper plates. The top plate is regulated to have constant temperature and is connected through a variable resistance to liquid helium bath which acts as the cold reservoir. Outside the sample space a common cryo-pumped vacuum provides protection against either conductive or convective heating in parallel, and radiative heating is controlled through the use of various concentric shields surrounding the sample space, the inner-most one being at the temperature of the top plate.

Mean pressure is measured by a Baratron gauge, using heads appropriate to the absolute pressure so as to maximize resolution. Temperature measurements rely on semiconductor resistance thermometers made of doped germanium. Within the fluid, cubes of neutron transmutation doped germanium, 250 μ on a side, are used to monitor temperature fluctuations.

The fluid has some special properties: its kinematic viscosity ν can be quite small when the density is large (i.e, near the critical point). The thermal expansion coefficient α is naturally large in the ideal gas limit, as its value is simply the inverse of the absolute temperature. The thermal diffusivity can be both very small and very large depending on the operating point in the pressure-temperature phase space. That is, near the critical point (at 5.2 K) the specific heat C_P diverges and so the thermal diffusivity $\kappa = k/\rho C_P$ vanishes, where k and ρ are, respectively, the thermal conductivity of the fluid and its density. In addition, α is thermodynamically related to the specific heat and so it also gets quite large near the critical point. Taken together, cryogenic helium gas presents a widely tunable fluid that allows both extremely large values of the principal control parameter Ra but also large ranges of it.

3 An Effective Diffusivity

Let us consider the simple diffusion equation

$$\frac{\partial T}{\partial t} = \kappa \frac{\partial^2 T}{\partial x_j \partial x_j}. \tag{3}$$

Under conditions of fully developed bulk turbulence (i.e. decoupled from any solid boundaries) we may consider that the turbulence can be modeled as a "fluid" having an effective diffusivity κ^{eff}. We may then consider Eq. (3), with an effective diffusivity κ^{eff} replacing the molecular value κ. The corresponding time scale, then, for turbulent diffusion is dimensionally given by L^2/κ^{eff} on some characteristic length scale L. On the other hand, we know for turbulent flows with characteristic length scales L and velocity scales u, the corresponding time scale is of order L/u.

Equating the two times scales gives $\kappa^{\mathrm{eff}} \sim uL$. Taking its ratio with the molecular value then gives

$$\frac{\kappa^{\mathrm{eff}}}{\kappa} \sim \frac{uL}{\kappa} = \mathrm{Pe}, \tag{4}$$

where Pe is the Péclet number. From Niemela et al. (2001) we know that $\mathrm{Pe} = 0.13\mathrm{Ra}^{0.50}$. We will return to consider this further below.

For a flow with mean velocity **U** sweeping along a heated surface we may also define an effective heat diffusivity starting with the diffusion-advection equation

$$\frac{\partial T}{\partial t} + U_j \frac{\partial T}{\partial x_j} = \kappa \frac{\partial^2 T}{\partial x_j \partial x_j}. \tag{5}$$

Decomposing velocity and temperature for turbulent flows into mean and fluctuating parts and averaging over the fluctuations (see Tennekes and Lumley 1997) we obtain for the heat transfer

$$q_j = \rho c_P \left(\overline{\theta u_j} - \kappa \frac{\partial T}{\partial x_j} \right), \tag{6}$$

where θ and u represent fluctuating components of temperature and velocity, respectively. By Reynolds' analogy we can define an eddy diffusivity for heat, κ_T as

$$\overline{\theta u_j} = -\kappa_T \frac{\partial T}{\partial x_j} \tag{7}$$

so that we may write for the total heat transfer Q in the vertical direction (for instance)

$$\frac{Q}{\rho c_p} = -(\kappa_T + \kappa) \frac{\partial T}{\partial x_3}. \tag{8}$$

Approximating the gradient in temperature as the average over the entire fluid layer, $H/\Delta T$, and taking as a definition of the dimensionless heat transport the Nusselt number Nu, where Nu is given by

$$\mathrm{Nu} = \frac{QH}{k\Delta T}, \tag{9}$$

we obtain from Eq. (8)

$$\mathrm{Nu} = \frac{\kappa^{\mathrm{eff}}}{\kappa}, \tag{10}$$

where

$$\kappa^{\text{eff}} = \kappa_T + \kappa. \tag{11}$$

It is interesting to note that Eq. (10) can be reconciled with Eq. (4) only if

$$\text{Nu} \sim \text{Ra}^{1/2}. \tag{12}$$

In fact, Eq. (4) resulted from our consideration of a hypothetical fully developed turbulence—i.e., without considering any boundary effects. We know that in the case that the diffusive boundary layers are artificially removed in RBC simulations (Lohse and Toschi 2003), Nu is indeed described by Eq. (12).

4 Time-Dependent Measurements of the Effective Diffusivity

Despite its academic interest, could we in fact measure an effective thermal diffusivity of convective turbulence directly? The thermal diffusivity of solids, or any purely conducting material, can be measured by applying a time-varying temperature at one surface and measuring the damped amplitude or phase of the resulting heat wave at some known distance along the direction of its propagation. The corresponding experiment then is to oscillate the temperature of the bottom boundary of a Rayleigh-Bénard cell and to then measure temperature at that frequency carefully at a known height within the layer. Certainly this will work when the fluid is quiescent, but the question is whether it will work when there are in addition turbulent eddies transporting heat.

The experiment performed was just this: at the top plate we retained a constant temperature as usual, while the bottom plate was subject to a sinusoidally oscillating heat flux with a non-zero mean value giving rise to an oscillation of the bottom plate temperature at the same frequency about some mean value larger than that of the top plate.

Let us consider only the oscillating part of the bottom plate temperature, which is of the form

$$T = T_0 \cos(\omega t). \tag{13}$$

A solution of Eq. (3) satisfying this boundary condition is

$$T = T_0 \exp\left(-\frac{z}{\delta_S}\right) \cos\left(\omega t - \frac{z}{\delta_S}\right), \tag{14}$$

where $\delta_S = (2\kappa/\omega)^{1/2}$ is the penetration depth. Equation (14) describes a wave-like phenomenon whose phase and amplitude depend on the thermal diffusivity of the medium and the frequency ω of the oscillation. Knowing the oscillation frequency, the position of the temperature sensor, and the amplitude of the temperature oscillation at the plate, the only unknown is the thermal diffusivity, which is then measured. With κ^{eff} in place of κ, Eq. (14) describes a heat wave that experiences damping and phase variation according to an effective penetration depth

$$\delta_{\text{eff}} = \sqrt{\frac{2\kappa^{\text{eff}}}{\omega}}. \tag{15}$$

If we also measure the dimensionless heat transfer or Nusselt number Nu simultaneously, we have two independent measurements of diffusivity: one through Nu, that requires no assumptions but only indirectly gives the diffusivity, and the other which is directly sensitive to the diffusivity but requires that we postulate a turbulent "fluid". To see how Nu infers the diffusivity, we refer to Eq. (9). Multiplying both numerator and denominator by the heat capacity of the fluid we obtain Eq. (10), where we identify $QH/\Delta T$ as the effective thermal conductivity of the fluid and κ^{eff} is then that value multiplied by the heat capacity.

The experiment is aided by the fact that metals at low temperature have high conductivity compared to the fluid, as was noted above, and also have nearly negligible heat capacity compared to the fluid. The resulting low thermal mass allows us to produce large amplitude, high frequency heat waves that can penetrate with detectable amplitude through the entire bulk region.

The frequency of the modulation, f_M (in Hz), was chosen to be both below and above the characteristic frequency of the largest scale circulation. The amplitude was also varied and a dimensionless form is given by

$$\Delta_M = (T_0)_{\text{rms}}/\langle\Delta T\rangle. \tag{16}$$

Here and elsewhere $\langle ... \rangle$ refers to averaging over integral periods of the modulation.

The experimental procedure consisted of applying sinusoidal heating at the bottom with a DC offset and then waiting for at least 200 cycle times of the large scale circulation to reach a statistical steady state before taking measurements. To obtain Nu conventionally, the temperature difference was averaged over integer numbers of the modulation cycle. From this both $<$ Ra $>$ and $<$ Nu $>$ could be computed.

Fluctuations in the temperature within the bulk were measured at the mid-height of the cell, and about 4.4 cm radially inward from the sidewall. This point was 25 cm above the bottom plate for $\Gamma = 1$ and 6.25 cm above the bottom plate for $\Gamma = 4$. Data were collected at a rate of 50 Hz using the off-balance signal from an audio frequency bridge circuit with lock-in detection. By Fourier analysis it was possible to measure the amplitude of the signal due specifically to the heat wave having frequency f_M.

Table 1 Experimental conditions and measurements

Γ	< Ra >	f_M (Hz)	Δ_M	κ (cm^2 s^{-1})	κ^{eff} (cm^2 s^{-1})	$\frac{\kappa}{\kappa^{\text{eff}}}$	< Nu >
1	3.4×10^9	0.01	0.05	4.03×10^{-2}	3.893	96.6	98.6
4	1.9×10^9	0.032	0.15	7.04×10^{-3}	0.583	82.9	81.6
1	4.5×10^{12}	0.04	0.28	7.93×10^{-4}	3.6099	4,535	975.6
1	1.0×10^{13}	0.025	0.22	4.79×10^{-4}	3.31	6,904	1,277

5 Results and Discussion

Table 1 shows experimental conditions and measurements made in both $\Gamma = 1$ and $\Gamma = 4$ cells. In particular, the last two columns show the ratio of measured diffusivities and the measured (time-averaged) Nusselt number. For the first two rows at low Ra, the values in the last two columns are in excellent agreement. This is remarkable given that in two very different-height systems Eq. (14) returns the same effective diffusivity that one would have inferred from the Nu-measurements. However, the situation changes for the last two rows, which correspond to the same $\Gamma = 1$ cell but at much higher Ra. In this case, the measured diffusivity is larger than we would have expected and its ratio with the molecular value is roughly five times that of the measured Nu.

The result is illustrated graphically in Fig. 1. Here we plot the various values of Nu corresponding to the last two columns of Table 1 for $\Gamma = 1$. In addition, values of Nu taken in the absence of modulation, namely those from Niemela and Sreenivasan (2003), are included. The dashed line represents $\text{Ra}^{1/2}$. One striking fact is that < Nu > and Nu are the same within experimental uncertainty for all Ra, even when the amplitude of modulation is of the same order as the average temperature difference.

The dashed line in Fig. 1 is meant to denote the expected slope of Nu_κ if the effective diffusivity were determined by Eq. (3). It is arbitrarily adjusted to connect through the two upper data points. It is tempting to think that at high Ra we are measuring for the most part pure bulk turbulence by the propagation of heat waves while at lower Ra we are sensitive to extended boundary layers above the heated plates.

The heat transfer measured by Nu or < Nu > does not require any assumptions about the fluid layer. It is simply the integration of the contributions of all features of the flow, even if the entire contribution is only from two thin boundary layers near the top and bottom plates at high Ra. On the other hand, the heat wave, at a first approximation, is assumed to propagate through a homogeneous medium. The fact that the thermal boundary layer is much smaller than wither the molecular or effective penetration depth (a fact that is true for all Ra investigated here) would seem to validate its use in RBC.

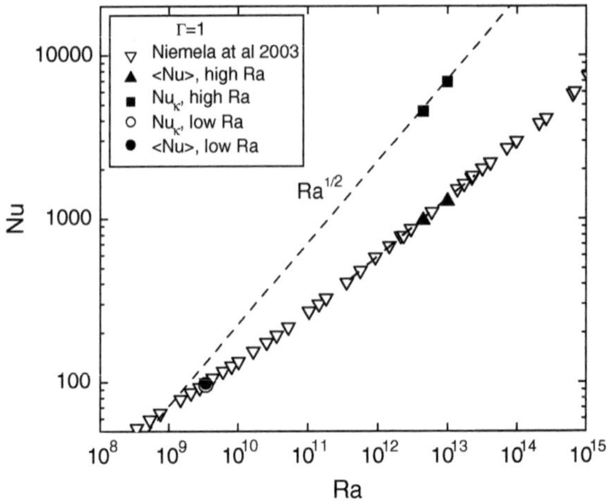

Fig. 1 Nu vs. Ra for $\Gamma = 1$ both with and without modulation. Included are values for Nu_κ as computed by the ratio of the effective to molecular diffusivity. Inverted open triangles, raw Nu from Niemela and Sreenivasan (2003) without modulation; open circles, Nu_κ for low Ra; solid squares, Nu_κ for high Ra; solid circles, $< Nu >$ for low Ra; solid triangles, $< Nu >$ for high Ra

It should be noted that Nu can be determined as the ratio between the half-height of the layer and the thermal boundary layer thickness. That is:

$$\mathrm{Nu} \simeq \frac{H}{2\delta}, \tag{17}$$

where δ is the thermal boundary layer thickness. On the other hand, Nu_κ is not dependent on the thermal boundary layer thickness at all, but rather on the ratio between δ_{eff} and δ_S, the effective and molecular Stokes layer thicknesses, respectively, so that

$$\mathrm{Nu}_\kappa \simeq \left(\frac{\delta_{\mathrm{eff}}}{\delta_S}\right)^2. \tag{18}$$

In Niemela and Sreenivasan (2008) it was proposed that the heat wave method "failed" at high Ra due to the emergence of a highly turbulent core region, which was assumed, for simplicity, to have an infinite conductivity. In this situation the amplitude of the heat wave would clearly cease to decrease with distance within the core region, and therefore the position of the sensor would no longer be a relevant parameter (i.e., it could be changed within the core region without affecting the measurement). Note that the conclusions are not really that different than the discussion above; namely, that at very high Ra there may exist a region of well-developed bulk turbulence completely decoupled from the boundaries.

Finally, the data here are few and additional conclusions would clearly benefit from distributed temperature measurements instead of one fixed sensor position and, of course, a larger coverage of Ra. Such experiments are currently underway.

Acknowledgments The author wishes to acknowledge the contribution of K.R. Sreenivasan to the work presented here and the Elettra Synchrotron Facility in Trieste for providing space and support for the experiment.

References

Lohse D, Toschi F (2003) Ultimate state of thermal convection. Phys Rev Lett 90:034502

Niemela JJ, Sreenivasan KR (2003) Confined turbulent convection. J Fluid Mech 481:355–384

Niemela JJ, Skrbek L, Sreenivasan KR, Donnelly RJ (2001) The wind in confined thermal convection. J Fluid Mech 449:169–178

Niemela JJ, Sreenivasan KR (2006) Turbulent convection at high Rayleigh numbers and aspect ratio 4. J Fluid Mech 557:411–422

Niemela JJ, Sreenivasan KR (2008) Formation of the "superconducting" core in turbulent thermal convection. Phys Rev Lett 100:184502

Tennekes H, Lumley JL (1997) A first course in turbulence. The MIT Press, Cambridge

Numerical Simulation of Sliding Drops on an Inclined Solid Surface

Marco Maglio and Dominique Legendre

Abstract We consider numerical simulations of drops sliding on an inclined solid surface. The simulations are performed using our in house research code JADIM based on the Volume of Fluid formulation of the mass and momentum equations. Special algorithms have been developed for the simulation of the hysteresis of the contact line as well as for the description of moving contact lines. The onset of motion is analyzed and the effect of the contact line hysteresis is studied. The critical angle of inclination, as well as the corresponding drop shape, are discussed and compared with previous experiments. The sliding velocity for a constant angle of inclination is also considered and compared with experiments. The different shapes observed in experiments (rounded, corner, cusp, or pearling drop) are recovered depending on both the fluid properties and the angle of inclination. The drop sliding velocity is then considered for larger values of the hysteresis.

1 Introduction

Motion of drops on surfaces is a phenomenon observed in everyday life as well as in many environmental or industrial applications: coating processes, combustion processes, pesticide and insecticide pulverization on cultivations, lab-on-a-chip devices, etc. In particular, small droplets have the capability to stick on non-horizontal surfaces. Despite its apparent simplicity, the behaviour of a drop on an inclined solid surface is far to be completely understood. It involves static, hysteresis, and dynamic contact line behaviours. Depending on the fluid properties, the hysteresis and the wall

M. Maglio · D. Legendre (✉)
Institut de Mécanique des Fluides de Toulouse, IMFT, Université de Toulouse, CNRS-INPT-UPS, 2 Allée du Professeur Camille Soula, Toulouse 31400 , France
e-mail: marco.maglio@imft.fr

D. Legendre
e-mail: dominique.legendre@imft.fr

L. Di G. Sigalotti et al. (eds.), *Computational and Experimental Fluid Mechanics with Applications to Physics, Engineering and the Environment*, Environmental Science and Engineering, DOI:10.1007/978-3-319-00191-3_3, © Springer International Publishing Switzerland 2014

inclination, different drop shapes (rounded, corner, or pearling drop) can be observed (Le Grand et al. 2005). In addition, its numerical modelling is still a great challenge. For example, up to now the different sliding regimes observed in the experiments have not been reproduced by numerical simulations.

From the numerical point of view, the description of a moving contact line cannot be performed by direct discretization of the equations. Indeed, the solution of the Navier-Stokes equations gives an infinite viscous dissipation at the moving contact line when a no-slip condition is applied on the wall (Huh and Scriven 1971). As a consequence, refining a grid induces a divergence of the viscous stress at the contact line (Afkhami et al. 2009).

In order to be predictive (i.e., with no adjustable parameters), numerical simulations of moving contact lines at the macroscopic scale must describe correctly the contact angle and the contact line speed. A full numerical simulation would consist in resolving all scales involved in the problem, i.e., from the macro scale L to the nano-metric scale associated with the effective slip length λ. For example, let us consider a millimeter-size droplet as studied in this work with an equivalent radius $a = (3V/4\pi)^{1/3} \approx 1$ mm, where V is the drop volume. The full resolution of all scales involved in this problem would require $N_{2D} \approx (R/\lambda)^2 \approx 10^{12}$ nodes for a 2D simulation and $N_{3D} \approx (R/\lambda)^3 \approx 10^{18}$ nodes for a 3D simulation. Such a grid size is obviously not compatible with the present computer's resources and/or requires an extremely long CPU time even on parallel computers. Consequently, it is clear that the simulation of the contact line hydrodynamics can not be performed up to the nano-scale level for a millimeter size drop. For $L = O(1)$mm, the grid size Δ is limited to some microns. In contrast, Molecular Dynamics simulations are able to describe the nano-scale effects (Blake 2006), but due to the limited computer resources, the size of the macro scale is limited and simulations are only possible for nano-drops (Winkels et al. 2012).

Different approaches can be used to overcome the singularity at the contact line (Bonn et al. 2009). The hydrodynamic models are based on analytical solutions of the interface shape, while the contact line velocity U_{cl} is obtained by matching methods. Typically, an inner region, whose characteristic length is imposed by the slip length λ, is matched to an outer region (the apparent region) of characteristic size L where no-slip occurs. Assuming Stokes flow at both scales, the apparent or dynamic contact angle θ_d is found to be a function of the wall contact angle θ_W (considered to be constant), the capillary number $\mathrm{Ca} = \mu_d U_{cl}/\sigma$, the logarithm of the scale ratio $\ln(L/\lambda)$, and the viscosity ratio q (Voinov 1976; Dussan 1976; Cox 1986). We have developed a sub-grid model of moving contact lines based on this matching relation between the molecular wetting at the nano-scale (not solved) and the macro-scale that is solved.

The paper is decomposed as follows. The problem considered in this paper is presented in Sect. 2. The numerical code JADIM as well as the numerical modelling developed for the static, hysteresis, and dynamic contact angle, are described in Sect. 3. Section 4 reports experimental tests concerning the onset of motion of a drop deposited on a inclined wall. Section 5 presents the simulation concerning the sliding regime and Sect. 6 contains the conclusions.

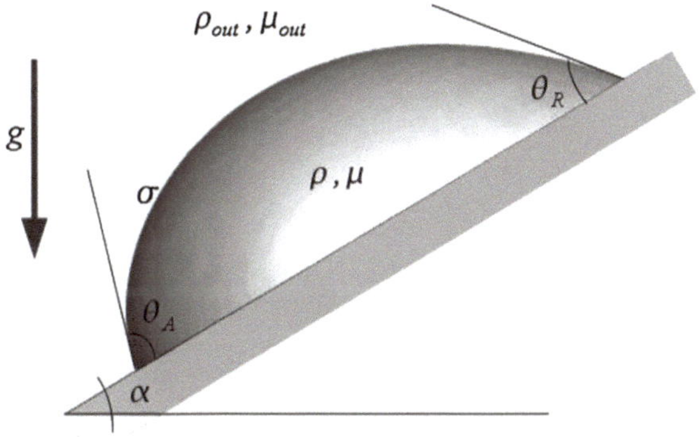

Fig. 1 Definition of parameters for a drop on an inclined surface

2 Statement of the Problem

We consider a drop of volume $V = 4\pi a^3/3$ (where a is the equivalent radius used as the characteristic length scale) located on a inclined surface (see Fig. 1). We denote by ρ_L, μ_L, and σ the drop density, the drop viscosity, and the surface tension of the interface between the drop and the air, respectively. Here ρ_G and μ_G denote the density and viscosity of the air. The surface wettability is characterized by the advancing θ_A and receding θ_R contact angles. The inclination is characterized by the angle α made by the surface and the horizontal. With the density and viscosity ratios, ρ_G/ρ_L and μ_G/μ_L, being much smaller than unity, the problem is completely characterized by the introduction of two additional dimensionless numbers, namely the Eötvös number $\mathrm{Eo} = \rho g a^2/\sigma$ and the Morton number $\mathrm{Mo} = \mu_L^4 g/\rho_L\sigma^3$. The values of the Ohnesorge number $\mathrm{Oh} = \mu_L/\sqrt{\rho_L\sigma a}$ are also specified.

Two series of numerical experiments are presented. In Sect. 4, we first focus on the angle and the drop shape at the onset of motion. At time $t = 0$, the drop is a hemispherical cap (the initial contact angle is $\theta_i = 90°$) and the wall is horizontal ($\alpha = 0$). Once the drop shape is stabilized, the wall is inclined at $t = t_0$, following a linear time evolution given by $\alpha(t) = \pi(t - t_0)/T$. The characteristic time T of the inclination is chosen to be much larger than the physical times of the problem, i.e. the visco-capillary time $t_\mu \approx \mu_L a/\sigma$ and the inertia-capillary time $t_i \approx \sqrt{\rho_L a^3/\sigma}$. In practice, T is chosen such that $T > 10\max(t_\mu, t_i)$. The angle at the onset of motion is denoted by α_c and corresponds to the angle when the advancing and receding points of the drop are both moving in the direction of sliding.

In Sect. 5, we consider the sliding velocity of a drop on a surface with a fixed angle of inclination α. The drop is initially a hemispherical cap making a contact angle $\theta_i = 50°$ with the wall. We record the time evolution of the drop velocity and determine the sliding velocity corresponding to the steady state once gravity balances capillary and viscous forces.

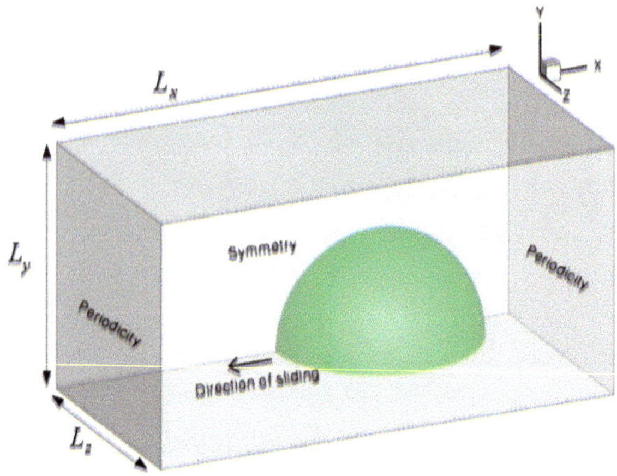

Fig. 2 Domain definition and initial layout of the simulations

For each simulation, a cap of initial radius $R_0 = a\,\left[2\,/\,(2 - 3\cos\theta_i + \cos^3\theta_i)\right]^{1/3}$ is deposited on the surface. We define x as the direction of sliding along the plane, y as the normal direction of the plane, and z as the lateral direction. Since the plane (x, y) is a symmetry plane for the problem, the computational domain shown in Fig. 2 is reduced to half the space as defined by $z \geq 0$. The dimensions of the computational domain are $L_x = 6.5\,a$ and $L_y = L_z = 2.25\,a$, where L_x, L_y, and L_z are the domain sizes along the x-, y-, and z-directions, respectively. A symmetry condition is thus imposed on the plane (x, y) and periodicity is enforced on both sides of the box in the sliding direction, allowing for long-term simulations in a reduced domain. For pearling drops, as described in Sect. 5, the length L_x of the domain was doubled, i.e. $L_x = 13\,a$.

3 Numerical Code

3.1 Volume of Fluid (VoF) Solver

The numerical simulations reported in this work were performed using the Volume of Fluid (VoF) solver developed in the JADIM code (Bonometti and Magnaudet 2007; Dupont and Legendre 2010). The one-fluid exact system of equations is obtained by introducing the distribution C, which is used to localize one of the two phases. In this study, we define C as $C = 1$ within the liquid drop, and $C = 0$ for the external fluid. The one-fluid function C makes possible the definition of the one-fluid variables $U = CU_L + (1 - C)U_G$ for the velocity, $P = CP_L + (1 - C)P_G$ for the pressure, $\rho = C\rho_L + (1 - C)\rho_G$ for the density, and $\mu = C\mu_L + (1 - C)\mu_G$ for the viscosity. The position of the interface is then given by the transport equation:

$$\frac{\partial C}{\partial t} + U \cdot \nabla C = 0. \quad (1)$$

The two fluids are assumed to be Newtonian and incompressible, with no phase change. Under isothermal conditions and in the absence of any surfactant, the surface tension is constant and uniform at the interface between the two fluids. Under these conditions, the velocity field U and the pressure P satisfy the classical one-fluid formulation of the Navier-Stokes equations:

$$\nabla \cdot U = 0, \quad (2)$$

$$\rho \left(\frac{\partial U}{\partial t} + U \cdot \nabla U \right) = -\nabla P + \nabla \cdot \Sigma + \rho g + F_\sigma, \quad (3)$$

where Σ is the viscous stress tensor, g is the gravity, and F_σ is the capillary force contribution given by

$$F_\sigma = \sigma \left(\nabla \cdot n \right) n \delta_I, \quad (4)$$

where σ is the surface tension, n denotes the outward unit vector normal to the drop surface, and δ_I is the Dirac distribution associated to the interface position.

The system of Eqs. (1)–(3) is discretized using the finite volume method and time is advanced through a third-order Runge-Kutta scheme for the viscous stresses. Incompressibility is satisfied at the end of each time step through a projection method. The overall algorithm is second-order accurate in both time and space. The volume fraction C and the pressure P are volume-centred, while the velocity components are face-centred. Due to the discretization of C, the interface is numerically represented by a surface of finite thickness and cells cut by the interface correspond to regions with $0 < C < 1$. One important aspect of our approach compared to the classical VoF or Level Set methods (Sussman et al. 1998; Scardovelli and Zaleski 1999; Sethian 1999) concerns the technique used to control the stiffness of the interface. In our approach no interface reconstruction or redistancing techniques are employed. The interface location and stiffness are both controlled by an accurate transport algorithm based on an FCT (Flux-Corrected-Transport) scheme (Zalesak 1979). This method leads to an interface thickness of about three grid cells by the implementation of a specific procedure to calculate the velocity used to transport C in flow regions of strong strain and shear (Bonometti and Magnaudet 2007).

The numerical description of the surface tension is one crucial point when we consider systems where capillary effects control the interface shape. This interfacial force is solved using the classical CSF (Continuum Surface Force) model (Brackbill et al. 1992):

$$F_\sigma = \sigma \nabla \cdot \left(\frac{\nabla C}{\|\nabla C\|} \right) \nabla C. \quad (5)$$

A classical problem connected to this formulation is the generation of spurious currents (Lafaurie et al. 1994; Popinet and Zaleski 1999). In order to decrease the intensity of spurious currents, a classical solution, introduced by Brackbill et al. (1992), is employed which consists in calculating the surface curvature from a smoothed density gradient, while the discretization of the delta function uses an non-smoothed density. The spurious currents have been characterized by Dupont and Legendre (2010). Their maximum magnitude is shown to evolve as $0.004\sigma/\mu$, in agreement with other codes that use the Brackbill's formulation.

3.2 Numerical Modelling of the Contact Angle

The aim of the numerical method is to handle static (either with or without hysteresis) and dynamic contact lines. The method has been developed to simulate the transition from a static contact line in the hysteresis range to a moving contact line and vice versa. The numerical scheme used in this study has been initiated by Dupont and Legendre (2010) for 2D and axisymmetric geometries. It has also been used to study droplet spreading in axisymmetry (Legendre and Maglio 2013). In this paper, the method is extended to 3D geometries.

The calculation of the capillary term requires the knowledge of the contact angle made by the interface at the wall. Indeed, the capillary contribution (5) in the momentum equation requires the knowledge of ∇C. Since $\nabla C/\|\nabla C\|$ is the unit vector normal to the interface, the boundary condition for ∇C is thus given directly by the value of the contact angle θ_W, as shown by the following relation:

$$\frac{\nabla C}{\|\nabla C\|} = \mathbf{n} = \sin\theta_W \mathbf{n}_{\parallel} + \cos\theta_W \mathbf{n}_{\perp}, \tag{6}$$

where $\mathbf{n}_{\parallel}$ and $\mathbf{n}_{\perp}$ are the unit vector components of the normal vector parallel and perpendicular to the wall.

The general method is divided into two steps. We first determine the value of the contact angle to be applied at the wall. This value is then imposed as a boundary condition using relation (6) for the calculation of the capillary contribution (5) in the momentum balance equation (3).

Static contact angle

The static equilibrium of a drop on an horizontal wall is characterized by the static contact angle θ_S between the interface and the wall. A simple force balance at the interface gives the Young-Dupré relation:

$$\sigma\cos\theta_S = -\sigma_{SG} - \sigma_{SL}, \tag{7}$$

where σ_{SL}, σ_{SG}, and σ are, respectively, the values of surface tension for the solid/liquid, solid/gas, and gas/liquid interfaces. The value of the static contact angle θ_S is a parameter of the simulation, characterizing the fluid-fluid-wall wettability. Simulations can also be performed by imposing a constant contact angle. This is the simplest situation in which relation (6) is used by imposing $\theta_W = \theta_S$. In Dupont and Legendre (2010), some simulations have been performed by imposing a static contact angle θ_S in order to compare with the dynamic models.

Dynamic contact line

As was outlined in the introduction, direct numerical simulations that resolve all scales involved in the problem are not possible. In particular, simulations of contact line hydrodynamics cannot be performed up to the nano-scale. A sub-grid description, as introduced by Dupont and Legendre (2010), must be implemented for macroscopic simulations of problems controlled by moving contact lines. This description, which is implemented in JADIM, aims to correctly reproduce the physics at the contact line. The model is based on considerations at the macroscopic scale based on the analytical derivation by Cox (1986) that connect the macroscopic region (imposed by the grid resolution Δ) to the inner region (imposed by the slip length λ). Thus, at the macroscopic scale, the wall condition seen by the fluid is a no-slip boundary condition:

$$U_W = 0. \tag{8}$$

This consists in imposing a zero numerical slip length $\lambda_N = 0$. At the macroscopic level, the interface shape characterized by the dynamic apparent contact angle θ_d is connected to the microscopic contact angle θ_S by means of the relation (Cox 1986):

$$g(\theta_d) - g(\theta_S) = \mathrm{Ca}\ln(\frac{L}{\lambda}), \tag{9}$$

for two fluids of arbitrary viscosity, where $g(\theta)$ is a function that simplifies to:

$$g(\theta) = \int_0^\theta \frac{x - \sin x \cos x}{2 \sin x} dx, \tag{10}$$

when the surrounding fluid is of much smaller viscosity (for example, the air). In practice, the functions $g(x)$ and $g(x)^{-1}$ can be approximated with good accuracy by means of a fitting polynomial (Dupont and Legendre 2010). When the condition $\theta_d < 3\pi/4$ is satisfied, Eq. (9) reduces to the well-known Cox-Voinov relation: $\theta_d^3 = \theta_S^3 + \mathrm{Ca}\ln(L/\lambda)$ (Voinov 1976; Cox 1986). In relation (9), L is imposed by the grid spacing and λ is the physical slip length. Due to the use of a staggered grid, where the VoF function C is located at the centre of the volume, while the velocities are face-centred, the interface is transported by the velocity a distance $\Delta/2$ and therefore one has $L = \Delta/2$. Several experiments have demonstrated that the apparent region is characterized by $L \approx 10\,\mu\mathrm{m}$ and that the characteristic slip length is $\lambda \approx 1$–$10\,\mathrm{nm}$ (Marsh et al. 1983; Ngan and Dussan 1989; Dussan et al. 1991; Shen and Ruth 1998). More recently, Rio (2005) and Le Grand et al. (2005) have shown that Eq. (9), in its simplified form $\theta_d^3 = \theta_S^3 + 9\mathrm{Ca}\ln(L/\lambda)$, provides a good description of their experiments. The length L is taken as the distance at which the measurement is taken and λ is used as an adjustable parameter. Thus, Rio (2005) showed that for two different measurement techniques the macroscopic length was $L \approx 30\,\mu\mathrm{m}$, using the Laser measurement, and $L \approx 200\,\mu\mathrm{m}$, using the optic circle technique. From these results, the microscopic length λ was found to be between $\lambda = 7\,\text{nm}$ and $\lambda = 15\,\text{nm}$, depending on the characteristics of the measurement and the model employed to fit the data. The slip length should be intended as the molecular slip, which is of the order of some molecules. Considering the value usually found for water-like liquids on solid substrates (Lauga et al. 2007), we have chosen $\lambda = 10^{-9}$ m in Eq. (9) for the simulations reported in this work.

Several authors have dealt with the "stress singularity" paradox by introducing the Navier slip condition, that gives a relation between the fluid velocity at the wall U_W and a numerical slip length λ_N:

$$U_W = \lambda_N \left(\frac{\partial U}{\partial y} \right)_W . \tag{11}$$

For example, this has been implemented by Renardy et al. (2001) in the case of a VoF scheme and by Spelt (2005) in a Level-Set code. Both of these methods impose a static contact angle θ_S at the wall, assuming that the microscopic contact angle is $\theta_W = \theta_S$. The dynamic contact angle is then obtained solving the full hydrodynamic problem up to a microscopic neighborhood. Unfortunately, due to the grid refinement limitation, these simulations use unrealistically large slip length values and, therefore, the slip length λ_N becomes in practice an adjustable parameter for the simulation (see Bonn et al. (2009) for a similar comment). The grid convergence of the simulations is then reached but an unphysical slip condition is required for this.

Hysteresis of contact angle

A number of numerical calculations has considered the modelling of the dynamic contact angle, while only few studies have implemented models for the hysteresis of the contact line of sessile drops. Some examples can be found in the literature (Dimitrakopoulos and Higdon 1999; Spelt 2005; Fang et al. 2008; Yokoi et al. 2009). All these methods consist in implementing the following conditions for the normal velocity U_{cl} of the contact line:

$$U_{cl} < 0 \quad \text{if} \quad \theta_d < \theta_R, \tag{12}$$

$$U_{cl} = 0 \quad \text{if} \quad \theta_R \leq \theta_d \leq \theta_A, \tag{13}$$

$$U_{cl} > 0 \quad \text{if} \quad \theta_A < \theta_d. \tag{14}$$

The method implemented here makes possible the transition from a static (resp. moving) to a moving (resp. static) contact line. The procedure is divided into two steps:

Step 1: For every cell containing the interface ($0 \leq C \leq 1$), the angle θ^* that cancels the local momentum balance Eq. (3) is determined by an iterative procedure using a simple Newton-Raphson scheme. The iteration is stopped when $\theta^{n+1} - \theta^n < \epsilon_\theta$ with a convergence limit of $\epsilon_\theta = 10^{-4}$.

Step 2. The value of θ^* is compared to the hysteresis range and two possible situations can be found:

(i) the contact angle is inside the hysteresis: $\theta_R \leq \theta^* \leq \theta_A$. The contact line is static and the contact angle is imposed as θ^* so that the momentum balance locally cancels. Thus, if the interface was immobile at the beginning of the time step, it remains immobile for the next time step; if the interface was previously in motion, the interface is stopped.

(ii) the contact angle is outside the hysteresis: $\theta^* < \theta_R$ or $\theta^* > \theta_A$. As a consequence, the contact line cannot be static. The value of the contact angle is then calculated using the dynamic model described above. The static contact angle is the advancing (resp. receding) angle when $\theta^* > \theta_A$ (resp. $\theta^* < \theta_R$).

3.3 *Numerical Validation and Grid Convergence*

The validation as well as the time and grid convergence of the numerical method used for the simulation of the dynamic contact angle have been extensively discussed in Dupont and Legendre (2010), Maglio (2012), and Legendre and Maglio (2013). Static, hysteresis, and dynamic situations have been considered for spreading drops, drops on inclined surfaces, and drops in a surrounding shear flow. In particular, a very satisfactory agreement has been found with experiments of spreading drops for both water drops and viscous drops. The inertia-capillary regime of spreading characterized by a contact line expansion varying as $t^{1/2}$ and the Tanner's evolution going as $t^{1/10}$ have been recovered. Comparisons of our model with other variants, i.e. static contact angle versus dynamic contact angle and no-slip condition versus slip condition, were also reported by Maglio (2012). These comparisons clearly stress the importance of considering a dynamic model for the simulation of moving contact lines. For the study reported in this paper, additional tests have been performed to ensure grid and time convergence for the onset of motion, as well as for the sliding drop velocity. They are discussed in the sections below.

4 The Onset of Motion

The angle of inclination α_c of the surface at the onset of motion can be obtained by considering the force balance acting on the drop. The volume V of the largest drop that can stick on the surface is given by the balance between the weight of the drop and the interfacial force acting along the contact line cl:

$$\rho g_L V \sin\alpha_c + \sigma \int_{cl} \cos\theta \sin\beta dl = 0, \tag{15}$$

where θ is the local contact angle and β is the angle between the unit normal pointing outwards the drop and the direction perpendicular to the sliding direction. By definition $\beta = 90°$ at the front of the drop and $\beta = -90°$ at the rear. Thus, the above relation reduces to

$$\rho_L V g \sin\alpha_c = \kappa_c \sigma (\cos\theta_R - \cos\theta_A), \tag{16}$$

where κ_c is the characteristic length κ at the onset of motion, where κ is given by

$$\kappa = -\frac{1}{\cos\theta_R - \cos\theta_A} \int_{cl} \cos\theta \sin\beta dl. \tag{17}$$

Furmidge (1962) found that relation (16) is in excellent agreement with the experiments, if κ is assumed to be the drop width w. As shown by relation (17), κ is not necessarily the drop width because it depends on the contact line shape. Nevertheless, for some particular drop shapes it is possible to demonstrate that $\kappa = w$. This is the case of a spherical cap drop with a small hysteresis range ($\theta_A \approx \theta_R$) and of a parallel sided drop with circular advancing and receding contact lines.

Using the equivalent radius a as the characteristic length, we can rewrite relation (16) in terms of the Eötvös number $\text{Eo} = \rho g a^2/\sigma$:

$$\sin\alpha_c = \frac{\kappa}{a}\frac{3}{4\pi}(\cos\theta_R - \cos\theta_A)\,\text{Eo}^{-1} \tag{18}$$

so that the critical contact angle varies as $(\cos\theta_R - \cos\theta_A)\,\text{Eo}^{-1}$ and depends on the contact line shape.

Much work, mostly experimental, has focused on the characterization of the onset of motion (Bikerman 1950; Rotemberg et al. 1984; Dussan 1985; Milinazzo and Shinbrot 1988; Extrand 1995; Dimitrakopoulos and Higdon 1999; Podgorski 2000; Le Grand et al. 2005). Considering a spherical cap drop with a small value of the contact angle hysteresis (i.e. $\theta_A \approx \theta_R$), Dussan (1985) showed that the onset of motion is described by

$$\sin\alpha_c = \frac{3}{\pi\, 2^{1/3}} \frac{(1+\cos\theta_A)^{1/2}}{(2+\cos\theta_A)^{1/3}\,(1-\cos\theta_A)^{1/6}} (\cos\theta_R - \cos\theta_A)\,\text{Eo}^{-1}. \tag{19}$$

This analytical solution is thus only valid for small values of the hysteresis, with $\theta_A - \theta_R$ being typically less than 10°. Relation (19) indicates that the characteristic length is $\kappa_c = 2^{5/3}(1+\cos\theta_A)^{1/2}\,(2+\cos\theta_A)^{-1/3}\,(1-\cos\theta_A)^{-1/6}\,a$. A plot of this relation as a function of θ_A reveals that the variation is moderate for $30° \leq \theta_A \leq 145°$ so that in a first approximation κ_c can be expressed using $\theta_A = 90°$:

$$\kappa_c \approx 2^{4/3}\,a. \tag{20}$$

Among the existing experimental studies, here we consider those by Podgorski (2000) and Le Grand et al. (2005), who reported well documented experiments of silicon oil drops deposed on different inclined surfaces. The corresponding experimental conditions are displayed in Fig. 3. We see that these experiments have focused on small hysteresis ranges. The purpose of our study is to compare our simulations

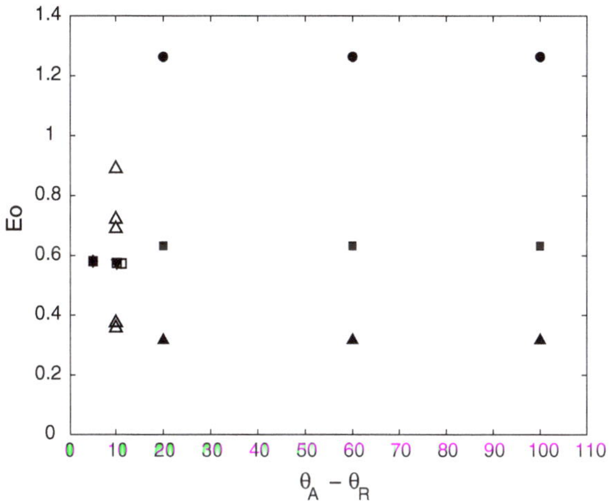

Fig. 3 Phase diagram $(\theta_A - \theta_R)$–Eo for the cases considered. Experiments: □ Le Grand et al. (2005), △ Podgorski (2000). Numerical simulations: ▲ Eo = 0.32 (FL1), ■ Eo = 0.63 (FL2), • Eo = 1.26 (FL3), ▼ Eo = 0.57 (47V100), ▼ Eo = 0.57 (47V100), and ◆ Eo = 0.58 (47V10)

with the experimental results for this small range of the contact angle hysteresis and then to extend the study to a larger hysteresis range. For the sake of direct comparison, we have considered two situations reported by Le Grand et al. (2005), corresponding to fluids 47V10 and 47V100, with Eötvös numbers of 0.58 and 0.57, and hysteresis ranges of (42.7°, 52.9°) and (45.5°, 50.5°), respectively. We have further extended the experimental studies by considering three larger contact angle hysteresis, i.e., $(\theta_A, \theta_R) = (80°, 100°), (60°, 120°)$, and $(40°, 140°)$, for Eo = 0.32, 0.63, and 1.26. The drop properties used in these simulations are listed in Table 1.

The grid and time convergence is here discussed for fluid FL1 in Table 1. Two regular grids M1 and M2 are considered. They are, respectively, made of $150 \times 50 \times 50$ and $300 \times 100 \times 100$ nodes in the x-, y-, and z-directions. The drop radius a corresponds to 22.5 and 45 nodes, respectively. The grid M1 is first used for discussing the effect of the time step. The simulations produced $\alpha_c = 31.5°$, $\alpha_c = 34.0°$, and $\alpha_c = 34.7°$ for time steps $\Delta t = 2 \times 10^{-5}$ s, 10^{-5} s, and 5×10^{-6} s, respectively.

Table 1 Onset of motion: parameters used for the simulations

Fluid	a (mm)	ρ kg/m^3	μ Pa.s	σ N/m	Eo (-)	Oh (-)	Mo (-)
FL1	0.595	655	0.01	7.2	0.32	0.19	4.0×10^{-4}
FL2	0.595	655	0.0168	3.6	0.63	0.45	3.2×10^{-2}
FL3	0.595	655	0.01	1.8	1.26	0.38	2.6×10^{-2}
47V100	1.127	964	0.103	20.9	0.57	0.68	1.3×10^{-1}
47V10	1.127	936	0.01	20.1	0.58	0.69	1.3×10^{-5}

A simulation performed using grid M2 and $\Delta t = 5 \times 10^{-6}$ s yielded $\alpha_c = 36.0°$, which differs only by ~3 % from the corresponding value obtained using the grid M1. Therefore, in order to minimize the computational time, the results reported in this section were obtained using $\Delta t = 1 \times 10^{-5}$ s and the mesh M1 ($150 \times 50 \times 50$).

4.1 Drop Shape During Surface Inclination

We first consider the case when Eo $= 0.63$ and (θ_A, θ_R)=(100°, 80°) (see Table 1) in order to describe the evolution of the shape and the contact angle during the inclination of the surface. Figure 4 shows the shape of the drop from the beginning of the simulation until the onset of motion for different inclinations of the surface. During the tilting of the surface, the drop re-arranges its shape while remaining pinned to the inclined surface. Figure 4h corresponds to the onset of motion and Fig. 4i shows when the drop is sliding.

Figure 5 represents the corresponding evolution of the contact angle at the front and rear points of the drop. As explained in Sect. 2, the surface was inclined after the drop stabilization. This corresponds to the vertical line at $\alpha = 0°$. Then, we can notice that the advancing angle at the front of the drop reaches first the limit value of hysteresis ($\theta_A = 100°$). The onset of motion is observed once the rear contact angle reaches the inferior limit given by the receding angle, here $\theta_R = 80°$. For this case, the sliding happens at 18°. This is confirmed by Fig. 6 (top), where the velocities for the front, rear, and centre of mass of the drop are displayed. We can see that the front contact line and the centre of mass start to move first. The rear stagnation point starts moving at a larger angle, which corresponds to the onset of motion. Figure 6 (bottom) shows the evolution of the contact angle along the contact line as a function of the local angle β between the normal of the contact line and the z-direction, for different inclinations of the surface. Due to the initial condition (the drop is initialized as a spherical cap), the contact angle is everywhere 90°. Then, the contact angle evolves progressively until it reaches the distribution corresponding to the beginning of the sliding, where the advancing and receding limits are clearly observed at the front and the rear of the drop, respectively.

4.2 Angle of Inclination at the Onset of Motion

The shapes of the drops at the onset of motion for all the cases considered are shown in Fig. 7. For each case, the shape is compared with the spherical one at the beginning of the simulation in order to outline the deformation. The value of the critical angle α_c is also shown in the figure. For the smallest Eötvös number Eo $= 0.31$, with the hysteresis of $\theta_A - \theta_R = 120° - 60°$ and $\theta_A - \theta_R = 140° - 40°$, the drop is found to remain pinned to the wall for the vertical inclination. When the Eötvös number increases, the critical angle α_c decreases, because the drop is more deformed by

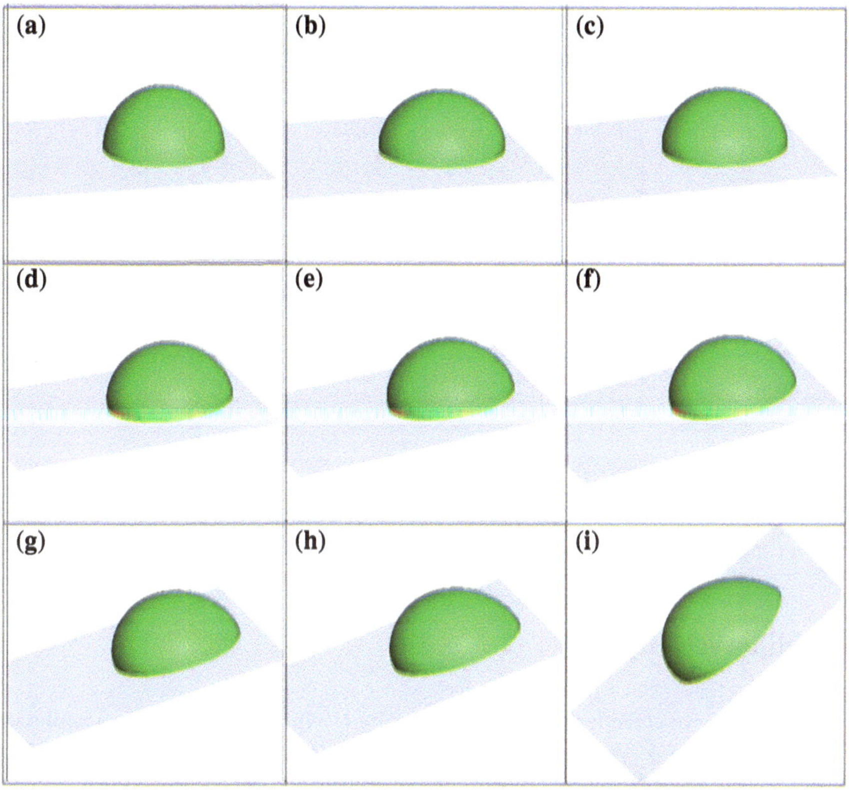

Fig. 4 Eo $= 0.63$ and $\theta_A - \theta_R = 100° - 80°$. Shapes of the drop during the tilting of the surface. **a** beginning of the simulation; **b** $\alpha = 0°$ (end of stabilization); **c** $\alpha = 2°$; **d** $\alpha = 6°$; **e** $\alpha = 9°$; **f** $\alpha = 12°$; **g** $\alpha = 15°$; **h** $\alpha = 18°$ (onset of motion); and **i** $\alpha = 39°$

gravity and reaches faster the shape of motion. As expected, we observe also that the hysteresis range increases the retention force of the drop on the surface.

The angle of inclination α_c at the onset of motion is shown in Fig. 8. Considering the equilibrium force balance (18), the figure shows $\sin\alpha_c$ as a function of $(\cos\theta_R - \cos\theta_A)\text{Eo}^{-1}$. The experimental values of Le Grand et al. (2005) and Podgorski (2000) are also shown for comparison. Surprisingly, all the results seem to collapse on the same evolution whatever are the value of the contact angle hysteresis and the Eötvös numbers considered. For $(\cos\theta_R - \cos\theta_A)\text{Eo}^{-1} \leq 1$, the critical angle is correctly described by relation (18), where the characteristic length $\kappa_c = 2^{4/3}\,a$ is deduced from the analytical relation (19) obtained by Dussan (1985). For larger values of $(\cos\theta_R - \cos\theta_A)\text{Eo}^{-1}$, the numerical results are always under the line representing relation (18) with $\kappa_c = 2^{4/3}\,a$. They correspond to lower values of κ_c, which can be explained by a more elongated shape of the drops. According to Fig. 8, a drop cannot slide on a wall regardless of its inclination if $(\cos\theta_R - \cos\theta_A)\text{Eo}^{-1} \gtrsim 2$.

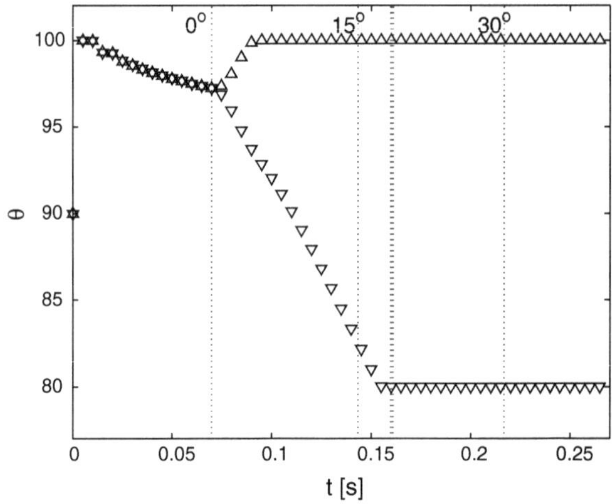

Fig. 5 Eo = 0.63 and $\theta_A - \theta_R = 100° - 80°$. Evolution of the advancing (△) and receding (▽) angles. The corresponding inclination of the surface is shown on the *top* of the graph. The angle at the onset of motion $\alpha_c = 18°$ is shown using a *larger dash line*

5 Sliding Velocity

In this section, we consider the terminal velocity U_T of a drop sliding on a wall with a fixed inclination α. We first compare our simulations with the experiments of Le Grand et al. (2005). Then, we discuss the effect of the contact angle hysteresis on the sliding velocity. The parameters used for the simulations correspond to fluids 47V100 and 47V10, while the drop size is taken from the experiments of Le Grand et al. (2005), (see Table 1). The corresponding contact angle hysteresis ranges (θ_A, θ_R) are (52.9°, 42.7°) and (50.5°, 45.5°), respectively. A spherical cap of volume $V = 6$ mm^3 with a contact angle $\theta_S = 50°$ is assumed as an initial condition (at $t = 0$). The numerical layout of the simulation is identical to the one presented in the previous section. The wall inclination α is fixed during the entire simulation. We have performed simulations for α in the range from α_c to 90°, where α_c is the angle at the onset of motion. We have checked for some cases (typically for the two fluids at $\alpha = 50°$) that the terminal velocity is similar when increasing progressively the wall inclination from 0° to α following the procedure described in the previous section.

In the simulations described here we have used $\Delta t = 1 \times 10^{-5}$ s and mesh M1 (150 × 50 × 50). The time and grid convergence has also been checked by varying the grid spacing and the time step on the drop velocity. For example, with mesh M1 the terminal velocity of a 47V100 drop is $U_T = 0.00204$ m s^{-1}, $U_T = 0.00188$ m s^{-1}, and $U_T = 0.00179$ m s^{-1} for $\Delta t = 1 \times 10^{-5}$ s, $\Delta t = 0.5 \times 10^{-5}$ s, and $\Delta t = 0.2 \times 10^{-5}$ s, respectively. For a finer grid, say M3, with 225 × 75 × 75 nodes and $\Delta t = 0.2 \times 10^{-5}$ s, the simulation gives $U_T = 0.00199$ m s^{-1}, which implies about a 1 % difference with the value obtained using grid M1.

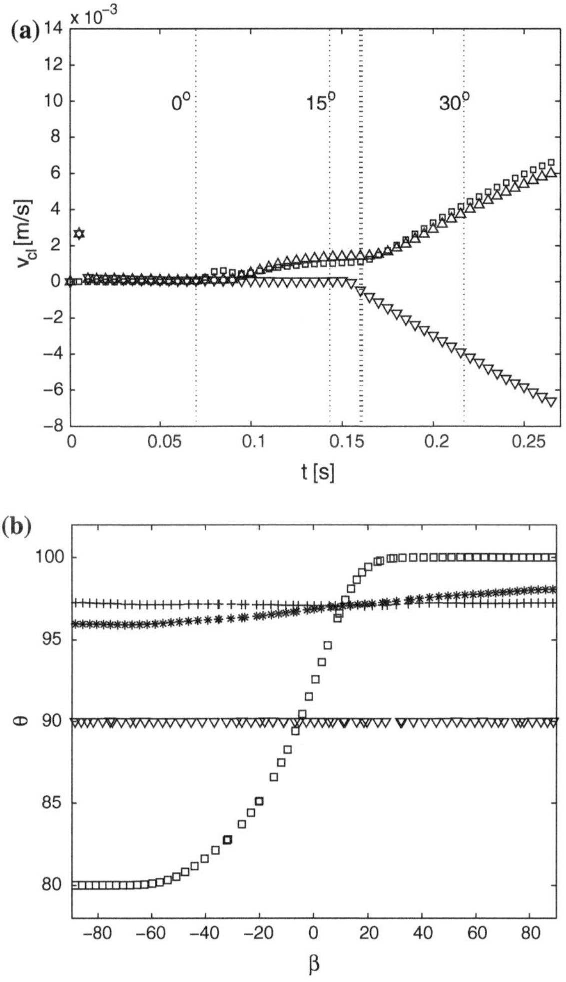

Fig. 6 Eo $= 0.63$ and $\theta_A - \theta_R = 100° - 80°$. (*Top*) Contact line velocities at the *front* ($\triangle$), at the rear ($\triangledown$) and at the centre of mass ($\square$). The corresponding inclination of the surface is shown on the *top* of the graph. The angle at the onset of motion $\alpha_c = 18°$ is shown using a *larger dash line*. (*Bottom*) Distribution of the contact angle along the contact line for different inclinations of the surface. $\triangledown$: $\alpha = 0°$; $+$: $\alpha = 9°$; $\square$: $\alpha = 19°$ (close to the starting point). Contact angle θ along the contact line orientation β

The drop shapes are depicted in Figs. 9, 10 for fluids 47V100 and 47V10, respectively. For both fluids the angles $\alpha = 15°$, $\alpha = 50°$, and $\alpha = 90°$ are shown. In accordance with the experiments of Le Grand et al. (2005), different shapes are observed. When increasing the drop velocity (by increasing α and decreasing the viscosity), the shape varies significantly and so we observe rounded, corner, cusp, and pearling drops. For all cases considered, the front contact line remains circu-

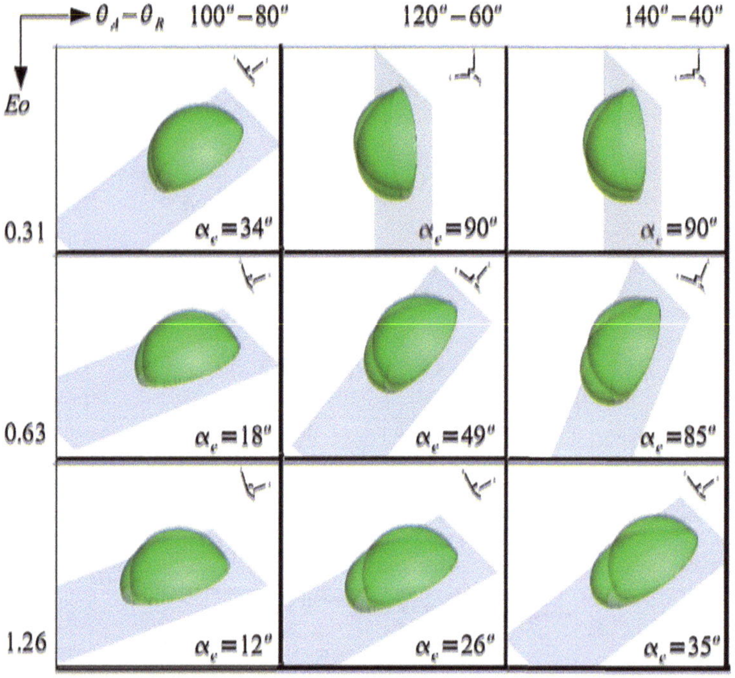

Fig. 7 Drop shapes at the onset of motion compared with the shape at the beginning of the inclination, for different Eötvös numbers and hysteresis ranges. For $\theta_A = 100°$: $\theta_R = 80°$ and Mo = 0.000421; for $\theta_A = 120°$: $\theta_R = 60°$ and Mo = 0.00321; and for $\theta_A = 140°$: $\theta_R = 40°$ and Mo = 0.02568

lar, the difference being mainly noticed at the rear contact line. For the smallest velocities ($\alpha = 15°$ and $\alpha = 50°$ for 47V100 and $\alpha = 15°$ for 47V10), the shape of both the advancing and the receding contact lines is circular. For intermediate velocities, the rear contact line is progressively deformed to form a corner ($\alpha = 50°$ for 47V10) and then a cusp ($\alpha = 90°$ for 47V100). For the largest velocity ($\alpha = 90°$ for 47V10), the pearling regime is observed. The pearling regime is characterized by the emission of droplets from the tip of the cusp. As was reported by Le Grand et al. (2005) (see their Fig. 6), the size of the first droplet is larger than the second one. The transition between the cusp regime and the pearling regime is observed between $\alpha = 60°$ and $\alpha = 70°$ for fluid 47V10. The corresponding capillary number is close to $\mathrm{Ca} = \mu U_T/\sigma \sim 0.01$, in agreement with the experiments of Le Grand et al. (2005).

The time evolution of the drop velocity U for some particular inclinations is shown in Fig. 11 for fluids 47V100 (left) and 47V10 (right), respectively. For both fluids, U increases at the beginning of the simulation, reaches a maximum around $t \sim 0.01$ s, and then decreases and stabilizes to a constant value corresponding to the drop terminal velocity U_T. As expected, the terminal velocity is found to

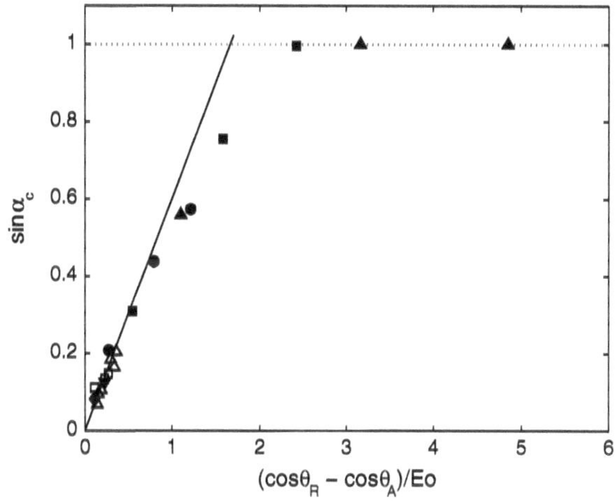

Fig. 8 Variation of the angle $\sin\alpha_c$ with $(\cos\theta_R - \cos\theta_A)\,\mathrm{Eo}^{-1}$. Experiments: □ Le Grand et al. (2005), △ Podgorski (2000). Numerical simulations: ▲ Eo = 0.32 (FL1), ■ Eo = 0.63 (FL2), • Eo = 1.26 (FL3), ▼ Eo = 0.57 (47V100), ▼ Eo = 0.57 (47V100), ◆ Eo = 0.58 (47V10),— $\sin\alpha_c \approx 0.6(\cos\theta_R - \cos\theta_A)\,\mathrm{Eo}^{-1}$ derived from relation (20) (Dussan 1985) with $\kappa_c = 2^{4/3}\,a$, ··· drop sticked on a vertical surface

increase with α. Figure 11 (right) shows that the velocity is not perfectly stabilized for the less viscous fluid 47V10 with the largest angles $\alpha = 70°$ and $\alpha = 90°$. This is due to the pearling observed at the rear of the drop. The inspection of the first part of the evolution reveals that the evolution of the centre of mass follows $V \sim g\,\sin(\alpha)\,t$. It corresponds to the volume acceleration inside the drop due to gravity. The drop deforms and the front and rear contact angles evolve from the initial value $\theta_S = 50°$ to the advancing and receding contact angle θ_A and θ_R, respectively. Thus, during this first phase of the evolution, the advanced and receding contact lines remain immobile and the drop sticks on the surface. Once the drop deformation has induced values for both the front and rear contact angle outside the hysteresis range (see previous section), the drop starts to slide until it reaches the terminal velocity U_T. The drop Reynolds number $\mathrm{Re} = \rho U_T a/\mu$, based on the terminal velocity, as obtained for $\alpha = 90°$, is Re = 0.035 and 2.8 for fluids 47V100 and 47V10, respectively, evidencing a viscous dominated situation. As a consequence, the contact line Reynolds number $\mathrm{Re}_{cl} = \rho U_{cl} L/\mu \sim \rho U_T \Delta/\mu$ is much smaller than unity, in agreement with the range of validity of the Cox's hydrodynamic model for the apparent contact angle [relation (9)].

The terminal velocity U_T results from the following force balance:

$$\rho g V \sin\alpha - \kappa\sigma(\cos\theta_R - \cos\theta_A) + F_D = 0, \tag{21}$$

where F_D is the drag force experienced by the drop. In F_D we can identify two different viscous contributions. The first corresponds to the bulk contribution F_D^{bulk},

Fig. 9 Drop shape when the terminal velocity is achieved for fluid 47V100 at different inclinations. First column: view normal to the wall. Second column: side view. *Top* $\alpha = 15^\circ$, middle $\alpha = 50^\circ$, *bottom* $\alpha = 90^\circ$

resulting from the viscous effects at the macro scale L. This effect comes from the internal motion and the viscous stress in the air. The second is the contact line contribution F_D^{cl}, which results from the viscous dissipation at the contact line. The drop terminal velocity can be simply derived from the force balance (21) when considering the two following assumptions. The dissipation at the contact line is negligible compared to the bulk dissipation, i.e. $F_{bulk} \gg F_{cl}$ and the drop motion is controlled by viscous effects, i.e. Re$<$ 1. Under such conditions, the drag force experienced by the drop can be written as:

$$F_D \approx F_{vi} \approx -C\mu\pi a U_T, \tag{22}$$

where C is a coefficient depending a priori on the drop shape. Note that for a spherical drop of radius a settling in air, the Stokes drag gives $C = 6$. Assuming that the drop shape during the sliding is similar to its shape at the onset of motion (i.e. $\kappa \approx \kappa_C$), we can write

$$\kappa\sigma(\cos\theta_R - \cos\theta_A) \approx \rho g V \sin\alpha_C$$

Fig. 10 Drop shape when the terminal velocity is achieved for fluid 47V10 at different inclinations. *First column* View normal to the wall. *Second column* Side view. *Top* $\alpha = 15°$, middle $\alpha = 50°$, *bottom* $\alpha = 90°$

so that the force balance gives the following relation between the capillary and the Eötvös number:

$$\mathrm{Ca} \approx \frac{4}{3C}\mathrm{Eo}(\sin\alpha - \sin\alpha_C). \tag{23}$$

The capillary number is plotted as a function of the Eötvös number in Fig. 12. For both fluids, the linear dependence shows that relation (23) makes possible the description of the sliding velocity. The experimental results of Le Grand et al. (2005) are also depicted for comparison. The figure confirms the good agreement with the predicted value of α_c corresponding to the onset of motion. Figure 12 shows very good agreement with the experiments for the more viscous fluid, 47V100 (left), while a significant deviation is found for fluid 47V10 (right). For fluid 47V100, the evolution is split up into two linear regimes. The first regime agrees with the experiments of Le Grand et al. (2005). Relation (23) fits both the simulation and

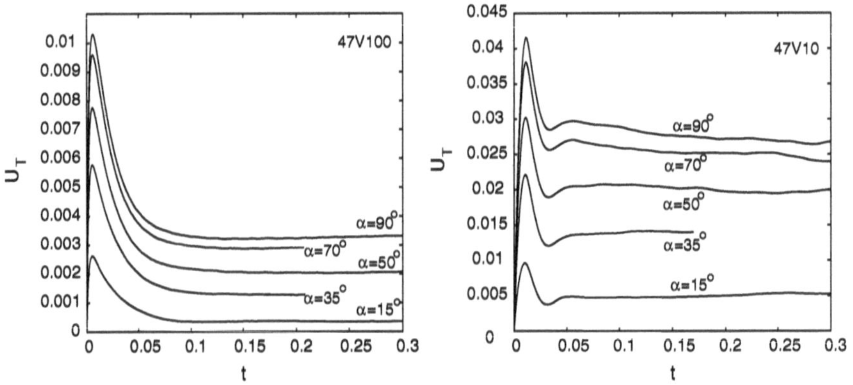

Fig. 11 The time evolution of the drop velocity V for different values of α. From *bottom to top* $\alpha = 15, 35, 50, 70$ and $90°$. (*left*) fluid V100, (*right*) fluid V10

the experiments with $C \sim 83$. On the other hand, the second regime corresponds to $C \sim 50$. This lower value is due to the change of the drop shape. The receding contact line shape generates a lower resistance in the force balance. These values of C show that the drag force experienced by a viscous drop sliding on an inclined wall is one order of magnitude bigger than the one experienced by a settling drop ($C = 6$). For fluid 47V10, relation (23) is also seen to fit the numerical simulation for $C \sim 83$, while the experiments are better described using $C \sim 131$. Our numerical results for fluids 47V100 and 47V10 follow a similar evolution ($C \sim 80-83$), while a significant difference is observed with the experiments made with the two fluids. One possible explanation of this discrepancy could be the contact line contribution

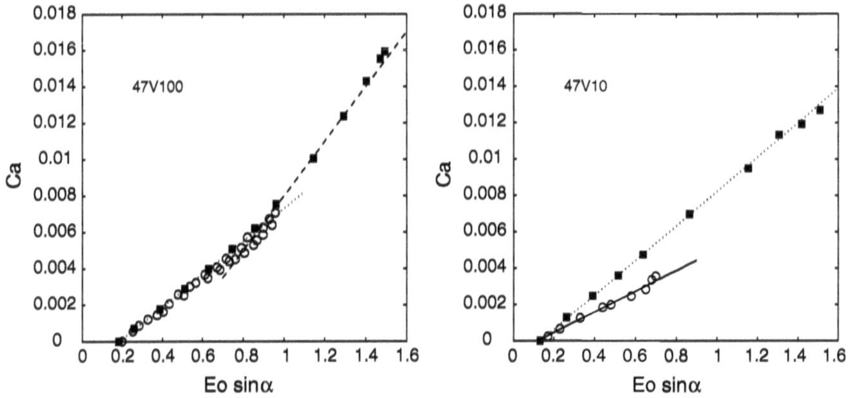

Fig. 12 Evolution of the drop capillary number as a function of the Eötvös number Eo_α. (o) experiments of Le Grand et al. (2005), ■ Numerical simulations. (*left*) fluid 47V100: $\cdots\cdot$ $Ca = 0.009\,Eo\,\sin\alpha$, - - - $Ca = 0.015\,Eo\,\sin\alpha$. (*right*) fluid 47V10: $\cdots\cdot$ $Ca = 0.0095\,Eo\,\sin\alpha$, —$Ca = 0.0057\,Eo\,\sin\alpha$

F_D^{cl} to the drag force. In our simulations, only the large scale dissipation is considered since we do not solve the hydrodynamics at the contact line. While Eq. (9) relates the value of the apparent contact angle to the velocity of the contact line, no sub-grid dissipation was introduced in our models. For the more viscous fluid, the agreement is very satisfactory, suggesting that the main dissipation occurs in the bulk, since F_D^{cl} is small compared to F_{bulk}. For the less viscous fluid, the contribution of F_D^{cl} is more important and could be of the same order of F_{bulk}, resulting in an underestimated drag force in our simulations since the models do not take into account the contribution of F_D^{cl}.

We finally consider the effect of the contact angle hysteresis on the drop sliding velocity. The simulations were performed for fluid 47V100 since the agreement with the experiments was very good. A surface inclination of $\alpha = 50°$ and contact angle hysteresis of (θ_A, θ_R)=(50.5°, 45.5°), (58°, 38°), (63°, 33°), and (68°, 28°) were considered. As expected, the simulations indicate that the sliding velocity decreases when increasing the hysteresis range. Figure 13 depicts the drop capillary number as a function of $\cos\theta_R - \cos\theta_A$. This plot shows that the decrease is linear and reveals that $(\cos\theta_R - \cos\theta_A)$ is again the pertinent parameter for taking into account the hysteresis.

6 Conclusions

We have reported numerical simulations of drops on a inclined solid surface. Both the onset of motion and the sliding regime have been considered. The simulations have compared satisfactorily with the experiments of Le Grand et al. (2005). In addition,

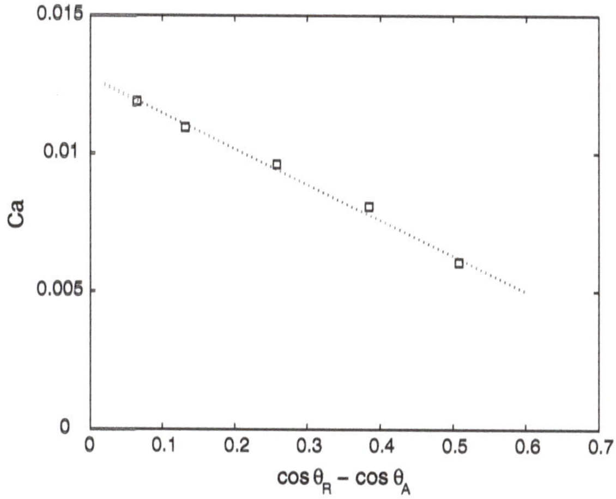

Fig. 13 Effect of the hysteresis on the drop velocity for fluid 47V100 and $\alpha = 50°$. Ca is plotted as a function of $\cos\theta_R - \cos\theta_A$. □ simulations, · · · · $y = 0.0128 - 0.013x$

we have extended the analysis of the effects of the contact angle hysteresis to values of the hysteresis range beyond those usually considered in current experiments. The critical angle of inclination, corresponding to the onset of motion, depends on the Eötvös number, the hysteresis, and the shape of the contact line. When the drop starts to slide, an almost rounded shape is observed in all cases, which contrasts with the very different shapes observed when the drop is sliding. All the experimental and numerical results are found to collapse on the same evolution as shown when $\sin\alpha_c$ is plotted as a function of $(\cos\theta_R - \cos\theta_A)\,\mathrm{Eo}^{-1}$. Regarding the sliding regime, the characteristic shapes observed in the experiments have been recovered by the simulations. Rounded, corner, cusp, and pearling drops have all been observed. The sliding velocity has been found to be in very good agreement with the experiments for the more viscous fluid, while a significant discrepancy has been seen for the less viscous fluid. One possible explanation for this discrepancy are the effects of dissipation at the contact line, which were not included in the present models. A proper inclusion of the effects of dissipation at the contact line is therefore needed to validate this hypothesis. Our study has also stressed the effect of the hysteresis on the sliding velocity. When increasing the hysteresis range, the sliding velocity is found to decrease as $\cos\theta_R - \cos\theta_A$, which appears to be the pertinent parameter for the description of the effects of hysteresis in the sliding motion. Future work will focus on considering the effects of the hysteresis on the different sliding regimes and, in particular, on the pearling regime.

References

Afkhami S, Zaleski S, Bussmann M (2009) A mesh-dependent model for applying dynamic contact angles to VOF simulations. J Comput Phys 228:5370–5389

Bikerman JJ (1950) Sliding of drops from surfaces of different roughnesses. J Colloid Interface Sci 5:349–359

Blake TD (2006) The physics of moving wetting lines. J Colloid Interface Sci 299:1–13

Bonn D, Eggers J, Indekeu J, Meunier J, Rolley E (2009) Wetting and spreading. Rev Mod Phys 81:739–805

Bonometti T, Magnaudet J (2007) An interface capturing method for incompressible two-phase flows. Validation and application to bubble dynamics. Int J Multiph Flow 33:109–133

Brackbill J, Kothe DB, Zemach C (1992) A continuum method for modeling surface tension. J Comput Phys 100:335–354

Cox RG (1986) The dynamics of the spreading of liquids on solid surfaces. Part 1: Viscous flow. J Fluid Mech 168:169–194

Dimitrakopoulos P, Higdon JJL (1999) On the gravitational displacement of three dimensional fluid droplets from inclined solid surfaces. J Fluid Mech 295:181–209

Dupont JB, Legendre D (2010) Numerical simulations of static and sliding drop with contact angle hysteresis. J Comput Phys 229:2453–2478

Dussan EB (1976) The moving contact line: the slip boundary condition. J Fluid Mech 77:665–684

Dussan EB (1985) On the ability of drops or bubbles to stick to non-horizontal surfaces of solids. Part 2. Small drops and bubbles having contact angle of arbitrary size. J Fluid Mech 151:1–20

Dussan EBV, Ramé E, Garoff S (1991) On identifying the appropriate boundary conditions at moving contact lines: an experimental investigation. J Fluid Mech 230:97–116

Extrand CW (1995) Liquid drops on an inclined plane: the relation between contact angles, drop shape, and retentive force. J Colloid Interface Sci 170:515–521
Fang C, Hidrovo C, Wang F, Eaton J, Goodson K (2008) 3-D numerical simulation of contact angle hysteresis for microscale two phase flow. Int. J. Multiph. Flow 34:690–705
Furmidge CGL (1962) I. The sliding of liquid drops on solid surfaces and a theory for spray retention. J Colloid Sci 17:309–324
Huh C, Scriven LE (1971) Hydrodynamic model of the steady movement of a solid/liquid/fluid contact line. J Colloid Interface Sci 35:85–101
Lafaurie B, Nardone C, Scardovelli R, Zaleski S, Zanetti G (1994) Modelling merging and fragmentation in multiphase flows with surfer. J Comput Phys 113:134–147
Lauga E, Brenner MP, Stone HA (2007) Microfluidics: the no-slip boundary condition. In: Tropea C et al (ed) Handbook of experimental fluid dynamics, Chap. 19. Springer, NY, pp 1219–1240
Le Grand N, Daerr A, Limat L (2005) Shape and motion of drops sliding down an inclined plate. J Fluid Mech 541:293–315
Legendre D, Maglio M (2013) Numerical simulation of spreading drops. Colloids and Surf A 432:29–37
Maglio M (2012) Numerical simulation of spreading, sliding and coalescing drops on surfaces. PhD thesis, Institut National Polytechnique de Toulouse, France
Marsh JA, Garoff S, Dussan EB (1983) Dynamic contact angles and hydrodynamics near a moving contact line. Phys Rev Lett 70:2778–2782
Milinazzo F, Shinbrot M (1988) A numerical study of a drop on a vertical wall. J Colloid Interface Sci 121:254–264
Ngan CG, Dussan EB (1989) On the dynamics of liquid spreading on solid surfaces. J Fluid Mech 209:191–226
Podgorski MT (2000) Ruissellement en conditions de mouillage partiel. PhD thesis, Université Paris 6, France
Popinet S, Zaleski S (1999) A front-tracking algorithm for accurate representation of surface tension. Int J Numer Meth Fluids 30:775–793
Renardy M, Renardy Y, Li J (2001) Numerical simulation of moving contact line problems using a volume-of-Fluid method. J Comput Phys 94:243–263
Rio E (2005) Gouttes, Flaques et Arches sèches: Des lignes de contact en présence d'un ècoulement. PhD thesis, Université Paris 6, France
Rotemberg Y, Boruvka L, Neumann W (1984) The shape of nonaxisymmetric drops on inclined planar surfaces. J Colloid Interface Sci 102:424–434
Scardovelli R, Zaleski S (1999) Direct numerical simulation of free surface and interfacial flow. Annu Rev Fluid Mech 31:567–603
Sethian J (1999) Level set methods and fast marching methods. Cambridge University Press, Cambridge
Shen C, Ruth DW (1998) Experimental and numerical investigations of the interface profile close to a moving contact line. Phys Fluids 10:789–799
Spelt PDM (2005) A level-set approach for simulations of flows with multiple moving contact lines with hysteresis. J Comput Phys 207:389–404
Sussman M, Fatemi E, Smereka P, Osher S (1998) A level set approach for computing solutions in incompressible two-phase flows. J Comput Phys 27:663–680
Voinov OV (1976) Hydrodynamics of wetting. Fluid Dyn 11:714–721
Winkels KG, Weijs JH, Eddi A, Snoeijer JH (2012) Initial spreading of low-viscosity drops on partially wetting surfaces. Phys Rev E 85:055301(R)
Yokoi K, Vadillo D, Hinch J, Hutchings I (2009) Numerical studies of the influence of the dynamic contact angle on a droplet impacting on a dry surface. Phys Fluids 21:072102
Zalesak S (1979) Fully multidimensional flux-corrected transport algorithms for fluids. J Comput Phys 31:335–362

Fluids in Cosmology

Jorge L. Cervantes-Cota and Jaime Klapp

Abstract We review the role of fluids in cosmology by first introducing them in General Relativity and then by applying them to a FRW model of the Universe. We describe how relativistic and non-relativistic components evolve in the background dynamics.We also introduce scalar fields to show that they are able to yield an inflationary dynamics at very early times (inflation) and late times (quintessence). Then, we proceed to study the thermodynamical properties of the fluids and, lastly, its perturbed kinematics. We make emphasis on the constrictions of parameters by recent cosmological probes.

1 Introduction

Modern cosmology is understood as the study of fluids and geometry in the Universe. This task involves the development of theoretical ideas about the nature of fluids and gravity theories, both to be compared with current observations that cosmic probes have been undertaking. The present understanding is condensed in the standard model of cosmology, that incorporates the material content of the standard model of particle physics and Einstein's theory of General Relativity (GR) with a cosmological constant. These two schemes, the fluid and gravity parts, have made predictions that have been tested and confirmed, albeit there are still some issues that remain open. Certainly, we have really no firm knowledge of what dark

J. L. Cervantes-Cota (✉) · J. Klapp
Departamento de Física, Instituto Nacional de Investigaciones Nucleares, ININ, Km 36.5, Carretera México-Toluca, La Marquesa 52750, Estado de México, Mexico
e-mail: jorge.cervantes@inin.gob.mx

J. Klapp
Departamento de Matemáticas, Cinvestav del Instituto Politécnico Nacional (I.P.N.), 07360 México, D. F, Mexico
e-mail: jaime.klapp@inin.gob.mx; jaime.klapp@hotmail.com

L. Di G. Sigalotti et al. (eds.), *Computational and Experimental Fluid Mechanics with Applications to Physics, Engineering and the Environment*, Environmental Science and Engineering, DOI:10.1007/978-3-319-00191-3_4,

matter and dark energy are, as well as their nature and detailed properties. Still we are confident of some specific roles that these dark components play in cosmology and astrophysics. Their influence is at least gravitational, as so far we know from cosmic measurements. This knowledge allows us to build a picture of fluids in the background and perturbed geometry in the history of the Universe and this is what we deal with in the present work.

The purpose of the present review is to provide the reader with a panorama of the role that fluids play in the standard model of cosmology. Tracking the recent history, in the late 1940s George Gamow (1946, 1948) predicted that the Universe should had begun from a very dense state, characterized by a huge density at very high temperatures, a scenario dubbed the *Big Bang*, that was conjectured by George Lemaître in the early 1930s. This scenario predicts that matter and light were at very high energetic states in thermal equilibrium and described by a Planckian blackbody. As the Universe expanded, it cooled down, and eventually matter and light decoupled. The image of the last scattering of light is a fingerprint of the initial state and remains today imprinted in the Cosmic Microwave Background Radiation (CMBR). Gamow's scenario predicted that this primeval radiation would be measured at a temperature of only a few Kelvin's degrees; since the expansion of the Universe cools down any density component.

The CMBR was for the first time measured by A. A. Penzias and R. W. Wilson in 1965 (Penzias and Wilson 1965). Later on, in the early 1990s Mather et al. (1990) and Smoot et al. (1992) measured further important properties of this radiation: its tiny anisotropies for large angular scales and its blackbody nature. The first property—also imprinted in the matter distribution—accounts for the perturbed fluids in the Universe that led to structure formation in the cosmos. The second property is a distinctive sign of the equilibrium thermodynamic properties of the primeval plasma—composed of photons, electrons, and baryons, plus decoupled (but gravitationally coupled) neutrinos, dark matter, and dark energy. The evolution and effects of these fluids is the main concern of the present review.

We begin our work by explaining the context of fluids in GR, and especially in cosmology. We then analyze the evolution of perfect fluids—since real fluids allow them to be described as such- and their background dynamics. We explain that the main cosmic components are baryons, photons, neutrinos, dark matter, and dark energy. We also introduce scalar fields since they are ubiquitous in modern cosmology because they enable to model different cosmic dynamics, from inflation (Guth 1981; Linde 1990) and dark energy (Caldwell et al. 1998; Copeland et al. 2006) to dark matter (Matos et al. 2000; Magaña and Matos 2012). Then, we proceed to study the thermodynamical properties of the fluids (as in Cervantes-Cota 2004) and, lastly, its perturbed kinematics. We make emphasis on the constraints of parameters as imposed by recent cosmological probes.

In this work, we use "natural" units $\hbar = c = k_B = 1$ and our geometrical sign conventions are as in Misner et al. (1973).

2 Fluids in General Relativity

The GR theory is based on the Einstein-Hilbert Lagrangian density

$$\mathscr{L} = \frac{1}{16\pi G}(R + L_m)\sqrt{-g}, \tag{1}$$

where R is the Ricci scalar, G the Newton constant, $g = |g_{\mu\nu}|$ is the determinant of the metric tensor, and L_m is the material Lagrangian that will give rise to the fluids. By performing the metric variation to this equation, one obtains the well known Einstein's field equations

$$\mathbf{R}_{\mu\nu} - \frac{1}{2} R\, \mathbf{g}_{\mu\nu} = 8\pi G \mathbf{T}_{\mu\nu}, \tag{2}$$

where $\mathbf{R}_{\mu\nu}$ is the Ricci tensor and $\mathbf{T}_{\mu\nu}$ is the stress energy–momentum tensor whose components are given by $T_{\mu\nu} \equiv -\frac{2}{\sqrt{-g}}\frac{\partial L_m \sqrt{-g}}{\partial g^{\mu\nu}}$. Tensors in Eq. (2) are symmetric which is a requirement of the theory. Being space-time 4D the imposed symmetry implies that Eq. (2) represents a collection of ten coupled partial differential equations. However, the theory is diffeomorphism invariant, and one adds to them a gauge condition, implying in general four extra equations to Eq. (2) that reduce the physical degrees of freedom. Thus, symmetries and gauge choice determine the fluid properties allowed by the theory.

The stress energy-momentum tensor $\mathbf{T}$ encodes the information of the fluid, and all kinds of energy types contribute to curve space-time: density, pressure, viscosity, heat, and other physical quantities. But before introducing them, one needs other elementary concepts.

Giving some reference frame, one defines the four-velocity $\mathbf{u} \equiv d\mathbf{x}/d\tau$ as the vector tangent to the worldline of a particle, with $\mathbf{x}$ being the local coordinates and τ the proper time along the worldline; its four-momentum is $\mathbf{p} = m\mathbf{u}$, where m is the rest mass of the particle. Now, given a space-time surface, $x^\alpha =$ const., one defines its associated one-form as $d\tilde{x}^\alpha$, to obtain the components $\mathbf{T}(d\tilde{x}^\alpha, d\tilde{x}^\beta) = T^{\alpha\beta}$, which is interpreted as the flux of momentum α, $p^\alpha = < d\tilde{x}^\alpha, \mathbf{p} >$, passing through the surface $x^\beta =$ const. In this way, T^{00} is the energy density, which is the flux of momentum ($p^0 =$ particle's energy) that crosses the surface $x^0 = t =$ const. and T^{0i} is the flux of energy that crosses the surface $x^i =$ const.; where latin labels run from 1 to 3 and greek labels from 0 to 3. Given the symmetry of the tensor, $T^{i0} = T^{0i}$, that is, energy fluxes are equal to momentum densities since mass equals relativistic energy. Finally, the components T^{ij} denote the momentum flux i crossing the surface $x^j =$ const., and again symmetry implies that $T^{ij} = T^{ji}$, avoiding a net intrinsic angular momentum.

The left-hand side of Eq. (2) is known as the Einstein tensor ($\mathbf{G}_{\mu\nu}$) and, giving the symmetries of the theory, it happens to fulfill the Bianchi identities, that is, its covariant derivative is null. This in turn implies, on the right-hand side, a conservation law for any fluid within this theory. The conservation law reads:

$$\mathbf{T}_{\mu\ ;\nu}^{\ \nu} = 0. \tag{3}$$

As we shall see, this equation is the most important since it encodes the thermodynamic laws of matter.

3 Fluids in Cosmology

The kinematical properties of a fluid element are determined by its velocity, acceleration, shear, and vorticity. All these quantities are defined in the space-time, and for convenience one uses comoving coordinates, that is Lagrangian coordinates that follow the flow motion. We refer the reader to standard gravity textbooks for details (Misner et al. 1973; Schutz 1985). One splits the space-time structure into surfaces of simultaneity to rest frame observers, with a projected metric on the surface $\mathrm{h}_{\mu\nu} = g_{\mu\nu} + u_\mu u_\nu$; where u^μ are the components of the four velocity $\mathbf{u}$. In this frame it is natural to define an expansion tensor, $\Theta_{\mu\nu} = \Theta_{(\mu\nu)} = \nabla_{(\mu}u_{\nu)}$, and the vorticity tensor, $\omega_{\mu\nu} = \omega_{(\mu\nu)} = \nabla_{[\mu}u_{\nu]}$, where ∇ operates on the projected 3D space. The trace of the expansion tensor is a scalar measure of the volume expansion, given by $\Theta = \nabla_\mu u^\nu$, and the shear tensor is the projected symmetric free-trace part of $\Theta_{\mu\nu}$, such that $\Theta_{\mu\nu} = \sigma_{\mu\nu} + \frac{1}{3}\Theta\, \mathrm{h}_{\mu\nu}$ (Ellis et al. 2012).

Accordingly, the energy-momentum tensor associated to the fluid can be separated into components parallel and orthogonal to the four velocity as:

$$T_{\mu\nu} = \rho u_\mu u_\nu + q_\mu u_\nu + q_\nu u_\mu + P\,\mathrm{h}_{\mu\nu} + \pi_{\mu\nu}, \tag{4}$$

where $\rho = T_{\alpha\beta}u^\alpha u^\beta$ is the energy density that includes rest masses and possibly the internal energy, such as the chemical energy; $P = \mathrm{h}^{\alpha\beta}T_{\alpha\beta}/3$ is the pressure; $q_\mu = -\mathrm{h}_\mu^\alpha T_{\alpha\nu}u^\nu$ is the momentum density or energy flux due to either diffusion or heat conduction; and $\pi_{\mu\nu} = [\mathrm{h}_{(\mu}^{\ \alpha}\mathrm{h}_{\nu)}^{\ \beta} - \frac{1}{3}\mathrm{h}_{\mu\nu}\mathrm{h}^{\alpha\beta}]T_{\alpha\beta}$ is the trace-free anisotropic stress tensor due to viscosity.

A perfect fluid is an inviscid fluid with no heat conduction, that is, $q_\mu = 0$ and $\pi_{\mu\nu} = 0$. It is analogous to an ideal gas in standard thermodynamics. In terms of the full metric, it is a standard practice to represent it as:

$$T^{\mu\nu} = (\rho + P)u^\mu u^\nu + P g^{\mu\nu}, \tag{5}$$

in comoving coordinates, $u^\mu = \delta_0^\mu$. Equation (5), is the energy-momentum tensor that correctly describes fluids in the background geometry of the Universe.

4 Fluids in the Standard Model of Cosmology

The Universe is described by its material components and geometry. The former is fed with microscopic or thermodynamic information about the fluids and the latter is determined by Eq. (2). In the following, we explain the features of the geometry and the properties of the fluids that have governed the evolution of the standard model of cosmology.

The *cosmological principle* states that the Universe is both spatially homogeneous and isotropic on large scales, and this imposes a symmetry on the possible fluids present in it. Any departure from this symmetry in the fluid would be reflected in the geometry through Eq. (2). The symmetry assertion is compatible with observations made of the all-sky cosmic microwave background radiation from the last twenty years, through the satellites COBE (Smoot et al. 1992) in the 1990s, the Wilkinson Microwave Anisotropy Probe (WMAP) (Bennett et al. 2013; Hinshaw et al. 2013) in the 2000s, and the Planck (Ade et al. 2013c) nowadays, although some large scale CMBR anomalies in the isotropy have been detected (Ade et al. 2013d) that require further investigation. On the other hand, homogeneity and isotropy have also been tested for the distribution of matter at large scales, see for instance Marinoni et al. (2012); Hoyle et al. (2013).

In GR, as in any other metric theory, symmetries of the physical system are introduced through the metric tensor. The homogeneous and isotropic space-time symmetry was originally studied by Friedmann, Robertson, and Walker (FRW) (Friedmann 1922, 1924; Robertson 1935, 1936a,b; Walker 1937). The symmetry is encoded in and defines the unique form of the line element:

$$ds^2 = g_{\mu\nu} dx^\mu dx^\nu = -dt^2 + a^2(t) \left[\frac{dr^2}{1 - kr^2} + r^2 (d\theta^2 + \sin^2\theta \, d\phi^2) \right], \quad (6)$$

where t is the cosmic time, r, θ, and ϕ are polar coordinates, and the constant curvature can be adjusted to take the values $k = 0, +1$, or -1 for a flat, closed, or open space, respectively. $a(t)$ is the unknown potential of the metric that encodes the size at large scales, and more formally, it is the *scale factor* of the Universe that measures how the model grows or shrinks as time evolves. Measurements show that it always grows, but a bounce in the very early or final stages is possible (De-Santiago et al. 2013).

The beautiful symmetric FRW solutions to the Einstein Eqs. (2) represent a cornerstone in the development of modern cosmology, since with them it is possible to understand the expansion of the Universe. Although in the first years of relativity, Einstein sought for a static solution – since observations seemed to imply that – it was soon realized by E. Hubble and others in the mid 1920's that the Universe is indeed expanding, following Hubble's law (Hubble 1929).

Using the FRW metric and a perfect fluid, the GR cosmological field equations are,

$$H^2 \equiv \left(\frac{\dot{a}}{a}\right)^2 = \frac{8\pi G}{3}\rho - \frac{k}{a^2} \tag{7}$$

and

$$\frac{\ddot{a}}{a} = -\frac{4\pi G}{3}(\rho + 3P), \tag{8}$$

where H is the *Hubble parameter* that has dimensions of inverse of time, and therefore, it encodes the model's expansion rate ; H^{-1} is proportional to the age of the Universe. Moreover, ρ and P are the density and pressure that enter in Eq. (5). Dots stand for cosmic time derivatives.

As explained above, the energy-momentum tensor is covariantly conserved, as shown by Eq. (3). In the present case, it implies the continuity equation,

$$\dot{\rho} + 3H(\rho + P) = 0. \tag{9}$$

Equations (7), (8), and (9) involve three unknown variables $(a,\ \rho,\ p)$ for three equations, but the system is not mathematically closed, since the equations are not all linearly independent, but just only two of them. Thus, an extra assumption has to be made to solve the system. The answer comes from the micro-physics of the fluids considered. For the moment let us assume a barotropic equation of state that is characteristic for different cosmic fluids, i.e., $w =$ const. so that

$$\frac{P}{\rho} = w = \begin{cases} \frac{1}{3} & \text{for radiation or relativistic matter,} \\ 0 & \text{for dust,} \\ 1 & \text{for stiff fluid,} \\ -1 & \text{for cosmological constant or vacuum energy,} \end{cases} \tag{10}$$

to integrate Eq. (9), yielding

$$\rho = \frac{M_w}{a^{3(1+w)}} \quad \text{or} \quad \frac{\rho_i}{\rho_{i0}} = \left(\frac{a_0}{a}\right)^{3(1+w_i)}, \tag{11}$$

where M_w is the integration constant and has different dimensions for different w-fluids. The equation on the right shows a different re-scaling of the integration constant, where the subscript i stands for the different i-fluids. Quantities with either a subscript or superscript "0" are evaluated at the present time. With this equation the system is mathematically closed and can be solved.

The system of ordinary differential equations described above needs a set of initial, or alternatively boundary conditions to be integrated. One has to choice a set of two initial values, say, $(\rho(t_*), \dot{a}(t_*)) \equiv (\rho_*, \dot{a}_*)$ at some (initial) time t_*, in order to determine its evolution. A full analysis of this assumption can be found in many textbooks (Misner et al. 1973; Weinberg 1972, 2008). In order to show some physical consequences of the early Universe, we assume that $k = 0$. This is consistent with data from recent cosmological probes, as we shall explain shortly. This tell us that curvature has not played a role for most of the age of the Universe. On the other hand, this can be justified as follows: from Eqs. (7) and (11) we may see that the expansion rate, given by the Hubble parameter, is dominated by the density term as $a(t) \to 0$, since $\rho \sim 1/a^{3(1+w)} > k/a^2$ for $w > -1/3$, that is, the flat solution fits very well the very beginning of times. Therefore, taking $k = 0$, Eq. (7) implies that

$$a(t) = [6\pi G M_w (1+w)^2]^{\frac{1}{3(1+w)}} (t - t_*)^{\frac{2}{3(1+w)}}$$
$$= \begin{cases} (\frac{32}{3}\pi G M_{\frac{1}{3}})^{1/4} \ (t-t_*)^{1/2} & \text{for } w = \frac{1}{3} \text{ radiation}, \\ (6\pi G M_0)^{1/3} \ (t-t_*)^{2/3} & \text{for } w = 0 \text{ dust}, \\ (24\pi G M_1)^{1/6} \ (t-t_*)^{1/3} & \text{for } w = 1 \text{ stiff fluid}, \end{cases} \tag{12}$$

and

$$a(t) = a_* e^{Ht} \qquad \text{for} \qquad w = -1 \text{ cosmological constant}, \tag{13}$$

where quantities with subscript "*" are integration constants, representing quantities evaluated at the beginning of times, $t = t_*$. It is thought that within a classical theory (as GR) this initial time is at most as small as the Planckian time ($t_{Pl} = 10^{-43}$s), since prior to it GR has to be modified to include quantum effects. To obtain Eq. (13), the argument given above to neglect k is not anymore valid, since here $\rho =$ const.; that is, from the very beginning it must be warranted that $H^2 \approx \frac{8\pi G}{3}\rho_* > k/a_*^2$, otherwise k cannot be ignored. Nevertheless if Λ is present, it will eventually dominate over the other decaying components, this is the so-called *cosmological no-hair theorem* (Chambers and Moss 1994). A general feature of all the above solutions is that they are expanding, at different Hubble rates, $H = \frac{2}{3(1+w)}\frac{1}{t}$ for Eqs. (12) and $H =$ const. for Eq. (13).

From Eq. (12) one can immediately see that at $t = t_*,\ a_* = 0$ and from Eq. (11), $\rho_* = \infty$, that is, the solution has a singularity at that time, at the beginning of the Universe. This initial cosmological singularity is precisely the Big Bang singularity. As the Universe evolves the Hubble parameter goes as $H \sim 1/t$, i.e., the expansion rate decreases, whereas the matter-energy content acts as an expanding agent [cf. Eq. (7)]. It decelerates the expansion, however, by decreasing asymptotically [cf. Eqs. (8) and (11)]. In this way, H^{-1} represents an upper limit to the longevity of the Universe; for instance, $H^{-1} = 2t$ for $w = 1/3$ and $H^{-1} = 3t/2$ for $w = 0$, t being the age of the Universe.

The exponential expansion (13) possesses no singularity (at finite times), being the Hubble parameter a constant. A fundamental ingredient of this inflationary solution is that the right-hand side of Eq. (8) is positive, $\ddot{a} > 0$, and this occurs when $\rho + 3p < 0$, that is, one does not have necessarily to impose the stronger condition $w = -1$, but it suffices that $w < -1/3$, in order to have a moderate inflationary solution; for example, $w = -2/3$ implies $a = a_* t^2$: a mild power-law inflation.

It is convenient to define dimensionless density parameters as $\Omega_i \equiv \frac{8\pi G \rho_i}{3H^2}$. With them, Eq. (7) can be expressed as the constraint:

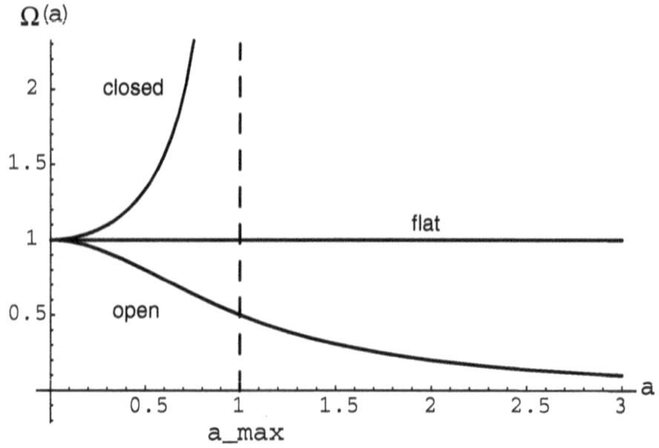

Fig. 1 The parameter Ω as a function of the scale factor, a, in a radiation dominated Universe (the dust model behaves similarly). For closed models, with $k = +1$, Ω diverges as the scale factor tends to its maximum value, whereas for open models, with $k = -1$, Ω tends asymptotically to zero as the Universe expands. Finally, for a flat metric, with $k = 0$, Ω always remains equal to one

$$\Omega \equiv \Omega_R + \Omega_M + \Omega_\Lambda = 1 + \frac{k}{a^2 H^2}, \tag{14}$$

where i labels the different components present in the Universe: R stands for the radiation components (photons, neutrinos, and relativistic particles), M for matter which is composed of dark matter (DM) and baryons, and Λ for a cosmological constant. The actual values of the density parameters (Ω_R, Ω_M, Ω_Λ) impose a value for the curvature term. If $\Omega > 1$, it turns out that k is greater than zero, meaning a Universe with a positive, closed curvature. If $\Omega < 1$, then $k < 0$, which corresponds to a negative, open curvature. Obviously, a critical value is obtained when $\Omega = 1$, then the spatial curvature is null, $k = 0$. The value of the energy density for which $\Omega = [\rho + \Lambda/(8\pi G)]/\rho_c = 1$ holds is known as the *critical density*, $\rho_c \equiv 3H^2/8\pi G$. The last term in Eq. (14) can be defined as $\Omega_k = -k/(a^2 H^2)$, and thus the Friedmann equation becomes a constraint for the density parameters, i.e., $\sum_i \Omega_i = 1$, and this expression holds at any time. It is worth mentioning that solutions $\Omega(a)$ are unstable in the presence of a curvature term (see Fig. 1). In fact, this is related to the flatness problem in the old cosmological picture: Why the Universe is nowadays close to a flat model? Inflation offered the solution to this issue.

In (background) cosmology, typical times and distances are determined mainly by the Hubble parameter, and in practice measurements are often related to *redshift*, as measured from stars, gas, etc. It is then useful to express the individual density parameters in terms of the redshift (z), $1 + z \equiv a_0/a(t)$, where a_0 is the scale factor at present and is set to unity by convention. Today $z_0 = 0$ and towards the early Universe the redshift grows. In terms of the redshift the density parameters are, from Eq. (11),

$$\Omega_i = \Omega_i^{(0)} (1 + z)^{3(1+w_i)}, \tag{15}$$

where w_i is the equation of state parameter for each of the fluids considered. Now, the Hubble parameter can be put in terms of the density parameters. In the standard model of cosmology, considering baryons, photons, neutrinos, cold dark matter (CDM), and a cosmological constant (Λ) – termed ΛCDM –, one has:

$$H^2 = H_0^2 \sum_i \Omega_i^{(0)} (1+z)^{3(1+w_i)}. \tag{16}$$

As defined above, the density parameter depends on $1/H^2$, so to avoid a bias with the expansion rate one defines the *physical* density parameter $\omega_i \equiv \Omega_i h^2$, where h is a dimensionless number given by the Hubble constant $H_0 \equiv 100h$ km s^{-1} Mpc^{-1}. The physical density parameters of matter are important since they are directly determined from CMBR experiments. The current best-fit values for the physical density parameters from Planck are (Ade et al. 2013a): $\omega_b = 0.022$, $\omega_{DM} = 0.120$, from which one computes the best fits

$$\Omega_b^{(0)} = 0.0492, \quad \Omega_{DM}^{(0)} = 0.267, \quad \Omega_\Lambda^{(0)} = 0.683, \quad h = 0.671. \tag{17}$$

The Universe at present is dominated by dark energy, which accounts for 68 % of the energy budget, dark matter for 27 %, and in minor proportion baryonic matter only for about 5 %, from which visible matter is made of. Photons and neutrinos contribute in a much less proportion at present. When one considers a curved model, the best fit for the curvature parameter is $\Omega_k^{(0)} = -0.01$ with an uncertainty of a few percent (Ade et al. 2013a).

Since the scale factor evolves as a smooth function of time, one is able to use it as a variable, instead of time, in such a way that $d/dt = a\,H\,d/da$. This change of variable helps to integrate the continuity equation for non-constant $w(a)$ to obtain:

$$\rho(a) = \rho_0 e^{-3\int [1+w(a)] da/a}. \tag{18}$$

If, for instance, one parameterizes dark energy through an analytic function of the scale factor, $w(a)$, one immediately obtains its solution in terms of

$$t = \int \frac{1}{\sqrt{8\pi G \rho(a)/3}} \frac{da}{a}. \tag{19}$$

From Eq. (19) one obtains the age of the Universe in terms of the redshift, H_0, and the density parameters:

$$t_0 = H_0^{-1} \int_0^\infty \frac{dz}{(1+z)H(z)}. \tag{20}$$

When combining different cosmological probes one obtains for the ΛCDM model an age of $t_0 = 13.81 \pm 0.06$ Gyr (Ade et al. 2013a).

In general, if dark energy is a function of the redshift, from Eq. (18) one can generalize the Friedmann equation to:

$$H(z)^2/H_0^2 = \Omega_M^{(0)}(1+z)^3 + \Omega_\gamma^{(0)}(1+z)^4 + \Omega_k^{(0)}(1+z)^2 + \Omega_{DE}^{(0)}\mathsf{f}(z), \quad (21)$$

where DE stands for dark energy, and

$$\mathsf{f}(z) = \exp\left[3\int_0^z \frac{1+w(z')}{1+z'}dz'\right]. \quad (22)$$

Different DE models can be directly parametrized through $w = w(z)$. The most popular one is perhaps the Chevalier-Polarski-Linder's (Chevallier and Polarski 2001; Linder 2003) formula $w = w_0 + w_a(z/(1+z))$, where w_0 and w_a are constants.

We would like to remark that the first strong evidence for the existence of dark energy, and hence for a present accelerated expansion of the Universe, came from fits of supernovae luminosity curves to data (Cervantes-Cota and Smoot 2011). Two different supernova groups (Riess et al. 1998, 1999; Perlmutter et al. 1999) found a clear evidence for Λ in the late 1990s. The presence of a cosmological constant makes the Universe not only expanding, but also accelerating and, in addition, its age is older, and not in conflict with the globular cluster ages (Jimenez et al. 1996; Richer et al. 2001). In the course of the years, various supernova groups have been getting more confident that the data is compatible with the presence of dark energy, dark matter, and a high value of the Hubble parameter. By moving a little beyond the standard model of cosmology and letting w be a constant (but not necessarily -1), one of the latest data released, the Union2 compilation (Amanullah et al. 2010), reports that the flat concordance ΛCDM model remains an excellent fit to the data, with the best fit to the constant equation-of-state parameter being $w = -0.997^{+0.050}_{-0.054}$ for a flat Universe, and $w = -1.035^{+0.055}_{-0.059}$ for a curved Universe. Also, they found that $\Omega_M^{(0)} = 0.270 \pm 0.021$ (including baryons and DM) for fixed $\Omega_k^{(0)} = 0$. That is, $\Omega_\Lambda^{(0)} = 0.730 \pm 0.021$. Using CMB Planck data, these numbers change a few percent, having little less DE and more DM, as shown by Eq. (17).

4.1 Fluids' Chronology

The standard model of cosmology is described by a set of periods in which different fluids dominated the dynamics. We first consider a period of inflation in which the Universe experienced an accelerated expansion rendering enough *e*-folds to explain the horizon and flatness problems of the old Big Bang theory (Cervantes-Cota 2004). This very early epoch is well described by an exponential expansion characterized by an equation of state $w = -1$. This is achieved through a scalar field that slowly rolls its potential, as we will see in Sect. 5.1. Eventually, the scalar field steps down the potential hill and begins to oscillate, to behave as a fluid of dust ($w = 0$) (Turner

1983). This period is thought to be short to let particle production and to heat the Universe in a period of reheating (Albrecht 1982; Dolgov and Linde 1982; Abbott et al. 1982) and/or preheating (Kofman et al. 1994, 1996), for a modern review see Allahverdi et al. (2010). This is needed since after inflation the Universe is cooled down exponentially and it is deprived of particles. The new produced particles, generically lighter than the scalar field mass, are relativistic ($T \gg m$, m being its rest mass), and therefore they are well described by $w = 1/3$. This epoch is important because it marks the beginning of the hot Big Bang theory. In this very early epoch particle physics theories (such as grand unification schemes) should describe the details of particle interactions to eventually reach the lower energies of the well tested standard model of particle physics. Then, the material content of the Universe consisted of a hot plasma with photons, protons, neutrons, neutrinos, electrons, and possibly other particles with very high kinetic energy. After some cooling of the Universe, some massive particles decayed and others survived (protons, neutrons, electrons, and DM) whose masses eventually dominated over the radiation components (photon, neutrinos, and possibly dark radiation; the latter being any other relativistic degree of freedom present at that epoch) at the *equality* epoch ($\rho_{\rm rel} = \rho_m$) at $z_{\rm eq} \sim 3402$ (Ade et al. 2013a). From this epoch and until recent e-folds of expansion ($z_{\rm DE} \sim 0.8$), the main matter component produced effectively no pressure on the expansion and, therefore, one can accept a model filled with dust, $w = 0$, to be representative for the energy content of the Universe in the interval $3402 < z < 0.8$. The dust equation of state is then representative of inert CDM. DM does not (significantly) emit light and therefore it is dark. Another possibility is that dark matter interacts weakly, which is generically called WIMP (Weakly Interacting Massive Particle); the neutralino being the most popular WIMP candidate. Another popular dark matter candidate is the axion, a hypothetical particle postulated to explain the conservation of the CP symmetry in quantum chromodynamics (QCD). Back to the Universe evolution, from $z \sim 0.8$ (Busca et al. 2013) until now the Universe happens to be accelerating with an equation of state $w \approx -1$, due to some constant energy that yields a cosmological constant, $\Lambda = 8\pi G\rho = \text{const.}$ The cosmological constant is the generic agent of an inflationary solution (see the $k = 0$ solution in Eq. 13). The details of the accelerated expansion are still unknown and it is possible that the expansion is due to some new fundamental field (e.g., quintessence) that induces an effective $\Lambda(t) \sim \text{const.}$ (see Sect. 5.2). We call (as M. Turner dubbed it) *dark energy* (DE) this new element. Dark energy does not emit light nor any other particle, and as known so far, it simply behaves as a (transparent) media that gravitates with an effective negative pressure. The physics behind dark energy or even the cosmological constant is unclear since theories of grand unification (or theories of everything, including gravity) generically predict a vacuum energy associated with fundamental fields, $< 0|T_{\mu\nu}|0 > = < \rho > g_{\mu\nu}$, that turns out to be very large. This can be seen by summing the zero-point energies of all normal modes of some field of mass m, to obtain $< \rho > \approx M^4/(16\pi^2)$, where M represents some cut-off in the integration, $M \gg m$. Then, assuming that GR is valid up to the Planck (Pl) scale, one should take $M \approx 1/\sqrt{8\pi G}$, which gives $< \rho > = 10^{71}$ GeV^4. This term plays the role of an effective cosmological constant $\Lambda = 8\pi G < \rho > \approx M_{Pl}^2 \sim 10^{38}\,\text{GeV}^2$, which must be added to Einstein's Eqs. (2), or directly to Eqs. (7) and (8), yielding an inflationary solution as given by Eq. (13). However, since the cosmological constant seems to dominate the dynamics of the Universe nowadays, one has that

$$\Lambda \approx 8\pi G\rho_0 = 3H_0^2 \sim 10^{-83}\,\mathrm{GeV}^2, \tag{23}$$

which is very small compared to the value derived above on dimensional grounds. Thus, the cosmological constraint and the theoretical expectations are rather dissimilar, by about 121 orders of magnitude! Even if one considers symmetries at lower energy scales, the theoretical Λ is indeed smaller, but never as small as the cosmological constraint: $\Lambda_{GUT} \sim 10^{21}$ GeV^2, $\Lambda_{SU(2)} \sim 10^{-29}$ GeV^2. This problem has been reviewed many decades ago (Weinberg 1989; Carrol et al. 1992) and still remains open.

5 Scalar Fields as Perfect Fluids

Scalar fields are ubiquitous in cosmology since they allow for modeling different cosmic dynamics, from inflation (Guth 1981; Linde 1990) and dark energy (Caldwell et al. 1998; Copeland et al. 2006) to dark matter (Matos et al. 2000; Magaña and Matos 2012). The full characterization of scalar fields is not describable in terms of perfect fluids, but its background dynamics allows for that. A scalar field with mass, m_ϕ, has an associated Compton wavelength, $\lambda_C = 1/m_\phi$. Thus, one can conceive the fluid picture as a collection of scalar particles with a typical size of λ_C. For if $\lambda_C = H_0^{-1}$ the corresponding scalar field mass is of course very light, $m_\phi = 10^{-33}$eV. If $\lambda_C < H^{-1}$ the particle is localizable within the Hubble horizon, otherwise its mass is too light and counts effectively as a massless particle.

A canonical scalar field (ϕ) is given by the Lagrangian density

$$\mathscr{L} = \frac{1}{2}\,\partial^\mu\phi\,\partial_\mu\phi - V(\phi), \tag{24}$$

where the first term accounts for the kinetic energy and $V(\phi)$ is its potential.

The energy-momentum tensor of the ϕ-field is

$$T_{\mu\nu}(\phi) = \frac{\partial\mathscr{L}}{\partial(\partial^\mu\phi)}\partial_\nu\phi - \mathscr{L}g_{\mu\nu} = \partial_\mu\phi\partial_\nu\phi - \frac{1}{2}\partial_\lambda\phi\partial^\lambda\phi\; g_{\mu\nu} + V(\phi)g_{\mu\nu}. \tag{25}$$

The field energy density and pressure are, by associating $\rho(\phi) = T_{00}(\phi)$ and $P(\phi) = T_{ii}(\phi)/a^2$ (no i-sum),

$$\rho(\phi) = \frac{1}{2}\dot{\phi}^2 + V(\phi) + \frac{1}{2a^2(t)}(\nabla\phi)^2 \approx \frac{1}{2}\dot{\phi}^2 + V(\phi),$$
$$P(\phi) = \frac{1}{2}\dot{\phi}^2 - V(\phi) - \frac{1}{6a^2(t)}(\nabla\phi)^2 \approx \frac{1}{2}\dot{\phi}^2 - V(\phi), \tag{26}$$

where the gradient terms (in comoving coordinates) are neglected. This typically occurs for the background cosmology and the reason for this is that the Universe is assumed to be sufficiently homogeneous within a horizon distance.

The equation of state associated to a scalar field is

$$w = \frac{P}{\rho} = \frac{\frac{1}{2}\dot{\phi}^2 - V(\phi)}{\frac{1}{2}\dot{\phi}^2 + V(\phi)}, \tag{27}$$

with w taking values in the interval $-1 \leq w \leq 1$.

The conservation of energy, Eq. (9), yields, using Eq. (26), the equation of motion for the ϕ-field,

$$\ddot{\phi} + 3H\dot{\phi} + V'(\phi) = 0, \tag{28}$$

where the prime stands for the scalar field derivative. The expansion term plays the role of a friction, whereas the potential contribution depends upon the scalar field model at hand.

In what follows, we present the main features of two applications of the scalar field dynamics: inflation and quintessence. We will refer the reader to recent reviews on these subjects for a more profound account of these topics, (cf. Mazumdar and Rocher 2011; Baumann 2012; Tsujikawa 2012).

5.1 Inflation

The scalar field responsible for the inflationary dynamics is dubbed the *inflaton*. There are hundreds of models of inflation and several theoretical aspects related to perturbations (see Sect. 7.1), non-Gaussianities, etc; for a recent review see Baumann (2012). The basics of the dynamics is as follows: the inflaton evolves from an initial value (ϕ_*) down the hill of the potential, but typically in a *slow roll-over* way, to a final state in which reheating takes place.

In order to get enough e-folds of inflation the scalar field should stay long time, compared to the cosmic time, in a potential 'flat region' where the potential is almost constant $V(\phi) \sim V(0)$. To construct such a flat curvature for the potential and to permit the ϕ-field to evolve slowly, one has to impose the slow roll-over conditions, namely, that $\ddot{\phi} \approx 0$. From Eq. (28), it implies that $\dot{\phi} \approx -V'/3H$, which in turn means that (Steinhardt and Turner 1984):

$$\frac{\ddot{\phi}}{3H\dot{\phi}} = -\frac{V''}{9H^2} + \frac{1}{48\pi G}\left(\frac{V'}{V}\right)^2 \ll 1, \tag{29}$$

or in terms of the dimensionless potential slow-roll parameters, $\epsilon \equiv 1/(16\pi G)(V'/V)^2 \ll 1$ and $\eta \equiv 1/(8\pi G)(V''/V) \ll 1$.

This condition also ensures that $\rho(\phi) \approx V(\phi) > \frac{1}{2}\dot{\phi}^2$, and so from Eq. (27) one has $w \approx -1$, which guarantees an accelerated expansion. However, if the initial conditions are such that at the outset $\frac{1}{2}\dot{\phi}^2 \gg V(\phi)$, then the solution takes the form, $\dot{\phi}^2 = \text{const.}/t^2$ and $\phi = \phi_0 - A\ \ln(1 + Bt)$, where A and B are constants. Then, the kinetic terms fall faster than the logarithmic decrease of a polynomial potential. Therefore, after some asymptotic time the Universe will be dominated by its potential and thus, inflation follows (Linde 1990). However, in other gravity theories the kinetic terms play an important role and could prevent the Universe from inflation (Cervantes-Cota and Dehnen 1995a,b).

The scalar field solution with $w \approx -1$, considered in Eq. (10), emulates a vacuum energy term or a cosmological constant. Given the slow roll-over of the ϕ-field this behaviour happens for a minimum of N e-folds of expansion in which the Hubble rate is effectively given by

$$H^2 = \frac{8\pi G}{3} V(\phi \approx \text{const.}). \tag{30}$$

In this way, $H \approx$ const. and the scale factor exhibits an exponential behaviour, as given by Eq. (13). Strictly speaking, during inflation ϕ is an increasing function of time, since $V' < 0$ in Eq. (28). However, under slow roll-over conditions its characteristic evolution time will be much greater than the cosmological time. Therefore, H will be a very slow, monotonically decreasing function of time.

Inflation lasts for a sufficient number of N e-folds to solve the horizon and flatness problems in cosmology, and this depends very much on the energy scale of inflation. In standard inflationary scenarios $N \sim 60$. This ensures that a possible curved model will look like a flat one for all the expansion history, including today (see Fig. 2 and compare with Fig. 1).

Among the multiple inflaton potentials considered in the literature, the most favoured models by the Planck CMBR temperature map fits (Ade et al. 2013b) are those having potentials with $V'' < 0$. Exponential potential models, the simplest hybrid inflationary models, and monomial potential models of degree $n \geq 2$ do not provide a good fit to the data. The most favoured models are Hill-top models, a simple symmetry breaking potential, natural inflation, R^2 inflation, and non-minimal coupled to gravity with a Mexican-hat potential; see Ade et al. (2013b) for details.

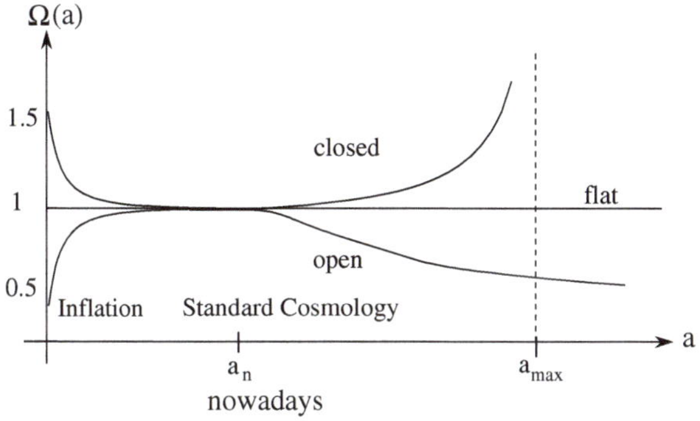

Fig. 2 The parameter Ω as a function of the scale factor, a, during inflation and thereafter in a radiation/matter dominated Universe. Inflation makes the space to look like as flat, even if it is initially curved. If there are enough e-folds of inflation to solve the horizon problem, it implies that the Universe nowadays is still flat. Later on, the behaviour is as in Fig. 1

5.2 Dark Energy: Quintessence

Dark energy is a generic name for an energetic "fluid" that has had little or no evolution in the past few giga-years of the cosmic expansion. Since then dark energy dominates the total density of the Universe over all other components (dark matter, baryons, photons, and neutrinos). During dark energy domination, the Hubble parameter, as given by Eq. (7), is basically a constant. Thus, a cosmological constant added to the gravitational theory is the simplest candidate for dark energy that fits the data from the different cosmic probes. There are at least seven independent observations that imply the presence of dark energy: the ages of some globular clusters surpasses the age of the Universe in models without dark energy (Jimenez et al. 1996; Richer et al. 2001); the supernovae best fits to distance moduli (Riess et al. 1998, 1999; Perlmutter et al. 1999); the dynamics of clusters of galaxies (Allen et al. 2004); the combination of the CMBR lensing deflection power spectrum with temperature and polarization power spectra (Sherwin et al. 2011); the measurements of the integrated Sachs-Wolf effect (Giannantonio et al. 2012); the measurements of Baryon Acoustic Oscillations (BAO) (Eisenstein et al. 2005); and the change of the Hubble rate behaviour from galaxy surveys (Busca et al. 2013).

Another possible candidate for dark energy is a canonical scalar field, dubbed quintessence (Caldwell et al. 1998). The equations governing the scalar field dynamics in a cosmological background are those displayed in Sect. 5. Basically, the FRW equations, i.e., Eqs. (7) and (8), are now fulfilled with the density and pressure terms given by Eqs. (26). To complete the whole picture, we add the rest of the known four material elements (dark matter, baryons, photons, and neutrinos) to the scalar field.

In a similar fashion to inflation, one demands that $V(\phi) > \frac{1}{2}\dot{\phi}^2$ has a flat potential and allows for an accelerated behaviour. One may again use the slow roll-over parameters (ϵ, η) to ensure an accelerated dynamics, but here we have the other four components that may spoil the exact accelerated dynamics. Still, this approach works well.

Originally, runaway potentials were considered, but nowadays there is a vast set of models that achieve the desired accelerated dynamics, including non-standard kinetic terms (Copeland et al. 2006; De-Santiago et al. 2013) or scalar fields interacting with matter (Aviles and Cervantes-Cota 2011), among many others. To avoid the over dominance of the scalar field during the early stages of the cosmic dynamics, one looks for scaling properties (of tracker nature) of the scalar field dynamics in which the field energy density (ρ_ϕ) evolves proportionally to the material fluid energy density (ρ_m) with $\rho_\phi < \rho_m$, and only until recently the scalar field turns to dominate. Depending on the evolution of the scalar-field equation of state, Eq. (27), quintessence models can be freezing or thawing (Caldwell and Linder 2005). The former class is when the scalar field gradually slows down to eventually freeze in a constant value. The latter class implies that the scalar field has recently started to change from a past constant value. These behaviours can in principle be tested (see Tsujikawa (2012) for a recent review on the subject).

6 Thermodynamics in the Early Universe

In the early Universe one considers a plasma of particles and their antiparticles, as was done originally by Gamow (1946), who first considered a physical scenario for the hot Big Bang model as a description of the beginning of the Universe. Later on, with the development of modern particle physics theories in the 70's it was unavoidable to think about a physical scenario which should include the "new" physics for the early Universe. It was also realized that the physics described by GR should not be applied beyond Planckian initial conditions, because there the quantum corrections to the metric tensor become very important, a theory which is still in progress.

After preheating/reheating, one assumes that the Universe is filled with a plasma of relativistic particles which include quarks, leptons, and gauge and Higgs bosons, all in thermal equilibrium at a very high temperature, T, with some gauge symmetry dictated by a particle physics theory.

Theoretically, one introduces some thermodynamic considerations necessary for the description of the physical content of the Universe, which we would like to present here. Assuming an ideal-gas approximation, the number density n_i of particles of type i, with a momentum p, is given by a Fermi or Bose distribution (Kolb and Turner 1990):

$$n_i = \frac{g_i}{(2\pi)^3} \int \frac{d^3 p}{e^{(E_i - \mu_i)/T} \pm 1}, \tag{31}$$

where $E_i = \sqrt{m_i^2 + p^2}$ is the particle energy, μ_i is the chemical potential, the sign $(+)$ applies for fermions and $(-)$ for bosons, and g_i is the number of spin states. One has that $g_i = 2$ for photons, quarks, baryons, electrons, muons, taus, and their antiparticles, but $g_i = 1$ for neutrinos because they are only left-handed. For the particles existing in the early Universe one usually assumes that $\mu_i = 0$: one expects that in any particle reaction the μ_i are conserved, just as the charge, energy, spin, and lepton and baryon number are. For a photon, which can be created and/or annihilated after some particle's collisions, its number density, n_γ, must not be conserved and its distribution with $\mu_\gamma = 0$, $E = p = \omega$, reduces to the Planckian one. For other constituents, in order to determine the μ_i, one needs n_i. Note from Eq. (31) that for large $\mu_i > 0$, n_i is large too. One does not know n_i in advance. However, the WMAP data constrains the baryon density at nucleosynthesis such that (Cyburt et al. 2005):

$$\eta \equiv \frac{n_B}{n_\gamma} \equiv \frac{n_{\text{baryons}} - n_{\text{anti-baryons}}}{n_\gamma} = 6.14 \pm 0.25 \times 10^{-10}. \tag{32}$$

The smallness of the baryon number density, n_B, relative to the photon's, suggests that n_{leptons} may also be small compared to n_γ. Therefore, one takes for granted that $\mu_i = 0$ for all particles. The ratio n_B/n_γ is very small, but not zero. The reason of why matter prevailed over antimatter is one of the puzzles of the standard model of cosmology called *baryogenesis* (Kolb and Turner 1990). There are some attempts to achieve baryogenesis at low energy scales, as low as few GeV or TeV (Dolgov 1992; Cohen et al. 1993; Cervantes-Cota and Dehnen 1995b; Trodden 1999; Bezrukov and Shaposhnikov 2008). Recent attempts to solve this problem are looking for prior to lepton asymmetry, *leptogenesis*, generated in the decay of a heavy sterile neutrino (Davidson et al. 2008), to then end with baryogenesis.

The above approximation allows one to treat the density and pressure of all particles as a function of the temperature only. According to the second law of thermodynamics, one has (Weinberg 1972):

$$dS(V,T) = \frac{1}{T}[d(\rho V) + P dV], \tag{33}$$

where S is the entropy in a volume $V \sim a^3(t)$, with $\rho = \rho(T)$ and $P = P(T)$ in equilibrium. Furthermore, the following integrability condition $\frac{\partial^2 S}{\partial T \partial V} = \frac{\partial^2 S}{\partial V \partial T}$ is also valid, which turns out to be

$$\frac{dP}{dT} = \frac{\rho + P}{T}. \tag{34}$$

On the other hand, the energy conservation law, Eq. (9), leads to

$$\frac{d}{dt}\left[\frac{a^3(t)}{T}(\rho + P)\right] = 0, \tag{35}$$

after using Eq. (34). Using Eq. (34) again, the entropy equation can be written as $dS(V,T) = \frac{1}{T} d[(\rho + P)V] - \frac{V}{T^2}(\rho + P)dT$. These last two equations imply that the entropy is a constant of motion:

$$S = \frac{a^3}{T}[\rho + P] = \text{const.}. \tag{36}$$

Moreover, the density and pressure are given by

$$\rho \equiv \int E_i n_i dp, \qquad P \equiv \int \frac{p^2}{3E_i} n_i dp. \tag{37}$$

For photons or ultra-relativistic fluids, $E = p$, and the above equations become $P = \frac{1}{3}\rho$, thus confirming Eq. (10) for $w = 1/3$. After integration of Eq. (34), it comes out that

$$\rho = bT^4, \tag{38}$$

where b is a constant of integration. In the real Universe there are many relativistic particles present, each of which contributes like Eq. (38). By including all of them, $\rho = \sum_i \rho_i$ and $P = \sum_i P_i$, where the summations are over all relativistic species, one has that $b(T) = \frac{\pi^2}{30}(N_B + \frac{7}{8}N_F)$, which depends on the effective relativistic degrees of freedom of bosons (N_B) and fermions (N_F). Therefore, this quantity varies with the temperature. Different i-species remain relativistic until some characteristic temperature $T \approx m_i$ and after this point N_{F_i} (or N_{B_i}) no longer contributes to $b(T)$. The factor 7/8 accounts for the different statistics of the particles [see Eq. (31)]. In the standard model of particle physics $b \approx 1$ for $T \ll 1\,\text{MeV}$ and $b \approx 35$ for $T > 300$ GeV (Kolb and Turner 1990). In particular, one accounts for the effective number of neutrinos (N_{eff}) in terms of photons' degrees of freedom as

$$\frac{\rho_\nu}{\rho_\gamma} = \frac{7}{8}\left(\frac{4}{11}\right)^{4/3} N_{\text{eff}}, \tag{39}$$

with $N_{\text{eff}} = 3.046$ for standard model neutrino species (Mangano et al. 2005). Extra neutrino-type relativistic species—dark radiation—should augment N_{eff}, as was recently suggested from measurements of different cosmological probes. Combining Planck with previous CMB data and Hubble Space Telescope measurements, it has been concluded that $N_{\text{eff}} = 3.6 \pm 0.5$ with a 95 % confidence level (Di Valentino et al. 2013).

For relativistic particles, we obtain from Eq. (31) that

$$n = cT^3, \quad \text{with } c = \frac{\zeta(3)}{\pi^2}(N_B + \frac{3}{4}N_F). \tag{40}$$

where $\zeta(3) \approx 1.2$ is the Riemann zeta function of 3. Nowadays, $n_\gamma \approx 411 T_{2.73}^3\,\text{cm}^{-3}$, where $T_{2.73} \equiv T_{\gamma 0}/(2.73\,\text{K})$. The precise measured value is $T_{\gamma 0} = 2.72548 \pm 0.00057°$ K (Fixsen 2009). The mean energy per photon is 6.34×10^{-4} eV which corresponds to a wavelength of 2 mm, and hence it is called cosmic "microwave" background radiation.

Using the relativistic equation of state given above ($w = 1/3$), From Eq. (36) it follows that $T \sim 1/a(t)$. From its solution in Eq. (12) one has

$$T = \sqrt[4]{\frac{M_{\frac{1}{3}}}{b}} \frac{1}{a(t)} = \sqrt[4]{\frac{3}{32\pi G b}} \frac{1}{(t - t_*)^{\frac{1}{2}}}, \tag{41}$$

which predicts a decreasing temperature behaviour as the Universe expands. Then, initially at the Big Bang, $t = t_*$ implies that $T_* = \infty$, and so the Universe was not only very dense but also very hot. As time evolves the Universe expands, cools down, and its density diminishes.

The entropy for an effective relativistic fluid is given by Eq. (36) together with its equation of state and Eq. (38), i.e., $S = \frac{4}{3}\, b\, (a\, T)^3 = \text{const}$. Combining this with Eq. (41), one can compute the value of $M_{\frac{1}{3}}$ to be $M_{\frac{1}{3}} = (\frac{3}{4}S)^{4/3}/b^{1/3} \approx 10^{116}$, since $b \approx 35$ and the photon entropy $S_0 = \frac{4}{3}\, b\, (a_0\, T_0)^3 \approx 10^{88}$ for $a_0 \to d_H(t_0) = 10^{28}$ cm and $T_{\gamma 0} = 2.73\,\text{K}$, as evaluated at the present time. One defines the entropy per unit volume, *entropy density*, to be $s \equiv S/V = \frac{4}{3}\frac{\pi^2}{30}(N_B + \frac{7}{8}N_F)T^3$, then at the present time $s \approx 7n_\gamma$. The nucleosynthesis bound on η, Eq. (32), implies that $n_B/s \approx 10^{-11}$.

We now consider particles in their non-relativistic limit ($m \gg T$). From Eq. (31) one obtains for both bosons and fermions that

$$n = g\left(\frac{mT}{2\pi}\right)^{3/2} e^{-m/T}. \tag{42}$$

The abundance of equilibrium massive particles decreases exponentially once they become non-relativistic. This situation is referred to as *in equilibrium annihilation*. Their density and pressure are given through Eqs. (37) and (42) by $\rho = nm$ and $P = nT \ll \rho$. Therefore, using these last two equations, the entropy for non-relativistic particles, given by Eq. (36), diminishes also exponentially during the in equilibrium annihilation. The entropy of these particles is transferred to that of the relativistic components by augmenting their temperature. Hence, the constant total entropy is essentially the same as the one given above, but the i-species contributing to it are just those which are in equilibrium and maintain their relativistic behaviour, that is, particles without mass such as photons.

Having introduced the abundances of the different particle types, we would like to comment on the equilibrium conditions for the constituents of the Universe as it evolves. This is especially important in order to have an idea of whether or not a given i-species disappears or decouples from the primordial brew. To see this, let us consider n_i when the Universe temperature, T, is such that (a) $T \gg m_i$, during the ultra-relativistic stage of some particles of type i and (b) $T \ll m_i$, when the particles i are non-relativistic, both cases in thermal equilibrium. From Eq. (40), one has that for the former case $n_i \sim T^3$ and the total number of particles, $\sim n_i a^3$, remains constant, whereas for the latter case, using Eq. (42), $n_i \sim T^{3/2} e^{-m_i/T}$, i.e., when the Universe temperature goes down below m_i, the number density of the i-species

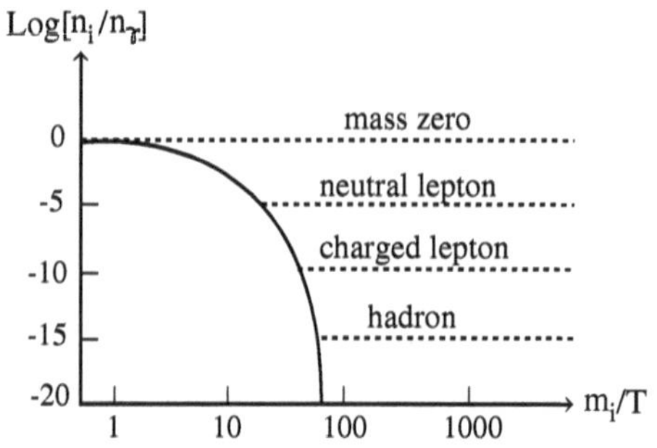

Fig. 3 Evolution of the particle density for different i-species. If a given i-species is in equilibrium, its abundance diminishes exponentially after the particle becomes non-relativistic (*solid line*). However, interactions of an i-species can freeze out, then it decouples from equilibrium and maintains its abundance (*dashed line*). Figure adapted from Kolb and Turner (1990)

significantly diminishes; it occurs an in equilibrium annihilation. Let us take as an example the neutron-proton annihilation. Then we have

$$\frac{n_n}{n_p} \sim \exp\left(\frac{m_p - m_n}{T}\right) = \exp\left(-\frac{1.5 \times 10^{10}\mathrm{K}}{T}\right), \tag{43}$$

which drops with the temperature from near 1 at $T \geq 10^{12}\,\mathrm{K}$ to about 5/6 at $T \approx 10^{11}\,\mathrm{K}$ and 3/5 at $T \approx 3 \times 10^{10}\,\mathrm{K}$ (Narlikar 2002). If this is valid forever, we then end up without massive particles and our Universe would have been consisted only of radiative components. However, our own existence prevents that! Therefore, eventually the in equilibrium annihilation had to be stopped. The quest is now to freeze out this ratio to $n_n/n_p \approx 1/6$ (due to neutron decays) until the time when nucleosynthesis begins (i.e., when n_n/n_p reduces to 1/7) in order to leave the correct number of hadrons and achieve later successful nucleosynthesis. The answer comes from comparing the Universe expansion rate, H, with the particle physics reaction rates, Γ. Hence, for $H < \Gamma$ the particles interact with each other faster than the Universe expansion rate, then equilibrium is established. For $H > \Gamma$ the particles cease to interact effectively, then thermal equilibrium drops out. This is only approximately true; a proper account of that involves a Boltzmann equation analysis. For that analysis numerical integration should be carried out in which annihilation rates are balanced with inverse processes (Steigman 1979; Kolb and Turner 1990). In this way, the more interacting the particles are, the longer they remain in equilibrium annihilation and, therefore, the lower their number densities are after some time, e.g., baryons vanish first, then charged leptons, neutral leptons, etc.; finally, the massless photons and neutrinos, whose particle numbers remain constant, as it was mentioned above (see Fig. 3). Note that if interactions of a given i-species freeze out while it is still relativistic, then its abundance will be significant at the present time and will account for dark radiation, as was recently suggested by Di Valentino et al. (2013).

It is worth mentioning that if the Universe would expand faster, then the temperature of decoupling, when $H \sim \Gamma$, would be higher and thus, the fixed ratio n_n/n_p would be greater and the ^{4}He abundance would be higher, leading to profound implications in the nucleosynthesis of light elements. Thus, the expansion rate cannot be arbitrarily modified during the equilibrium era of some particles. Furthermore, if a particle species is still highly relativistic ($T \gg m_i$) or highly non-relativistic ($T \ll m_i$), when decoupling from the primordial plasma occurs, it maintains an equilibrium distribution; the former being characterized by $T_r a$ = const. and the latter by $T_m a^2$ = const. [cf. Eq. (46)].

There are also some other examples of decoupling, such as neutrino decoupling: during nucleosynthesis there exist reactions, e.g. $\nu\bar{\nu} \longleftrightarrow e^+e^-$, which maintain neutrinos efficiently coupled to the original plasma ($\Gamma > H$) until about 1 MeV, since $\Gamma/H \approx T^3\,\mathrm{MeV}^{-3}$. The reactions are no longer efficient below 1 MeV and therefore neutrinos decouple and continue evolving with a temperature $T_\nu \sim 1/a$. Then, at $T \gtrsim m_e = 0.51\,\mathrm{MeV}$ the particles in equilibrium are photons (with $N_B = 2$) and electron-positron pairs (with $N_F = 4$), which contribute to the entropy with $b(T) = (11/2)(\pi^2/30)$. Later, when the temperature drops to $T \ll m_e$, the reactions are again no longer efficient ($\Gamma < H$) and, after the $e^\pm$ pair annihilation, there will be only photons in equilibrium with $b(T) = 2(\pi^2/30)$. Since the total entropy, $S = (4/3)b(aT)^3$, must be conserved, a decrease of $b(T)$ must be balanced with an increase of the radiation temperature so that $T_\gamma/T_\nu = (11/4)^{1/3}$, which should remain so until today, implying the existence of a cosmic background of neutrinos with a present temperature of $T_{\nu_0} = 1.95\,\mathrm{K}$. This cosmic relic has not been measured yet.

Another example is the gravitation decoupling, which should be also present if gravitons were in thermal equilibrium at the Planck time and then decouple. Today, the temperature background should be characterized at most by $T_{\mathrm{grav}} = (4/107)^{1/3}$ $\mathrm{K} \approx 0.91\,\mathrm{K}$.

For the matter dominated era we have stressed that effectively one has $P = 0$. Next we will see the reason for this. First, consider an ideal gas (such as atomic hydrogen) with mass m, then $\rho = nm + \frac{3}{2}nT_m$ and $P = nT_m$. From Eq. (35), one equivalently obtains that

$$\frac{d}{da}(\rho a^3(t)) = -3P\,a^2(t), \tag{44}$$

which after substitution of ρ and P, as given above, becomes

$$\frac{d}{da}\left(nma^3(t) + \frac{3}{2}n\,T_m a^3(t)\right) = -3\,n\,T_m a^2(t), \tag{45}$$

where $nma^3(t)$ is a constant. This equation yields

$$T_m a^2(t) = \mathrm{const.}, \tag{46}$$

so that the matter temperature drops faster than the radiation temperature as the Universe expands [cf. Eq. (41)]. Now, if one considers both radiation and matter, one has that $\rho = nm + \frac{3}{2}nT_m + bT_r^4$ and $P = nT_m + \frac{1}{3}bT_r^4$. The source of the Universe expansion is proportional to $\rho + 3P = nm + \frac{9}{2}nT_m + 2bT_r^4$, where the first term dominates over the second, precisely because T_m decreases very rapidly. The third term diminishes as $\sim 1/a^4$, whereas the first does it as $\sim 1/a^3$. After the time of density equalization, $\rho_m = \rho_r$, the matter density term is greater than the others and this explains why one assumes a zero pressure for that era.

From now on, when we refer to the temperature, T, it should be related to the radiation temperature. The detailed description of the Universe thermal evolution for the different particle types, depending on their masses, cross-sections, etc., is well described in many textbooks, going from the physics known in the early 1970s (Weinberg 1972) to the late1980s (Kolb and Turner 1990), and therefore it will not be presented here. However, we notice that as the Universe cools down a series of spontaneous symmetry–breaking (SSB) phase transitions are expected to occur. The type and/or nature of these transitions depend on the specific particle physics theory considered. Among the most popular ones are the Grand Unification Theories (GUT's), which bring together all known interactions except for gravity. One could also regard the standard model of particle physics or some extensions of it. Ultimately, when constructing a cosmological theory, one should settle the energy scale that one wants to describe physically. For instance, at a temperature between 10^{14} and 10^{16} GeV a transition to the $SU(5)$ GUT should take place, if this theory would be valid, in which a Higgs field breaks this symmetry to $SU(3)_C \times SU(2)_W \times U(1)_{HC}$, a process through which some bosons acquire their masses. Due to the gauge symmetry, there are color (C), weak (W), and hypercharge (HC) conservation, as the subscripts indicate. Later on, when the Universe evolved to around 150 GeV the electroweak phase transition took place in which the standard model Higgs field broke the symmetry $SU(3)_C \times SU(2)_W \times U(1)_{HC}$ to $SU(3)_C \times U(1)_{EM}$; through this breaking fermions acquired their masses. At this stage, there were only color and electromagnetic (EM) charge conservation, due to the gauge symmetry. Afterwards, around a temperature of 200 MeV (Aoki et al. 2009) the Universe should undergo a transition associated to the chiral symmetry-breaking and color confinement from which baryons and mesons were formed out of quarks. Subsequently, at approximately 10 MeV (Dolgov 2002) the synthesis of light elements (nucleosynthesis) began and lasted until temperatures below 100 keV, when most of the today observed hydrogen, helium, and some other light elements abundances were produced. So far the nucleosynthesis represents the earliest scenario tested in the standard model of cosmology. After some thousands of years ($z \sim 3402$ (Ade et al. 2013a)), the Universe became matter dominated, over the radiation components. At about 380,000 years ($z \sim 1090$ (Jarosik et al. 2011; Ade et al. 2013a)) recombination took place, that is, the hydrogen ions and electrons combined to form neutral hydrogen atoms, then matter and electromagnetic radiation decoupled from each other. At this moment, the (baryonic) matter structure began to form. Since that moment, the surface of last scattering of the CMBR evolved as an imprint of the early Universe. This is the light that Penzias and Wilson (1965) first measured, and that, later on, was measured in more detail by BOOMERANG (de Bernardis et al. 2000), MAXIMA (Hanany et al. 2000), COBE (Smoot et al. 1992), WMAP (Bennett et al. 2013), and now Planck (Ade et al. 2013a), among other probes.

7 Perturbed Fluids in the Universe

In the previous sections, we have outlined how the evolution of a homogeneous Universe can be described by means of a few equations and simple concepts such as the ideal perfect fluids. The next step is to introduce in this scenario small inhomogeneities that can be treated as first order perturbations to those equations, the goal being the description of the structures we see today in the Universe. This perturbative approach is sufficient to accurately explain the small temperature anisotropies ($\Delta T/T \sim 10^{-5}$) observed in the CMBR today, but it can only describe the distribution of matter today at those scales that are still in the linear regime. At the present epoch, scales smaller than $\sim$30 Mpc h^{-1} (Reid et al. 2010) have already entered the non linear-regime ($\Delta\rho/\rho >> 1$) due to the fact that matter tends to cluster under the effects of gravity. These scales can therefore be described only by means of numerical or semi-numerical approaches (Carlson et al. 2009).

The approach is quite straightforward. It involves a differential equation for the density perturbation of each individual constituent: scalar fields in inflation, or baryons, radiation, neutrinos, DM, and DE (usually treated as cosmological constant) in later times, and in general it needs to be solved numerically. In the context of the metric theories of gravity, and in particular GR, the metric is treated as the general expansion term $g^{(0)}_{\mu\nu}$ plus a perturbation $h_{\mu\nu}$:

$$g_{\mu\nu} = g^{(0)}_{\mu\nu} + h_{\mu\nu}, \tag{47}$$

with $h_{\mu\nu} << g^{(0)}_{\mu\nu}$, where $^{(0)}$ indicates unperturbed homogeneous quantities.

Inhomogeneities in the distribution of the components of the Universe are a source of scalar perturbations of the metric. Nevertheless, vector or tensor perturbations can modify the metric as well. The standard cosmological model does not predict vector perturbations that would introduce off-diagonal terms in the metric tensor. These perturbations would produce vortex motions in the primordial plasma, which are expected to rapidly decay. Models with topological defects or inhomogeneous primordial magnetic fields instead predict a consistent fraction of vector perturbations (Seljak and Zaldarriaga 1997; Turok et al. 1998; Kim and Naselsky 2009).

On the other hand, the standard cosmological model predicts the production of gravitational waves during the epoch of inflation, when the Universe expanded exponentially. Gravitational waves induce tensor perturbations $h^T_{\mu\nu}$ on the metric of the type:

$$h^T_{\mu\nu} = a^2 \begin{pmatrix} 0 & 0 & 0 & 0 \\ 0 & h_+ & h_\times & 0 \\ 0 & h_\times & -h_+ & 0 \\ 0 & 0 & 0 & 0 \end{pmatrix}$$

where h_+ and $h_\times$ are the polarization directions of the gravitational wave. This tensor is traceless, symmetric, and divergentless, i.e. it perturbs the time space orthogonally to the direction of propagation of the wave. The amplitudes of these tensor perturbations are expected to be small compared to the scalar ones, and therefore negligible in a first approximation as far as we are interested in studying the perturbations of the metric tensor. Nevertheless, these waves are expected to leave an imprint in the polarization of the CMBR, and their eventual detection would unveil an extremely rich source of information about an epoch of the Universe that is very hardly observable otherwise.

It is important to underline that choosing to model the metric perturbations corresponds to choosing a *gauge*, i.e. a specific coordinate system in which the metric tensor is represented. Changing the coordinate system, of course, do not change the physics, but can remarkably vary the difficulty of the calculations and ease the understanding of the physical meaning of the different quantities. In order to solve the perturbed equations one chooses convenient gauges for the different expansion epochs and depending on whether the formalism is theoretical or numerical, as we will see below.

The presence of weak inhomogeneous gravitational fields introduces small perturbations in the metric tensor. The most general perturbation to the FRW metric is:

$$ds^2 = a^2(\eta)\left[-(1+2A)\,d\eta^2 - B_i dx^i d\eta + [(1+2D)\delta_{ij} + 2E_{ij}]dx^i dx^j\right], \tag{48}$$

where η and x^i are comoving coordinates in which the expansion factor $a(\eta)$ is factored out. Different choices of them imply different gauges. We refer to Mukhanov et al. (1992); Ma and Bertschinger (1994, 1995) and Lyth and Liddle (2009) for an account of the physical meaning of the metric potentials and a full treatment of the perturbations.

In correspondence to the above metric perturbations, the energy-momentum tensor is also perturbed. One has:

$$\begin{aligned} T^0_0 &= -(\rho + \delta\rho), \\ T^0_i &= (\rho + P)(v_i - B_i), \\ T^i_0 &= -(\rho + P)v^i, \\ T^i_j &= (P + \delta P)\delta^i_j + \pi^i_j, \end{aligned} \tag{49}$$

where $v^i = dr^i/dt$ is the velocity in local orthonormal coordinates [$dt = a\,(1 + A)\,d\eta$; $dr^i = a\,dx^i$] and π^i_j are the anisotropic stresses; if they are null the perturbed fluid is also a perfect fluid. Anisotropic stresses are important before last scattering, when the primordial plasma was coupled. Later on, when structure formation begins they are set to zero.

A convenient gauge choice is given through two scalar functions $\Phi(\eta, x^i)$ and $\Psi(\eta, x^i)$ as (Mukhanov et al. 1992):

$$ds^2 = a^2(\eta)\left[-[1 + 2\Phi(\eta, x^i)]\,d\eta^2 + [1 + 2\Psi(\eta, x^i)]dx_i dx^i\right], \tag{50}$$

where the perturbed part of the metric tensor is:

$$h_{00}(\eta, x^i) = -2\Phi(\eta, x^i), \quad h_{0i}(\eta, x^i) = 0, \quad h_{ij}(\eta, x^i) = a^2\delta_{ij}(2\Psi(\eta, x^i)). \tag{51}$$

This metric is just a generalization of the well-known metric for a weak gravitational field usually described in the textbooks (e.g. Chap. 18 of Misner et al. (1973)) for the case of a static Universe [$a(\eta) = 1$]. The function Φ describes Newton's gravitational field, while Ψ is the perturbation of the space curvature. The above gauge is the *Newtonian conformal gauge*, which has the advantage of having a diagonal metric tensor $g_{\mu\nu}$ in which the coordinates are totally fixed with no residual gauge modes and therefore with a straightforward interpretation of the functions introduced.

Another example of a gauge that is particularly popular in the literature is the *synchronous gauge*, defined by:

$$ds^2 = a^2(\eta)[-d\eta^2 + (\delta_{i,j} + h_{i,j})\,dx^i\,dx^j], \tag{52}$$

which is especially used in numerical codes for calculations of the anisotropies and inhomogeneities in the Universe. It behaves well numerically by choosing that observers fall freely without changing their spatial coordinates.

The full perturbed equations are obtained by substituting the above expressions, for the chosen gauge, into the Einstein equations. Alternatively, one may obtain the continuity equation from the time ($\mu = 0$) component of Eq. (3) and the Euler equation from its space sector ($\mu = i$). Here we do not write down the perturbed equations for any particular gauge, but rather refer the reader to standard textbooks (Lyth and Liddle 2009), where these equations are fully described.

7.1 Perturbations During Inflation

The primeval fluctuations are thought to be present at the very beginning of time, at the inflationary epoch. The perturbations are produced by quantum fluctuations of the ϕ-field during the accelerated stage. These fluctuations are usually studied in the *comoving gauge* in which the scalar field is equal to its perturbed value at any given time during inflation and therefore, the perturbation information resides in the metric components (see Mukhanov et al. (1992); Cervantes-Cota (2004) and Lyth and Liddle (2009) for reviews on the subject).

To understand how perturbations evolve it is necessary to introduce the concept of horizon (Cervantes-Cota and Smoot 2011). There are two types of horizons in cosmology: the *causal* or particle horizon (d_H) and the *event* horizon (d_e). The former determines the region of space which can be connected to some other region by causal physical processes, at most through the propagation of light with $ds^2 = 0$. For the radiation cosmological era, one has that $d_H(t) = 2t = H^{-1}$ and for the matter era one has $d_H(t) = 3t = 2H^{-1}$; H^{-1} is sometimes called the Hubble horizon. During inflation (under an exponential expansion of the Universe) $d_H(t) = H^{-1}(e^{Ht} - 1)$ ($H =$ const.) and hence, the causal horizon grows exponentially. The event horizon, on the other hand, determines the region of space which will keep in causal contact (again complying with $ds^2 = 0$) after some time; that is, it delimits the region from which one can ever receive (up to some time $t_{\rm max}$) information about events taking place now (at time t). For the matter/radiation dominated eras $d_e \to \infty$ as $t_{\rm max} \to \infty$. However, during inflation one has that $d_e = H^{-1}(1 - e^{-(t_{\rm max}-t)H}) \approx H^{-1}$, which implies that any observer will see only those events that take place within a distance $\leq H^{-1}$. In this respect, there is an analogy with black holes, from whose surface no information can get away. Here, in an exponentially expanding Universe, observers encounter themselves in a region which is apparently surrounded by black holes (Gibbons and Hawking 1977; Linde 1990), since they receive no information located farther than H^{-1}.

Now, we turn back to the perturbation discussion. During the de Sitter stage the generation of perturbations, which is a causal microphysical process, is localized in regions of the order of $d_e = H^{-1}$ in which the microphysics operates coherently. At this time, the wavelength of inhomogeneities grows exponentially (as the causal horizon does) and eventually they cross outside the event horizon. Much later on, they re-enter into the event horizon, at the radiation and matter dominated epochs, to yield an almost scale invariant density perturbation spectrum (Harrison-Zel'dovich, $n_S = 1$), as is required for structure formation and measured by different cosmological probes.

It was shown that the amplitude of inhomogeneities produced corresponds to the Hawking temperature in the de Sitter space, $T_H = H/(2\pi)$. In turn, this means that perturbations with a fixed physical wavelength of size H^{-1} are produced throughout the inflationary era. Accordingly, a physical scale associated to a quantum fluctuation, $\lambda_{\rm phys} = \lambda a(t)$, expands exponentially and once it leaves the event horizon, it behaves as a metric perturbation; its description is then classical, general relativistic. If inflation lasts for enough time, the physical scale can grow as much as a galaxy or horizon-sized perturbation. The field fluctuation expands always with the scale factor and after inflation, it evolves according to t^n ($n = 1/2$ radiation or $n = 2/3$ matter). On the other hand, the Hubble horizon evolves after inflation as $H^{-1} \sim t$. This means that it will come a time at which field fluctuations cross inside the Hubble horizon and re-enters as density fluctuations. Thus, inflation produces a gross

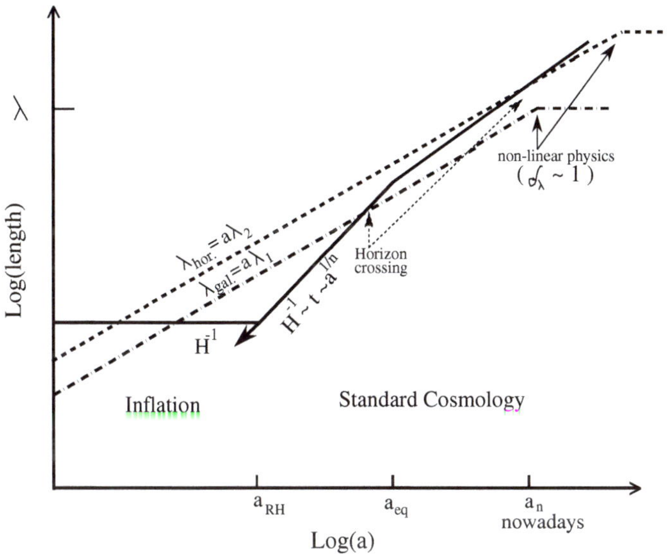

Fig. 4 Quantum perturbations were initially subhorizon-sized. During inflation they grow exponentially ($\lambda_{\text{phys.}} = \lambda a(t)$), whereas the event horizon remains almost constant. Then, eventually they cross outside H^{-1} and evolve as classical perturbations. Later on, they re-enter the event horizon to produce an almost scale invariant, Harrison-Zel'dovich density perturbation spectrum. In the figure are depicted two physical perturbations scales: galaxy and horizon-sized. Figure adapted from Kolb and Turner (1990)

spectrum of perturbations, the largest scale ones being originated at the start of inflation with a size H_i^{-1}, and the smallest ones with size H_f^{-1} at the end of inflation (see Fig. 4).

The power spectra for scalar (S) and tensor (T) perturbations are given by:

$$P_S(k) \approx \left(\frac{H^2}{16\pi^3 \dot{\phi}_c^2} \right) \Bigg|_{k=aH}, \qquad P_T(k) \approx \left(\frac{H^2}{4\pi^2 m_{Pl}^2} \right) \Bigg|_{k=aH}, \tag{53}$$

where $\dot{\phi}_c$ is the classical scalar field velocity. The equations are evaluated at the horizon crossing ($k = aH$) during inflation. Each of the k-modes generate an anisotropy pattern in the CMBR that was measured for scalar perturbations by the COBE (Smoot et al. 1992) and later probes. The Planck satellite may have the chance to detect the ratio of tensor to scalar amplitudes $r \equiv C_l^T / C_l^S < 0.12$ (95% limits) (Ade et al. 2013a), since the tensor modes modulate CMBR photons coming from last scattering.

The power spectra above give rise to the observed curvature and tensor power spectra in terms of the wavenumber (k) in a power law manner (Lyth and Liddle 2009):

$$P_{\mathscr{R}}(k) = A_S \left(\frac{k}{k_0}\right)^{n_S - 1 + \frac{1}{2} dn_S/d\ln k \, \ln(k/k_0)}, \qquad P_t(k) = A_t \left(\frac{k}{k_0}\right)^{n_t} \tag{54}$$

that has been determined by recent CMBR probes, such as Planck to give a best fit of $n_S = 0.96$ and $dn_S/d\ln k \approx -0.0090$ (Ade et al. 2013a). One should also have a tensor spectral index n_t that has not been measured yet.

These scalar and metric perturbations are small, but still very important. We discuss in the next section how to include them so that the information contained can be recognized and exploited.

7.2 Perturbations Inside the Horizon

We explained that in the early Universe baryons were tightly coupled to photons in an expanding background. Baryonic and dark matter potential wells provoked the local collapse of density fluctuations up to a certain point, at which the radiation pressure was big enough to pull out the matter apart and smooth the potential wells. These oscillations of the plasma are in fact *acoustic waves*. As we know, any wave can be decomposed into a sum of modes with different wave numbers, $k = 2\pi/\lambda$. Since these modes are in the sky, their wavelengths are measured as angles rather than as distances. Accordingly, instead of decomposing the wave in a Fourier series, what is normally done is to decompose the wave in terms of spherical harmonics, $Y_{lm}(\hat{n})$, where $\hat{n}$ is the direction of a measured photon. The angular power spectrum can be expanded in Legendre polynomials, since there is no preferred direction in the Universe and only the angular separation θ is relevant. A mode l plays the same role of the wavenumber k, thus $l \approx 1/\theta$. We are interested in the temperature fluctuations that are analyzed experimentally in pairs of directions $\hat{n}$ and $\hat{n}'$, where $cos(\theta) = \hat{n}\cdot\hat{n}'$. We then average these fluctuations, obtaining the multipole expansions:

$$\frac{\Delta T}{T} = \sum_{l=1}^{\infty} \sum_{m=-l}^{l} a_{lm}(\mathbf{x}, \eta) Y_{lm}(\hat{n}), \qquad P_S(\theta) = \sum \frac{(2l+1)}{4\pi} C_l P_l(\cos\theta), \tag{55}$$

where $P_S(\theta)$ is the angular power spectrum, P_l are the Legendre polynomials, and the C_l are estimated as averages of the a_{lm} over m. All this information can be used to determine the cosmological parameters Ω_i. We will not discuss here the detailed calculations nor the curve that must be adjusted to obtain the best fit values for such parameters. The peak of the fundamental mode appears at approximately

$$l \simeq \frac{200}{\sqrt{\Omega^{(0)}}}. \tag{56}$$

BOOMERANG (de Bernardis et al. 2000) and MAXIMA (Hanany et al. 2000) were two balloon-borne experiments designed to measure the anisotropies at scales smaller than the horizon at decoupling ($\theta_{\text{hor-dec}} \sim 1°$), hence measuring the acoustic features of the CMBR. The sensitivity of the instruments allowed for a measurement of the temperature fluctuations of the CMBR over a broad range of angular scales. BOOMERANG found a value of $l = 197 \pm 6$ and MAXIMA-1 found a value of $l \approx 220$. This implies that the cosmological density parameter $\Omega^{(0)} \approx 1$ (see Eq. (14)), suggesting that the Universe is practically flat, $\Omega_k^{(0)} \approx 0$. These two experiments provided the first strong evidence for a flat Universe from observations. Happily, this result was expected from inflation since an accelerating dynamics effectively flattens the curvature of the event horizon, which we later identify with our Universe (see Fig. 2). These results were confirmed by WMAP in a series of data releases in the last decade, as well as by other cosmological probes: the Universe is flat or pretty close to be flat. The problem in the exact determination of the curvature is because the CMBR anisotropies show strong degeneracies among the cosmological parameters (Bond et al. 1997; Zaldarriaga et al. 1997). However, the satellite Planck offers results on the density parameters with uncertainties less than a percent level, $\Omega_k^{(0)} = -0.0105$ (Ade et al. 2013a).

Since baryons and photons were in thermal equilibrium until recombination, also called *last scattering* (ls), the acoustic oscillations (BAO) were also imprinted in the matter perturbations, as they were in the CMBR anisotropies. The sound horizon, at the moment when the baryons decoupled from the photons, plays a crucial role in the determination of the position of the baryon acoustic peaks. This time is known as the *drag epoch* which happens at $z_d = 1/a_d - 1$. The sound horizon at that time is defined in terms of the effective speed of sound of the baryon-photon plasma, $c_s^2 \equiv \delta p_\gamma/(\delta\rho_\gamma + \delta\rho_b)$,

$$r_s(z_d) = \int_0^{\eta_d} d\eta\, c_s(\eta) = \frac{1}{3}\int_0^{a_d} \frac{da}{a^2 H(a)\sqrt{1+(3\Omega_b/4\Omega_\gamma)a}}\,. \tag{57}$$

Note that the *drag epoch* does not coincide with the last scattering. In most scenarios $z_{\text{d}} < z_{ls}$ (Hu and Sugiyama 1996). The redshift at the drag epoch can be computed with a fitting formula that is a function of $\omega_m = \Omega_m^{(0)} h^2$ and $\omega_b = \Omega_b^{(0)} h^2$ (Eisenstein and Hu 1998). The WMAP team, and recently Planck, computed these quantities for the ΛCDM model, obtaining $z_d = 1059.29 \pm 0.65$ and $r_s(z_d) = 147.53 \pm 0.64$ Mpc (Ade et al. 2013a).

BAO can be characterized by the angular position and the redshift (Seo and Eisenstein 2003; Amendola and Tsujikawa 2010):

$$\theta_s(z) = \frac{r_s(z_d)}{(1+z)\,d_A(z)}, \tag{58}$$

$$\delta z_s(z) = r_s(z_d)\, H(z), \tag{59}$$

where $d_A(z) = \frac{1}{H_0|\Omega_k|^{1/2}(1+z)} \sin_k\left(|\,\Omega_k\,|^{1/2} \int_0^z \frac{dz'}{H(z')}\right)$ is the proper (not comoving) angular diameter distance to the redshift z, with $\sin_{\text{k}} = \sin$ for $\Omega_k < 0$ and $\sin_{\text{k}} = \sinh$ for $\Omega_k > 0$; where $H(z)$ is determined by Eq. (21). The angle $\theta_s(z)$

corresponds to the direction orthogonal to the line-of-sight, whereas $\delta z_s(z)$ measures the fluctuations along the line-of-sight. Observations of these quantities are encouraging to determine both $d_A(z)$ and $H(z)$. However, from the current BAO data is not simple to independently measure these quantities. This will certainly happen in forthcoming surveys (Schlegel et al. 2011). Therefore, it is convenient to combine the two orthogonal dimensions to the line-of-sight with the dimension along the line-of-sight to define (Eisenstein et al. 2005):

$$D_V(z) \equiv \left((1+z)^2 d_A(z)^2 \frac{z}{H(z)} \right)^{1/3}, \tag{60}$$

where the quantity $D_M \equiv d_A/a = (1+z)d_A(z)$ is the comoving angular diameter distance. The BAO signal has been measured in large samples of luminous red galaxies from the SDSS (Eisenstein et al. 2005). There is a clear evidence (3.4σ) for the acoustic peak at a scale of $100h^{-1}$ Mpc. Moreover, the scale and amplitude of this peak are in good agreement with the prediction of the ΛCDM as confirmed by the WMAP and Planck data. One finds that $D_V(z = 0.35) = 1370 \pm 64$ Mpc, and more recently new determinations of the BAO signal has been published (Carnero et al. 2012) in which $\theta_s(z = 0.55) = 3.90° \pm 0.38°$ and $w = -1.03 \pm 0.16$ for the equation of state parameter of the dark energy, or $\Omega_M^{(0)} = 0.26 \pm 0.04$ for the matter density, when the other parameters are fixed. One also defines the BAO distance $d_z \equiv r_s(z_d)/D_V(z)$, which has been measured by surveys. For instance, an analysis of the BOSS survey gives $d(0.57) = 13.67 \pm 0.22$ (Anderson et al. 2012), which is the current most precise determination of the BAO scale.

Measuring the BAO feature in the matter distribution at different redshifts will help break the degeneracy that exists in the determination of the cosmological parameters. By combining line-of-sight with angular determinations of the BAO feature one will constrain even more the parameter space. Furthermore, a complete combination of BAO, the full matter power spectrum, direct $H(z)$ measurements, supernovae Ia luminosities, and CMBR data will certainly help envisage the true nature of the mysterious, dark Universe.

8 Outlook

We have reviewed the role that fluids have played in the entire history of the Universe. Their components are relatively simple and behave as perfect fluids, at least at the background level. The fluids' evolution is as follows: first, scalar fields governed a very early inflationary dynamics with an equation of state $w \sim -1$. After inflation, the Universe was deprived of particles and it had a very low temperature. Then, reheating/preheating took place to give rise to the hot Big Bang era, governed by a radiation period with $w = 1/3$. But the density of radiation and/or relativistic particles (photons, neutrinos) decayed faster than that of non-relativistic particles (protons, neutrons, DM) and eventually matter dominated over the relativistic components in a dust ($w = 0$) period of the evolution. More recently, but still seven billion years ago, dark energy with $w \sim -1$ entered to dominate the dynamics and to inflate the Universe again.

Real fluids are in a perturbed state, and the five main components of the Universe (photons, neutrinos, baryons, dark matter, and dark energy) are not the exception. The plasma that composed the hot Big Bang era oscillated with the well-known kinematics of perturbed fluids, and as a consequence anisotropies in the CMB and inhomogeneities in the matter distribution left a unique fingerprint that we measure at present. On the other hand, if dark energy is the simplest candidate, the cosmological *constant*, its perturbations are null, since it is simply a geometrical term in the Einstein's equations. But if it is a fluid, perturbations are to be computed to understand their effect on structure formation.

Cosmological and astrophysical observations, since the early 1990s, have been playing a main role in the cosmological science, which was governed mainly by exact solutions and mathematical analyses. Indeed, we have just entered in a high precision era in which the observations demand to construct new theoretical observables, and vice versa. In the coming years, we expect not only to learn more about the fluids in cosmology, such as dark matter and dark energy, but also about the left-hand side of Einstein's equations: is GR correct? or, are modified gravity schemes more properly fitted to the cosmic kinematics? These are quests that challenge our present knowledge and that should be answered in the coming years.

Acknowledgments This work was partially supported by the Consejo Nacional de Ciencia y Tecnología of México (CONACyT) under the project CONACyT-EDOMEX-2011-C01-165873.

References

Abbott LF, Farhi E, Wise MB (1982) Particle production in the new inflationary cosmology. Phys Lett B 117:29–33

Ade PAR, Aghanim N, Armitage-Caplan C, Arnaud M, Ashdown M, et al. (2013a) Planck 2013 results. XVI. Cosmological parameters, p 67. arXiv:1303.5076v1 [astro-ph.CO].

Ade PAR, Aghanim N, Armitage-Caplan C, Arnaud M, Ashdown M, et al. (2013b) Planck 2013 results. XXII. Constraints on inflation, p 43. arXiv:1303.5082v1 [astro-ph.CO].

Ade PAR, Aghanim N, Armitage-Caplan C, Arnaud M, Ashdown M, et al. (2013c) Planck 2013 results. I. overview of products and scientific results, p 44. arXiv:1303.5062v1 [astro-ph.CO].

Ade PAR, Aghanim N, Armitage-Caplan C, Arnaud M, Ashdown M, et al. (2013d) Planck 2013 results.XXIII. Isotropy and statistics of the CMB, p 42. arXiv:1303.5083v1 [astro-ph.CO].

Albrecht A, Steinhardt PJ, Turner MS, Wilczek F (1982) Reheating an inflationary universe. Phys Rev Lett 48:1437–1440

Allahverdi R, Brandenberger R, Cyr-Racine F-Y, Mazumdar A (2010) Ann Rev Nucl Part Sci 60:27. [arXiv:1001.2600 [hep-th]].

Allen SW, Schmidt RW, Ebeling H, Fabian AC, van Speybroeck L (2004) Constraints on dark energy from Chandra observations of the largest relaxed galaxy clusters. Mon Not R Astron Soc 353:457–467

Amanullah R, Lidman C, Rubin D, Aldering G, Astier P, et al. (2010) Spectra and Hubble space telescope light curves of six type Ia supernovae at $0.511 \leq z \leq 1.12$ and the Union2 compilation. Astrophys J 716:712–738
Amendola L, Tsujikawa S (2010) Dark energy: theory and observations. Cambridge University Press, Cambridge
Anderson L, Aubourg E, Bailey S, Bizyaev D, Blanton M, et al. (2012) The clustering of of galaxies in the SDSS-III baryon oscillation spectroscopic survey: Baryon acoustic oscillations in the Data Release 9 Spectroscopic Galaxy Sample. Mon Not R Astron Soc 427:3435–3467
Aoki Y, Borsányi S, Dürr S, Fodor Z, Katz SD, et al. (2009) The QCD transition temperature: results with physical masses in the continuum limit II. J High Energy Phys 06:088
Aviles A, Cervantes-Cota JL (2011) Dark matter from dark energy-baryonic matter couplings. Phys Rev D 83:023510
Baumann D (2012) TASI lectures on inflation, p 159. arXiv:0907.5424v2 [hep-th].
Bennett CL, Larson D, Weiland JL, Jarosik N, Hinshaw G, et al. (2013) Nine-year Wilkinson microwave anisotropy probe (WMAP) observations: final maps and results, p 177. arXiv:1212.5225v3 [astro-ph.CO].
de Bernardis P, Ade PAR, Bock JJ, Bond JR, Borrill J, et al. (2000) A flat universe from high-resolution maps of the cosmic microwave background radiation. Nature 404:955–959
Bezrukov FL, Shaposhnikov M (2008) The standard model Higgs boson as the inflaton. Phys Lett B 659:703–706
Bond JR, Efstathiou G, Tegmark M (1997) Forecasting cosmic parameter errors from microwave background anisotropy experiments. Mon Not R Astron Soc 291:L33–L41
Busca NG, Delubac T, Rich J, Bailey S, Font-Ribera A, et al. (2013) Baryon acoustic oscillations in the Ly$alpha$; forest of BOSS quasars. Astron Astrophys 552:A96
Caldwell RR, Dave R, Steinhardt PJ (1998) Cosmological imprint of an energy component with general equation of state. Phys Rev Lett 80:1582–1985
Caldwell RR, Linder EV (2005) The limits of quintessence. Phys Rev Lett 95:141301
Carlson J, White M, Padmanabhan N (2009) Critical look at cosmological perturbation theory techniques. Phys Rev D 80:043531
Carnero A, Sanchez E, Crocce M, Cabre A, Gaztanaga E (2012) Clustering of photometric luminous red galaxies II: cosmological implications from the baryon acoustic scale. Mon Not R Astron Soc 419:1689–1694
Carrol S, Press W, Turner E (1992) The cosmological constant. Ann Rev Astron Astrophys 30:499–542
Cervantes-Cota JL (2004) An introduction to standard cosmology. In: Bretón N, Cervantes-Cota JL, Salgado M (eds) Lecture notes in physics: the early universe and observational cosmology, vol 646. Springer, Berlin, pp 53–107
Cervantes-Cota JL, Dehnen H (1995) Induced gravity inflation in the SU(5) GUT. Phys Rev D 51:395–404
Cervantes-Cota JL, Dehnen H (1995) Induced gravity inflation in the standard model of particle physics. Nucl Phys B 442:391–409
Cervantes-Cota JL, Smoot G (2011) Cosmology today-a brief review. Am Inst Phys Conf Ser 1396:28–52
Chambers CM, Moss IG (1994) Cosmological no-hair theorem. Phys RevLett 73:617–620
Chevallier M, Polarski D (2001) Accelerating universes with scaling dark matter. Int J Modern Phys D 10:213–223
Cohen AG, Kaplan DB, Nelson AE (1993) Progress in electroweak baryogenesis. Ann Rev Nuc Part Sci 43:27–70
Copeland EJ, Sami M, Tsujikawa S (2006) Dynamics of dark energy. Int J Modern Phys D 15:1753–1935
Cyburt RH, Fields BD, Olive KA, Skillman E (2005) New BBN limits on physics beyond the standard model from ^{4}He. Astropart Phys 23:313–323
Davidson S, Nardi E, Nir Y (2008) Leptogenesis. Phys Rep 466:105–177

De-Santiago J, Cervantes-Cota JL, Wands D (2013) Cosmological phase space analysis of the $F(X)$-$V(\phi)$ scalar field and bouncing solutions. Phys Rev D 87:023502
Di Valentino E, Melchiorri A, Mena O (2013) Dark radiation candidates after Planck, p 6. arXiv:1304.5981v1 [astro-ph.CO].
Dolgov AD (2002) Big bang nucleosynthesis. Nucl Phys B Proc Suppl 110:137–143
Dolgov AD, Linde AD (1982) Baryon asymmetry in the inflationary universe. Phys Lett B 116:329–334
Dolgov AD (1992) Non-GUT baryogenesis. Phys Rep 222:309–386
Eisenstein DJ, Hu W (1998) Baryonic features in the matter transfer function. Astrophys J 496:605–625
Eisenstein DJ, Zehavi I, Hogg DW, Scoccimarro R, Blanton MR (2005) Detection of the baryon acoustic peak in the large-scale correlation function of SDSS luminous red galaxies. Astrophys J 633:560–574
Eisenstein DJ, Zehavi I, Hogg DW, Scoccimarro R, Blanton MR, et al. (2005) Detection of the baryon acoustic peak in the large-scale correlation function of SDSS luminous red galaxies. Astrophys J 633:560–574
Ellis GFR, Maartens R, MacCallum MAH (2012) Relativistic cosmology. Cambridge University Press, Cambridge
Fixsen DJ (2009) The temperature of the cosmic microwave background. Astrophys J 707:916–920
Friedmann A (1922) Über die Krümmung des Raumes. Zeitschrift für Physik A 10:377–386
Friedmann A (1924) Über die Möglichkeit einer Welt mit konstanter negativer Krümmung des Raumes. Zeitschrift für Physik A 21:326332
Gamow G (1946) Expanding universe and the origin of elements. Phys Rev 70:572–573
Gamow G (1948) The origin of elements and the separation of galaxies. Phys Rev 74:505–506
Giannantonio T, Crittenden R, Nichol R, Ross AJ (2012) The significance of the integrated Sachs-Wolfe effect revisited. Mon Not R Astron Soc 426:2581–2599
Gibbons GW, Hawking SW (1977) Cosmological event horizons, thermodynamics, and particle creation. Phys Rev D 15:2738–2751
Guth AH (1981) Inflationary universe: a possible solution to the horizon and flatness problems. Phys Rev D 23:347–356
Hanany S, Ade PAR, Balbi A, Bock J, Borrill J, et al. (2000) MAXIMA-1: a measurement of the cosmic microwave background anisotropy on angular scales of $10' - 5°$. Astrophys J 545:L5–L9
Hinshaw G, Larson D, Komatsu E, Spergel DN, Bennett CL, et al. (2013) Nine-year Wilkinson microwave anisotropy probe (WMAP) observations: cosmological parameter results, p 32. arXiv:1212.5226v3 [astro-ph.CO].
Hoyle B, Tojeiro R, Jimenez R, Heavens A, Clarkson C, et al. (2013) Testing homogeneity with galaxy star formation histories. Astrophys J 762:L9–L13
Hu W, Sugiyama N (1996) Small-scale cosmological perturbations: an analytic approach. Astrophys J 471:542–578
Hubble EP (1929) A relation between distance and radial velocity among extra-galactic nebulae.Proceedings of the National Academy of Science of the United States of. America 15:168–173
Jarosik N, Bennett CL, Dunkley J, Gold B, Greason MR, et al. (2011) Seven-year Wilkinson microwave anisotropy probe (WMAP) observations: sky maps, systematic errors, and basic results. Astrophys J Suppl 192:14 (15 pp).
Jimenez R, Thejll P, Jorgensen U, MacDonald J, Pagel B (1996) Ages of globular clusters: a new approach. Mon Not R Astron Soc 282:926–942
Kim J, Naselsky P (2009) Cosmological Alfvén waves in the recent CMB data, and the observational bound on the primordial vector perturbation. J Cosmol Astropart Phys 7:041
Kofman L, Linde A, Starobinsky AA (1994) Reheating after inflation. Phys Rev Lett 73:3195–3198
Kofman L, Linde A, Starobinsky AA (1996) Nonthermal phase transitions after inflation. Phys Rev Lett 76:1011–1014
Kolb EW, Turner MS (1990) The early universe. Addison-Wesley, Frontiers in physics
Linde AD (1990) Particle physics and inflationary cosmology. CRC Press, Boca Raton

Linder EV (2003) Exploring the expansion history of the universe. Phys Rev Lett 90:091301
Lyth DH, Liddle AR (2009) The primordial density perturbation: cosmology, inflation and the origin of structure. Cambridge University Press, Cambridge
Ma CP, Bertschinger E (1994) A calculation of the full neutrino phase space in cold + hot dark matter models. Astrophys J 429:22–28
Ma CP, Bertschinger E (1995) Cosmological perturbation theory in the synchronous and conformal Newtonian gauges. Astrophys J 455:7–25
Magaña J, Matos T (2012) A brief review of the scalar field dark matter model. J Phys Conf Ser 378:012012
Mangano G, Miele G, Pastor S, Pinto T, Pisanti O, Serpico PD (2005) Relic neutrino decoupling including flavor oscillations. Nucl Phys B 729:221–234
Marinoni C, Bel J, Buzzi A (2012) The scale of cosmic isotropy. J Cosmol Astropart Phys 10:036
Mather JC, Cheng ES, Eplee RE Jr, Isaacman RB, Meyer SS, et al. (1990) A preliminary measurement of the cosmic microwave background spectrum by the Cosmic Background Explorer (COBE) satellite. Astrophys J 354:L37–L40
Matos T, Guzman FS, Urena-Lopez LA (2000) Scalar field as dark matter in the universe. Class Quantum Gravity 17:1707–1712
Mazumdar A, Rocher J (2011) Phys Rept 497:85. [arXiv:1001.0993 [hep-ph]].
Misner CW, Thorne TS, Wheeler AJ (1973) Gravitation. W. H. Freeman & Co, New York
Mukhanov VF, Feldman HA, Brandenberger RH (1992) Theory of cosmological perturbations. Phys Rep 215:203–333
Narlikar JV (2002) Introduction to cosmology. Cambridge University Press, Cambridge
Penzias AA, Wilson RW (1965) A measurement of excess antenna temperature at 4080 Mc/s. Astrophys J 142:419–421
Perlmutter S, Aldering G, Goldhaber G, Knop RA, Nugent P, et al. (1999) Measurements of Ω and Λ from 42 high-redshift supernovae. Astron J 517:565–586
Reid BA, Percival WJ, Eisenstein DJ, Verde L, Spergel DN, et al. (2010) Cosmological constraints from the clustering of the Sloan Digital Sky Survey DR7 luminous red galaxies. Mon Not R Astron Soc 404:60–85
Richer HB, Brewer J, Fahlman GG, Gibson BK, Hansen BM, et al. (2001) The Lower main sequence and mass function of the globular cluster Messier 4. Astron J 574:L151–L154
Riess AG, Filippenko AV, Challis P, Clocchiatti A, Diercks A, et al. (1998) Observational evidence from supernovae for an accelerating universe and a cosmological constant. Astron J 116:1009–1038
Riess AG, Kirshner RP, Schmidt BP, Jha S, Challis P, et al. (1999) BVRI light curves for 22 type Ia supernovae. Astron J 117:707–724
Robertson HP (1935) Kinematics and world-structure. Astrophys J 82:284–301
Robertson HP (1936a) Kinematics and world-structure II. Astrophys J 83:187–201
Robertson HP (1936b) Kinematics and world-structure III. Astrophys J 83:257–271
Schlegel D, Abdalla F, Abraham T, Ahn C, Allende Prieto C, et al. (2011) The bigBOSS experiment, p 212. arXiv:1106.1706v1 [astro-ph.IM].
Schutz BF (1985) A first course In general relativity. Cambridge University Press, Cambridge
Seljak U, Zaldarriaga M (1997) Signature of gravity waves in the polarization of the microwave background. Phys Rev Lett 78:2054–2057
Seo HJ, Eisenstein DJ (2003) Probing dark energy with baryonic acoustic oscillations from future large galaxy redshift surveys. Astrophys J 598:720–740
Sherwin BD, Dunkley J, Das S, Appel JW, Bond JR, et al. (2011) Evidence for dark energy from the cosmic microwave background alone using the atacama cosmology telescope lensing measurements. Phys Rev Lett 107:021302
Smoot GF, Bennett CL, Kogut A, Wright EL, Aymon J, et al. (1992) Structure in the COBE differential microwave radiometer first-year maps. Astrophys J 396:L1–L5
Steigman G (1979) Cosmology confronts particle physics. Ann Rev Nucl Part Sci 29:313–338

Steinhardt PJ, Turner MS (1984) Prescription for successful new inflation. Phys Rev D 29:2162–2171
Trodden M (1999) Electroweak baryogenesis. Revf Modern Phys 71:1463–1500
Tsujikawa S (2012) Quintessence: a review, p 20. arXiv:1304.1961 [gr-qc].
Turner MS (1983) Coherent scalar-field oscillations in an expanding universe. Phys Rev D 28:1243–1247
Turok N, Pen UL, Seljak U (1998) Scalar, vector, and tensor contributions to CMB anisotropies from cosmic defects. Phys Rev D 58:023506
Walker AG (1937) On Milne's theory of world-structure. Proc Lond Math Soc 42:90–127
Weinberg S (1972) Gravitation and cosmology: principles and applications of the general theory of relativity. John Wiley, New York
Weinberg S (1989) The cosmological constant problem. Rev Modern Phys 61:1–23
Weinberg S (2008) Cosmology. Oxford University Press, Oxford
Zaldarriaga M, Spergel D, Seljak U (1997) Microwave background constraints on cosmological parameters. Astrophys J 488:1–13

Fluid Mechanics and Systems Biology for Understanding the Cosmic Distribution of Life: A Review

Julián Chela-Flores

Abstract Due to progress in instrumentation both in cryogenics and in space exploration, the 20th century witnessed the extension of fluid mechanics applications in two novel systems. While the major aim for the first of these two cases—low temperature physics—was to understand the underlying microscopic theory, in the second case of fluid mechanics in the outer Solar System the major problem was, and still is, one of instrumentation, rather than theory. This second kind of environments may provide hints regarding the central problem of astrobiology, namely the search for life outside our own planet. The Galileo Mission (1995–2003) allowed closer probing of the Jovian satellite Europa, both with imaging techniques, as well as with spectroscopy of its icy surface over a deep ocean that is covered with chemical elements. Other examples of oceans are found in Ganymede and Callisto, two other icy Galilean moons, but possibly these oceans are not in contact with a silicate core, as in the cases of the life-friendly world: the Earth. In addition, Europa, with possibly the same internal geological structure as our planet, is also potentially a life-friendly world. These appealing phenomena are currently the source of plans for the next European mission to Europa that will provide a baseline for the search of life. For this purpose knowledge of our oceans will guide us in the search of life in other solar system oceans. These possibilities have encouraged underlining technologically feasible proposals for delivering small missiles ("penetrators") with appropriate instrumentation. Whenever compatible with the available payloads, one objective of these instruments has been to identify bioindicators. We are interested essentially in understanding the surficial sulfur stains of Europa's icy surface. Although not included in the most recent approved mission for Europa, penetrators remain a valid alternative

J. Chela-Flores (✉)
The Abdus Salam International Centre for Theoretical Physics (ICTP),
Strada Costiera 11, 34014 Trieste, Italy
e-mail: chelaf@ictp.it

J. Chela-Flores
Fundación Instituto de Estudios Avanzados, IDEA, Carretera Nacional
Hoyo de la Puerta, Valle de Sartenejas, Baruta, Caracas 1015-A, Estado Miranda, Venezuela

L. Di G. Sigalotti et al. (eds.), *Computational and Experimental Fluid Mechanics with Applications to Physics, Engineering and the Environment*, Environmental Science and Engineering, DOI:10.1007/978-3-319-00191-3_5,

in lunar research that we have shown to be relevant to the basis of astrobiology. In this context we have argued that already existing miniaturized mass spectrometers are particularly relevant. The arguments of this work bring together fluid mechanics, systems biology, and feasible cutting-edge technology.

1 Introduction: Novel Applications of Fluid Mechanics

Generally fluid mechanics is understood as the response of fluids to forces exerted upon them. The fluids that first concerned this discipline were restricted to those that were easily observable, mainly liquid water. Interest in the field goes back to Classical Greece, to the well-known work of Archimedes (c. 290–280 BC-212/211 BC).

Since those early times significant changes have taken place in fluid mechanics, but we shall dwell especially on relatively recent events. For we will not concern ourselves with the details of the development by Leonhard Euler and Daniel Bernoulli in the 18th century, or with the work of G. G. Stokes and William Thomson in the 19th century, or even the definite steps forward taken by Ludwig Prandtl at the beginning of last century.

Instead, we wish to highlight briefly scientific disciplines in which fluid mechanics has been fundamental and those that are closely related to the main objectives of the science of astrobiology (the reader will find in Sect. 5 the relevant references). This is a relatively new science that studies the origin, evolution, distribution, and destiny of life in the universe. Astrobiology is flourishing in the present and our opinion is that it will continue to flourish at a faster pace in the future, due to the many space agencies including the European Union, the United States, Russia, Japan, the Popular Republic of China, India, and to these larger efforts other countries are beginning to join forces, including our own country for some time now 1999–2005 (Chela-Flores et al. 2000; Falcón and Loyo 2007). But let us begin firstly by returning to fluid mechanics on Earth. One evident example that is relevant to astrobiology is physical oceanography (as in the new environments provided by the icy satellites off Jupiter): this sub-discipline of oceanography is concerned with the properties of seawater including temperature, density and pressure, movement (waves, currents, and tides), and the interactions between the ocean water and its overlying atmosphere. In Sect. 5 we shall return to this topic in relation with the plumes that may reach the icy surface of Europa, the Galilean satellite of Jupiter. Oceanography is a wider discipline, since it deals with topics beyond fluid mechanics, including chemical oceanography, marine geology, and marine ecology.

Secondly, once again fluid mechanics is particularly relevant for atmospheric science (meteorology, climatology, and aeronomy). These disciplines are concerned with composition, structure, and dynamics of the Earth's atmosphere (Vallis 2006). Fluid mechanics is also needed in aeronomy, since this sub-discipline of atmospheric science studies the physics and processes of the upper atmosphere, information of which may be measurable in the middle term in worlds around other stars.

In Sect. 2 there is a brief description, by way of illustration, of how progress in instrumentation has extended the range of applications of fluid mechanics to include quantum fluid phenomena (superfluidity). But later on we shall underline how for astrobiology more relevant are oceanography, as well as atmospheric science. These two sciences are and, due to a series of possible space probes, will be increasingly more relevant in extra-terrestrial conditions, and in the short term, in an extra-solar context.

2 A 20th Century Application of Fluid Mechanics

Fluid mechanics has ventured into new pathways, of which one originated from improved low temperature instrumentation and the other was due to the exploration of the Solar System and will be introduced in the next section. But we underline here that technology has found applications for fluids firstly, at extremely low temperatures, and secondly additional applications arose in locations out of this world. Both of these unusual venues for fluids have concerned our research in the past. In the first case of "extreme fluids" the major difficulty was to propose the correct theory. In the second case the main issue was of a different kind, once the extraterrestrial fluids (oceans) were identified, the question was not a theoretical one, but the question that was called for was one of identifying, developing, testing, and challenging space agencies for approval of the appropriate instrumentation.

Returning to the first case, the development of advanced cryogenic technology at the beginning of the 20th century (in 1908) allowed to liquefy helium at (4.2 K) into a state that is called helium I. Special attention was paid to liquid helium when it cooled to near absolute zero (0 K [−273.15 °C]) in both of the stable isotopes of helium: ^{3}He and ^{4}He. It was in 1938 when an unusual set of properties was shown to occur in liquid ^{4}He underneath a critical temperature. Hence, liquid helium I assumes different properties and we called this new state of condensed matter helium II, a true "superfluid". (One of the properties that first gave this liquid its name was the capability of displacing itself without viscosity). The major problem that raised by the discovery of superfluid ^{4}He was to find its theoretical bases at a microscopic level. (Subsequently, in 1972, it was shown that the phenomenon also occurs in the second stable isotope ^{3}He at temperatures that were even lower than in the liquid ^{4}He).

Quantum mechanics gives a general understanding of superfluidity, since for ^{4}He, the liquid state consists of atoms with null total spin angular momentum. Consequently, the distribution between their possible states is given by Bose statistics. Neglecting interactions between the ^{4}He atoms, Bose condensation takes place (but the subsequent introduction of interactions does not change significantly the microscopic explanation). With the development of field theory, an alternative approach to fluid mechanics of superfluid ^{4}He was suggested amongst various attempts (Chela-Flores 1975). As the temperature is lowered this field theoretic approach allows the subsequent estimate of the increment of the fraction of condensed atoms

(Chela-Flores 1976). The theory even allows an understanding of diffraction data, both X-rays, as well as neutron diffraction (Chela-Flores 1977). Although not discussed in these three papers, Bose condensation essentially applies also to the case of ^{3}He superfluidity, where Fermi statistics are required for the ^{3}He fermions. The insight that led to this further understanding was based on the earlier theory of superconductivity, where the fermions pair in structures known as Cooper pairs that have integral spin for which the correct statistics is, once again, that of Bose.

3 An Additional Application of Fluid Mechanics

We wish to underline that, once again, it is instrumentation, rather than theory (as in the above case of superfluidity of ^{4}He) that now takes the central position of our enquiries for the new venue of the extraterrestrial fluids. Indeed, with the advent of advanced space technology the exploration of the outer Solar System was possible in the three decades that went from 1973 till 2003. Gradually it became evident that large bodies of liquid water were present in our cosmic neighborhood. Evidence began to emerge during the first steps of exploration of the possible presence of large oceans on the moons of the giant planets: Jupiter and Saturn.

The science of oceanography was untested in these novel environments, a situation that began to change at the very end of last century, as we will briefly refer to in Sect. 1. But within the 20th century the Galilean moons Europa, Ganymede, and Callisto were shown to be very likely the host of oceans of liquid water. These steps forward in the exploration of the outer Solar System have been a gradual process:

- Pioneers 10 and 11 were the first Jovian flybys: Pioneer 10 (1972) flew by Jupiter in December 1973. This was a major achievement for the period, since it was the first such mission. Pioneer 11 (1973) passed by Jupiter in December 1974.
- Voyager 1 went past Jupiter on March 5, 1979. Voyager 2 traveled more slowly and went by Jupiter on July 9, 1979.
- A decade later the Galileo mission built its success on the heritage from the much more modest missions mentioned above. Galileo was placed into Earth orbit in 1989, but from 1995–2003 the Galileo mission successfully explored the Jovian System, providing strong evidence for satellites, where life as it is known to have emerged on Earth, may have also have taken its initial steps providing an opportunity to identify a "second Genesis" using the suggestive phrase of Christopher McKay. (For a detailed discussion of the consequences of life on Europa, the reader should consult "A second Genesis: Stepping-stones towards the intelligibility of nature" (Chela-Flores 2009), especially Chaps. 8–12 and the Glossary, p. 199 for the original use of the suggestive phrase for the origin of life in an extraterrestrial context).

The Galileo mission has added insights, such as the presence on Europa of some form of 'ice tectonics'. The Jet Propulsion Laboratory, which handled Galileo for NASA, has released some images that suggest that part of the surface is understood in terms of shifting plates of ice. From all the information gathered from Voyager and Galileo, reasonable guesses have been put forward regarding possibly a substantial amount of liquid water between the Europan silicate crust and its icy surface.

4 New Paths for Fluid Mechanics in the 21st century

A preliminary proposal for a return mission to Europa and the Jupiter system was entitled LAPLACE. In February 2009 NASA and ESA took a preliminary decision to support a Jupiter mission with the name of the Europa-Jupiter System Mission (EJSM) replacing temporarily and extending our original Laplace proposal (Grasset et al. 2009).

The Jovian System exploration was reformulated by ESA as a European-led single spacecraft mission to the Jovian system, namely, the JUpiter ICy moon Explorer, JUICE (Dougherty et al. 2011). The timeline is launch in 2022, and arrival at the Jupiter system in 2030. The new mission is based on the design of the Jupiter Ganymede Orbiter, which is the ESA flight element of EJSM-Laplace Mission. Indeed, since three of the Galilean satellites are thought to host internal oceans, the JUICE mission will study the moons as potential habitats for life.

In this context an appropriate technology concerns the micro-penetrator. These instruments consist of small projectiles that can be delivered at high velocity to reach just beneath the surface of planets or their satellites for probing samples of surficial chemical elements, amongst other investigations. This type of instrumentation (the penetrators) has a long history of feasible technological development by several space agencies.

Although the limited payload constraints does not include penetrators in the JUICE mission, it is forcing a choice between penetrators and landers. Some advantages of the penetrator approach are nevertheless evident and remain a valid instrument for studying our origins in lunar research (Chela-Flores 2012). The low mass of these instruments, combined with their agility in deployment, makes them worthy complements to orbiter missions launched without landers. We have attempted to describe the feasibility of this technology both on the surface of Europa (Gowen et al. 2011), or on the Moon (Smith et al. 2012). The Europa's stained icy surface has been the focus of recent search for possible biomarkers. The science of biogeochemistry presents us the tantalizing option of inferring from the sulfur surficial patches tests of biogenic chemical elements. Several Earth-bound regions are good analogues of what may be happening in recent geologic times on Europa. These regions are on the Canadian Arctic (Damhnait et al. 2012) and in the Antarctic (Chela-Flores 2011).

Fluid mechanics provides a rationale for the mechanisms that could bring biomarkers from the seafloor to the icy surface. The original intention of some oceanographers was to understand the special changes in the Europa's surface in the Conamara Chaos Region (cf., Sect. 5). With the Galileo Mission we were able to retrieve detailed images of Europa's frozen and stained surface. One of the most intriguing and possibly significant was the Conamara Chaos.

The surface morphology can be understood in terms of oceanic plumes bearing chemical elements, including sulfur from hydrothermal sources in the oceanic bottom. A little beyond the present time technology will allow us to probe the atmospheres of planets beyond the Solar System, where the atmospheric sciences will be set in a new context (cf., Sect. 8) and this, once again, will allow fluid mechanics to explore a novel physical context, namely, the atmospheric structure of hot giants, Neptunes and super-Earths (Segura et al. 2010).

Several other instrumentation issues are also relevant. For instance, laser-induced breakdown spectroscopy (LIBS). This has been a technique for the analysis of elements by retrieving a unique elemental fingerprint spectrum. Since chemical elements are known to emit light of a given frequency when excited to sufficiently high temperatures, LIBS suggests itself for detecting all elements in a given target. There are advantages when planning the exploration of the Solar System.

LIBS shows potential for development instrumentation with characteristics typical of LIBS, but in addition rapid *in situ* analysis is possible with little or no sample preparation and the feasibility of automated spectroscopic analysis (Multari et al. 2010). But as in the case of the penetrators the payload constraint of, for instance the JUICE Mission, does exclude some of these relevant instruments.

5 Buoyant Plumes from the Underlying Seafloor

These are possibilities that can eventually be tested in the laboratories of fluid mechanics. The original intention was to simulate the circulation of Europa by solving the magneto-hydrodynamic equations of motion for a stratified incompressible conduction fluid in a rotating frame of reference. The argument was centered on the fact that the tidal forces can implement oceanic motions in the oceanic annulus generated by the other Galilean moons, by hydrothermal venting from crustal heating and by the intense Jovian magnetosphere. In addition, there will be oceanic stratification influenced by large-scale ocean circulation driven by hydrothermal venting from below and conductive cooling from above.

Such stratification is expected to determine the height of the hydrothermal plume rise, which if the conditions are given could be comparable with the ocean depth and modify the surficial ice. This was a phenomenon that could be observed by the space probes around Europa. This anomalous ice morphology began to be studied in terms of fluid mechanics around the time when Galileo Mission data from the Jovian System was available (Thomson and Delaney 1996).

Five years later, it was demonstrated that these plumes could indeed bear sufficient energy to alter the morphology of the surficial ice, as observed by the Galileo probe around the Conamara Region (Thomson and Delaney 2001). In detail, Thomson and Delaney interpreted this region as melt-through structures formed by oceanic plumes that rise to the base of the ice shell-surface from magmatically heated regions deep in the seafloor. But what is most interesting from our point of view is that these mechanisms of plume delivery from the hydrothermal vents could be the source of biomarkers.

On Earth the origin of life may have occurred around hydrothermal vents, where chemosynthetic bacteria may have flourished. In a similar manner, traces of living organisms could be part of the supply of the stained ice, where sulfur is a main chemical component. Mass spectrometry is the appropriate instrument that could decide whether the internal source of sulfur is of biogenic origin, as the fractionation produced by living organisms can be radically different from that produced by inorganic means (Dudeja et al. 2012).

6 Fluid Mechanics and a System-Level Understanding of Exolife

Systems biology has been a remarkable step forward in the life sciences, especially after we have learnt how to handle large data banks. The first steps in this direction were in the area of molecular biology with the genome and proteome projects. One specific area of impact has been molecular medicine. We have suggested extending systems biology to all areas of the life sciences, especially regarding "exolife" life, namely life elsewhere in the Universe, which is the main topic of the new science of astrobiology (Chela-Flores 2013a).

In fact, systems astrobiology is forced upon us, since our objective is not to reduce problems to first principles, but more modestly our main objective is to attempt defining a set of parameters that may lead to identifying the condition for the presence of complex life on an exo-world (exo-planets and exo-moons). The relevant parameters include amongst many others: an anomalous fraction of oxygen, the star class hosting the Earth-like planet, the age, the metallicity of the star, the position of the exo-planet in the habitability zone of its star, and the possible presence of an exo-moon.

Life in the Universe will emerge from statistical analysis of large data banks that are now rapidly beginning to accumulate. Our combined assumptions of convergence and the cosmos as a complex system imply that all the Earth-like exo-planets that will be in the habitable zone of their corresponding star will have an identifiable bioindicator (anomalous production of biogenic gases).

The signs of life are predicted to be a biologically produced atmosphere, largely fractionated towards one of the biogenic gases (in the case of the Earth the large fractionation triggered by biosystems is the 21 % of oxygen). Such atmospheres would not be the result of natural accretion processes in the processes that give origin to the planets, but instead, the emergence of the biogenic atmospheres would be the result of the innate phenomenon of life that the laws of biochemistry will allow in brief geologic times.

Systems astrobiology is analogous to systems biology, but it has to wait for its full implementation until after we have gathered enough data from the sector of our Galaxy. The practical reason why systems biology is a promising frontier for the future of astrobiology is that it is not easy to have access to information on these planets, except through the now incipient data banks of observable geophysical data, such as methane and oxygen atmospheres, as well as information on the presence of liquid water beyond the present data that has already been searched. In view of the large rate of data retrieval systems astrobiology needs to be formulated at present to prepare for its most convenient management and interpretation.

Since we are discussing how fluid mechanics has found a new area of application in the astrobiology of moons of our solar system, we shall dwell with some care on the question of the relevance of the moons for favouring the origin of life. The potential detection of exo-moons has raised the possibility of bringing the distribution of life in the cosmos closer to reality. The bases of exomoonology are the initial success of the CoRoT mission that was the first space mission designed to search for exo-planets similar to the Earth itself. It was launched with a Soyuz-Fregat rocket in December 2006. CoRoT is the French Space Agency (CNES) mission containing a small space telescope in a terrestrial orbit at a height of 900 km.

7 Distribution of Life in Other Solar Systems: Kepler Worlds

On the other hand, the Kepler Mission, unlike CoRoT, is in a solar orbit. It was launched on March 7, 2009 from Cape Canaveral Air Force Station in Florida. It has a capability to scan some 150,000 stars in the local neighborhood of our Galaxy for extrasolar planets (Kipping 2009a). Its main objective is to search for exo-planets, especially Earth-like planets. At the time of writing, Kepler now has selected out of the 150,000 stars a set of 2326 candidate transiting planets.

The search for exo-planets can be viewed as the first step in an eventual discovery of life as a complex cosmic system. Following the lines outlined above, we expect that a rationalization of life will eventually emerge from the data banks of a very large number of stars in our galactic sector. The geophysical data, rather than data banks of biological information, will provide a gradual emergence of the living phenomenon. The geophysical (atmospheric) bio-indicators point towards ecosystems that have evolved around stars producing measurable biomarkers in our galactic sector. Subsequently, with better missions and with improved instrumentation, this identification of life as a complex system can be extended from a sector of the Galaxy now being probed to other more distant parts of the Universe. It will be at that stage that the methods of computational biology are necessary.

8 The Moon's Influence on the Emergence of Habitability

The presence of an exo-moon would stabilize the magnetic axis of the exo-Earth and hence discard oscillations in the range 0–80° that would constrain the evolution of life from small anaerobic to large complex life capable of photosynthesis (Kipping 2009b; Chela-Flores 2013b). Although no exo-moon has been discovered so far they are in principle detectable with the Kepler data and, indeed, hints of an exo-moon-forming region around exo-planets have been reported (Heller and Barnes 2012). On Earth the stability of the terrestrial magnetic axis is a well-known factor for the evolution of complex multicellular life. The Moon has stabilized the axis of rotation of the Earth, so that its axis of rotation stays in the same direction.

This has had a profound effect on Darwinian evolution, since drastic climatic changes would restrict the survival to only small, robust organisms to survive (Batalha et al. 2012). We have been presented with a remarkable discovery of several oceans in the moons of the outer solar system. The knowledge we are gathering from the moons of our solar system to which fluid mechanics has made a contribution (cf., Sect. 5), will in turn serve to understand the role of exo-moon in the emergence of life in systems of habitable environments around other stars.

The more challenging possibilities that we have to face include the example of a Neptune around an M2 star with a widely separated Earth-like Moon (Kipping et al. 2012). If a moon happens to be leading the planet, as it passes by, it will pull the planet across the face of the star a little faster than average. If it happens to be following, it will hold the planet back. Whether the moon is leading, or trailing, the silhouette of the planet and moon will be wider than that of a planet alone. The planet-moon system will block more of the star's light.

If the moon is directly in between the planet and the visual range of Kepler, on the other hand, or if it is between the planet and the star, more starlight will reach Kepler's sensors – and the moon itself will not be visible. After the planet passes around the star several times the changes in speed caused by a moon can be compared with an average speed, and so that moons that are completely hidden on one pass can have a chance to show themselves on the next.

It should be kept in mind that the feasible detection of exo-moons will add additional parameters for the emergence of habitability on their exo-planets, as it has happened in our own local environment (cf., Sect. 6). The Moon has been a stabilizing factor for the axis of rotation of the Earth. In the case of Mars, for instance, the lack of large satellites has allowed axis obliquity change. Consequently, the ice at the poles could in some moonless exo-planets be displaced to the equator. But the Moon has helped stabilize the Earth, so that its axis of rotation stays in the same direction, leading to less climatic change than if the Earth resembled the moonless planet Venus. The emergence of more complex multi-cellular organisms has been favored compared to a planet where drastic climatic change would allow only small, robust organisms to survive.

With the advent of exomoonology (Kipping et al. 2012), the new batch of data to arrive will be particularly relevant for adding yet another factor in defining habitability and life, as suggested in a systems astrobiological approach. We will face with the Kepler data and the HEK Project a selection of data for discriminating those Kepler worlds that have more favorable options for habitability if they have companion satellites.

9 Insights from a System-Level Understanding

The systems biology approach should also give us insights into one of its branches, namely astrobiology, whose major problem is to understand habitability in alternative abodes for life. With its geophysical/astronomical data, astrobiology can follow up the tracks of genetics and biochemistry for solving fundamental problems that were intrinsic to these disciplines of the life sciences such as protein, or proteome folding. By having the option of focusing on how systems properties emerge in astrobiology, we can raise the question whether habitability can be interpreted as an emergent phenomenon. We suggest basing such an approach on different forthcoming projects:

9.1 The Kepler Mission

This NASA mission is already producing valuable data related to over 2000 candidates for exo-planets (at the time of writing).

9.2 The FINESSE Mission

The NASA Mission FINESSE, Fast INfrared Exoplanet Spectroscopy Survey Explorer (Swain 2010) is to be launched in 2016. It is also a source of data in the near future. It would measure the spectra of stars and their planets.

FINESSE will analyze the planetary atmospheric components using a space telescope to survey more than 200 planets around other stars. This mission attempts to find the fraction of biogenic gases in exo-planet atmospheres and how the Solar System fits into the family of planets in the galactic neighborhood focused by the Kepler mission. FINESSE science objectives overlap the topic of our interest, since firstly, they intend to measure fundamental parameters in the exo-atmospheres to allow knowing the physical and chemical processes of their atmospheres.

Secondly, the science objectives once again overlap with one of the atmospheric science sub-disciplines—climatology—concerned with the weather in the same layers of the atmosphere over given periods of time. The second relevant FINESSE science objective is to trace the composition and temperature change with longitude and time. It is expected that the details of the day side-night side differences will allow the mission to determine insights into the exo-planet climate. A project now in its first steps, the "Hunt for Exomoons with Kepler", (the HEK project mentioned in Sect. 8) aims at distilling the entire list of known transiting planet candidates found by Kepler.

This effort is pursued in order to track down the most promising candidates for hosting at least an all-important moon, whose interaction with the host planet is relevant for the pathway along which life evolves.

9.3 The TESS Mission

With Transiting Exoplanet Survey Satellite Mission, TESS (Foust 2012) the Kepler search for exo-planets will be extended to additional G, K type of stars up to the 12 magnitude, including over two million stars and M type (red-dwarfs) to about one thousand up to 30 parsecs.

9.4 The EChO Mission

With the Exoplanet Characterisation Observatory Mission, EChO (Tinetti et al. 2012), exo-moons down to $0.33R_{\oplus}$ would be detectable for our target stars, providing a complementary set of information from what is being searched from the Kepler data (Kipping et al. 2012). In addition, ECho will be able to analyze the atmospheres of super-Earths in the habitable zones of their host stars. One of their objectives is to measure the spatial (vertical and horizontal) and temporal variability of the thermal/chemical atmospheric structure of hot giants, Neptunes, and super-Earths orbiting bright stars.

10 Discussion and Conclusions

From the point of view of the comparatively recent science of astrobiology (Chela-Flores 2011), we have aimed to illustrate a novel area of application of the time-honored discipline of fluid mechanics. Since ancient times fluid mechanics has been relevant in a context of our civilization. An extraordinary new venue for fluid mechanics emerged early in the 20th century for macroscopic quantum phenomena of the quantum liquids.

These relatively new applications became even broader, due to the technological revolution in instrumentation that we are going through at present. This on-going revolution is to be materialized with the forthcoming extension of aeronomy from its present Solar System constraints to planetary systems around other stars, since as mentioned in Sect. 2 aeronomy is concerned with the physics and processes of the upper atmosphere. Now we are in a position to anticipate that the upper exo-atmospheres will be measurable with the coming step forward in instrumentation with the missions FINESSE, EChO, and TESS (cf., Sect. 9).

An underlying hypothesis in the previous work (Chela-Flores 2013a) has been evolutionary convergence, namely, independent evolution of similar genetic or morphological features. Assuming both biochemistry (Pace 2001) and biology (Dawkins 1983) to be universal sciences, evolutionary convergence has been assumed to be possible, even in other lines of biological evolution elsewhere in the universe (Conway-Morris 1998, 2003; Chela-Flores 2007). For a more careful detailed discussion of evolutionary convergence we refer the interested reader to Chap. 12 in "The Science of Astrobiology" (Chela-Flores 2011).

The eventual verification of the validity of the fluid mechanics theory that has been applied to the internal ocean of Europa (cf., Sect. 5) does not have to wait for long-term technological developments. We originally proposed with our JPL co-workers instruments of the kind of cryobots and coupled hydrobots that may penetrate the icy cover to probe directly the oceanic phenomena that were to be modelled by fluid mechanics (Horvath et al. 1997). However, it is clear now that the surficial probing that can be performed with the help of the micro-penetrators would suffice for extracting most of the relevant information from the upper layers of the icy Europan surface (Gowen et al. 2011).

Finally, a point that we would like to highlight is that the new venues for the science of astrobiology have been suggested by older approaches that come from the life and physical sciences. Indeed, systems chemistry is a physical science clearly outlined (Anderson 1972), in which an interdisciplinary approach focuses on complex interactions in chemical systems, using a new point of view, holism rather than reductionism, where collective phenomena are the main ingredient in basic research of chemical systems. It attempts to produce a more holistic understanding of biochemistry, especially the question of folding in proteins.

On the other hand, systems biology is a life science in which an interdisciplinary approach focuses on complex interactions in biological systems, using a new point of view. It attempts to produce a more holistic understanding of biology, especially genetics. The new approach aims to construct a network of interacting processes that can be related to the information sciences (Buchanan et al. 2010). A major aim is to discover emergent properties of a system that would be understood by focusing on its complex interactions and relying on the information sciences. These computational techniques have given rise to systems astrobiology, where the new space science is also considered as a branch of biology (Chela-Flores 2013a).

Instead of applying the new methodology of systems biology to genetics, it is applied to other biologically relevant questions, namely the origin, evolution, distribution, and destiny of life in the Universe. The distribution of systems of habitable worlds with their biomarkers will be testable in the short term with forthcoming space missions mentioned above. This would justify subsequent use of quantitative systems biology methods that are already available in other branches of biology.

References

Anderson PW (1972) More is different-broken symmetry and nature of hierarchical structure of science. Science 177:393–396

Batalha NM, et al. (2012) Planetary candidates observed by Kepler III: analysis of the first 16 months of data, arXiv:1202.5852v1 [astro-ph.EP].

Buchanan M, Caldarelli G, De los Rios P, Rao F, Vendruscolo M (eds) (2010) Networks in cell biology. Cambridge University Press, Cambridge

Chela-Flores J (1975) Gauge theory of superfluidity. J Low Temp Phys 21:307–319

Chela-Flores J (1976) Condensate fraction of liquid helium four. J Low Temp Phys 23:775–783
Chela-Flores J (1977) Atomic order in liquid helium II. J Low Temp Phys 28:213–228
Chela-Flores J (2007) Testing the universality of biology. Int J Astrobiol 6:241–248
Chela-Flores J (2009) A second genesis: stepping-stones towards the intelligibility of nature. World Scientific Publishers, Singapore http://www.ictp.it/ chelaf/ss220.html
Chela-Flores J (2011) The science of astrobiology a personal point of view on learning to read the book of life 2nd edn, Book series: cellular origin, life in extreme habitats and astrobiology. Springer, Dordrecht. Chapters 3 and 8, http://www.ictp.it/ chelaf/ss220.html
Chela-Flores J (2012) A case for landing on the moon's farside to test nitrogen abundances. Int J Astrobiol 11:61–69
Chela-Flores J (2013a) From systems chemistry to systems astrobiology: life in the universe as an emergent phenomenon. Int J Astrobiol 12:8–16
Chela-Flores J (2013b) Habitability on Kepler worlds: are moons relevant? In: de Vera J-PP, Seckbachi J (eds) Habitability on other planets and satellites - the quest for extraterrestrial life. Book series: Cellular origin, life in extreme habitats and astrobiology. Springer, Dordrecht (in press).
Chela-Flores J, Lemarchand GA, Oro J (2000) Astrobiology: origins from the big bang to civilisation. Kluwer Academic Publishers, Dordrecht
Conway-Morris S (1998) The crucible of creation. Oxford University Press, Oxford
Conway-Morris S (2003) Life's solution inevitable humans in a lonely universe. Cambridge University Press, Cambridge
Damhnait F, Gleeson RT, Pappalardo MS, Anderson SE, Grasby RE, et al. (2012) Biosignature detection at an Arctic analog to Europa. Astrobiology 12:135–150
Dawkins R (1983) Universal Darwinism. In: Bendall DS (ed) Evolution from molecules to men. Cambridge University Press, London, pp 403–425
Dougherty MK, Grasset O, Bunce E, Coustenis A, Titov DV, et al. (2011) JUICE (JUpiter ICy moon Explorer): a European-led mission to the jupiter system. EPSC Abstracts, EPSC-DPS Joint Meeting, vol 6. EPSC-DPS 2011–1343-1.
Dudeja S, Bhattacherjee AB, Chela-Flores J (2012) Antarctica as model for the possible emergence of life on Europa. In: Hanslmeier A, Kempe S, Seckbach J (eds) Earth and other planets in view of biogenesis. Book series: cellular origin and life in extreme habitats and astrobiology, Springer, Dordrecht pp 407–419.
Falcón N, Loyo Y (eds) (2007) Orígenes del Universo. Consejo de Desarrollo Científico y Tecnológico de la Universidad de Carabobo, la Vida y la Inteligencia.
Foust J (2012) Exoplanet explorers. The space review. http://www.thespacereview.com/article/2003/1
Gowen RA, Smith A, Fortes AD, Barber S, Brown P, et al. (2011) Penetrators for in situ sub-surface investigations of Europa. Adv Space Res 48:725–742
Grasset O, Lebreton JP, Blanc M, Dougherty M, Erd C, et al. (2009) The jupiter ganymede orbiter as part of the ESA/NASA Europa jupiter system mission (EJSM). EPSC Abstracts 4: EPSC2009-784, European planetary Science Congress.
Heller R, Barnes R (2012) Exomoon habitability constrained by illumination and tidal heating. Astrobiology J. 13:18–46
Horvath J, Carsey F, Cutts J, Jones J, Johnson E, et al. (1997) Searching for ice and ocean biogenic activity on Europa and Earth. In: Hoover RB (ed) Instruments, methods and missions for investigation of extraterrestrial microorganisms, The International Society for Optical Engineering, Bellingham, Washington USA, Proceedings SPIE 3111: 490–500 http://www.ictp.trieste.it/chelaf/searching_for_ice.html
Kipping DM (2009a) Transit timing effects due to an exomoon. MNRAS 392:181–189
Kipping DM (2009b) Transit timing effects due to an exomoon II. MNRAS 396:1797–1804
Kipping DM, Bakos GÁ, Buchhave LA, Nesvorný D, Schmitt A (2012), The hunt for exomoons with Kepler (HEK). I. Description of a new observational project. Astrophys J 750:115 (19 pp).

Multari RA, Cremers DA, Dupre JM, Gustafson JE (2010) The use of laser-induced breakdown spectroscopy for distinguishing between bacterial pathogen species and strains. Appl Spectrosc 64:750–759

Pace NR (2001) The universal nature of biochemistry. Proc Natl Acad Sci U S A 98:805–808

Segura A, Walkowicz L, Meadows VS, Kasting JF, Hawley S (2010) The effect of a strong stellar flare on the atmospheric chemistry of an Earth-like planet orbiting an M dwarf. Astrobiology 10:751–771

Smith A, Crawford IA, Gowen RA, Ambrosi R, Anand N, et al. (2012) LunarNet A proposal in response to an ESA M3 call in 2010 for a medium size mission. Exp Astron 33:587–644

Swain MR (2010) Finesse-a new mission concept for exoplanet spectroscopy. Bull Am Astron Soc 42:1064

Thomson RE, Delaney JR (1996) Circulation of an Europan ocean: what we need to know. Paper Presented at the Europa Ocean Conference, 12–14. At San Juan Capistrano Research Institute, San Juan Capistrano, California, USA, p 75 November 1996.

Thomson RE, Delaney JR (2001) Evidence for a weakly stratified Europan ocean sustained by seafloor heat flux. J Geophys Res 106: E6 12355–12365.

Tinetti G, et al. (2012) EChO-exoplanet characterisation observatory. Exp Astron 34:311–353

Vallis GK (2006) Atmospheric and oceanic fluid dynamics: fundamentals and large-scale circulation. Cambridge University Press, Cambridge

The Impact of Computational Fluid Mechanics on Cancer Research

Dimas C. Belisario and Leonardo Di G. Sigalotti

Abstract This chapter presents an overview of recent contributions that show how fluid mechanics is drastically changing cancer research. The review will mainly focus on the computational modelling of fluid-mediated processes related to cancer dynamics, spanning different representation scales from cells to organs. Fluid mechanics seems to act as a fundamental organizing principle in many aspects of cancer, including its growth, progression, metastasis, and therapy. On the other hand, it is clear that fluid-dynamics modelling can make a huge contribution to many areas of experimental cancer investigation since there is now a wealth of data that requires systematic analysis. The relevance of microfluidics in the isolation, detection, molecular characterization, and migration of tumour cells is also discussed. In the last part of the chapter, future challenges and perspectives are briefly outlined.

1 Introduction

Cancer cannot be defined as just one disease, but rather as a broad group of more than 200 diseases. From the biological point of view, it is a complex phenomenon that can be characterized by a small set of hallmarks that point to a cascade of events from the molecular to the organismal level (Hanahan and Weinberg 2011). At the molecular level, cancer arises through a series of genetic mutations, which allow cells to grow and divide uncontrollably. An alteration of the DNA molecule can

---

D. C. Belisario (✉)
Dipartimento di Oncologia, Università degli Studi di Torino, Via Santena 5/BIS, 10126 Torino, Italy
e-mail: caro.belisario@gmail.com

L. Di G. Sigalotti
Centro de Física, Instituto Venezolano de Investigaciones Científicas, IVIC, Apartado 20632, Caracas 1020-A, Venezuela
e-mail: leonardo.sigalotti@gmail.com

L. Di G. Sigalotti et al. (eds.), *Computational and Experimental Fluid Mechanics with Applications to Physics, Engineering and the Environment*, Environmental Science and Engineering,
DOI:10.1007/978-3-319-00191-3_6,

disrupt the genes and produce faulty proteins, causing the cells to become abnormal (or malignant) and lose their restraints on growth.

In healthy individuals, the immune system can recognize abnormal cells and destroy them before they get a chance to divide. However, some mutant cells may escape detection and survive to form a tumour (or neoplasm), which looks like a small ball of cells and feeds on oxygen and nutrients that diffuse to its surface. As cells in the core of the tumour become starved of oxygen (hypoxic cells), they release substances (growth factors) that stimulate the growth of new blood vessels; a process called angiogenesis (Dvorak et al. 1988). These angiogenic growth factors activate receptors present on endothelial cells in pre-existing blood vessels. The activated endothelial cells begin to release enzymes (proteases) that allow them to escape from the parent vessel walls. These then proliferate into the surrounding matrix and form solid sprouts connecting neighbouring vessels, which extend towards the tumour, supplying it with blood (Leung et al. 1989; Hanahan and Folkman 1996). For a while, the tumour grows as a cohesive ball of cells with smooth edges. However, eventually some rogue cells break away from the growing tumour and invade the adjacent tissue. This is a key process in the growth of most cancers and an escape route for metastasis—the formation of secondary tumours owing to spreading of cancer cells to more distant parts of the body through the lymphatic system or bloodstream. Metastasis is the main cause of deaths due to cancer (Sporn 1996). For example, as a cause of mortality in the United States, metastatic cancer is second only to heart disease, with one out of four deaths being from cancer.

Cancer invasion occurs through several important steps, involving the interplay between the cells themselves and their microenvironment (Liotta and Kohn 2001): reduction in or loss of cell-cell adhesion, cell adhesion to the extracellular matrix (the surrounding connective tissue), secretion of enzymes that digest the extracellular matrix, and movement (migration) of the cancer cells coupled with their proliferation. Cancer cells experience both self-adhesion (cell-cell adhesion) and cell-matrix adhesion, while cell movement through the surrounding tissue may occur through diffusion with no preferred direction and by directed motion due to the breakdown of the extracellular matrix components (Hanahan and Weinberg 2000; Friedl and Wolf 2003; Weinberg 2007).

Notwithstanding decades of research in cancer biology and medicine, our present ability to predict and treat metastatic cancer is still very limited. The main difficulty to reliably forecast the risk of cancer metastasis for individual patients stems from the fact that cancer itself is the result of a complex interplay between a large number of factors. While biological data continue to pile up at an enhanced rate, a major obstacle to progress lies precisely on how to handle this overwhelming flow of data. As a result of this difficulty, cancer research has commenced to undergo radical changes towards a more quantitative approach, where mathematical models are slowly making their way out as predictive tools using the parameters and information from state-of-the-art experiments. Integrating mathematics, physics, and mechanics with genomic investigations of cancer and its therapy opens a window towards a novel multidisciplinary approach, which encompasses biomathematics and computation, cancer biology, bioengineering, and imaging (Suresh 2007; Michor et al. 2011).

Through the use of mathematical modelling and simulation software, this new approach to cancer research has the potential to predict prognosis, optimize surgical and pharmacological treatments for various cancers, and ultimately guide the design of novel therapeutics (Quaranta et al. 2005). For an extensive review on the novel mathematical tools applied to the modelling of cancer onset, evolution, and growth the reader is referred to Bellomo et al. (2008).

A branch of physics and engineering which is transforming the fight against cancer is fluid mechanics. As advocated by Koumoutsakos et al. (2013) in a recent review on the subject: *after a century of rapid advances in theory, numerical methods, hardware, and software, the fluid mechanics community has developed a powerful arsenal of multiscale imaging, analysis, and simulation tools that are highly suitable for the investigation of transport processes in cancer.* Fluid mechanics has been recognized to play an important role in most aspects of cancer, including tumour inception, growth, metastasis, and therapy. In this chapter, we review the most important contributions that project fluid mechanics as an essential organizing principle for cancer, spanning spatial scales from the gene to the organ and timescales of microseconds, as in gene mutations, to decades, as is pertinent to metastasis. We shall primarily focus on progress achieved in numerical simulation models of aspects of cancer that interface with fluid mechanics and discuss how significant future progress in the area is promising to change dramatically both the way experimental oncology is going on and our understanding of the processes involved from cancer initiation to metastasis and from the molecular to the patient level.

2 The Microscopic Level

The starting point of cancer is the generation of a neoplastic cell through phenotypic alterations, resulting from genetic mutations. However, this concept has not yet been well addressed through mathematical modelling, which so far has mainly focused on angiogenesis and invasion. After the onset of neoplasia, the characterization of the system suggests the identification of three natural scales, which are also connected to different stages of the disease, i.e., processes on the cellular scale (microscopic level) are triggered by signals stemming from the sub-cellular level and these have an impact on the macroscopic scale (organism), when tumours grow and spread.

On the microscopic (cellular) level, fluid-dynamic models have been proposed to simulate the effects of cell-cell interactions. These interactions are fundamental at all stages of tumour formation, whether they are among abnormal cells and host cells, or among abnormal cells themselves. If tumour cells skip recognition and suppression by the action of the immune system, the tumour may evade apoptosis or co-opt host cells, allowing progressive growth. During invasion and metastasis, alterations in cell-cell adhesion between individual tumour cells are key to driving the process. Existing experimental data suggests that tumour cell-adhesion to the endothelium under hydrodynamic shear rate—the change in flow velocity within the micro-capillaries—is a critical step that results in circulation-mediated metastasis (Liang et al. 2008, 2010; Fu et al. 2012).

The mechanism of cell-cell adhesion—a non-local interaction between two cells through transmembrane receptor binding—has naturally suggested the use of discrete cell approaches, which retain the finite cell size and permit incorporation of molecular interactions and/or forces that act between cells. A drawback of these approaches is the significant computational time required to simulate large populations. Therefore, it is desirable to augment such methodologies with continuous models that capture the dynamics of population-level behaviour. The past decade has witnessed the development of a wide variety of discrete models of increasing sophistication that incorporate cell adhesion, which can be classified into two major classes: lattice-based and lattice-free approaches. Examples of the former class include many cellular automata models (Deutsch and Dormann 2005; Moreira and Deutsch 2005) and discrete-continuum techniques (Anderson 2005; Anderson et al. 2006), where the discrete cells are allowed to interact with each other and surrounding continuous fields representing the extracellular matrix densities and growth factor concentrations. In particular, this latter approach has primarily been applied to models of tumour cell invasion, where some models have incorporated the effects of cell-adhesion, cell-migration, and phenotypic mutations (Anderson et al. 2006). These have suggested that invasive fingering is essentially driven by environmental heterogeneity. A spatially extended approach of the lattice-based class is the Cellular Potts Model, which has been adapted and applied to cell populations (Graner and Glazier 1992; Glazier and Graner 1993) and to simulation models of solid tumour growth (Turner and Sherratt 2002) and angiogenesis (Bauer et al. 2007).

In contrast to the above grid-imposed models, lattice-free models allow individual cells to move freely through continuous space. In a number of models of this type, cells are given variable, yet predefined, shapes such as deformable ellipsoids of fixed volume (Dallon and Othmer 2004; Palsson 2008). In more refined models, cells are allowed to shift between spheroidal and polyhedral shapes (Schaller and Meyer-Hermann 2005), or adopt continuously deforming shapes according to their interactions with neighbours and the environment (Newman 2005). Models have also been proposed in which individual cells are described as fluid-elastic structures in which their membrane is represented by a deformable boundary immersed in a fluid (Rejniak 2007; Dillon et al. 2008). In these models, adhesive forces are again represented by force balances that describe the movement and deformation of cells, while channels at their membranes permit the influx of fluid into them required for growth.

Hybrid models aimed at studying the adhesive rolling of leukocytes over a P-selectin coated surface in parabolic shear flow in microchannels, where the immersed boundary method is used for cell deformation coupled with a Monte Carlo simulation for receptor/ligand interaction, have reproduced the characteristic "stop-and-go" motion of rolling leukocytes and the "tear-drop" shape of adherent leukocytes as observed in experiments (Pappu et al. 2008). A software environment capable of simulating blood flows on cellular scale inside microfluidic devices have been recently proposed, where the blood is modelled as a suspension of liquid blood plasma, immersed blood cells, and magnetic beads (Gusenbauer et al. 2011). The blood flow is represented on a fixed grid by solving the lattice-Boltzmann equations, while the boundary of each suspended object is represented by a set of discrete Lagrangian immersed boundary points that do not need to lie on the fluid grid. A direct application of this model in biomedicine is the use of self-organized magnetic bead chains to isolate circulating tumour cells employing lab-on-a-chip technologies (see Sect. 5 below).

3 Continuous Macroscopic Level

The body of literature devoted to models which link the cellular scale to the macroscopic tissue scale has increased at a high rate during the last few years. We foresee that this trend will continue as cancer research in the immediate near future will focus on refining and improving the existing models, allowing us not only to understand but also diagnose and treat cancer beyond our present technical abilities. While discrete models permit the straightforward incorporation of many intra-, extra-, and inter-cellular processes, they can require a formidable number of cells to describe the transition from the cellular to the tissue level, making the problem computationally intractable. On the other hand, discrete models often resist a thourough analytical investigation that can shed light on generic properties of the system under study. Both of these difficulties can be relaxed by considering continuum-scale models based on fluid-dynamic simulations with genetic and molecular elements, where cells are represented through their density at the tissue level and where relevant aspects of cancer such as tumour inception, growth, metastasis, and therapy that have direct relevance to flow-mediated processes can be thouroughly analyzed. In most of these models, events at the cellular scale are accounted for by the particular choice of terms and parameter functions that enter the governing evolution equations.

3.1 Tumour Onset, Growth, and Invasion

A cell becomes cancerous when a set of mutations is accumulated in its genome. These mutations are linked to oncogenes—genes that have the potential to cause cancer—and to tumour suppressor genes, which in contrast prevent a cell from becoming cancerous. The combination of thousands of mutant genes across different cell lines enables uncontrolled tumour growth. As the cancerous cells accumulate genetic mutations, the rate of mutations increases as the molecular mechanisms of genome maintenance are lost (Negrini et al. 2010). One outcome of this series of mutations is an increase in the proliferation rate and a decrease in the death rate of the cells, giving rise to a tumoral mass consisting of distinct cell types intertwined with the extracellular matrix (Egeblad et al. 2010). However, even a fast growing clump of tumour cells cannot grow beyond a certain size, since there is a balance between cells inside the tumour consuming nutrients and nutrient diffusion into the tumour.

Once cells have formed a tumour mass, its sustained metabolic activity requires oxygen and nutrients, which in the avascular stage (i.e., tumours without blood vessels), are provided by diffusion through the surrounding perfused tissue. At this stage, the tumour has a volume that usually never exceeds 1 mm^3 and consists of an inner zone of necrotic cells surrounded by an intermediate zone of quiescent (or dormant) cells, owing to the lack of oxygen and nutrients, and an outer zone of proliferative cells (Koumoutsakos et al. 2013). Tumour substances (angiogenic growth factors), generated by the hypoxic zone near the necrotic one, induce blood vessel growth. During this step, one sees at a macroscopic scale capillary sprouts from existing vasculature moving towards the tumour to feed it and allow its further growth. In particular, sprouting and intussusceptive angiogenesis entail flow-related processes. In the former case, new blood vessels sprout from the existing vasculature and grow to form a new vascular network, characterized by intermittent and low-shear-stress conditions inside the vessel (Song and Munn 2011). The initiation of blood flow leads to active vessel remodelling, maturation, and differentiation into venules and arterioles. In contrast, intussusceptive angiogenesis is the process of transcapillary pillar formation inside existing vessels that result in the formation of new vessels (Styp-Rekowska et al. 2011). It involves three different steps: microvascular growth, arborization, and branching remodelling (Djonov et al. 2003). Its initiation possibly involves the imbalance of forces experienced by endothelial cells due to blood flow, cell-cell adhesion, and the extracellular matrix (Davies 2005).

Tumour vasculature shows increased vascular density and branching patterns, distorted and enlarged vessels, and highly convoluted segments (Goel et al. 2011; Narang and Varia 2011). The presence of large inter-cellular spaces renders the vessels leaky, allowing for enhanced macromolecule transport between the lumen and the extracellular space, offers ways for tumour cells to enter the vasculature, and leads to an increase of the interstitial vasculature (Narang and Varia 2011). The vascular shear rate has been found to influence vascular lumen formation as well as proliferation and migration of endothelial cells (Yamane et al. 2010), while pulsatile flow has been shown to stimulate angiogenesis in an *in vitro* environment (Cullen et al. 2002). In tumour-associated vasculature, the highly tortuous vessels increase the resistance to blood flow. The leakage of blood plasma leads to an increase in the interstitial pressure, causing vessel occlusion and acute hypoxia, which in turn leads to the persisting release of vascular endothelial growth factor. In response, angiogenesis continues, the network structures changes, and maturation is prevented, promoting vascular leakage. Once the tumour has acquired its own blood supply (vascular stage), peripheral tumour cells can escape via the circulatory system (migration) and set up secondary tumours elsewhere in the body (metastasis). After angiogenesis and metastasis, the patient is left with multiple tumours in different parts of his/her body that are very difficult to detect and even more difficult to treat. From a clinical point of view angiogenesis and vascular tumour growth together with metastasis is what cause the patient to die, and modelling and understanding these different stages is crucial for cancer therapy. However, a recent clinical study has reported a high degree of regression of a nonmelanoma skin tumour, particularly a basal cell carcinoma with a high microvessel density, after photodynamic therapy (Cabrera et al. 2012). While the tumour destruction was induced by the diffusion of cytotoxic agents from the irradiated zone to the neighbourhood of the tumour zone, this may represent a case where the process of angiogenesis may play a beneficial role in the regression of contiguous untreated tumours.

3.2 Fluid-Dynamic Models

In general, computational models at the macroscopic scale are formulated in terms of mass balance equations for the cellular components, coupled to a system of reaction-diffusion equations for the concentration of extracellular chemicals, which can be written in the form (Bellomo et al. 2008):

$$\frac{\partial(\rho_i\phi_i)}{\partial t} + \nabla\cdot(\rho_i\phi_i\mathbf{v}_i) = \Gamma_i, \qquad i = 1, 2, \ldots, n, \tag{1}$$

$$\frac{\partial c_k}{\partial t} + \nabla\cdot(c_k\mathbf{v}_l) = \nabla\cdot(Q_k\nabla c_k) + \Lambda_k, \qquad k = 1, 2, \ldots, m, \tag{2}$$

where ρ_i and ϕ_i denote, respectively, the density and concentration of the ith cellular component (i.e., cells, extra-cellular matrix, or extra-cellular fluid), $\mathbf{v}_i$ is the mass velocity vector of the ith population, c_k are the concentrations of the chemicals and nutrients, and $\mathbf{v}_l$ is the velocity of the liquid (blood). The term Γ_i in Eq. (1) is a source/sink term for each component, including production (cell birth) or destruction (cell death) terms. Tumours constantly produce waste products, mainly water, and a multitude of chemical factors. In particular, when a cell dies, its membrane ruptures releasing its content, which is mostly re-usable organic material. In Eq. (2), Q_k is the diffusion coefficient of the kth chemical factor and Λ_k is a source term for the particular nutrient or chemical. In the language of fluid mechanics ϕ_i is just the volume fraction of the ith constituent so that the tumour is modelled as a multiphase material. The sum of the volume fractions over all constituents must therefore equal one.

In order to close the above system of equations, an equation for the velocity components ($\mathbf{v}_i$) must be specified. Depending on the choice of this equation, macroscopic models can be defined as phenomenological or mechanical models. Phenomenological models are based on a diffusion equation for cell movement, i.e.,

$$\mathbf{v}_i = -D_i\nabla\phi_i, \tag{3}$$

where D_i is the diffusion coefficient. If this quantity is a positive constant, cell movement will be described by linear diffusion. However, in several models, the motion of cells is described by non-linear diffusion, where $D_i = D_i(\phi, c)$ (Thompson and Byrne 1999; Sherratt and Chaplain 2001). While these models are suitable for describing some interplay between cells such as contact inhibition, they cannot really account for the influence of an elastic membrane. However, they have been successfully used for evaluating the efficacy of therapy or resection in the case of brain tumours (gliomas) (Swanson and Alvord 2002), or the influence of acidity (Gatenby and Gillies 2004). Alternatively, phenomenological models can specify biased movement such as chemotaxis (Chaplain 1996)—the characteristic movement or orientation of a microorganism or cell in response to a chemical concentration gradient either towards or away the chemical stimulus—or haptotaxis (Anderson 2005)—the directional motility or outgrowth of cells towards or along a gradient of chemoattractants or adhesion sites in the extracellular matrix. An extension of the model combining diffusion and haptotactic movement predicted that heterogeneity of the extracellular matrix affects cancer invasion (Perumpanani and Byrne 1999).

As tumour cells proliferate, they push into the surrounding tissue and cause pressure to build. This pressure, along with other mechanical interactions, have very important implications on tumour growth and progression. Incorporation of the physical forces that influence cell motion requires complementing Eqs. (1) and (2) with the momentum balance equations (Bellomo et al. 2008):

$$\rho_i \phi_i \left(\frac{\partial \mathbf{v}_i}{\partial t} + \mathbf{v}_i \cdot \nabla \mathbf{v}_i \right) = \nabla \cdot \mathbb{T}_i + \phi_i \mathbf{f}_i + \mathbf{F}_i, \qquad i = 1, 2, ..., n, \tag{4}$$

for each constituent. Here $\mathbb{T}_i$ is the stress-tensor, $\mathbf{f}_i$ is the body force acting on the ith constituent, and $\mathbf{F}_i$ is the interaction force with the other constituents. In order to close this system of equations we need to specify constitutive equations that relate the forces to the level of stress and compression. For instance, as a cell undergoes mitosis and divides into two cells, these will generate a pressure on neighbouring cells, causing an increase in tumour size. If cells are assumed to behave as a fluid, the simplest constitutive equation for the stress can be written as

$$\mathbb{T}_i = -\sigma_i \mathbb{I}, \tag{5}$$

where σ_i is the response of the cells to compression and $\mathbb{I}$ is the unit tensor. Here the implicit assumption is made that cells behave as elastic liquids.

In many instances the filtration of organic liquids through tumours has been simulated by modelling the tumour as a growing and deformable porous medium. If cells move as an elastic fluid within a rigid extracellular matrix, Eqs. (1) and (2) can be closed using Darcy's law

$$\mathbf{v}_i = -K \nabla \sigma_i, \tag{6}$$

where K is the permeability of the matrix. Modifications of this equation for a deformable porous medium and for mass exchange between the constituents are given in De Angelis and Preziosi (2000) and Chaplain et al. (2003). Darcy models have been considered by several authors in simulations of tumour growth (Cristini et al. 2003), of fluid flow in solid tumours (Soltani and Chen 2011), and more recently for describing cancer-therapeutic transport in the lung (Erbertseder et al. 2012). The former models have predicted interstitial velocities in very good agreement with experimental results, while the latter has described the flow, transport, and reaction processes of a therapeutic agent in the pulmonary circulation in healthy and cancerous pulmonary tissue. In this case, the phase moving within the tissue continuum consists of two components, namely the interstitial fluid and the therapeutic agent. While it is assumed that the fluid phase is incompressible, the movement of the dissolved drug molecules in the interstitial tissue of the lung is modelled using a single-phase, two-component approach in a rigid, porous medium. With the additional assumption that the flow within the tissue is creeping, the flow velocity of the interstitial fluid can be very well described by Darcy's law (Baxter and Jain 1989; Baish et al. 1997; Erbertseder et al. 2012).

Alternatively, the cell-matrix medium can be viewed as a viscous fluid (Stokes flow), where the stress depends on the viscosity (Friedman and Hu 2007), or as a viscoelastic fluid (Holmes and Sleeman 2000). Other models treat the tumour tissue as a mixture of cells living in a porous medium made of extracellular matrix and filled with extracellular liquid (Graziano and Preziosi 2007). Darcy's law can be used to model both fluid flow and cell motion, where the latter is treated as a granular material flowing in the porous extracellular matrix scaffold. For example, the case of a multicell spheroid can be modelled as a growing poro-elastic medium using Eqs. (1) and (2) coupled to a variant of Eq. (4) for the interstitial pressure p, where the inertial terms are neglected and the stress-tensor of the mixture ($i = tc$, tumour cells; $i = l$, extracellular fluid with chemicals and nutrients) is given by $\mathbb{T}_m = -[p + \sigma_{tc}]\mathbb{I}$, and a composite velocity equation (Bellomo et al. 2008). This scheme has been used together with experimental data to show the cell-size reduction by solid stress inside tumour spheroids (Ambrosi and Mollica 2002; Roose et al. 2003). Combining Darcy's law with Stokes flow gives a further constitutive relation, known as the Brinkman equation. Models based on Darcy-Stokes flow have been used to study tumour morphology and stability (Zheng et al. 2005; Pham et al. 2011). However, models based solely on Stokes flow has been found to be more consistent with experimental data from *in vitro* three-dimensional multicellular tumour spheroids (Pham et al. 2011).

A number of illustrative mechanical models describing the growth of avascular tumours are reviewed in Roose et al. (2007), and details of some of the existing models can be found in the references therein. Multiscale mechanical models designed for simulating the growth of both avascular and vascular tumours, including environmental conditions, distribution of oxygen, elastic membrane response, membrane degradation, and the dynamics of the motion of the tumour have also started to appear in the literature (Mantzaris et al. 2004; Plank et al. 2004; Macklin et al. 2009; Bresch et al. 2010). While some of these models can be applied to investigate the therapeutic benefits of anti-invasive agents, they provide the basis of a numerical platform for more refined tumour growth simulations. On the other hand, the process of invasion of adjacent tissue by cancer cells has been recently modelled by assuming that cells migrate through a combination of diffusion and haptotaxis as well as undergoing proliferation and by incorporating the effects of cell-cell and cell-matrix adhesion (Chaplain et al. 2011). Multiscale models describing the growth of *in vitro* multicellular tumour spheroids and *in vivo* avascular tumour nodules that incorporate heterogeneous population of cells, drug diffusion, drug pharmacokinetics, cell-cycle-phase transitions, and the diffusion of multiple nutrients are being used to formulate effective therapeutic strategies by understanding the interactions between drugs and the heterogeous microenvironments in growing tumours (Venkatasubramanian et al. 2008).

Although macroscopic models based on a fluid-dynamic approach are deeply influencing modern cancer research in that they exhibit the general behaviour of tumour growth, angiogenesis, and invasion, they fail to examine details of the phenomena occurring at the single cell level. In particular, this makes detailed modelling of processes such as angiogenesis difficult because calculating average cell density fails to include the spatial structure of the vascular network. Moreover, it is not completely clear if invasion and metastasis are driven by average population behaviour, or instead by cells which deviate from the mean. For example, it is quite possible that individual rogue cells drive the macroscopic processes of invasion or metastasis. However, their individual behaviour is certainly not captured by a continuum approach. Although the development of multiscale approaches is very recent, future practical models must be based on some modular approach where at a certain scale the processes have to be consistent with the lower and higher scales. In this framework, the overall system can be regarded as a network of several interacting subsystems, each developed at a specific scale, while interactions between contiguous systems need to deal with compatibility (and possibly boundary) conditions at each specific scale. A brief outline of these issues is given in Sect. 6. The interested reader is referred to Bellomo et al. (2008), where perspectives of such a complexity multiscale theory is amply discussed.

4 Models of Vascular Transport and Angiogenesis

The blood flow in microvessels, whose diameters are $\sim$100 μm or less, is called the microcirculation (Sugihara-Seki and Fu 2005). Microvessels have irregular interconnections that form a network in tissues and are responsible for the exchange of materials between blood and surrounding tissues. Research on the flow through the neoplastic vacuslature of solid tumours has been largely motivated by the desire to understand the role of fluid convection in the treatment of cancer by therapeutic monoclonal antibodies. A key problem in this kind of treatment is the low transport rates into the main body of the tumour across the vasculature, which leads to low and ineffective concentrations of the therapeutic macromolecules. It is a common observation that interstitial fluid pressure is higher in both human and experimental solid tumours than in normal tissue (Heldin et al. 2004). Enhanced interstitial pressure is the result of a richly developed and highly permeable vascular network, combined with facilitated transendothelial fluid transfer (Boucher and Jain 1992; Lee et al. 1994). Clinically, a high interstitial pressure is marked by a reduced delivery and uptake of anticancer drugs (or macromolecules) and, hence, lack of therapeutic effects. Therefore, the analysis of blood flow and transport processes in the growing networks requires accurate modelling of blood flow in microvessels, solute transport, and angiogenesis.

Physically, blood is a suspension of red blood cells, white blood cells (leukocytes), and platelets in plasma. It is an incompressible Newtonian fluid with viscosity of about 1.2 cP at 37° C. Red blood cells are the most abundant, with a volume fraction of 40–45%, and therefore they strongly influence the rheology of blood. Because of their flexible viscoelastic membranes, they can easily pass through capillaries with diameters less than their major diameters at rest ($\sim 8\,\mu$m). In fact, the minimum diameter of a cylindrical tube that will allow a normal red cell to pass through intact is as narrow as $\sim 2.8\,\mu$m (Halpern and Secomb 1989). Leukocytes are generally spherical with a mean diameter of $\sim$6–8 μm and are much less deformable than red cells. Despite their relatively small numbers, leukocites can contribute significantly to blood flow resistance (Schmid-Schönbein et al. 1981). The rheological properties of blood flowing in microvessels have been extensively studied by *in vitro* experiments, using a suspension of red cells flowing through capillary tubes (Sugihara-Seki and Fu 2005).

Accurate numerical simulations of blood flow in microvessels must certainly include detailed models of blood cells as well as the glycocalyx layer attached to the the endothelial surface. The dimensional irregularities of vessel diameters is another important factor. The Reynolds number of the blood flow in microvessels is $\ll 1$, so that in general non-linear convective acceleration terms $(\mathbf{v} \cdot \nabla \mathbf{v})$ in the momentum-balance equations describing the plasma flow and the cell motion can be neglected (Sugihara-Seki and Fu 2005). Since the plasma is known to be an incompressible Newtonian fluid, its motion is governed by the Navier-Stokes equations

$$\rho \frac{\partial \mathbf{v}}{\partial t} = -\nabla p + \mu \nabla^2 \mathbf{v}, \tag{7}$$

along with the continuity equation

$$\nabla \cdot \mathbf{v} = 0, \tag{8}$$

where $\mathbf{v}$ is the velocity vector, p is the pressure, ρ is the density, and μ is the dynamic viscosity of the plasma. Early simulations aimed at modelling the flow of red cells in narrow tubes under axisymmetry, the flow fields around cells and shear stress on the cell membrane, and flow resistance due to irregularities of vessel lumen as well as the effects of glycocalyx and leukocytes are reviewed in Sugihara-Seki and Fu (2005), and described in full detail in the references therein. More recent simulations using continuum-based models have shown that coupling of solid components and fluid flow in these models poses a number of challenging problems (Pozrikidis 2005; Noguchi and Gompper 2005; Liu and Liu 2006; Skotheim and Secomb 2007; Wu and Aidun 2010; Fedosov et al. 2012). For example, computational complexity can be reduced by coupling discrete models of red cells with mesoscopic methods for flow discretization such as the lattice Boltzmann method, multiparticle collisional dynamics, and dissipative particle dynamics (Dupin et al. 2008). Numerical simulations have indicated that the effect of leukocyte adhesion to the vessel walls on flow depends strongly on the number of adherent leukocytes and the vessel diameter (Pappu et al. 2008). Owing to many similarities in the process of leukocyte and circulating tumour cell adhesion, models developed for leukocytes can also be applied to circulating tumour cells during the metastasis process.

Microvessel walls consist mainly of endothelial cells. Vascular endothelium is the principle barrier to, and regulator of, material exchange between circulating blood and the body tissues. The ultrastructural pathways and mechanisms whereby endothelial cells and the cleft between the cells modulate microvessel permeability to water and solutes have been a classical question in microvessel transport since the early 1950s. If capillary walls act like semi-permeable membranes, fluid motion across them depends on the net imbalance between the osmotic absorption pressure of the plasma proteins and the capillary hydraulic pressure generated by the heart beating (Levick and Michel 2010). Most existing models of transport through the inter-endothelial clefts are based on continuum approaches. However, it was suggested that more suitable analyses should be based on the molecular nature of the fluid because of the sizes of the mean intermolecular distances ($\sim$0.3 nm) and the cleft width ($\sim$18 nm) (Sugihara-Seki et al. 2008). The development of multiscale computational models (Praprotnik and Delle Site 2008), coupling, for example, the molecular structure of the glycocalyx with a continuum description of the flow, is highly suitable in this context. Moreover, solute transport from the vasculature to the cells has been largely modelled as passively transported elements with a flux proportional to the drug concentration. Solute transport inside the tumour was recently analyzed using computational models of diffusion based on high-resolution images (Baish et al. 2011).

On the other hand, tumour-induced angiogenesis has been modelled using both continuum and discrete models (Qutub and Mac Gabhann 2009). In a more recent continuum approach, the extracellular population is modelled by a density function that resolves the vascular branching patterns (Bergdorf et al. 2010). Cell-based and lattice-based discrete models are described in Bauer et al. (2009) and Chaplain (2000), respectively, while a hybrid modelling where a discrete tip-cell representation is coupled to a continuum description of the blood vessels is given in Milde et al. (2008) and Travasso and Corvera Poiré (2011). A model for sprouting angiogenesis based on Poiseuille flow inside a network of connected pipes can be found in McDougall et al. (2002), which was successively extended to account for the variability in blood viscosity and evolving capillary vessels that can dilate and constrict to study the transport of therapeutic agents inside the growing vasculature (Stephanou et al. 2006) and combined with a continuum model of tumour growth (Macklin et al. 2009). An in-depth report on recent simulation models of vascularized tumours is given in Lowengrub et al. (2010), and references therein.

5 Microfluidics in Cancer

Microfluidics typically deals with the manipulation of fluids that are geometrically constrained to a submillimeter scale. Such small scales offer a number of advantages including cost effectiveness, low consumption of reagents, cellular separations and detections with high resolution and sensitivity, and other less obvious features of fluids in microchannels, such as laminar flow (Whitesides 2006). The early development of microfluidics in life science applications has been mainly focused on the analysis of biomolecules from small volumes of fluids (typically nanolitres or less). However, the use of microfluidics in manipulating and analyzing individual cells has notably increased in recent years. Its application to biological systems is compelling because it allows manipulation at the single or even subcellular level. Recently, there has been a push towards applying microfluidic tools to specific biological research areas so that development of these engineering approaches can be better guided.

Microfluidic devices allow for a lab-on-a-chip array to simplify single cell analysis by providing a microenvironment that is of micrometer dimension and containing nanomoles of reagent/media. They also allow for controlled placement of cells and precise delivery factors (Chao and Ros 2008). One conventional system that is commonly used as a model to study cell migration is the transwell Boyden chamber, in which a porous membrane with pore size of $\sim$5–10 μm is placed between cells and chemoattractant so that cells are attracted to move across the membrane (Karnoub et al. 2007). Rapid advances in microtechnology have made microfluidic devices easy to design and construct. For instance, polydimethylsiloxane (PDMS) membrane stamps are typically molded off through soft lithography and other rapid prototyping techniques (Xia and Whitesides 1998). Refinements in the fabrication process, such as e-beam lithography, makes it possible to construct channels on the submicron or even nanoscale, which in theory would be able to constrain a fluid volume down to a femtolitre (billionth of microlitre) range (Qin et al. 2010). Recent microfluidic approaches in studying cellular migration are reviewed in Huang et al. (2011).

Another application of microfluidic-based devices is in the isolation, detection, and molecular characterization of circulating tumour cells. Efficient methods for the isolation and characterization of circulating tumour cells can also contribute to a much better understanding of the metastatic process. The development of passive microfluidic cell separation biochips, which can isolate circulating tumour cells from whole blood without the use of antibodies or magnetic beads, is revolutionizing disease detection, diagnosis, and prognosis as cancer cells can be obtained from blood (termed liquid biopsy) rather than via the needle aspiration tumour biopsy, which is invasive, painful, and cannot be performed on a regular basis (Lim 2012). Recent overviews of various methods for circulating tumour cell isolation, detection, and molecular characterization can be found in Hou et al. (2011), Lianidou and Markou (2011), and Yu et al. (2011). Isolation of cells with differential deformabilities remains a great challenge (Wirzt et al. 2011; Gossett et al. 2012). Microfabrication-assisted technology, using microscale arrays of round or rectangular posts, channels, or other simple patterns, has the potential to solve this problem. For instance, a mechanical separation chip, which employs artificial microbarriers in combination with hydrodynamic forces to separate deformable from stiff cells, has been used to demonstrate the separation of: (i) an artificial mixture of two breast cancer cell types (MDA-MB-436 and MCF-7) with distinct deformabilities and matastatic potentials, and (ii) a heterogeneous breast cancer cell line (SUM149), into enriched flexible and stiff sub-populations (Zhang et al. 2012). The flexible phenotype is associated with overexpression of multiple genes involved in cancer cell motility and metastasis.

Microfluidic devices have also been used for studying metastatic cancer cell invasion. Much of the initial work in applying microfluidics to metastasis has focused on studying how cancer cells respond to concentration gradients of chemicals suspected to drive cell motion. For example, with the aid of a PDMS-based device it has been possible to monitor 3D migration of the invasive MDA-MB-231 (mammary carcinoma) cells across extracellular matrix-coated microgaps with real-time light microscopy and map out their migration paths (Chaw et al. 2007). This not only permits to quantify the percentage of migrated cells, but also to obtain information on migration of individual cells. Microdevices for cell isolation and enumeration from blood have also been presented by several other authors (Cheng et al. 2007; Vickerman et al. 2008; Tan et al. 2009). Today, most of these devices has the potential to be used for routine monitoring of cancer development and cancer therapy in a clinical setting. Recently, a microfluidic optical stretcher have been used to study mechanical properties of cells from the inside (Lautenschläger et al. 2009). This helps investigate how the cytoskeleton, cell mechanics, and cell motility may be related, so that we may better understand how to develop therapies that hinder movement of metastatic cells.

6 Future Challenges and Perspectives

In this chapter we have reviewed recent modelling aspects of cancer fluid mechanics at different representation scales. In particular, model simulations of how cellular changes affect macroscopic distributions are especially important when examining sustained angiogenesis, tissue invasion, and metastasis. Although these models have been successful in describing macroscopic evolution properties of cancer, it is well known that they occur through genetic mutations and evolutionary selection; a link that has not yet been fully modelled. On the other hand, while all macroscopic models either assume that cells move through a diffusion-like process or act as an elastic fluid, only discrete models have the ability to track the behaviour of single cells. Therefore, macroscopic continuum models should be derived from the underlying cellular models by suitable asymptotic methods linking inter-cellular distances to those typical of the tissue level. This necessity has recently given rise to multiscale modelling constructs, where the dynamics at the cellular scale is coupled with the continuum mechanics of solid tumours. However, as is the case with very complex systems, all of the components cannot be usually included if we wish to develop practical models.

One possible solution to the above difficulty that has been envisaged is to make use of the so-called theory of modules proposed by Hartwell et al. (1999), where the whole system is decomposed into subsystems (or modules) such that the identification of each module is related to the expression of specific biological functions. However, this modular approach must overcome a number of challenges before becoming a workable and practical multiscale modelling framework. For instance, the analysis of large interacting systems as occurs if the numerous signalling pathways or interacting cytokines are incorporated, which will unavoidably lead to a significant increase in system size and complexity. Other difficult features involve the processes of angiogenesis and metastasis, where a detailed modelling of branching, anastomosis, vascular normalization as well as active cell migration to blood vessels, intravasation, extravasation, and distant site colonization in metastatic spread are completely ignored by present models. In addition, factors as cell geometry, diffusion terms, chemotaxis and haptotaxis, and cell invasion as an active and coordinated process represent future challenges that must be accurately modelled if we want to reproduce cancer in the computer and convert such complexity multiscale models into powerful tools for the diagnosis, prediction, and therapy of cancer.

Acknowledgments D. C. Belisario acknowledges the organizers of the I Workshop of the Venezuelan Society of Fluid Mechanics for inviting me to write part of this chapter.

References

Ambrosi D, Mollica F (2002) On the mechanics of a growing tumour. Int J Eng Sci 40:1297–1316

Anderson ARA (2005) A hybrid mathematical model of solid tumour invasion: the importance of cell adhesion. Math Med Biol 22:163–186

Anderson ARA, Weaver AM, Cummings PT, Quaranta V (2006) Tumor morphology and phenotypic evolution driven by selective pressure from the microenvironment. Cell 127:905–915

De Angelis E, Preziosi L (2000) Advection diffusion models for solid tumours in vivo and related free-boundary problems. Math Models Methods Appl Sci 10:379–408

Baish JW, Netti PA, Jain RK (1997) Transmural coupling of fluid flow in microcirculatory network and interstitium in tumors. Microvasc Res 53:128–141

Baish JW, Stylianopoulos T, Lanning RM, Kamoun WS, Fukumura D, et al. (2011) Scaling rules for diffusive drug delivery in tumour and normal tissues. Proc Nat Acad Sci USA 108:1799–1803

Bauer AL, Jackson TL, Jiang Y (2007) A cell-based model exhibiting branching and anastomosis during tumor-induced angiogenesis. Biophys J 92:3105–3121

Bauer AL, Jackson TL, Jiang Y (2009) Topography of extracellular matrix mediates vascular morphogenesis and migration speeds in angiogenesis. PLoS Comput Biol 5:e1000445

Baxter LT, Jain RK (1989) Transport of fluid and macromolecules in tumors: I. Role of interstitial pressure and convection. Microvasc Res 37:77–104

Bellomo N, Li NK, Maini PK (2008) On the foundations of cancer modelling: selected topics, speculations, and perspectives. Math Models Methods Appl Sci 18:593–646

Bergdorf M, Milde F, Koumoutsakos P (2010) Continuum models of mesenchymal cell migration and sprouting angiogenesis. In: Deisboeck T, Stamatakos GS (eds) Multiscale cancer modeling. CRC, Boca Raton, pp 213–235

Boucher Y, Jain RK (1992) Microvascular pressure is the principal driving force for interstitial hypertension in solid tumours: Implications for vascular collapse. Cancer Res 52:5110–5114
Bresch D, Colin T, Grenier E, Ribba B, Saut O (2010) Computational modeling of solid tumor growth: the avascular stage. SIAM J Sci Comput 32:2321–2344
Cabrera H, Castro J, Grassi HC, Andrades EDJ, López-Rivera SA (2012) The effect of photodynamic therapy on contiguous untreated tumor. Dermatol Surg 38:1097–1099
Chao T-C, Ros A (2008) Microfluidic single-cell analysis of intracellular compounds. J Royal Soc Interface 5:S139–S150
Chaplain MAJ (1996) Avascular growth, angiogenesis, and vascular growth in solid tumors: the mathematical modelling of the stages of tumor development. Math Comput Model 23:47–87
Chaplain MA (2000) Mathematical modelling of angiogenesis. J Neuro-Oncol 50:37–51
Chaplain MAJ, Graziano L, Preziosi L (2003) Mathematical modelling of the loss of tissue compression responsiveness and its role in solid tumour development. IMA J Math Appl Med Biol 23:197–229
Chaplain MAJ, Lachowicz M, Szymańska Z, Wrzosek D (2011) Mathematical modelling of cancer invasion: the importance of cell-cell adhesion and cell-matrix adhesion. Math Models Methods Appl Sci 21:719–743
Chaw KC, Manimaran M, Tay FE, Swaminathan S (2007) Matrigel coated polydimethylsiloxane based microfluidic devices for studying matastatic and non-metastatic cancer cell invasion and migration. Biomed Microdevices 9:597–602
Cheng SY, Heilman S, Wasserman M, Archer S, Shuler ML, Wu M (2007) A hydrogel-based microfluidic device for the studies of directed cell migration. Lab on a Chip 7:720–725
Cristini V, Lowengrub JS, Nie Q (2003) Nonlinear simulation of tumor growth. J Math Biol 46:191–224
Cullen JP, Sayeed S, Sawai RS, Theodorakis NG, Cahill PA, et al. (2002) Pulsatile flow-induced angiogenesis. Arterioscler, Thromb Vasc Biol 22:1610–1616
Dallon JC, Othmer HG (2004) How cellular movement determines the collective force generated by the Dictyostelium discoideum slug. J Theor Biol 231:203–222
Davies JA (2005) Mechanisms of morphogenesis. Elsevier, Philadelphia
Deutsch A, Dormann S (2005) Cellular automaton modeling of biological pattern formation: characterization, applications, and analysis. Birkhäuser, Boston
Dillon R, Painter KJ, Owen MR (2008) A single-cell based model of cellular growth using the immersed boundary method. In: Koo CB, Li Z, Li P (eds) Moving interface problems and applications in fluid dynamics (Contemporary Mathematics, AMS) pp 1–16
Djonov V, Baum O, Burri P (2003) Vascular remodeling by intussusceptive angiogenesis. Cell and Tissue Res 314:107–117
Dupin MM, Halliday I, Care CM, Munn LL (2008) Lattice Boltzmann modelling of blood cell dynamics. Int J Comput Fluid Dyn 22:481–492
Dvorak HF, Nagy J, Dvorak J, Dvorak A (1988) Identification and characterization of the blood vessels of solid tumors that are leaky to circulating macromolecules. Am J Pathol 133:95–109
Egeblad M, Nakasone ES, Werb Z (2010) Tumors as organs: complex tissues that interface with the entire organism. Dev Cell 18:884–901
Erbertseder K, Reichold J, Flemisch B, Jenny P, Helmig R (2012) A coupled discrete/continuum model for describing cancer-therapeutic transport in the lung. PLoS One 7:e31966
Fedosov DA, Fornleitner J, Gompper G (2012) Margination of white blood cells in microcapillary flow. Phys Rev Lett 108:028104
Friedl P, Wolf K (2003) Tumour-cell invasion and migration: diversity and escape mechanisms. Nat Rev Cancer 3:362–374
Friedman A, Hu B (2007) Bifurcation for a free boundary problem modeling tumor growth by Stokes equation. SIAM J Math Anal 39:174–194
Fu Y, Kunz R, Wu J, Dong C (2012) Study of local hydrodynamic environment in cell-substrate adhesion using side-view μPIV technology. PLoS One 7:e30721

Gatenby RA, Gillies RJ (2004) Why do cancers have high aerobic glycolysis? Nat Rev Cancer 4:891–899

Glazier JA, Graner F (1993) Simulation of the differential adhesion driven rearrangement of biological cells. Phys Rev E 47:2128–2154

Goel S, Duda D, Xu L, Munn L, Boucher Y, et al. (2011) Normalization of the vasculature for treatment of cancer and other diseases. Physiol Rev 91:1071–1121

Gossett DR, Tse HTK, Lee SA, Ying Y, Lindgren AG, et al. (2012) Hydrodynamic stretching of single cells for large population mechanical phenotyping. Proc Nat Acad Sci USA 109:7630–7635

Graner F, Glazier JA (1992) Simulation of biological cell sorting using a two-dimensional extended Potts model. Phys Rev Lett 69:2013–2016

Graziano L, Preziosi L (2007) Mechanics in tumour growth. In: Mollica F, Rajagopal KR, Preziosi L (eds) Modeling of biological materials. Birkhäuser, Boston, pp 267–328

Gusenbauer M, Cimrak I, Bance S, Exl L, Reichel F, Oezelt H, Schrefl T (2011) A tunable cancer cell filter using magnetic beads: cellular and fluid dynamic simulations. arXiv:1110.0995v1 [physics.flu-dyn]

Halpern D, Secomb TW (1989) The squeezing of red blood cells through capillaries with near-minimal diameters. J Fluid Mech 203:381–400

Hanahan D, Folkman J (1996) Patterns and emerging mechanisms of the angiogenic switch during tumorigenesis. Cell 86:353–364

Hanahan D, Weinberg RA (2000) The hallmarks of cancer. Cell 100:57–70

Hanahan D, Weinberg RA (2011) Hallmarks of cancer: the next generation. Cell 144:646–670

Hartwell HL, Hopfield JJ, Leibner S, Murray AW (1999) From molecular to modular cell biology. Nature 402:C47–C52

Heldin C-H, Rubin K, Pietras K, Östman A (2004) High interstitial fluid pressure—an obstacle in cancer therapy. Nat Rev Cancer 4:806–813

Holmes MJ, Sleeman BD (2000) A mathematical model of tumour angiogenesis incorporating cellular traction and viscoelastic effects. J Theor Biol 202:95–112

Hou HW, Lee WC, Leong MC, Sonam S, Vedula SRK, Lim CT (2011) Microfluidics for applications in cell mechanics and mechanobiology. Cell Mol Bioeng 4:591–602

Huang Y, Agrawal B, Sun D, Kuo JS, Williams JC (2011) Microfluidics-based devices: new tools for studying cancer and cancer stem cell migration. Biomicrofluidics 5:013412

Karnoub AE, Dash AB, Vo AP, Sullivan A, Brooks MW et al (2007) Mesenchymal stem cells within tumour stroma promote breast cancer metastasis. Nature 449:557–563

Koumoutsakos P, Pivkin I, Milde F (2013) The fluid mechanics of cancer and its therapy. Ann Rev Fluid Mech 45:325–355

Lautenschläger F, Paschke S, Schinkinger S, Bruel A, Bell M, Guck J (2009) The regulatory role of cell mechanics for migration of differentiating myeloid cells. Proc Nat Acad Sci USA 106:15696–15701

Lee I, Boucher Y, Demhartner TJ, Jain RK (1994) Changes in tumour blood flow, oxygenation and interstitial pressure induced by pentoxifylline. Br J Cancer 69:492–496

Leung D, Cachianes G, Kuang W, Goeddel D, Ferrara N (1989) Vascular endothelial growth factor is a secreted angiogenic mitogen. Science 246:1306–1309

Levick JR, Michel CC (2010) Microvascular fluid exchange and the revised starling principle. Cardiovasc Res 87:198–210

Liang S, Slattery MJ, Wagner D, Simon SI, Dong C (2008) Hydrodynamic shear rate regulates melanoma-leukocyte aggregation, melanoma adhesion to the endothelium, and subsequent extravasation. Ann Rev Biomed Eng 36:661–671

Liang S, Hoskins M, Dong C (2010) Tumor cell extravasation mediated by leukocyte adhesion in shear rate dependent on IL-8 signaling. Mol Cell Biomech 7:77–91

Lianidou ES, Markou A (2011) Circulating tumor cells in breast cancer: detection systems, molecular characterization, and future challenges. Clin Chem 57:1242–1255

Lim CT (2012) Microfluidics for cancer cell detection and diagnosis. IEEE Life Sci Newsl (June 2012)

Liotta LA, Kohn EC (2001) The microenvironment of the tumour-host interface. Nature 411:375–379
Liu Y, Liu WK (2006) Rheology of red blood cell aggregation by computer simulation. J Comput Phys 220:139–154
Lowengrub JS, Frieboes HB, Jin F, Chuang YL, Li X, et al. (2010) Nonlinear modelling of cancer: bridging the gap between cells and tumours. Nonlinearity 23:R1–R91
Macklin P, McDougall S, Anderson ARA, Chaplain MAJ, Cristini V, Lowengrub J (2009) Multiscale modelling and nonlinear simulation of vascular tumour growth. J Math Biol 58:765–798
Mantzaris NV, Webb S, Othmer HG (2004) Mathematical modeling of tumor-induced angiogenesis. J Math Biol 49:111–187
McDougall SR, Anderson ARA, Chaplain MAJ, Sherratt JA (2002) Mathematical modelling of flow through vascular networks: implications for tumour-induced angiogenesis and chemotherapy strategies. Bull Math Biol 64:673–702
Michor F, Liphardt J, Ferrari M, Widom J (2011) What does physics have to do with cancer? Nat Rev Cancer 11:657–670
Milde F, Bergdorf M, Koumoutsakos P (2008) A hybrid model for three-dimensional simulations of sprouting angiogenesis. Biophys J 95:3146–3160
Moreira J, Deutsch A (2005) Pigment pattern formation in zebrafish during late larval stages: a model based on local interaction. Dev Dyn 232:33–42
Narang AS, Varia S (2011) Role of tumor vascular architecture in drug delivery. Adv Drug Delivery Rev 63:640–658
Negrini S, Gorgoulis VG, Halazonetis TD (2010) Genomic instability: an evolving hallmark of cancer. Nat Rev Mol Cell Biol 11:220–228
Newman TJ (2005) Modeling multicellular systems using subcellular elements. Math Biosci Eng 2:613–624
Noguchi H, Gompper G (2005) Shape transitions of fluid vesicles and red blood cells in capillary flows. Proc Nat Acad Sci US 102:14159–14164
Palsson E (2008) A 3-D model used to explore how cell adhesion and stiffness affect cell sorting and movement in multicellular systems. J Theor Biol 254:1–13
Pappu V, Doddi SK, Bagchi P (2008) A computational study of leukocyte adhesion and its effect on flow pattern in microvessels. J Theor Biol 254:483–498
Perumpanani AJ, Byrne HM (1999) Extracellular matrix concentration exerts selection pressure on invasive cells. Euro J Cancer 35:1274–1280
Pham K, Frieboes HB, Cristini V, Lowengrub J (2011) Predictions of tumour morphological stability and evaluation against experimental observations. J Royal Soc Interface 8:16–29
Plank MJ, Sleeman BD, Jones PF (2004) A mathematical model of tumor angiogenesis, regulated by vascular endothelial growth factor and the angiopoietins. J Theor Biol 229:435–454
Pozrikidis C (2005) Numerical simulation of cell motion in tube flow. Ann Biomed Eng 33:165–178
Praprotnik M, Delle Site L (2008) Multiscale simulation of soft matter: from scale bridging to adaptive resolution. Ann Rev Phys Chem 59:545–571
Qin D, Xia Y, Whitesides GM (2010) Soft lithography for micro- and nanoscale patterning. Nat Protoc 5:491–502
Quaranta V, Weaver AM, Cummings PT, Anderson ARA (2005) Mathematical modeling of cancer: the future of prognosis and treatment. Clin Chim Acta 357:173–179
Qutub A, Mac Gabhann F (2009) Multiscale models of angiogenesis. IEEE Eng Med Biol Mag 28:14–31
Rejniak KA (2007) An immersed boundary framework for modelling the growth of individual cells: an application to the early tumour development. J Theor Biol 247:186–204
Roose T, Netti PA, Munn LL, Boucher Y, Jain RK (2003) Solid stress generated by spheroid growth estimated using a linear poroelasticity model. Microvasc Res 66:204–342
Roose T, Chapman SJ, Maini PK (2007) Mathematical models of avascular tumor growth. SIAM Rev 49:179–208

Schaller G, Meyer-Hermann M (2005) Multicellular tumor spheroid in an off-lattice Voronoi-Delaunay cell model. Phys Rev E 71:051910
Schmid-Schönbein GW, Sung K-P, Tözeren H, Skalak R, Chien S (1981) Passive mechanical properties of human leukocytes. Biophys J 36:243–256
Sherratt JA, Chaplain MAJ (2001) A new mathematical model for avascular tumour growth. J Math Biol 43:291–312
Skotheim JM, Secomb TW (2007) Red blood cells and other nonspherical capsules in shear flow: oscillatory dynamics and the tank-treading-to-tumbling transition. Phys Rev Lett 98:078301
Soltani M, Chen P (2011) Numerical modeling of fluid flow in solid tumours. PLoS One 6:e20344
Song JW, Munn LL (2011) Fluid forces control endothelial sprouting. Proc Nat Acad Sci USA 108:15342–15347
Sporn MB (1996) The war on cancer. Lancet 347:1377–1381
Stephanou A, McDougall SR, Anderson ARA, Chaplain MAJ (2006) Mathematical modelling of the influence of blood rheological properties upon adaptive tumour-induced angiogenesis. Math Comput Model 44:96–123
Styp-Rekowska B, Hlushchuk R, Pries AR, Djonov V (2011) Intussusceptive angiogenesis: pillars against the blood flow. Acta Physiol 202:213–223
Sugihara-Seki M, Fu BM (2005) Blood flow and permeability in microvessels. Fluid Dyn Res 37:82–132
Sugihara-Seki M, Akinaga T, Itano T (2008) Flow across microvessel walls through the endothelial surface glycocalyx and the interendothelial cleft. J Fluid Mech 601:229–252
Suresh S (2007) Biomechanics and biophysics of cancer cells. Acta Biomater 3:413–438
Swanson KR, Alvord EC Jr (2002) Virtual brain tumours (gliomas) enhance the reality of medical imaging and highlight inadequacies of current therapy. Br J Cancer 86:14–18
Tan SJ, Yobas L, Lee GYH, Ong CN, Lim CT (2009) Microdevice for the isolation and enumeration of cancer cells from blood. Biomed Microdevices 11:883–892
Thompson KE, Byrne WF (1999) Modelling the internalization of labelled cells in tumour spheroids. Bull Math Biol 61:601–623
Travasso RDM, Corvera Poiré E (2011) Tumor angiogenesis and vascular patterning: a mathematical model. PLoS One 6:e19989
Turner S, Sherratt JA (2002) Intercellular adhesion and cancer invasion: a discrete simulation using the extended Potts model. J Theor Biol 216:85–100
Venkatasubramanian R, Henson MA, Forbes NS (2008) Integrating cell cycle progression, drug penetration and energy metabolism to identify improved cancer therapeutic strategies. J Theor Biol 253:98–117
Vickerman V, Blundo J, Chung S, Kamm R (2008) Design, fabrication and implementation of a novel multi-parameter control microfluidic platform for three-dimensional cell culture and real-time imaging. Lab on a Chip 8:1468–1477
Weinberg RA (2007) The biology of cancer. Garland Science, New York
Whitesides GM (2006) The origins and the future of microfluidics. Nature 442:368–373
Wirzt D, Konstantopoulos K, Searson PC (2011) The physics of cancer. The role of physical interactions and mechanical forces in metastasis. Nat Rev Cancer 11:512–522
Wu JS, Aidun CK (2010) Simulating 3D deformable particle suspensions using lattice Boltzmann method with discrete external boundary force. Inter J Numer Methods Fluids 62:765–783
Xia Y, Whitesides GM (1998) Soft lithography. Angew Chem 37:550–575
Yamane T, Mitsumata M, Yamaguchi N, Nakazawa T, Mochisuki K, et al. (2010) Laminar high shear stress up-regulates type IV collagen synthesis and down-regulates MMP-2 secretion in endothelium: A quantitative analysis. Cell and Tissue Res 340:471–479
Yu M, Scott S, Toner M, Maheswaran S, Haber DA (2011) Circulating tumor cells: approaches to isolation and characterization. J Cell Biol 192:373–382
Zhang W, Kai K, Choi DS, Iwamoto T, Nguyen YH, et al. (2012) Microfluidics separation reveals the stem-cell-like deformability of tumor-initiating cells. Proc Nat Acad Sci USA 109:18707–18712
Zheng X, Wise SM, Cristini V (2005) Nonlinear simulation of tumor necrosis, neo-vascularization and tissue invasion via an adaptive finite-element/level-set method. Bull Math Biol 67:211–259

Growth of Bubbles in Reservoirs and its Consequences on the Foam Formation

Abel López-Villa and Abraham Medina

Abstract The effect of partial confinement on the shape and volume of bubbles generated by injection of gas at a constant flow rate, into a highly viscous liquid is studied numerically and experimentally. By using the Boundary Element Method, numerical solutions of the Stokes equations for the viscous liquid yield the evolution of the surface of a bubble. These solutions and experiments show that cylindrical, conical, and pipe walls with periodic corrugations, concentric with the gas injection orifice in the horizontal bottom of the liquid, may strongly affect the shape and volume of the bubbles. Thus, the presence of walls could be used to control the size of the generated bubbles without changing the gas flow rate. A well-known scaling law for the volume of the bubbles generated by injection of gas at a high flow rate in a highly viscous, unconfined liquid is extended to take into account the presence of cylindrical or conical walls around the injection orifice. In addition, we study numerically the thickness film that is formed between the free surface of a bubble and the cylindrical walls in both cases.

1 Introduction

The growth and detachment of bubbles, generated by the continuous injection of gas into a quiescent liquid, has long been studied in conditions where the viscosity of the liquid does not play an important role. Results of these studies are of interest in the metallurgical and chemical industries, for example, where liquids of low viscosity, such as liquid metals and aqueous solutions, need be handled. Bubbles in

A. López-Villa (✉) · A. Medina
SEPI ESIME Azcapotzalco, Instituto Politécnico Nacional (I.P.N.), Avenida de las Granjas 682, Colonia Santa Catarina, 02250 Azcapotzalco, México, D. F., Mexico
e-mail: abelvilla77@hotmail.com

A. Medina
e-mail: abraham_medina_ovando@yahoo.com

L. Di G. Sigalotti et al. (eds.), *Computational and Experimental Fluid Mechanics with Applications to Physics, Engineering and the Environment*, Environmental Science and Engineering, DOI:10.1007/978-3-319-00191-3_7,

these liquids can be used to modify the concentrations of different substances and promote chemical reactions between them, to clean liquids from impurities captured by adhesion or diffusion processes, and for many other purposes (López-Villa et al. 2011).

The generation and dynamics of bubbles in highly viscous liquids is also of interest, but it has been scarcely studied. Thus, while many aspects of the dynamics of bubbles in unbounded viscous liquids are well understood, the formation and detachment of bubbles in confined systems has received less attention. Bubbles in highly viscous liquids are commonly found when dealing with polymers in their liquid phases, in the flows of lava, and in processes of oil extraction from production pipelines, among others. The latter example is the main motivation of the present work, which is of great interest in the so-called gas lift technique of enhanced oil recovery, where bubbles formed by injecting gas in oil extraction pipes help pumping the oil.

Scaling laws show that the volume of the bubbles generated by gas injection at high flow rates into highly viscous liquids increases as the power 3/4 of the flow rate and it is independent of the diameter of the injection orifice. The simplest way to control the size of the bubbles in a given liquid is, therefore, to act on the flow rate of gas. This possibility, however, is limited in the application at hand, because the flow rate of gas to be injected in the confined space of an extraction pipe is often determined by other requirements of the gas lift technique. The confinement of the bubble brings to the front elements of the generation process such as the viscous drag of the bubbles and the shear stress in the vicinity of the walls, which are disregarded in inviscid analyses but offer a clue to the solution of the size-control problem. In this Chapter, we show that the shape of the tube in the vicinity of the injection orifice, or the use of properly shaped injection nozzles, may cause substantial distortion of the growing and shape bubbles and modify their volume at detachment. The extent of the confinement can be gradually increased by decreasing the radius of the cylinder, the amplitude and frequency of the corrugations, or the angle of the cone, which allows for quantification of the wall effect on the evolution and size of the bubbles. This size is determined numerically and experimentally, and scaling laws that are extensions of well-known laws for unconfined liquids are proposed and validated.

The numerical treatment was made using the Boundary Element Method (BEM). Boundary integral equations are a classical tool for the analysis of boundary value problems for partial differential equations. The term "Boundary Element Method" (BEM) denotes any method for the approximate numerical solution of these boundary integral equations. The approximate solution of the boundary value problem obtained by BEM has the distinguishing feature that it is an exact solution of the differential equation in the domain and is parameterized by a finite set of parameters living on the boundary (Costabel 1986).

The chapter is organized as follows. In Sect. 2, a formulation of the problem in terms of the dimensionless equations for the motion of the liquid, the dimensionless boundary conditions, and the evolution of the free surface is given. In Sect. 3, a brief description of the BEM is presented. In Sect. 4, we present the results of the numerical solutions and discuss the evolution of the free surface during the growth

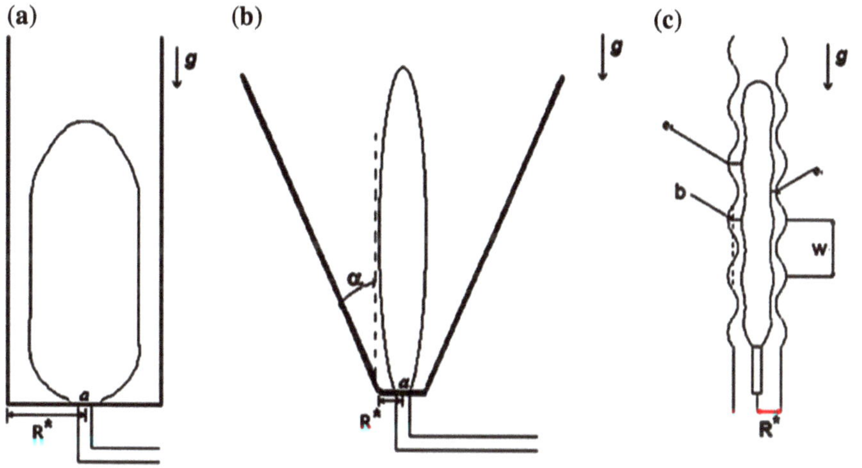

Fig. 1 Two-dimensional projections of the reservoirs, concentric to the gas injection needles

of the bubble at a constant gas flow rate in conical, cylindrical, and cylindrical corrugated containers. In Sect. 5, we describe a set of experiments that were carried out to validate the numerical results. Finally, Sect. 6 summarizes the main findings and limitations of this work.

2 Equations for Bubble Growth in a Confined Liquid

Suppose that a constant flow rate Q of an incompressible gas of negligible density and viscosity is injected into a liquid, of density ϱ and viscosity μ, initially at rest in a reservoir under the action of gravity. The gas is injected through a circular orifice of radius a at the centre of the base (of radius R^*) of the reservoir. The lateral wall of the reservoir may be cylindrical, conical, making an angle θ to the vertical, or periodically ribbed cylindrical, as sketched in Fig. 1. The gas accumulates in a bubble attached to the base of the reservoir. The volume of this bubble increases with time until it detaches and begins to ascend in the liquid, being replaced by a new attached bubble. The effect of the inertia is assumed to be negligible on the motion induced in the liquid by the growth and displacement of the bubbles. A sufficient condition for the effect of the inertia to be negligible is that $Re = \rho Q/\mu R_b \ll 1$ (Wong et al. 1998; Higuera 2005; Ajaev and Homsy 2006). Here R_b is the characteristic radius of the detaching bubble, or of its upper cap, which has to be found.

In particular, the model used here is valid for bubble formation in a highly viscous liquid in confined axi-symmetric geometries (López-Villa et al. 2011). The continuity and Navier-Stokes equations have the following dimensionless forms, respectively

$$\nabla \cdot \mathbf{v} = 0, \tag{1}$$

$$0 = -\nabla p + \nabla^2 \mathbf{v} - Bo\mathbf{i}. \tag{2}$$

Here the distances and times are scaled with the radius of the orifice a and the viscous time $\mu a/\sigma$; where p is the pressure, $\mathbf{v}$ is the velocity field, with $\mathbf{v} = \mathbf{v}^*\sigma/\mu$, $\mathbf{i}$ is the normal vector pointing in the upward vertical direction (x is the vertical coordinate), and Bo is the Bond number given by

$$Bo = \frac{\rho g a^2}{\sigma}, \tag{3}$$

where ϱ is the liquid density, g is the acceleration due to gravity, a is the radius of the orifice of gas injection, and σ is the surface tension. When the bubble grows, we assume that the dimensionless gas flow rate is constant. The dimensionless capillary number is

$$Ca = \frac{\mu Q}{\sigma a^2}, \tag{4}$$

where Q is the constant flow rate of gas.

When a bubble is formed in the liquid, a free surface of the form $f(\mathbf{x}, t) > 0$ exists, and therefore Eqs. (1) and (2) must be solved with the boundary conditions on the surface of the ith bubble

$$\frac{Df_i}{Dt} = 0, \tag{5}$$

$$-p\mathbf{n}_i + \tau' \cdot \mathbf{n}_i = (\nabla \cdot \mathbf{n}_i - p_{g_i})\mathbf{n}_i. \tag{6}$$

In the inner cylinder's surface, at $r = R^*$, and at infinity we have that

$$\mathbf{v} = 0, \tag{7}$$

because the fluid does not move there. Moreover, the pressure far from the bubble must satisfy

$$p + Box = constant, \tag{8}$$

where $p_g(t)$ is the uniform pressure of the gas in the bubble, which must be found by assuming that the volume V of the bubble increases linearly with time at a rate equal to Q for the attached bubble. In dimensionless variables, this condition reads

$$\frac{dV}{dt} = Ca. \tag{9}$$

In these equations $D/Dt = \partial/\partial t + \mathbf{v}\cdot\nabla$ is the material derivative at points on the surface of the bubble, $\boldsymbol{n}=\nabla f/|\nabla f|$ is a unit vector normal to the surface of the bubble, $\tau' = \nabla\mathbf{v}+(\nabla\mathbf{v})^T$ is the dimensionless viscous stress tensor, and x are r are the

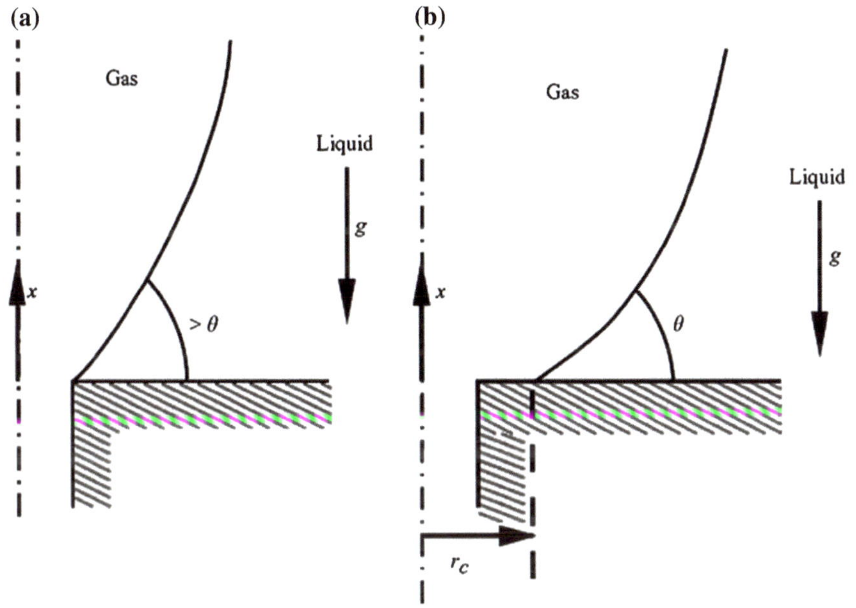

Fig. 2 Details of the contact line

distances along the axis of the reservoir and normal to it, respectively. Equation (5) means that the surface of the bubbles is a fluid surface separating the liquid from the gas, and therefore there is no mass exchange through it, while Eq. (6) is the balance of the stress acting on the surface of the bubbles.

If the gas pressure in each bubble, p_{gi}, is known, Eqs. (1) and (2) together with the boundary conditions (5–9) determine the velocity fields and the fluid pressure, and in particular the velocity on the surface of each bubble. In order to determine p_{gi} and complete the formulation of the conditions to be used, the volumes of the released bubbles ($i > 1$) are considered to be constant and equal to their values at the instant of release.

An additional condition is needed at the contact line between the growing bubble and the solid. Here the contact line will be taken to coincide with the edge of the orifice when the angle of the liquid–gas surface with the horizontal bottom is larger than the contact angle (i.e. when $-n_{x0}<\cos\theta$, where n_{x0} is the vertical component of the unit normal $\boldsymbol{n}_0$ to the attached bubble), and to spread away from the orifice, with the liquid–gas surface making a constant contact angle with the solid bottom ($-n_{x0} = \cos\theta$), otherwise (Higuera 2005; see Fig. 2).

The line of contact of the bubble ($i = 1$) adhered to the reservoir base is a circle which may either coincide with the edge of the orifice or move to a position $r > 1$ to be determined. Figure 2 illustrates both possibilities. In the first case (Fig. 2a), the radius of the contact line coincides with the hole edge, $r = 1$. In the second case (Fig. 2b), the angle formed by the bubble surface shape and the base of the reservoir must match the contact angle θ, which is a property of the liquid and the material of the shell.

Then the problem contains five dimensionless parameters, which are the Bond number *Bo*, a capillary number *Ca*, the dimensionless radius of the base of the reservoir $R = R * /a$, the semi-angle of the conical base α, and the contact angle θ between the liquid and the base (see Fig. 2). In the case when the walls of the tubes are corrugated, the mean is the tube radius, i. e., only for this case we have a further parameter: the frequency parameter *w*, which will be introduced later in Sect. 4.

The set of equations given above satisfy the boundary conditions of quite flow at infinity, the non-slip conditions on the walls, and the quasi-static pressure balance. The evolution of the free surface (bubble shapes) is given by the solution of Eq. (5), which is here solved using a fourth order Runge-Kutta scheme, after the hydrodynamic problem, imposed by Eqs. (1) and (2), is solved by using the BEM method (López-Villa et al. 2011). Figure 1 shows a schematic of the shape of the reservoir containing a viscous liquid.

3 Numerical Solution Method

We seek for axi-symmetric solutions of Eqs. (1–9). We use a standard Boundary Element Method (e.g., Pozrikidis 1992, 2002) to solve them along with a fourth order Runge-Kutta method to calculate the evolution of the free surface *f* given by Eq. (5).

3.1 History

After three decades of development, the Boundary Element Method (BEM) has found a firm footing in the arena of numerical methods for partial differential equations. Compared to more popular numerical methods, such as the Finite Element Method (FEM) and the Finite Difference Method (FDM), both of which can be classified as domain methods, the BEM distinguishes itself as a boundary method, meaning that the numerical discretization is conducted at reduced spatial dimension. For example, for problems in three-space dimensions, the discretization is performed on the bounding surface only, while in two-space dimensions, the discretization is on the boundary contour only. This reduced dimension leads to smaller linear systems, less computer memory requirements, and more efficient computation. This effect is most pronounced when the domain is unbounded. Unbounded domains need be truncated and approximated by domain methods. The BEM, on the other hand, automatically models the behaviour at infinity without the need of deployment of a mesh to approximate it. In the modern day industrial settings, mesh construction is the most intensive labour and also the most costly portion of numerical modelling, particularly for the FEM. Without the need of dealing with the interior mesh, the BEM is more cost effective in mesh preparation. For problems involving moving boundaries, the adjustment of the mesh is much easier with the BEM; hence it is again the preferred tool. With these advantages, the BEM is indeed an essential part in the repertoire of the modern day computational tools (Alexander et al. 2005).

One can view BEM as the numerical implementation of boundary integral equations based on Green's formula, in which the piecewise element concept of the FEM is utilized for the discretization.

3.2 BEM for Axi-Symmetric Domains

In this section, we describe the standard BEM for three-dimensional flow in an axi-symmetric domain. Our goal is to reduce the boundary integral equation to a one-dimensional equation, or a system of one-dimensional equations, over the trace of the boundaries in an azimuthal plane. First, the Green's functions of Stokes flow represent solutions to the continuity equation $\nabla \cdot \mathbf{v} = 0$ and the singularly forced Stokes equation

$$- \nabla P + \mu \nabla^2 \mathbf{v} + g\delta(\boldsymbol{x} - \boldsymbol{x}_0) = 0, \tag{10}$$

where $\delta(\mathbf{x} - \boldsymbol{x}_0)$ is the Dirac delta function in three dimensions, g is an arbitrary constant, and $\boldsymbol{x}_o$ is an arbitrary point. Introducing the Green's function G in three dimensions, we may write the solution of Eq. (10) in the form

$$\boldsymbol{u}_i = \frac{1}{8\pi\mu} \boldsymbol{G}_{ij}(\boldsymbol{x} - \boldsymbol{x}_0) g_i. \tag{11}$$

Here $\boldsymbol{x}$ is the observation or field point. Physically, Eq. (11) expresses the velocity field due to a concentrated point force of strength g placed at point $\boldsymbol{x}_0$, and may be identified with the flow produced by the slow settling of a small particle. In the literature of boundary integral methods, the Green's function may appear under the names *fundamental solution* or *propagator* (Pozrikidis 1992).

It is convenient to classify the Green's functions into three categories depending on the topology of the flow domain. First, we have the free-space Green's function for infinite unbounded flow; second, the Green's functions for infinite or semi-infinite flow that are bounded by a solid surface; and third, the Green's function for internal flow that is completely confined by solid surfaces. The Green's functions in the second and third categories are required to vanish over the internal or external boundaries of the flow. As the observation point $\boldsymbol{x}$ approaches the pole $\boldsymbol{x}_0$, all Green's functions exhibit a singular behaviour and, to leading order, behave like the free-space Green's function. The Green's functions for infinite unbounded or bounded flow are required to decay at infinity at a rate equal to or lower than that of the free-space Green's function. We find that

$$\frac{\partial \boldsymbol{G}_{ij}(\boldsymbol{x} - \boldsymbol{x}_0)}{\partial \boldsymbol{x}_i} = \mathbf{0}. \tag{12}$$

for the divergence of Eq. (11).

Integrating Eq. (12) over a volume of fluid that is bounded by the surface D and using the divergence theorem, we find

$$\int_D G_{ij}(\boldsymbol{x}-\boldsymbol{x}_0)\, n_i(\boldsymbol{x})\, dS(\boldsymbol{x}) = 0, \tag{13}$$

independently of whether the pole $\boldsymbol{x}_o$ is located inside, right on, or outside D.

The vorticity, pressure, and stress fields associated with the flow given by Eq. (11) may be represented by the corresponding forms:

$$\omega_i = \frac{1}{8\pi\mu}\Omega_{ij}(\boldsymbol{x}-\boldsymbol{x}_0)\, g_j, \tag{14}$$

$$\boldsymbol{P} = \frac{1}{8\pi\mu}p_j(\boldsymbol{x}-\boldsymbol{x}_0)\, g_j, \tag{15}$$

$$\tau_{ik} = \frac{1}{8\pi\mu}T_{ijk}(\boldsymbol{x}-\boldsymbol{x}_0)\, g_j, \tag{16}$$

where Ω_{ij}, p_j, andT_{ijk} are the vorticity tensor, the pressure vector, and the stress tensor associated with the Green's function. In particular, the stress tensor T_{ijk} is defined as

$$\boldsymbol{T}_{ijk}(\boldsymbol{x}-\boldsymbol{x}_0) = -\delta_{ik}p_j(\boldsymbol{x}-\boldsymbol{x}_0) + \frac{\partial G_{ij}(\boldsymbol{x}-\boldsymbol{x}_0)}{\partial x_k} + \frac{\partial G_{ij}(\boldsymbol{x}-\boldsymbol{x}_0)}{\partial x_i}. \tag{17}$$

Note that $T_{ijk} = T_{kji}$ as required by the symmetry of the stress tensor τ. When the flow domain is infinite, we require that Ω_{ij}, p_j, andT_{ijk} vanish as the observation point is moved to infinity.

First, for axi-symmetric flow with no swirling motion, we observe that in cylindrical coordinates, none of the boundary variables is a function of the azimuthal angle ϕ. This reduces the number of variables.

In the problem of bubble generation, the contours are the surfaces of the bubbles and the solid surfaces of the reservoir, which are all surfaces of revolution. The unknowns involved in the formulation of the boundary element are the velocities of the fluid particles that define the surfaces of the bubbles and the stresses on the wall of the vessel.

We introduce the dimensionless driving pressure $P = p + Box$, which allows us to write Eq. (2) as

$$-\nabla P + \nabla^2\mathbf{v} = \nabla\cdot\tau = 0, \tag{18}$$

where $\tau = -P\mathbf{I}+\tau'$ is a modified stress tensor, where the fluid pressure is replaced by the driving pressure. The stress of the liquid on the surface of the ith bubble, given by the left-hand side of Eq. (6), is therefore

$$-p\mathbf{n}_i + \tau'\cdot\mathbf{n}_i = -P\mathbf{n}_i + \tau'\cdot\mathbf{n}_i + Box\mathbf{n}_i.$$

We will use the notation $\boldsymbol{f} = \tau \cdot \mathbf{n}$ for the modified stress on the limiting contour of the liquid, where $\mathbf{n}$ is the normal to the contour directed towards the liquid. With this notation $-p\mathbf{n}_i + \tau' \cdot \mathbf{n}_i = \boldsymbol{f} + Box\mathbf{n}_i$ and so the boundary condition (6) on the bubble surface i takes the form

$$\boldsymbol{f} = (\nabla \cdot \mathbf{n}_{\mathrm{i}} - Box - p_{\mathrm{gi}})\mathbf{n}_i \tag{19}$$

In addition, we use the Green's functions for axi-symmetric flow, which are the solutions of the Stokes equations (10) in unlimited space for the action of forces concentrated on a circumference of radius $\boldsymbol{r}_0$ centred at point $\boldsymbol{x}_0$ on the axis of symmetry. These forces can be either parallel or perpendicular to the axis of symmetry, which gives rise to two distinct solutions whose velocity and pressure distributions denoted as $\boldsymbol{G}^x(\boldsymbol{x}, \boldsymbol{x}_0)$ and $P^x(\boldsymbol{x}, \boldsymbol{x}_0)$ for the axial force, and $\boldsymbol{G}^r(\boldsymbol{x}, \boldsymbol{x}_0)$ and $P^r(\boldsymbol{x}, \boldsymbol{x}_0)$ for a radial force. Here, $\boldsymbol{x}_0 = (x_0, r_0)$ and $\boldsymbol{x} = (x, r)$ (a generic point) in cylindrical coordinates as was defined above. The problems that must be solved to determine the Green's functions are

$$\nabla \cdot \boldsymbol{G}^x = 0, \qquad 0 = -\nabla P^x + \nabla^2 \boldsymbol{G}^x + 8\pi\delta(x, x_0)\mathbf{e}_{\mathrm{x}}, \tag{20}$$

and

$$\nabla \cdot \boldsymbol{G}^{\mathbf{r}} = 0, \qquad 0 = -\nabla P^{\mathbf{r}} + \nabla^2 \boldsymbol{G}^r + 8\pi\delta(x, x_0)\mathbf{e}_{\mathrm{r}}, \tag{21}$$

with the conditions $(\boldsymbol{G}^x, P^x) \to 0$ and $(\boldsymbol{G}^r, P^r) \to 0$ at infinity. In these equations $\mathbf{e}_x$ and $\mathbf{e}_r$ are the unit vectors parallel and perpendicular to the axis of symmetry, δ is the Dirac function, and the 8π factor is introduced by convention. We will also use the notation T^r and T^x for the stress tensor of solutions (20) and (21). These solutions are known and are given in Appendix A.

Since $\nabla \cdot \mathbf{v} = \nabla \cdot \boldsymbol{G}^x = 0$, we find that $\nabla \cdot (\boldsymbol{G}^x \cdot \tau - \mathbf{v} \cdot T^x) = \boldsymbol{G}^x \cdot (\nabla \cdot \tau) - \mathbf{v}(\nabla T^x)$. Using Eqs. (18) and (20) [i.e. $\nabla \cdot \tau = 0$ and $\nabla \cdot T^x = 8\pi\delta(x, x_0)\mathbf{e}_{\mathbf{x}}$] on the right part of the above equality, integrating the result over the volume occupied by the liquid, and using the Gauss theorem for flow in an axi-symmetric domain by transforming the integral of the left part of the equality in a surface integral and this in a line integral on the meridional section of the bubbles and the walls of the tube denoted by C, we obtain

$$\begin{aligned} &-\int_C \boldsymbol{G}^x(\boldsymbol{x}, \boldsymbol{x}_0) \cdot \boldsymbol{f}(\boldsymbol{x})\, r(\boldsymbol{x})\, dl(\boldsymbol{x}) + \int_C \mathbf{v}(\boldsymbol{x}) \cdot \boldsymbol{T}^x(\boldsymbol{x}, \boldsymbol{x}_0)\, \boldsymbol{n}(\boldsymbol{x})\, r(\boldsymbol{x})\, dl(\boldsymbol{x}) \\ &= 8\pi r_0 v_x(\boldsymbol{x}_0), \end{aligned} \tag{22}$$

where $\boldsymbol{f}(\boldsymbol{x})$, $r(\boldsymbol{x})$, and $dl(\boldsymbol{x})$ are defined in a point $\boldsymbol{x}$ on the contour C, where $r(\boldsymbol{x})$ and $dl(\boldsymbol{x})$ are, respectively, the distance from this point to the axis of symmetry and the arc element on the boundary. Similarly

$$- \int_C \boldsymbol{G}^r(\boldsymbol{x}, \boldsymbol{x}_0) \cdot \boldsymbol{f}(\boldsymbol{x})\, r(\boldsymbol{x})\, dl(\boldsymbol{x}) + \int_C \mathbf{v}(\boldsymbol{x}) \cdot \boldsymbol{T}^r(\boldsymbol{x}, \boldsymbol{x}_0)\, \boldsymbol{n}(\boldsymbol{x})\, r(\boldsymbol{x})\, dl(\boldsymbol{x})$$
$$= 8\pi r_0 v_r(\boldsymbol{x}_0). \tag{23}$$

In the derivation of Eqs. (22) and (23) we have assumed that the circumference on which the force is applied is concentrated in the volume occupied by the liquid. Otherwise, the right-hand sides of these equations would vanish. The second integral on the right side of Eqs. (22) and (23) diverges when the point $\boldsymbol{x}_0$ belongs to the contour C. A detailed calculation, by deforming the contour in the vicinity of $\boldsymbol{x}_0$ (see, for example, Pozrikidis 1992) shows that for $\boldsymbol{x}_0 \in C$,

$$\mathbf{v}_j(\boldsymbol{x}_0) = -\frac{1}{4\pi} \int_C G_k^j(\boldsymbol{x}, \boldsymbol{x}_0)\, f_k(\boldsymbol{x}) \frac{r}{r_0} dl + \frac{1}{4\pi} \int_C^{PV} v_k(\boldsymbol{x})\, T_{kl}^j(\boldsymbol{x}, \boldsymbol{x}_0)\, n_l(\boldsymbol{x}) \frac{r}{r_0} dl, \tag{24}$$

where PV indicates the principal value of the integral and subscript notation has been used, with $(j, k, l) = x$ or r, to write the equation in a more compact form.

Equation (24) is a ratio between speeds and stresses on the contour C of the surface domain occupied by the liquid. If the stresses $\boldsymbol{f}$ are known at all points of C, this equation is then employed to calculate the velocity of the liquid in C. Similarly, if $\boldsymbol{v}$ is known in C the equation will then allow for the calculation of $\boldsymbol{f}$ (except for an undetermined constant P (Pozrikidis 2002).

In the problem of bubble generation, $\mathbf{v} = 0$ on the solid surfaces wetted by the liquid. There, the liquid velocity on the bubbles is not known, however, the modified stress is obtained by using Eq. (19). Moreover, the stress given by Eq. (19) would then be known if p_{gi} is known. In this case, the solution of Eq. (24) determines the velocity of the liquid on the surfaces of the bubbles and the stresses on the solid surfaces. With p_{gi} unknown, the stress on the bubble i is the sum of a known stress $(\nabla \cdot \mathbf{n}_i - Box)\mathbf{n}_i$, and a uniform normal stress, $-p_{gi}\mathbf{n}_i$.

Given the linearity of Eq. (24), the velocity on the surfaces of the bubbles and the stress on the solid surfaces take the form

$$\mathbf{v} = \mathbf{v}^0 + \sum_i \mathbf{v}^i(p_{gi}) \quad \text{and} \quad \boldsymbol{f} = \boldsymbol{f}^0 + \sum_i \boldsymbol{f}^i(p_{gi}), \tag{25}$$

where v^0 and $\boldsymbol{f}^0$ are the velocity and the stress calculated for $p_{gi} = 0$ in Eq. (19) and v^i and $\boldsymbol{f}^i$ are the velocity and the stress calculated for $\boldsymbol{f} = 0$ on all surfaces of the bubbles except for bubbles i, where $\boldsymbol{f} = -\boldsymbol{n}_i$.

Equation (24) provides the solution of Eq. (23) in terms of the pressures of the gas in the bubbles, p_{gi}. We now need to establish appropriate equations to calculate these pressures. These equations express the conditions (9) and that the volumes of the released bubbles ($i > 1$) are constant. In terms of the velocity of the liquid, as given by Eq. (24), the time rate of change of the volume of bubble j is

$$\frac{dV_j}{dt} = 2\pi \int_{C_j} \mathbf{v} \cdot \boldsymbol{n}_j r\, dl = a_{j0} + \sum_i a_{ji} p_{gi}, \tag{26}$$

where C_j is the contour of the bubble. The coefficients a_{j0} and a_{ji} are easily calculated from the velocities $\mathbf{v}^0$ and $\mathbf{v}^i$ and satisfy the linear equations

$$a_{10} + \sum_i a_{1i} p_{gi} = Ca \quad \text{and} \quad a_{j0} + \sum_i a_{ji} p_{gi} = 0 \quad \text{for } j > 1, \tag{27}$$

from which the pressures p_{gi} can be calculated.

To complete the formulation of the problem, we must solve Eq. (5) for the surface of each bubble moving with the local velocity of the liquid. The position $\boldsymbol{x}(t)$ of a fluid particle on the surface of a bubble satisfies the equation

$$\frac{d\boldsymbol{x}}{dt} = \mathbf{v}(\boldsymbol{x}, t), \tag{28}$$

where $\mathbf{v}(\boldsymbol{x}, t)$ is the solution of Eqs. (24–27) at point $\boldsymbol{x}$ on the surface and at time t.

To solve numerically the integral equation (24), the contours of the bubbles and solid surfaces must be discretized. This is done using N_i nodes distributed at the same distance on the bubble boundary i and N_d nodes at the base of the reservoir, distributed non-uniformly in space with increasing separations from the axis of symmetry. The integrals in Eq. (24) are calculated using a Gaussian integration allocation with six points in the interval between each pair of nodes.

This surface is discretized with a finite number of nodes that move as material particles. Numerical tests conducted with different numbers of nodes show that a discretization of 120 nodes gives sufficient resolution.

A value of $\theta = 45°$ was used for the contact angle in the calculations. Numerical computations with other values of θ show that the effect of the contact angle on the volume of the bubbles is small provided that θ is smaller than about 90°.

4 Numerical Results

With the numerical method described above, we have obtained a set of important results. These results are here compared qualitatively with experiments that validate the reliability of the method.

4.1 Cylindrical Reservoir

A few numerical calculations were carried out to study the growth and detachment of a bubble in a cylindrical reservoir. Figure 3 shows a bubble which is about to detach from the base of the reservoir for $Bo = 0.2$, $Ca = 10$, and dimensionless radii of the reservoir $R = 5, 4$, and 3.5. Figure 4 shows the variation of the volume of detachment, V_f, and the aspect ratio of the bubble, Γ (defined as the ratio of L_f to the maximum diameter of the bubble), with R.

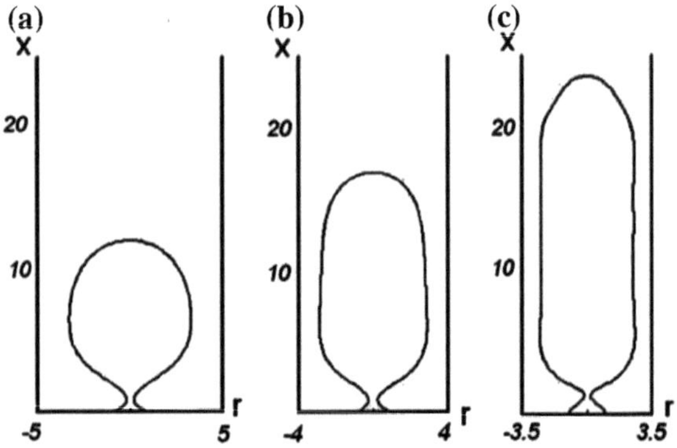

Fig. 3 Meridional sections of bubbles that are about to detach from the base of a cylindrical reservoir for $Bo = 0.2$, $Ca = 10$, and $R = 5$ **(a)**, 4 **(b)**, and 3.5 **(c)**

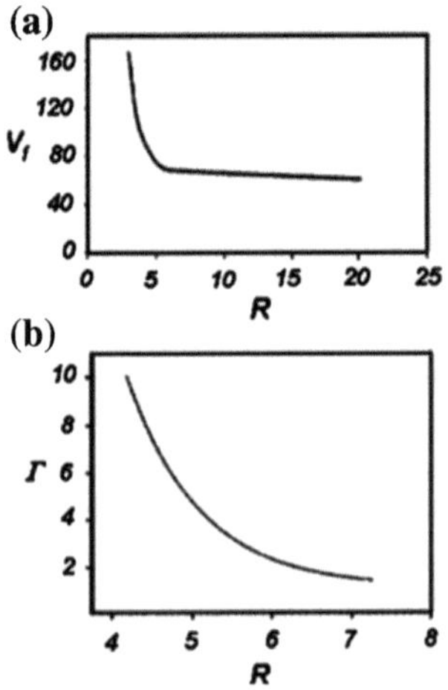

Fig. 4 **a** Volume V_f and **b** aspect ratio Γ of a detaching bubble from the base of a cylindrical reservoir as functions of R for $Bo = 0.2$ and $Ca = 10$

The finite radius of the reservoir affects only the high-flow-rate regime for the values of Bo and R used here. The decrease of Γ with increasing R in Fig. 4b is in qualitative agreement with the estimate $L_f/R \sim Ca/(BoR^4)$ for columnar bubbles. The decrease of V_f in Fig. 4a also agrees with previous estimates, according to which the ratio of the volume of a columnar bubble to the volume of a bubble detaching in an infinite reservoir is of order $(Ca/Bo)^{1/4}/R$ for Ca/Bo large compared to R^4. Figure 5 shows V_f as a function of Ca for $Bo = 0.2$ and three values of R. The nearly linear increase of V_f agrees with the estimate $V_f \sim Ca/Bo$. Note that $V_f \sim (Ca/Bo)^{3/4}$ for a bubble in an infinite reservoir. The numerical calculations also predict (results not shown here) that the centre of mass of a columnar attached bubble rises linearly with time during the growth of the bubble, and that the velocity

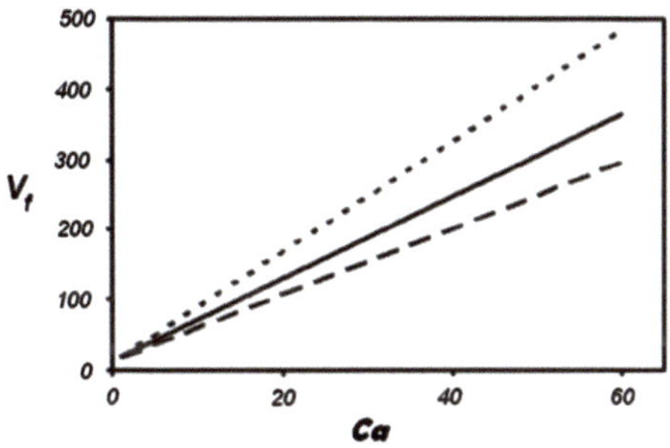

Fig. 5 Volume of a detaching bubble from the base of a cylindrical reservoir as a function of *Ca* for $Bo = 0.2$ and $R = 4$ (*dotted*), 5 (*solid*), and 6 (*dashed*)

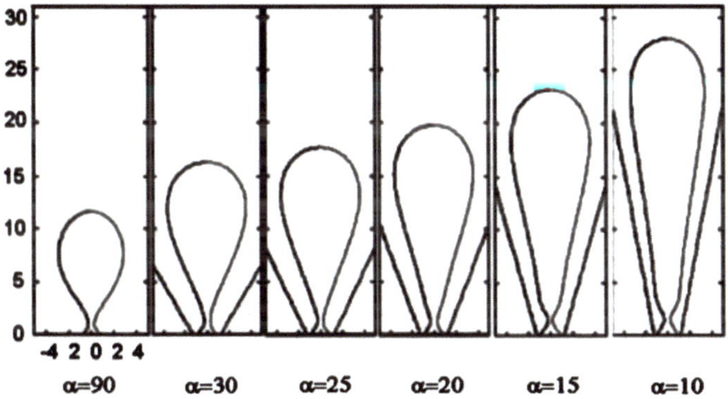

Fig. 6 Meridional sections of detaching bubbles growing from the base of conical reservoirs for $Bo = 0.2$, $Ca = 50$, $R = 1.2$, and varying values of α

of the centre of mass is nearly constant, except in the early stages of the process, when the bubble is still small compared to the radius of the tube.

4.2 Conical Reservoirs

Figure 6 shows the shape of a bubble which is about to detach from the base of a conical reservoir for $Bo = 0.2$, $Ca = 50$, $R = 1.2$, and different values of the semi-angle of the cone α, and Fig. 7 shows the volume V_f and the aspect ratio Γ of the detaching bubble as functions of α for $R = 1.2$ and different values of *Bo* and *Ca*. We see that the volume of the bubble always increases when the angle of the cone decreases, the effect being more pronounced for small values of the Bond number, for which the bubble is larger and therefore more easily affected by the wall of the reservoir. Figure 7a displays an important result of this work, namely, that at low Bond numbers and high capillary numbers the volume of the bubbles can be easily controlled through the angle of the cone without having to change the flow rate. This is a desirable aspect in some applications.

Fig. 7 **a** Volume V_f and **b** aspect ratio Γ of a detaching bubble from the base of a conical reservoir as functions of α, for $R = 1.2$ and $(Bo, Ca) = (0.2, 10)$ (*solid*), (2, 10) (*dashed*), and (2, 20) (*dotted*). The symbols in **b** show experimentally obtained values of Γ for $Bo = 0.0176$ (▲) and $Bo = 0.15$ (♦), with $Ca = 50.78$, and $R = 1.2$

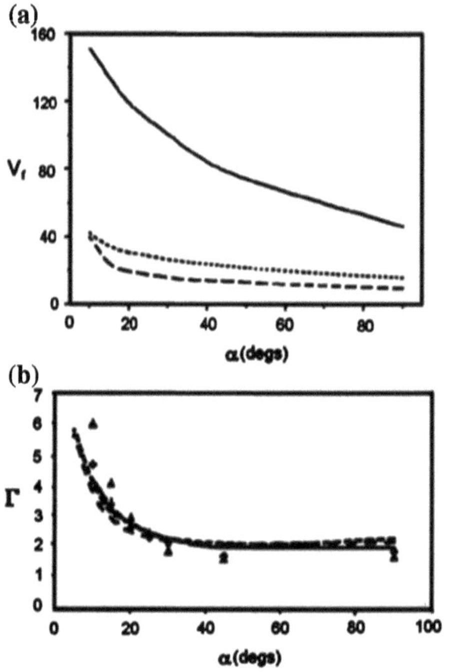

These results can be rationalized by means of a straightforward extension of the estimations of the previous section for the high-flow-rate regime in cylindrical reservoirs. Figure 6 shows that the bubbles in conical reservoirs are columnar for moderately small values of α, with a cap that increases linearly with its height above the bottom of the reservoir. (See also Fig. 7b; the bubble is slender for α smaller than about 30°).

4.3 Corrugated Pipes

For the corrugated case, the numerical calculations were mainly aimed at understanding the effects of corrugation on the bubble shape. When using different number of nodes for the tube walls, we observe that, as in the case of smooth walls, the number of nodes is not important.

Figure 8 shows the bubble shapes for different pipe corrugation wave lengths, in the viscous case, and Fig. 9 shows the same in the inviscid approximation, where $Ca = 0.1$ and $Bo = 0.2$ in Fig. 9a and $Ca = 1$ and $Bo = 0.2$ in Fig. 9b. We see that the film thickness and the bubble volume are small in the case of small capillary number.

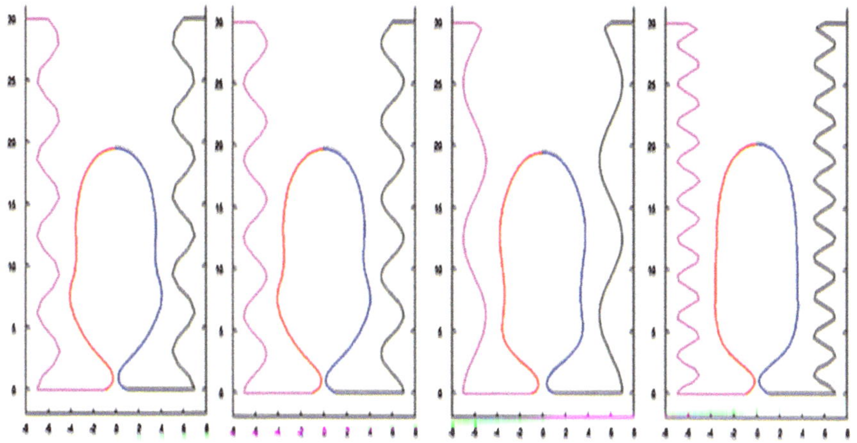

Fig. 8 Bubble shapes with the same amplitude and varying wave length of the corrugations for a viscous liquid with $Ca = 25$ and $Bo = 0.2$

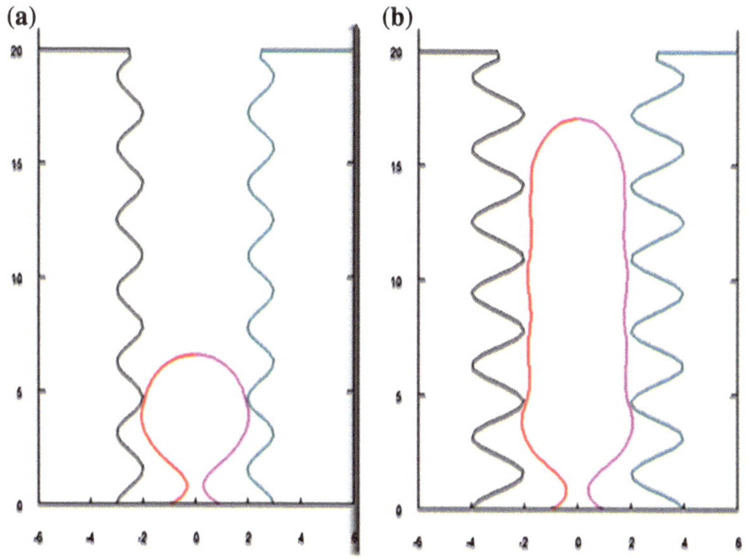

Fig. 9 Bubbles growing in a near inviscid liquid **a** $Ca = 0.1$, $Bo = 0.2$ and **b** $Ca = 1$, $Bo = 0.2$. Notice that the thickness of the film is very thin

In Figs. 10 and 11 we see the process of bubble detachment. These result may have possible applications to understand the foams formed in viscous fluids in porous media (Kovscek et al. 1995; Yan et al. 2006).

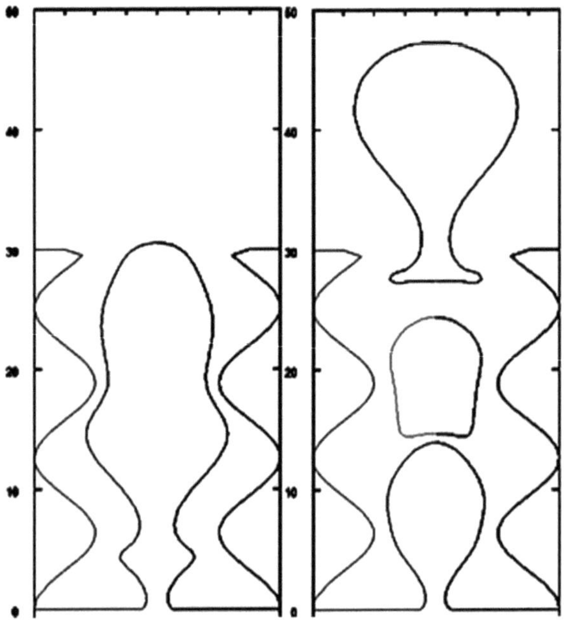

Fig. 10 Details of bubble detachment and the effect of the tube wall

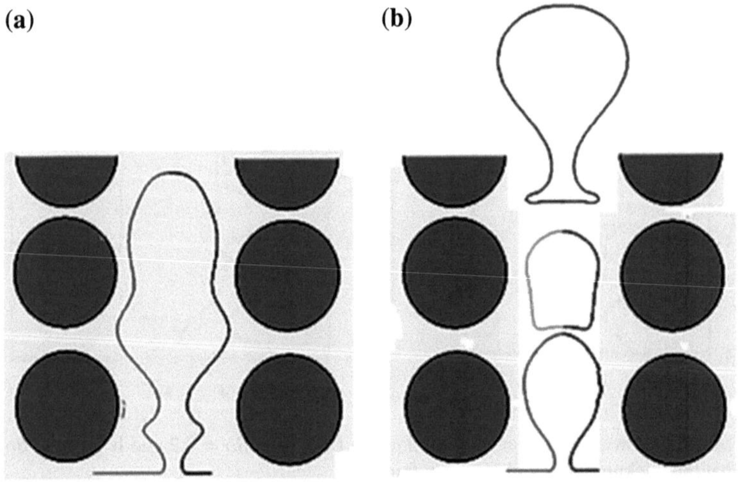

Fig. 11 Comparison with a porous medium

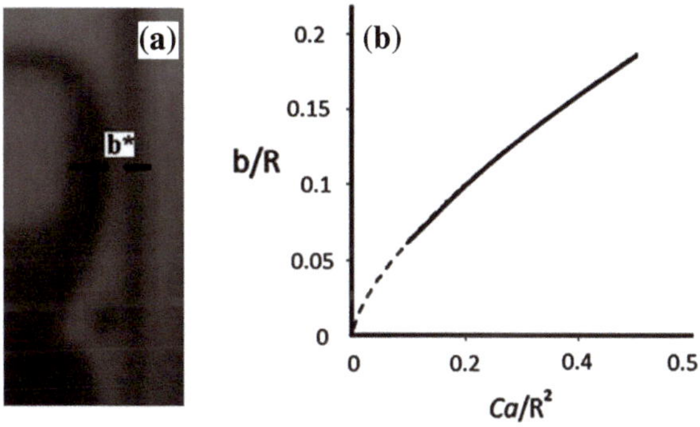

Fig. 12 **a** Bubble in a vertical tube. Notice the existence of a film of thickness b^*. **b** Plot of the scaled dimensionless film thickness *b/R* as a function of the scaled capillary number Ca/R^2

4.4 The Film Thickness

In Fig. 12a, we see that during growth of the bubble in the pipe there is an annular film of thickness b^*. In a classical study, Bretherton (1961) showed, by using the lubrication theory, that the dimensionless thickness of the film $b = b^*/a$ scales as

$$\frac{b}{R} \sim \left(\frac{Ca}{R^2}\right)^{\frac{2}{3}}, \tag{29}$$

which is valid whenever $Ca \to 0$ and R and the Bond number are also small. Figure 12b shows a numerically obtained curve that obeys this scaling. In this plot the continuous curve was obtained with our numerical solution method, while the dashed part of the curve shows the trend given by Eq. (29), which was numerically inaccessible. In spite of this, we clearly see that $b \to 0$ as $Ca \to 0$. Physically, the condition $Ca \to 0$ means that the bubble in an inviscid liquid is touching the inner solid wall.

Figure 13a shows some bubble profiles: in this case they were obtained for low capillary numbers and it is evident that the film thickness tends to zero for small values of Ca and R. It can also be seen that the profiles show some “corrugations”. This occurs because they become unstable when the height of the tube is very large compared to its radius. In our case, the height of the tube is 30 times its radius. Also, it is observed experimentally that when $Ca << 1$ the profiles of even small bubbles may become unstable.

Moreover, very different results are obtained when the film thickness is computed for highly viscous liquids, i.e., for $Ca >> 1$ in the limit of low Bond numbers. In this case, the film thickness tends to a constant value when the capillary number increases (see Fig. 13b). Figure 14 shows how $b \to$ constant for $Ca >> 1$. In dimensional terms the actual thickness of the annular film, $b* \to 1.5a^*$, i.e., the lower value of b^* is 1.5 times the radius of the gas injection orifice. Physically, this condition is attained for highly viscous liquids or at very large gas flow rates.

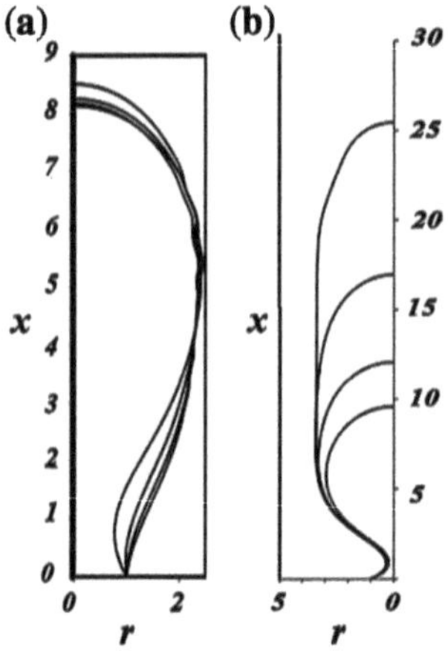

Fig. 13 Bubble shapes in cylinders filled with (**a**) liquids of low viscosity and (**b**) with highly viscous liquids. In (**a**) the height and the film thickness between the bubble and the wall diminishes when $Ca \rightarrow 0$ ($Ca = 0.4, 0.3, 0.2$, and 0.1). In (**b**) the film thickness $b \rightarrow$ constant, for $Ca >> 1$. The larger bubble corresponds to $Ca = 3$. The other cases are $Ca = 20$, *10*, and *5*. The dimensionless pipe radius is $R = 5$

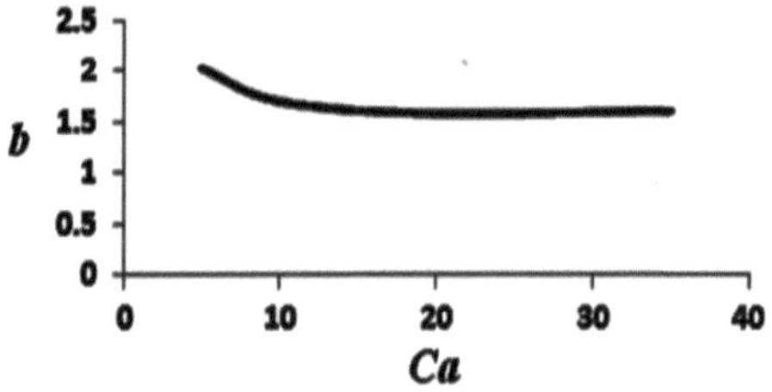

Fig. 14 Thickness of the annular film, b, as a function of the capillary number, Ca. Notice that $b \rightarrow constant$ for $Ca >> 1$.

5 Experiments

A series of experiments were carried out to study the growth and detachment of bubbles in highly viscous liquids. Glycerine and silicone oil have been used in the experiments. The properties of glycerine at 25° are: density $\varrho = 1260$ kg/m^3, viscosity $\mu = 7.9 \times 10^{-1}$ Ns/m^2, and surface tension $\sigma = 6.3 \times 10^{-2}$ N/m, while those of the silicone oil at the same temperature are: density $\varrho = 971$ kg/m^3 and viscosity $\mu = 9.71 \times 10^{-1}$ Ns/m^2. In each experiment, a large open container with a horizontal bottom was filled with liquid up to a height of 100 mm. At the bottom of the container a circular orifice of radius $a = 0.3$ mm was drilled.

A glass tube of inner radius $R^* = 3.2$ mm was positioned vertically and concentrically with the orifice to form a cylindrical reservoir. Conical reservoirs of various angles were formed by carefully inserting cones made of acetate sheet concentrically

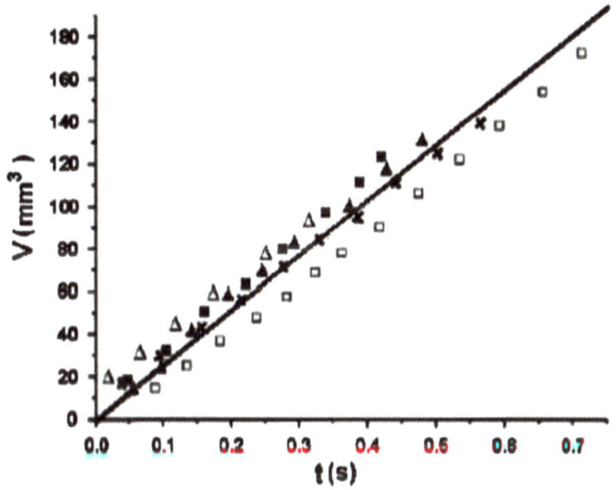

Fig. 15 Volume of a bubble attached to the base of a conical reservoir filled with glycerine as a function of time during the growth of the bubble for $Bo = 0.0176$, $R = 1.2$, and different values of the semi-angle of the cone: $\alpha = 10°$(□), 20°(x), 25°(■), 30°(▲), and 90°(△)

with the orifice. Air was pumped through a capillary tube 40 cm long and 0.6 mm of inner diameter, which ends at the orifice in the bottom of the container. We found in a previous work (Corchero et al. 2006) that a length of 40 cm suffices to make the pressure drop in the air line large compared to the pressure variations in the bubble during the growth process and therefore ensures a constant flow rate in our experiments, which was one of the premises of the numerical work. To check that the flow rate is constant, the evolution of the attached bubble was video recorded; the contour of the bubble was extracted from the video images using a standard algorithm (Russ 2002) implemented in a home-made code. The volume of the bubble, $V(t)$, and the height of its centre of mass, $x_{CM}(t)$, were computed assuming that the bubble is axi-symmetric. Some sample plots of V as a function of time for a bubble growing in glycerine within conical reservoirs of various angles are shown in Fig. 15. The approximate linear variation of V with time shows that the flow rate is nearly constant and independent of the angle of the cone. The value of the flow rate determined by fitting a straight line to the experimental data of Fig. 15 is $Q = 364.5$ mm^3/s. The same procedure was used to measure the flow rate of air injected into silicone oil and in cylindrical containers. The flow rate was found to be nearly constant in all cases.

In addition, Fig. 16 shows the shapes of bubbles in glycerine that are about to detach from the injection orifice in conical reservoirs of various angles. Here $\alpha = 90°$ corresponds to a bubble detaching in an infinite reservoir. The shape of the bubbles begins to differ significantly from this case when α becomes smaller than about 30°. Coalescence between previously detached bubbles can be seen in some of the images. The presence of the conical wall of the reservoir increases the drag of the ascending bubbles, decreasing their velocity and apparently promoting coalescence. We plan to analyze this important aspect of the generation of bubbles in a future work.

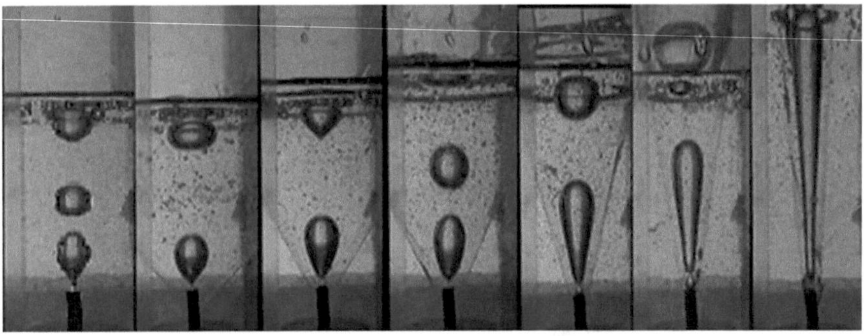

Fig. 16 Growth of bubbles at different angles of inclination α

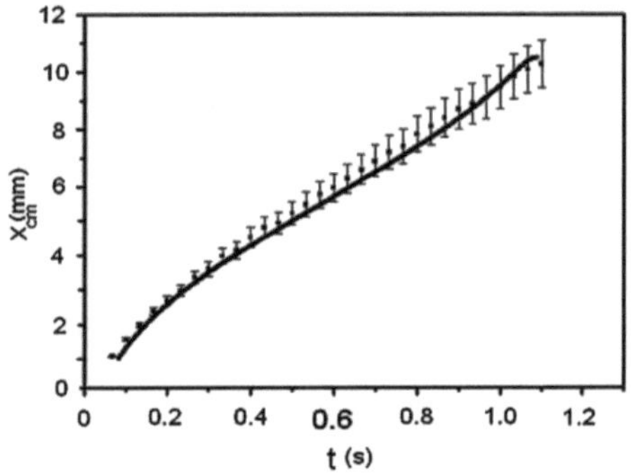

Fig. 17 Centre of mass of a bubble attached to the base of a conical reservoir as a function of time during the growth of the bubble for $Bo = 0.04$, $Ca = 70.23$, $R = 1.2$, and $\alpha = 15°$. Symbols are experimental results and the solid curve shows the results of the numerical computation

The gas flow rate in this sequence of experiments is that measured from Fig. 15. Values of the dimensionless parameters are $Bo = 0.0176$, $Ca = 50.78$, and $R = 1.2$.

The aspect ratio Γ of the bubbles in Fig. 16 and in other companion experiments was also extracted from the images and included in Fig. 7b (triangles and diamonds), where it is compared to the numerical results obtained for similar dimensionless parameters. The comparison is reasonably good, though the experimental values of Γ increase with decreasing α slightly faster than predicted numerically, and become larger for small values of α. We think that the difference is due to the vertical momentum carried by the injected gas, which was not taken into account in the numerical calculations. The evolution of the centre of mass of the bubble is shown in Fig. 17, where it is also compared to the numerical results. The nearly linear increase of x_{CM} with time must be compared to the $x_{CM} \propto t^{1/3}$ evolution expected for a round bubble growing in an infinite reservoir (Davidson and Schuler 1960). The difference clearly shows the effect of the conical wall.

In the experiments with cylindrical reservoirs only silicone oil was used because glycerine tends to produce small bubbles that linger in the reservoir for a long time and

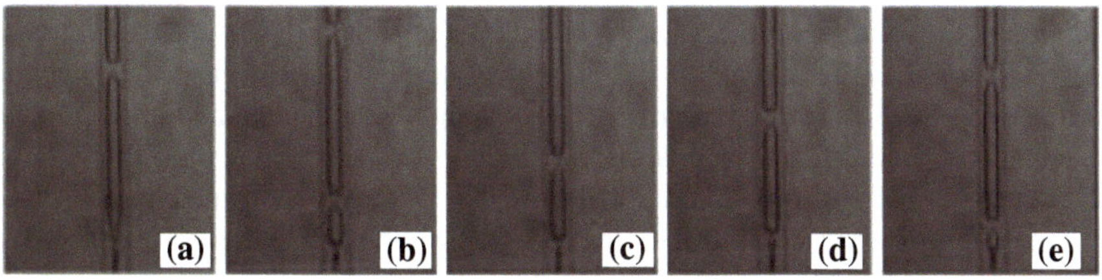

Fig. 18 Five equispaced images spanning the period of growth of a bubble attached to the base of a cylindrical reservoir for $Bo = 0.04$, $Ca = 209.94$, and $R = 10.66$. The period of bubbling is 1.33 s

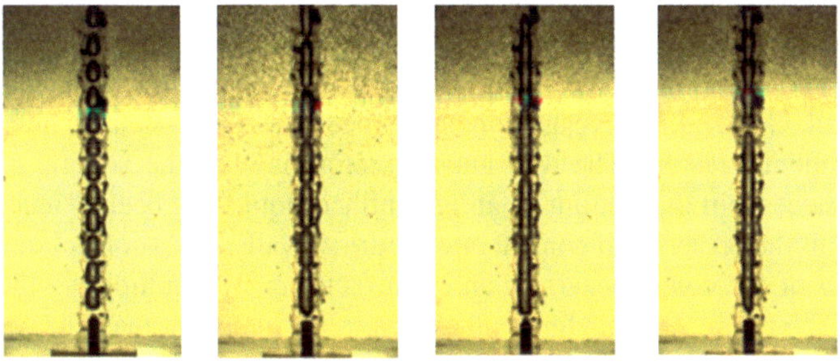

Fig. 19 Dimension of profile bubbles formed inside a tube with periodic corrugations $c = 7.33$, dimensionless radius $R^* = 3.7$, $Bo = 0.2$, and different capillary numbers

interfere with the observation of the bubble attached to the orifice. Figure 18 shows five equispaced images in time that span the cycle of growth and detachment of a bubble. In this experiment, the flow rate of gas measured from the video recording is $Q = 419.59$ mm^3/s and the period of bubbling is 1.33 s. Values of the dimensionless parameters are $Bo = 0.04$, $Ca = 209.94$, and $R = 10.66$. Figure 19 shows profile bubbles formed inside a tube with periodic corrugations $c = 7.33$, dimensionless radius $R^* = 3.7$, $Bo = 0.2$, and different capillary numbers.

6 Conclusions

The growth of a bubble due to the injection of gas at a constant flow rate through an orifice in the horizontal base of a container filled with highly viscous liquids has been investigated numerically and experimentally for the case in which nearby solid walls partially confine the space where the bubble is allowed to grow. Conical, cylindrical, and corrugated cylindrical walls coaxial with the injection orifice have been used in order to allow for an easy control of the extent of the confinement by simply changing the angle of the cone, the radius of the cylinder, or the frequency in the corrugated case.

Numerical solutions of the Navier-Stokes equations for the liquid and the evolution equation for the free surface of the bubble show that the wall near the injection orifice may have an important effect on the shape of the bubble and its volume at detachment. Calculations for small Bond numbers ($Bo = 0.2$) and moderately large capillary numbers (of the order of 10) show that vertically elongated bubbles with volumes significantly larger than those of the round bubbles, generated in the absence of walls, are obtained when the radius of the cylindrical wall is smaller than about six times the radius of the orifice, or when the semi-angle of the cone is smaller than about 30°. The calculated distributions of forces on the surface of the bubble and the wall suggest that buoyancy, viscous drag, and viscous friction with the wall all play a role in the dynamics of the bubbles.

Experiments have been carried out with two different viscous liquids to explore wide ranges of the Bond and capillary numbers by keeping the inertial effects of the liquid small. Good qualitative agreement has been found between the numerical and the experimental results. The well-known scaling law for the volume of a bubble at detachment from the bottom of an unconfined liquid has been extended to take into account the presence of conical or cylindrical walls. For a conical reservoir, the semi-angle of the cone appears as an extra factor $\alpha^{-1/4}$ multiplying the standard $(Ca/Bo)^{3/4}$ scaling. For a cylindrical reservoir, the exponent may change from 3/4 to 1 when the radius of the cylinder decreases.

The results of this work may have implications for the methods of enhanced oil recovery, where properly shape injection nozzles may help optimizing the volume of the bubbles generated in oil production pipes without changing the flow rate of the gas or the injection of foams in both homogeneous and fractured reservoirs.

Acknowledgments We acknowledge the I.P.N. for partial support through projects SIP20131821 and SIP20131821-IPN. We also acknowledge the CONACyT for partial support through the project SENER-CONACyT 146735.

Appendix A. The Green's functions for Axi-symmetric flow.

This Appendix lists Green functions for axi-symmetric flow generated by a unit ring force located at (x_0, r_0) and pointing in the direction $\mathbf{e}_\alpha$ with $\alpha = r, x$. Define

$$
\begin{aligned}
Z &= x - x_0,\\
L &= \sqrt{Z^2 + (r + r_0)},\\
D &= \sqrt{Z^2 + \sqrt{Z^2 + (r + r_0)}},\\
S &= \sqrt{Z^2 + r^2 + r_0^2},\\
m &= \frac{2(r r_0)^{\frac{1}{2}}}{L},
\end{aligned}
$$

and the elliptic integrals

$$K(m) = \int_0^{\frac{\pi}{2}} \frac{d\theta}{\sqrt{1 - m^2 sen^2\theta}},$$
$$E(m) = \int_0^{\frac{\pi}{2}} \sqrt{1 - m^2 sen^2\theta d\theta}$$

Then

$$G_x^x = 4\frac{r}{L}\left(K + E\frac{Z^2}{D^2}\right),$$
$$G_r^x = 2\frac{Z}{L}\left(K - E\frac{S^2 - r^2}{D^2}\right),$$
$$G_r^x = 2\frac{rZ}{r_0 L}\left(-K + E\frac{S^2 - 2r_0^2}{D^2}\right),$$
$$G_r^r = 2\frac{1}{r_0 L}\left[k\left(S^2 + Z^2\right) - E\left(L^2 + \frac{Z^2 S^2}{D^2}\right)\right],$$
$$T_{x\,x}^x = 8\frac{rZ^3}{D^2}\left(K - E\frac{4S^2}{D^2}\right),$$
$$T_{x\,r}^x = T_r^{x_x} = -4\frac{Z^2}{D^2 L}\left[K\frac{S^2 - 2r^2}{L^2} - E\left(1 + \frac{8r_0^2\left(2r_0^2 - S^2\right)}{D^2 L^2}\right)\right],$$
$$T_{r\,r}^x = -4\frac{rZ}{L}\left[K\left(\frac{1}{r^2} + \frac{2Z^2}{D^2 L^2}\right) - \frac{E}{D^2}\left(6 - S^2\left(\frac{1}{r^2} + \frac{8Z^2}{D^2 L^2}\right)\right)\right],$$
$$T_{x\,x}^r = -4\frac{rZ^2}{r_0 D^2 L}\left[K\frac{2r_0^2 - S^2}{L^2} + E\left(1 + \frac{8r_0^2\left(2r^2 - S^2\right)}{D^2 L^2}\right)\right],$$
$$T_{x\,r}^r = T_x^{r_r} = -4\frac{Z}{r_0}\left[K\left(\frac{Z^2 S^2}{D^2 L^2} - 2\right) + \frac{E}{D^2}\left(2S^2 - Z^2 - \frac{16r^2 r_0^2 Z^2}{D^2 L^2}\right)\right],$$
$$T_{r\,r}^r = -4\frac{r}{L}\left[\frac{K}{r_0}\left(\frac{Z^2\left(S^2 - 2r_0^2\right)}{D^2 L^2} - \frac{r^2 - r_0^2 - 2Z^2}{r^2}\right)\right.$$
$$\left.+\frac{E}{D^2}\left(\frac{8r_0 Z^2\left(S^2 - 2r^2\right)}{D^2 L^2} + \frac{r^2\left(r^2 + r_0^2\right) - S^2\left(r_0^2 + 2Z^2\right)}{r^2 r_0}\right)\right].$$

References

Ajaev VS, Homsy GM (2006) Modeling shapes and dynamics of confined bubbles. Annu Rev Fluid Mech 38:277–307

Bretherton FP (1961) The motion of long bubbles in tubes. J. Fluid Mech. 10:166–188

Cheng AH-D, Cheng DT (2005) Heritage and early history of the boundary element method. Eng Anal Boundary Elem 29:268–302

Corchero G, Medina A, Higuera FJ (2006) Effect of wetting conditions and flow rate on bubble formation at orifices submerged in water. Colloids Surf A Physicochem Eng Aspects 290:41–49

Costabel M (1986) Principles of boundary element methods. In: Lectures given at the first graduate summer course in computational physics finite elements in physics lausanne 1–10 Sept 1986, p 28

Davidson JF, Schuler BOG (1960) Bubble formation at an orifice in a viscous liquid. Trans Inst Chem Eng 38:144–154

Higuera FJ (2005) Injection and coalescence of bubbles in a very viscous liquid. J Fluid Mech 530:369–378

Hirasaki GJ, Lawson JB (1985) Mechanisms of foam flow in porous media: apparent viscosity in smooth capillaries. SPE J 25:176–190

Kovscek AR, Tretheway DC, Persoff P, Radke CJ (1995) Foam flow through a transparent rough-walled rock fracture. J Petrol Sci Eng 13:75–86

López-Villa A, Medina A, Higuera FJ (2011) Bubble growth by injection of gas into viscous liquids in cylindrical and conical tubes. Phys Fluids 23:102102

Pozrikidis C (1992) Boundary integral and singularity methods for linearized viscous flows. Cambridge University Press, Cambridge

Pozrikidis C (1997) Introduction to theoretical and computational fluid dynamics. Oxford University Press, Oxford

Pozrikidis C (2002) A practical guide to boundary element methods with the software library BEMLIB. Chapman & Hall/CRC, Boca Raton

Russ JC (2002) The Image processing handbook. CRC Press, Boca Raton

Wong H, Rumschitzki D, Maldarelli C (1998) Theory and experiment on the low-Reynolds-number expansion and contraction of a bubble pinned at a submerged tube tip. J Fluid Mech 356:93–124

Yan W, Miller CA, Hirasaki GJ (2006) Foam sweep in fractures for enhanced oil recovery. Colloids Surf A Physicochem Eng Aspects 282–283:348–359

Theoretical Physics of Granular Fluids and Solids

Leonardo Trujillo and Leonardo Di G. Sigalotti

Abstract Here we present a brief introduction to some theoretical ideas for granular matter. We start by reviewing the physical properties and constraints of granular materials. We then outline some approaches towards a thermodynamics for granular materials. We analyze the grain flow as a fluid mechanical phenomenon, with a brief introduction to the kinetic theory of inelastically colliding hard particles. We present a nonlinear theory of elasticity for granular solids. Finally, we briefly discuss the problem of formulating continuous field equations in discrete particulate systems and non–local constitutive relations.

1 Introduction

One grain of sand is a solid, but a lot of grains together can behave like a solid or a liquid. Their physical behaviour involves complex nonlinear phenomena, including non-equilibrium static configurations, energy dissipation, nonlinear elastic response, and peculiar flow dynamics (Guyon et al. 1990; Nagel 1992; Mehta and Barker 1994; Jaeger et al. 1996; de Gennes 1998, 1999; Kadanoff 1999; Rajchenbach 2000; Herrmann 2002; Aranson and Tsimring 2006). Classical examples are sand, powders, sugar, salt, and gravel. No limitations are imposed on the size of the particles, which may range from the nanometer scale, as in pigments and aerosols, to the scale of

L. Trujillo (✉) · L. Di G. Sigalotti
Centro de Física, Instituto Venezolano de Investigaciones Científicas, IVIC,
Apartado Postal 20632, Caracas 1020-A, Venezuela
e-mail: leonardo.trujillo@gmail.com

L. Di G. Sigalotti
e-mail: leonardo.sigalotti@gmail.com

L. Trujillo
The Abdus Salam International Centre for Theoretical Physics, ICTP,
Strada Costiera 11, 34014 Trieste, Italy

L. Di G. Sigalotti et al. (eds.), *Computational and Experimental Fluid Mechanics with Applications to Physics, Engineering and the Environment*, Environmental Science and Engineering,
DOI:10.1007/978-3-319-00191-3_8,

mined or quarried materials and rocks. The mechanical response of granular packings to external perturbations plays a major role in numerous scientific endeavours (such as soil mechanics and geophysics), as well as in industry (oil exploration, structural stability, product formulation in pharmacology and domestics, granular composites, heterogeneous materials, etc.) (Ottino and Khakhar 2001). Investigations in granular materials gather teams from civil, chemical and mechanical engineers, food technologists, powder metallurgists, materials scientists, pharmaceutical scientists, applied mathematicians, and physicists. Beyond their practical importance, granular materials interest physicists because they are an unusual form of matter with interesting properties that are not yet fully understood. So far, there is no accepted set of universal governing equations describing the physics of granular materials (Kadanoff 1999; Aranson and Tsimring 2006). Important contributions have been made to improve our understanding of many new aspects using modern tools from statistical dynamics to fluid mechanics. Foundational issues in granular materials are offering a great challenge for theoretical physicists. On the other hand, a precise model for granular dynamics would be particularly useful for optimizing industrial devices and processes. In addition, a mathematical model for granular materials, expressed in the form of governing equations with proper boundary conditions and/or initial conditions would be very useful for computational simulations of complex practical problems in science and engineering.

The question "which are the theoretical foundations of granular physics?" is not a mere philosophical one because it is not well understood how to formulate concisely a chain of logical arguments that can reveal what are the basic assumptions, how a theory for granular materials can be constructed part-by-part, what are its strengths and limitations, how *many* theoretical scenarios one can imagine for the gas, liquid or solid state coexistences, plastic flow, and glassy behaviour, and how, at least in principle, accurate calculations can be performed to decide unambiguously how the initial configurations evolve in time.

Here we present a brief introduction to some theoretical ideas for granular matter. In Sect. 2 some characteristic features, complexities, and constraints about granular materials are enumerated. The theoretical landscape is outlined through three main subjects: *granular thermodynamics* (Sect. 3); *granular hydrodynamics* (Sect. 4); and *granular elasticity* (Sect. 5). In Sect. 6, we address the problem of a formal justification of continuous matter fields for particulate discrete systems and constitutive relations. Section 7 summarizes the conclusions.

2 Some Features on Granular Materials

What are granular materials? They are materials made of sets of discrete macroscopic solid particles (granules). In order to characterize a granular material let us first consider the following attributes and physical constraints (Guyon et al. 1990; Nagel 1992; Mehta and Barker 1994; Jaeger et al. 1996; de Gennes 1998, 1999; Kadanoff 1999; Rajchenbach 2000; Herrmann 2002; Aranson and Tsimring 2006):

- *Size*: The lower size limit for grains in granular materials is about one micron (de Gennes 1998, 1999). However, it is believed that the physics of granular media can also be applied to much bigger objects. An interesting example are the rings of Saturn, which are composed of a myriad of individual particles that continually collide (Esposito 2010). The size of Saturn ring particles may extend over kilometers, i.e., from fine dust to embedded moonlets (Borderies et al. 1985). A sand grain (the archetypal granular particle) is of the order of 10^8 times more massive and voluminous than, say, a water molecule. This difference is not a particularly fundamental one as far as the microscopic description of particles motion is concerned since both types of particles can be treated according to the laws of classical mechanics. However, grain size has an important bearing on the applicability of the continuum hypothesis as will be discussed below. In the context of granular materials the size matters because the size of a grain affects the entire static or dynamic collective behaviour. Perhaps the best known of these effects is the so-called Brazil nut effect, whereby larger particles placed into a mechanically agitated system rise to the surface rather than sinking to the bottom (Kudrolli 2004).
- *Shape*: Real granular systems are composed of heterogeneous constituents with a rather intricate morphology, such as rocks, rice, flakes, etc. Sand is composed of finely divided rock and mineral particles, which under the effect of the air flow their shapes become well rounded as, for example, in sand dunes. Up to now, the mechanical response of a granular material and the shape of its constituents remain an open problem, making systematic studies infeasible. Idealized models for granular media consist of a packing formed by spherical particles in three dimensions (3D) or disks in two dimensions (2D). While the former model allows for efficient simulations of large systems, it is often insufficient when the static behaviour of such systems is considered. For example, results of experiments on a 2D model system, made of discrete square cells submitted to a point load, have shown that the collective response of the pile contradicts the standard elastic predictions and supports a diffusive description of stress transmission (Da Silva and Rajchenbach 2000). On the other hand, in vibro-fluidized systems complex collective behaviours and patterns emerge when the interactions between grains have shape anisotropy as occurs, for example, in a system of vertically vibrated rods, which can range from a nematic-like gas phase to single and multiple rotating vortices (Aranson and Tsimring 2006).
- *Many-body classical system:* Granular materials are macroscopic systems made of a large number of interacting particles. The term macroscopic implies that classical mechanics must be employed to provide an accurate description of the system. The dynamics of a granular material is governed by Newton's equations of motion for the centre-of-mass coordinates $\mathbf{r}_i$ and Euler angles ϕ_i of its constituent particles i (with $i = 1, \ldots, N$):

$$m_i \frac{\partial^2 \mathbf{r}_i}{\partial t^2} = \mathbf{F}_i(\mathbf{r}_j, \mathbf{v}_j, \phi_j, \omega_j),$$

$$\hat{I}_i \frac{\partial^2 \phi_i}{\partial t^2} = \mathbf{M}_i(\mathbf{r}_j, \mathbf{v}_j, \phi_j, \omega_j), \quad (j = 1, \ldots, N), \tag{1}$$

where $\mathbf{F}_i$ and $\mathbf{M}_i$ are, respectively, the force and torque acting on particle i of mass m_i and tensorial moment of inertia $\hat{I}_i$, $\mathbf{v}_j$ are the linear velocities, and ω_j denote the angular velocities.

- *Non-thermal system:* Granular materials can be used to explore the non-thermal and non-equilibrium behaviour found in microscopic systems because they are not affected by thermal fluctuations and typically exist far from equilibrium. Under suitable conditions, a non-cohesive granular system can be maintained in a gas-like state of colliding particles by continuous vibration of the container. This picture has led several authors to use kinetic theory and thermodynamical concepts to describe granular media. In contrast to a molecular gas in equilibrium, where the mean kinetic energy (mean-square velocity) of a molecule is proportional to the gas temperature (thermodynamic temperature), the natural equilibrium state of a granular material is a static configuration due to the inelastic nature of particle collisions. Therefore, a stationary fluidized state needs a constant flow of energy into the system. This steady state driven by the energy flux can be assumed as a condition of thermal equilibrium. On the other hand, in analogy to a molecular gas, we can define a granular temperature T_g, which is proportional to the mean kinetic energy E of the particles' velocity (Goldhirsch 2008). Let us remark that this generalized notion of temperature is introduced for pure theoretical convenience to take advantage of the existing thermodynamical analogy. In fact, the definition of thermodynamic variables for non-equilibrium states is theoretically straightforward. The thermodynamics of non-equilibrium states has always been a matter of debate. Alternatives for a thermodynamic formulation for granular materials have been proposed by Edwards and Oakeshott (1989), Herrmann (1993), and Hong and Hayakawa (1997). Certainly, the problem of a thermodynamic formulation for granular materials deserves a deeper analysis (cf. Kadanoff 1999)

Here we can anticipate that granular materials represent a good framework to explore the frontiers of many-body classical mechanics. However, at the lower limit the problem of solving the dynamics of a mechanical system becomes intractable for $N \geq 3$ (c.f., Poincaré's three-body problem). This is one reason to consider pairwise interactions between particles in computational simulations based on Molecular Dynamics techniques, thus limiting the study of multiple collisions and contacts. Therefore, we need to resort to a statistical approach of the problem. However, statistical mechanics is well understood for systems where the thermodynamical limit is satisfied (i.e., when the number of particles is around the Avogrado's number $N \sim 10^{23}$). On the other hand, classical statistical mechanics is founded on the postulate of equal *a priori* probability which is valid for systems that are in thermodynamic equilibrium. Both conditions are not satisfied by granular materials. An inevitable consequence of these properties is a plethora of theoretical complexities, where some methodological questions arise: What should a theory for granular physics deliver? and What would be the postulates for such a theory? At least it seems that a statistical description cannot be avoided. Classical statistical mechanics addresses this via ensemble averaging. Given the discrete nature of granular materials an alternative approach is to implement space-time averaging directly, starting with an equation governing the motion of a centre-of-mass point in an inertial frame (Murdoch and Bedeaux 1994). On the other hand, continuum field theories have been developed to model reproducible macroscopic behaviour. Such theories involve fields whose values can be related to those of experimental measurements. As a granular system implies a discrete distribution of matter and experiments inherently involves probing a system at specific scales of length and time (e.g., in a monochrome CCD camera data are limited by the maximum resolution of pixels and maximum recording rate), it is clearly of interest to link field values of continuum mechanics with space-time averages of fundamental discrete entities.

Where is located the physics of granular media among the research fields in physics? Granular physics is a very old subject dating back to the 17th century. The work of Coulomb, published in 1776, is considered to be one of the earliest scientific studies of granular media. His report: *"Essai sur une application des règles de Maxims et Minimins à quelques problèmes de statique relatifs à l'architecture"*, published in *Mémoires de Mathématique et de Physique de l'Academie Royale des Sciences*, pages 343–382, addresses the issue of the origin of the static angle of repose, defined as the maximum angle that a sandpile can build up while remaining stable against gravity-driven avalanches. Once the sample begins to flow, the moving sand expands under shear so that the grains move out of each other's way. This dilatancy was noted by Reynolds in 1885 in his work *"On the Dilatancy of Media Composed of Rigid Particles in Contact"*, published in the *Philosophical Magazine*. Another major contribution was made by Bagnold in his 1941 classical book: *"The Physics of Blown Sand and Desert Dunes"*, which remains a basic reference even 72 years after its publication. However, the study of granular materials has been ignored by physicists until 1987 when the papers: *"Self Organized Criticality: A Universal Explanation for 1/f Noise"* in *Physical Review Letters* by Bak, Tang and Wiesenfeld (see Nagel (1992) for a detailed account of this problem), and *"Why the Brazil Nuts are on Top: Size Segregation of Particulate Matter by Shaking"* by Rosato and collaborators (Kudrolli (2004) presents a review about the segregation problem in granular materials) appeared. These two papers *were* at least partly responsible for the subsequent wealth of attention paid by physicists to granular systems, inspiring experimental and theoretical investigations of the fascinating nonlinear dynamics exhibited by these deceptively simple seemingly systems. The importance of fundamental research in granular materials has been appreciated very early by engineers but physicists have joined in more recently. Frequently, granular matter is considered to be a new type of condensed matter, as fundamental as liquids and solids. An inspiring statement by P.G. de Gennes was *"Granular matter, in 1998, is at the level of solid state physics in 1930"* (de Gennes 1998, 1999).

Now the physics of granular materials is broad and interdisciplinary in scope, focusing on collective phenomena of many-body systems, with statistical physics and nonlinear dynamics as the central foundations. On the other hand, the contemporary field of "ill"-condensed matter physics: Soft Matter, includes granular materials together with colloids, complex fluids, liquid crystals, and polymers. Granular physics also covers fluid dynamics, rheology, and tribology. Experiments in granular physics range from easily accessible table-top experiments to studies in geomorphology to astronomical observations.

3 New Kinds of Thermodynamics

A first step towards a mathematical theory for granular materials is the quest to answer the following question: Can one construct a thermodynamics for compact, slowly moving grains? Thermal energy is unimportant for the dynamics of granular media. Therefore, the concept of "temperature" could be superfluous. However, there are a number of contexts in which a generalized notion of temperature is important. Many attempts have been made to extend the concepts of thermodynamics to non-equilibrium systems—such as those exhibiting spatio-temporal chaos or weak turbulence and aging such as glasses (Cugliandolo et al. 1997). In this context we can define an "effective temperature" for stationary non-equilibrium systems. As we have argued above, in vibro-fluidized systems an analogous "effective temperature"—the granular temperature—(which is proportional to the mean kinetic energy of the particles velocity) may also be defined for granular materials.

Since 1989, Edwards and co-workers (e.g., Edwards and Oakeshott 1989) have proposed an alternative description of granular materials in complete analogy with statistical mechanics. In their formulation, the granular temperature does not appear. The basic assumption is that the volume V of a powder has the role played by the energy of a classical statistical system. Thus, instead of the Hamiltonian of the system, they introduced a function W, which specifies the volume of the system in terms of the positions of the individual grains. Then, physical observables are obtained by averaging over the usual equilibrium distribution at the corresponding volume, energy, etc., but restricting the sum to the *blocked configurations* defined as those in which every grain is unable to move. This definition leads immediately to an *entropy* S_{edw} given by the logarithm of the number of blocked configurations of given volume:

$$S_{edw} := \lambda \ln \left[\int \delta(V - W) \mathrm{d}(\text{all configurations}) \right], \tag{2}$$

where δ is the Dirac delta function and λ is the analog of the Boltzmann's constant. Associated with this entropy are the state variables such as the "compactivity":

$$X^{-1} := \frac{\partial}{\partial V} S_{edw}(V), \tag{3}$$

and the *Edwards temperature*, T_{edw}:

$$T_{edw}^{-1} := \frac{\partial}{\partial E} S_{edw}(E). \tag{4}$$

The compactivity characterizes the static system, i.e., in the close-packed limit $X = 0$ and in the limit of low density $X = \infty$.

Another proposal of thermodynamics for moving granular materials was introduced by Herrmann (1993). As the starting point, the conservation of energy was modified to include the effects due to dissipation by the interaction between colliding inelastic grains at a mesoscopic level (first law):

$$\Delta I = \Delta E_{int} + \Delta D + \Delta W, \tag{5}$$

where, for a given time interval, ΔE_{int} is the variation of the internal energy in a thermodynamical sense due to variations in the kinetic and potential energy, ΔD is the energy dissipated by the system, and ΔI is the energy that was injected to the system in order to maintain a stationary energy flux state. The variation ΔI is related to some kind of work performed on the system as, for example, gravity on a inclined plane, or on the plate of a mechanical shaker, and ΔW is the work done to allow changes of (bulk) volume in the system. The excess of dissipated energy is given by

$$\Delta \mathscr{D} := \Delta D - \Delta I. \tag{6}$$

This excess of dissipated energy can be associated to the internal pressure acting on collisions as a form of compression $\mathscr{P}$ and their conjugated variable $\mathscr{C}$:

$$\delta \mathscr{D} = \mathscr{P} \delta \mathscr{C}. \tag{7}$$

The quantity $\mathscr{C}$ was named "contactopy", in analogy to the entropy, and has dimensions of volume (contact volume). In this sense, the contactopy is similar to the Edwards' approach. From the point of view of classical mechanics, this contactopy is congruent with the "dissipation function" introduced by Lord Rayleigh to describe dissipational effects in dynamical systems. For instance, an analog to the second law of thermodynamics is that *any change of state at constant internal energy E_{int} should decrease the contactopy*:

$$\Delta \mathscr{C} \leq 0. \tag{8}$$

In addition, a granular potential can be defined as

$$G := E_{int} + \mathscr{P}\mathscr{C}. \tag{9}$$

Therefore, the definition for granular temperature (at constant compression) is given by

$$T_g = \frac{\partial G}{\partial S}, \tag{10}$$

where S is the Boltzmann's entropy. The granular temperature controls the granular canonical ensemble with the granular free energy F_g, defined as the Legendre transformation of the granular potential:

$$F_g = G - T_g S. \tag{11}$$

In equilibrium F_g should have a minimum.

An analogy between the configurational statistics of weakly excited 2D granular materials in a vibrating bed and the Fermi statistics was done by Hong and Hayakawa (1997). They defined the entropy as $S = \ln W$, with W being the total number of ways of arranging N particles into a system. Since excluded volume grains are not allowed to occupy the same states, the statistics is given by a Fermi-like statistics:

$$W = \prod_i \left[\frac{\Omega!}{N_i!(\Omega - N_i)!} \right], \tag{12}$$

where N_i is the number of particles at the i-th row and Ω is the degeneracy of each row (i.e., $\Omega = L/d$, where L is the width of the container and d the diameter of the grain). The density profile as a function of the height z was calculated via the maximization of S, giving a Fermi-like distribution

$$\phi(z) = \frac{N_i}{\Omega} = \frac{1}{(1 + \exp[\beta(z - \mu)])}, \tag{13}$$

where $\beta \to mgd/T$ at the low temperature limit and μ is the Fermi-like energy measured in units of d and given by the initial number of layers measured from the bottom.

In the context of granular elasticity Jiang and Liu (2003) introduced the elastic free energy potential

$$f = \frac{\rho}{m} f_1 + \frac{1}{2} K_b u_{nn}^2 + K_a u_{kl}^0 u_{kl}^0 + \rho g z, \tag{14}$$

where K_b and K_a are the constant compressional and shear moduli, respectively, u_{nn} is the trace of the strain tensor u_{ij}, u_{ij}^0 is its traceless part, i.e., $u_{ij}^0 = u_{ij} - u_{nn}\delta_{ij}/3$, g is the standard value of Earth's gravitational acceleration, and z denotes the height coordinate. The bulk and shear elastic moduli are given by

$$K_b = \widetilde{K}_b \delta^b, \; K_a = \widetilde{K}_a \delta^a, \tag{15}$$

with $\widetilde{K}_b, \widetilde{K}_a > 0$ for $\delta \geq 0$ and $\widetilde{K}_b, \widetilde{K}_a = 0$ for $\delta = 0$ so that the elastic moduli remain finite. The exponents a and b are related to the type of contact between the grains. That is, when $a = b = 0$ linear elasticity is recovered, whereas $a = b = 1/2$ implies Hertz contacts. This formulation provides a much better approximation to granular elastic behaviour in which we can specify any type of contact by suitably choosing the exponents a and b. Using the Jiang-Liu model for granular elasticity we can calculate a stability criterion for the granular packing and yield. In this way, the stability criterion demands that the elastic energy must be a convex function of the state variables, i.e., $\partial f/\partial\delta < 0$, and satisfy the cross convexity condition

$$\left(\frac{\partial^2 f}{\partial \delta^2}\right)\left(\frac{\partial^2 f}{\partial u_s^2}\right) \leq \left(\frac{\partial^2 f}{\partial \delta \partial u_s}\right)^2,$$

with deformation δ and shear strain u_s.

4 Hydrodynamic Theory for Granular Materials

A granular medium behaves in many ways like an ordinary fluid. Insofar as possible, individual grains are treated as the molecules of a *granular fluid*. One example is the motion of sand on a vibrating plate. At sufficiently high vibrations the individual grains randomly jump up and down colliding among them. Other examples are the displacements inside a shear-cell or the flows of grains driven by gravity down inclines. These observations have inspired several authors to use continuum balance equations for mass, momentum, and energy analogous to fluid dynamics (Haff 1983). Whereas classical fluids are well described by the Navier-Stokes equations, no constitutive laws can indeed reproduce the diversity of behaviours observed for granular materials. Here, we restrict the discussion to rigid dry grains only and do not consider soft particles, cohesive and friction effects, and the interaction with the surrounding fluid.

Under the action of gravity, a fluidized state can be reached when the granular material is vibrated in the vertical direction by means of a harmonic displacement of the bottom plate of the form $A(t) = A_0 \sin(\omega t)$, where A_0 is the maximum amplitude and $\omega = 2\pi f$ is the circular frequency of oscillation. For most dissipative systems perturbed by external forces, energy balance can be described by the expression $dE/dt = I - D$, where E is the total energy of the system, I is the power induced by the external force, and D is the power dissipated by the system. Note that in the absence of dissipative processes and external forces the total energy of the system is a constant because both I and D vanish identically. If the medium attains a statistically stationary state, the time averages of I and D, say $\langle I\rangle$ and $\langle D\rangle$, must be equal. In other words, a granular system in a steady state far from equilibrium requires a constant flux of external energy (*thermostat*) to balance the dissipated power (Herrmann 1993; Aumaître et al. 2001). In analogy with a molecular gas, we may introduce the definition of *granular temperature* as being proportional to the mean kinetic energy of the grains. However, this temperature bears no relation with the internal (thermodynamic) temperature of the grains and gives only a measure of the particle velocity fluctuations. We refer the reader to Goldhirsch (2008), Baldassarri et al. (2005), and Baxter and Olafsen (2007), where a complete discussion is given on the definition of temperature in granular materials.

Since granular systems are dissipative and far from equilibrium we cannot expect *a priori* that the granular temperature can be used to establish (a) the direction of the heat flux and (b) a thermalization criterion in the sense of the thermodynamic temperature. The former point is clearly evidenced by the temperature gradients that appear in a granular system lying between a vibrating wall (*heat source*) and a reflecting wall (Brey et al. 2000). On the other hand, recent experiments have shown that fluidized granular systems behave as thermal baths satisfying the fluctuation-dissipation relation (Baldassarri et al. 2005). Because of the inelastic nature of grain collisions we cannot expect equipartition of the granular energy. In fact, the non-equipartition of the granular energy has been observed in binary mixtures of grains differing in size and/or density (Barrat and Trizac 2002; Feitosa and Menon 2002). Moreover, the numerical simulations carried out by Barrat et al. (2004) have also shown that the classical Green-Kubo relations are satisfied by each component of a granular mixture on its own. They observed that when the concentration of one of the components reaches the dilute limit, it acts as a *thermometer* without affecting the system considerably. In this case the thermometer measures its own temperature and not that of the surrounding granular medium. These findings imply that we must be careful in establishing direct analogies between a thermal bath and the fluidized state of granular materials.

On the other hand, further recent experiments have revealed that the granular temperature is not uniform through a granular layer (Olafsen and Urbach 1998; Prevost et al. 2004; Olafsen and Urbach 2005). In particular, the experiments of Olafsen and Urbach (1998) have shown that as the rate of injection of energy to the system is decreased by decreasing the amplitude of the vibrations, the system cools progressively down and favours the growth of small localized transient clusterings. By cooling of the granular system we mean a decrease of the particle velocity fluctuations and hence a decrease of the system kinetic energy. As long as the size of the perturbation is further decreased, a phase transition occurs leading to the nucleation of a single localized cluster of motionless particles surrounded by a granular fluid. Similar results were also found by Prevost et al. (2004), who observed the homogeneous nucleation of a crystal in a dense layer of steel spheres. It comes from these experiments that the coexisting phases are characterized by different granular temperatures that do not satisfy the zeroth law of thermodynamics. Therefore, the assumption of thermal equilibrium for a granular medium must be seriously questioned.

The development of statistical descriptions for granular systems involves averaging over the 'microscopic' laws for the particles' motion to obtain 'macroscopic' balance equations for the hydrodynamic fields. A basis for the derivation of granular hydrodynamic equations (analogous to the Navier–Stokes equations) and detailed expressions for the constitutive relations (transport coefficients) is provided by the kinetic theory of gases conveniently modified to account for inelastic binary collisions. We refer the reader to the book by Brilliantov and Pöschel (2004), which provides an excellent introduction of the current knowledge about granular gases. The balance laws for a granular fluid can be obtained on the basis of a mean field kinetic equation, like the Boltzmann or Enskog–Boltzmann equation. The only "non-classical" term is the collision rate of dissipation per unit volume per unit time due to inelastic collisions. Many calculations are based on the kinetic theory of nonuniform gases (Chapman and Cowling 1970), using the Chapman–Enskog procedure. Jenkins and Mancini (1987, 1989) introduced a remarkable extension of the kinetic theory of nonuniform gases to bidisperse granular mixtures. In the context of the so–called Revised Enskog Theory (RET) for multicomponent mixtures (López de Haro et al. 1983), they derived balance laws and constitutive relations for a dense binary mixture of smooth, nearly elastic particles by assuming a Maxwellian velocity distribution. They investigated in detail the use of the theory to a steady rectilinear shearing flow induced by the relative motion of parallel boundaries in the absence of external forces. Important modifications in these models were introduced by Arnarson and Willits (1998),Willits and Arnarson (1999), Alam et al. (2002), and Arnarson and Jenkins (2004). It is important to notice that there has also been an attempt to extend the kinetic theory of multicomponent mixtures to systems consisting of an arbitrary number of inelastic particles (Garzó et al. 2007a, b).

The starting point here are the conservation equations for the mixture density, momentum, and energy:

$$\frac{\mathrm{D}\rho}{\mathrm{D}t} = -\rho\nabla\cdot\mathbf{u}, \tag{16}$$

$$\rho\frac{\mathrm{D}\mathbf{u}}{\mathrm{D}t} = -\nabla\cdot\hat{\mathbf{P}} + \sum_i n_i\mathbf{F}_i, \tag{17}$$

$$\rho\frac{d}{2}\frac{\mathrm{D}T}{\mathrm{D}t} = \frac{d}{2}T\nabla\cdot\mathbf{J} - \nabla\cdot\mathbf{Q} - \hat{\mathbf{P}}:\nabla\mathbf{u} + \sum_i \mathbf{J}_i\cdot\mathbf{F}_i - \mathscr{D}, \tag{18}$$

where ρ is the total mixture mass density, $\mathbf{u}$ is the mass average velocity of the mixture, $\hat{\mathbf{P}}$ is the pressure tensor, n_i is the number density of species i, $\mathbf{F}_i$ is the external force acting on the particle, T is the mixture *granular temperature*, $\mathbf{J}$ is the diffusive mass-flux, $\mathbf{Q}$ is the mixture energy flux, and $\mathscr{D}$ is the total inelastic dissipation rate. Here $\mathrm{D}(\cdot)/\mathrm{D}t = \partial(\cdot)/\partial t + \mathbf{u}\cdot\nabla(\cdot)$, is the substantial derivative (or material derivative), the symbol : denotes full tensor contraction, and d is the dimension ($d = 2, 3$). Equations (16–18) are rigorous consequences of the Boltzmann–Enskog kinetic equation and must be supplemented with constitutive relations for $\hat{\mathbf{P}}$, $\mathbf{J}$, $\mathbf{Q}$, and $\mathscr{D}$.

4.1 Kinetic Theory and Granular Temperature

Here we shall give a brief account of the kinetic theory for smooth inelastic hard particles. To do so we consider a binary granular mixture with unequal granular temperatures and derive explicit expressions for the inelastic dissipation rate and the granular temperature ratio of the species.

4.1.1 Mean Fields

As a mechanical model for granular fluids we consider a binary mixture of slightly inelastic, smooth particles (disks/spheres) with radii r_i ($i = A, B$) and masses m_i in two and three dimensions ($d = 2, 3$). The system contains a number N_i of particles of species i in a volume V. The volume has a constant regular shape, and no particles are allowed to flow across the surface so that the total number of particles is constant.

Mean values are calculated in terms of the single particle velocity distribution function $f_i(\mathbf{c}, \mathbf{r}, t)$ for each species. By definition $f_i(\mathbf{c}, \mathbf{r}, t)\mathrm{d}\mathbf{c}\mathrm{d}\mathbf{r}$ is the number of particles which, at time t, have velocities in the interval $\mathrm{d}\mathbf{c}$ centered at $\mathbf{c}$ and positions lying within a volume element $\mathrm{d}\mathbf{r}$ centered at $\mathbf{r}$. The information that there is a number N_i of particles of species i in the volume V of the system is expressed by means of the normalization condition

$$\int \mathrm{d}\mathbf{c}\mathrm{d}\mathbf{r}\, f_i(\mathbf{c}, \mathbf{r}, t) = N_i(\mathbf{r}, t). \tag{19}$$

If the particles are uniformly distributed in space, so that f_i is independent of $\mathbf{r}$, then the number density $n_i(\mathbf{r}, t)$ of species i is

$$n_i(\mathbf{r}, t) = \int \mathrm{d}\mathbf{c}\, f_i(\mathbf{c}, \mathbf{r}, t). \tag{20}$$

The total number density n is the sum over both species, $n = n_A + n_B$. The mass density of each species ρ_i is defined by the product of n_i and m_i, and the total mixture density is $\rho = \rho_A + \rho_B = \rho_A\phi_A + \rho_B\phi_B$, where ϕ_i is the d-dimensional volume fraction for species i: $\phi_i = \Omega_d n_i r_i^d/d$, where Ω_d is the surface area of a d-dimensional unit sphere. The mean value of any quantity $\psi_i = \psi_i(\mathbf{c})$ of a particle species i is given by

$$\langle\psi_i(\mathbf{c})\rangle \equiv \frac{1}{n_i}\int \mathrm{d}\mathbf{c}\psi_i(\mathbf{c}) f_i(\mathbf{c}). \tag{21}$$

The mean velocity of species i is $\mathbf{u}_i = \langle\mathbf{c}_i\rangle$ and the mass average velocity (i.e., the barycentric velocity) of the mixture is defined by

$$\mathbf{u} \equiv \frac{1}{\rho}(\rho_A\mathbf{u}_A + \rho_B\mathbf{u}_B). \tag{22}$$

The peculiar velocity $\mathbf{C}_i$ of particle i is its velocity relative to the barycentric velocity, so that

$$\mathbf{C}_i \equiv \mathbf{c}_i - \mathbf{u}, \tag{23}$$

and the diffusion velocity $\mathbf{v}_i$ of species i is its relative mean motion

$$\mathbf{v}_i \equiv \langle\mathbf{C}_i\rangle = \mathbf{u}_i - \mathbf{u}. \tag{24}$$

The granular temperature of species i is proportional to its mean kinetic energy and obeys the relation

$$T_i \equiv \frac{1}{d} m_i \langle \mathbf{C}_i \cdot \mathbf{C}_i \rangle = \frac{1}{d} m_i \langle C_i^2 \rangle, \tag{25}$$

while the temperature of the mixture is

$$T \equiv \frac{1}{n}(n_A T_A + n_B T_B). \tag{26}$$

Moreover, the coefficient of restitution for collisions between particles i and j is denoted by e_{ij}, with $e_{ij} \leq 1$ and $e_{ij} = e_{ji}$.

4.1.2 The Boltzmann–Enskog Equation for a Mixture

The distribution functions $f_i(\mathbf{c}_i, \mathbf{r}, t)$ for the two species are determined from the set of nonlinear Boltzmann–Enskog equations

$$\left(\frac{\partial}{\partial t} + \mathbf{c}_i \cdot \nabla + \frac{\mathbf{F}_i}{m_i} \cdot \frac{\partial}{\partial \mathbf{c}_i}\right) f_i(\mathbf{c}_i, \mathbf{r}, t) = \sum_j J_{ij}\big[\mathbf{c}_i | f_i(\mathbf{c}_i), f_j(\mathbf{c}_j)\big]. \tag{27}$$

The Boltzmann–Enskog collision operator $J_{ij}\big[\mathbf{c}_i | f_i, f_j\big]$ describing the scattering of pairs of particles is

$$J_{ij}\big[\mathbf{c}_1 | f_i(\mathbf{c}_1), f_j(\mathbf{c}_2)\big] \equiv g_{ij} r_{ij}^{d-1} \int\int \mathrm{d}\mathbf{c}_2 \mathrm{d}\hat{\boldsymbol{\sigma}} \Theta(\hat{\boldsymbol{\sigma}} \cdot \mathbf{c}_{21})(\hat{\boldsymbol{\sigma}} \cdot \mathbf{c}_{21}) \times \left[\frac{1}{e_{ij}^2} f_i(\mathbf{c}_1') f_j(\mathbf{c}_2') - f_i(\mathbf{c}_1) f_j(\mathbf{c}_2)\right], \tag{28}$$

where g_{ij} is the radial distribution function, $r_{ij} = r_i + r_j$, $\hat{\boldsymbol{\sigma}}$ is the unit vector directed from the centre of the particle of type i to the centre of particle j separated at contact by r_{ij}, and $\Theta(x)$ is the Heaviside step function ($\Theta(x) = 0$, for $x < 0$ and $\Theta(x) = 1$, for $x > 0$). The post–collisional velocities $\mathbf{c}_i'$ are given in terms of the pre–collisional velocities $\mathbf{c}_i$ by

$$\mathbf{c}_i' = \mathbf{c}_i + M_{ji}(1 + e_{ij})(\hat{\boldsymbol{\sigma}} \cdot \mathbf{c}_{ji})\hat{\boldsymbol{\sigma}}, \tag{29}$$

where $M_{ji} = m_j/m_{ij}$, $m_{ij} = m_i + m_j$, and $\mathbf{c}_{ji} \equiv \mathbf{c}_j - \mathbf{c}_i$ is the relative velocity between particles.

Also, in the Boltzmann–Enskog collision operator we have used the Enskog assumption for dense gases, i.e.,

$$f_{i,j}^{(2)}(\mathbf{c}_1, \mathbf{c}_2) \to g_{ij} f_i^{(1)}(\mathbf{c}_1) f_j^{(1)}(\mathbf{c}_2), \tag{30}$$

for the complete pair velocity distribution function $f_{i,j}^{(2)}(\mathbf{c}_1, \mathbf{c}_2)$.

4.1.3 Analysis Based on the Boltzmann–Enskog Equation

To find the equation for $\langle \psi_i \rangle$, as defined by relation (21), we multiply both sides of Eq. (27) by ψ_i and then integrate over all velocities $\mathbf{c}$ to get

$$\frac{\partial}{\partial t}\langle n_i \psi_i \rangle = -\nabla \cdot \langle n_i \mathbf{c}_i \psi_i \rangle + n_i \frac{\mathbf{F}_i}{m_i} \cdot \left\langle \frac{\partial \psi_i}{\partial \mathbf{c}_i} \right\rangle + \sum\nolimits_j \left[\chi_{ij}(\psi_i) - \nabla \cdot \boldsymbol{\theta}_{ij}(\psi_i) \right], \quad (31)$$

where $\chi_{ij}(\psi_i)$ is the collisional source integral

$$\chi_{ij}(\psi_i) \equiv g_{ij} r_{ij}^{d-1} \int\int\int \mathrm{d}\mathbf{c}_1 \mathrm{d}\mathbf{c}_2 \mathrm{d}\hat{\boldsymbol{\sigma}} \Theta(\hat{\boldsymbol{\sigma}} \cdot \mathbf{c}_{21})(\hat{\boldsymbol{\sigma}} \cdot \mathbf{c}_{21}) \left[\psi_i' - \psi_i \right] \quad (32)$$
$$\times \left[1 + (1/8) r_{ij}^2 (\hat{\boldsymbol{\sigma}} \cdot \nabla)^2 + \cdots \right] f_i(\mathbf{c}_1) f_j(\mathbf{c}_2),$$

and $\theta_{ij}(\psi_i)$ is the collisional flux integral

$$\theta_{ij}(\psi_i) \equiv \frac{1}{2} g_{ij} r_{ij}^{d} \int\int\int \mathrm{d}\mathbf{c}_1 \mathrm{d}\mathbf{c}_2 \mathrm{d}\hat{\boldsymbol{\sigma}} \hat{\boldsymbol{\sigma}} \Theta(\hat{\boldsymbol{\sigma}} \cdot \mathbf{c}_{21})(\hat{\boldsymbol{\sigma}} \cdot \mathbf{c}_{21}) \left[\psi_i' - \psi_i \right] \quad (33)$$
$$\times \left[1 + (1/8) r_{ij}^2 (\hat{\boldsymbol{\sigma}} \cdot \nabla)^2 + \cdots \right] f_i(\mathbf{c}_1) f_j(\mathbf{c}_2).$$

When $\psi_i(\mathbf{c}_i) \to \psi_i(\mathbf{C}_i)$, the balance equation may be written in the form

$$\frac{\partial}{\partial t}\langle n_i \psi_i \rangle = -\nabla \cdot \langle n_i \mathbf{c}_i \psi_i \rangle + \left\langle n_i \frac{\partial \mathbf{C}_i}{\partial t} \cdot \frac{\partial \psi_i}{\partial \mathbf{C}_i} \right\rangle \quad (34)$$
$$+ \sum\nolimits_j \left[\chi_{ij}(\psi_i) - \nabla \cdot \boldsymbol{\theta}_{ij}(\psi_i) - \boldsymbol{\theta}_{ij} \left(\frac{\partial \psi_i}{\partial \mathbf{C}_i} \right) : \nabla \mathbf{u} \right].$$

To investigate transport processes, we must solve the Boltzmann–Enskog equation (27), with given initial conditions, to obtain the velocity distribution function. Some rigorous properties of any solution of Eq. (27) can be obtained from the fact that in any particle collision there are dynamical quantities that are rigorously conserved. For a granular fluid the independent conserved properties are mass and momentum. The transformation (29) conserves momentum but, when $e < 1$, it does not conserve energy. Therefore, in a binary collision the change of kinetic energy $\Delta E \neq 0$. In the lowest–order approximation we assume that the fluid has a local Maxwell–Boltzmann (Maxwellian) distribution given by

$$f_i(\mathbf{c}_i) \approx f_i^{(0)}(\mathbf{c}_i) = \pi^{d/2} n_i \left(\frac{m_i}{2T_i} \right)^{d/2} \exp\left[-\frac{m_i}{2T_i} (\mathbf{c}_i - \mathbf{u})^2 \right]. \quad (35)$$

4.1.4 Energy Balance

Taking $\psi_i = \frac{1}{2}m_i\mathbf{C}_i^2$ and using the balance law (34), we obtain the balance of fluctuation energy (granular temperature) for each species:

$$\frac{\partial}{\partial t}(n_i T_i)+\nabla\cdot(n_i\mathbf{u}_i T_i) = -\nabla\cdot\mathbf{Q}_i-\hat{\mathbf{P}}_i : \nabla\mathbf{u}+\mathbf{J}_i\cdot\mathbf{F}_i + \frac{\rho_i}{\rho}\mathbf{v}_i\cdot(\nabla\cdot\hat{\mathbf{P}}-n\mathbf{F})+\mathscr{D}_i. \quad (36)$$

The Chapman–Enskog method is a perturbation method based on a power series expansion of $f_i(\mathbf{c}_i)$ in terms of the Knudsen number $\mathrm{K}_n = l/L$, where l is the mean free path and L a characteristic length. The method provides different expressions for the pressure tensor $\hat{\mathbf{P}}_i$, the energy (heat) flux $\mathbf{Q}_i$, and the kinetic energy dissipation rate $\mathscr{D}_i$. The species stress tensor has at the Navier-Stokes level, the standard Newtonian form

$$\hat{\mathbf{P}}_i = p_i\hat{\mathbf{I}} + \mu_i(\nabla\mathbf{u} + \nabla\mathbf{u}^T), \quad (37)$$

where p_i is the partial pressure of species i, μ_i is the viscosity of species i, and $\hat{\mathbf{I}}$ denotes the unit tensor. The equation of state for the partial pressure of species i can then be written as: $p_i = n_i Z_i T_i$, where $Z_i := (1 + \sum_{j=l,s} K_{ij})$ is the *compressibility* factor of species i and $K_{ij} := \phi_j g_{ij}\left(1 + R_{ij}\right)^d/2$, with g_{ij} being the radial distribution function and $R_{ij} = r_i/r_j$ the size-ratio. The energy flux is given by

$$\mathbf{Q}_i = -\kappa_i\sqrt{T_i}\nabla T_i, \quad (38)$$

where κ_i is the analog of the thermal conductivity of species i. Finally, the rate of kinetic energy dissipation of species i is given by

$$\mathscr{D}_i = \sum_j \frac{\pi^{\frac{d-1}{2}}}{d\Gamma(d/2)} g_{ij} r_{ij}^{d-1} m_i n_i n_j M_{ji}\left[M_{ji}(1-e_{ij}^2)\left(\frac{2T_i}{m_i}+\frac{2T_j}{m_j}\right)\right.$$
$$\left. +4(1+e_{ij})\frac{T_i - T_j}{m_i + m_j}\right]\left(\frac{2T_i}{m_i}+\frac{2T_j}{m_j}\right)^{1/2}. \quad (39)$$

In the elastic limit $e_{ij} = 1$, the above constitutive relations reduce to the "classical" theory of non-uniform dense gases (Chapman and Cowling 1970). On the other hand, keeping the next order gradient terms corresponds to the *Burnett* or *super-Burnett* hydrodynamics (García-Colín et al. 2008; Serero et al. 2008). Let us mention that a complete treatment of the energy flux $\mathbf{Q}_i$ should include a "non-Fourier" term which relates the energy flux with the density gradient ∇n. This term does not have an analogue in the hydrodynamics of molecular fluids and should be included in the calculation of temperature and density profiles of a vertically vibrated granular system. Early studies of granular mixtures, as referred to above, used a nearly elastic formalism. Recently, an important step was taken by Serero et al. (2007) and Garzó et al. (2009), where it has been proposed a new Sonine approach for the calculation of higher order expansions to treat strong inelasticity. These contributions open a new perspective to increase the accuracy of hydrodynamics calculations for granular binary mixtures.

4.1.5 Temperature Ratio

We can split Eq. (39) into two terms:

1. The inter-species collisional dissipation rate $\mathscr{D}_i^I$

$$\mathscr{D}_i^I := \frac{\sqrt{2}d}{\sqrt{\pi}\Omega_d} \sum_j g_{ij} R_{ij}^d (1+R_{ji})^{d-1} M_{ji}^2 (1-e_{ij}^2) \frac{\phi_i \phi_j}{r_i^{d+1}} \frac{T_i^{3/2}}{m_i^{1/2}} \left(1+\frac{m_i T_j}{m_j T_i}\right)^{3/2} \tag{40}$$

2. The exchange collisional dissipation rate $\mathscr{D}_i^E$

$$\mathscr{D}_i^E := \frac{\sqrt{2}d}{\sqrt{\pi}\Omega_d} g_{ik} R_{ik}^d (1+R_{ki})^{d-1} M_{ik} M_{ki} (1+e_{ik}) \frac{\phi_i \phi_k}{r_i^{d+1}} \frac{T_i^{3/2}}{m_i^{1/2}} \left(1-\frac{T_k}{T_i}\right) \left(1+\frac{m_i T_k}{m_k T_i}\right)^{3/2} \tag{41}$$

Hence, we could assert that the *exchange* term $\mathscr{D}_i^E$ is a clear consequence of the non-equipartition assumption. Note that $\sum_{i=A,B} \mathscr{D}_i^E = 0$. On the other hand, with the assumption of equipartition of granular energy ($T_A = T_B = T$) we have that $\mathscr{D}_i^E = 0$ and so Eq. (40) reduces to:

$$\mathscr{D}_i^I := \frac{\sqrt{2}d}{\sqrt{\pi}\Omega_d} \sum_j g_{ij} R_{ij}^d (1+R_{ji})^{d-1} M_{ji}^2 (1-e_{ij}^2) \frac{\phi_i \phi_j}{m_i^{1/2} r_i^{d+1}} \frac{T_i^{3/2}}{m_i^{1/2}} \left(\frac{T}{M_{ji}}\right)^{3/2}. \tag{42}$$

We can conclude that physically the lack of energy equipartition is determined by the different dissipation rates given by Eqs. (40) and (41). The way how the energy is injected into the two-species system determines together with the dissipation rates the functional form of the temperature ratio. To illustrate this point, let us consider the case of a *uniformly heated* granular system. This idealized situation has been used to model rapid granular flows. To study the system we start from the *stochastic* equation of motion for a particle

$$m_i \frac{\mathrm{d}\mathbf{c}_i}{\mathrm{d}t} = \mathbf{F}_i^D + m_i \hat{\xi}_i. \tag{43}$$

In this equation $\mathbf{F}_i^D$ is the force due to inelastic collisions and $\hat{\xi}_i$ is the random acceleration due to external forcing, which is assumed to be Gaussian white noise and uncorrelated for different particles

$$\langle \xi_{i\alpha}(t)\xi_{j\beta}(t')\rangle = \xi_0^2\delta_{ij}\delta_{\alpha\beta}\delta(t-t'), \tag{44}$$

where Greek indices denote Cartesian components and ξ_0^2 is the strength of the correlation which is proportional to the rate of the energy input per unit mass. The average is done over the noise source. When we consider the associated forcing term in the Boltzmann–Enskog equation, the balance equation for species temperatures satisfies, in the stationary state, the relation

$$m_i\xi_0^2 = \mathscr{D}_i. \tag{45}$$

Using this relation it is straightforward to derive the corresponding equation for the temperature ratio T_A/T_B (with $r_A > r_B$):

$$c_1\left(\frac{T_A}{T_B}\right)^{3/2} + c_2\left(1+\frac{m_B}{m_A}\frac{T_A}{T_B}\right)^{3/2} + c_3\left(1+\frac{m_B}{m_A}\frac{T_A}{T_B}\right)^{1/2}\left(\frac{T_A}{T_B}-1\right) + c_4 = 0, \tag{46}$$

where the coefficients are given by

$$\begin{aligned}
c_1 &= 2^{d-1}(1-e^2)R_{BA}^d\phi_A g_{AA}\left(\frac{m_B}{m_A}\right)^{3/2},\\
c_2 &= \sqrt{2}(1-e^2)(1+R_{BA})^{d-1}(\phi_B M_{BA}^2 - \phi_A R_{BA}^d M_{AB}^2)g_{AB},\\
c_3 &= 2\sqrt{2}(1+e)(1+R_{BA})^{d-1}M_{BA}(\phi_B M_{BA} - \phi_A R_{BA}^d M_{AB})g_{AB},\\
c_4 &= -2^{d-1}(1-e^2)R_{BA}^d\phi_B g_{BB}.
\end{aligned}$$

We can suppose that this theoretical construction could describe, at least in a first-order approximation, the bulk of a vibrated granular system. Under this assumption any further analysis should be restricted to regions far from the boundaries.

5 Theory of Elasticity for Granular Materials

The mechanical properties of granular systems are largely affected by the interaction between the grains. The nonlinear elastic nature of contact surfaces are responsible for many observable features as the formation of three-dimensional force chain networks (Jaeger et al. 1996), the increase of material stiffness with depth (Landau and Lifshitz 1970), the scattering effects on wave propagation (Jia 2004), and the structure signature of granular phases (Corwin et al. 2005). It is well-known that conventional linear elasticity fails to explain such complexities because it does not provide a satisfactory description of the relations between stress and strain in a granular system. This can be seen from the expression of the free energy for an elastic medium

$$f_{el} = \frac{1}{2}K_b u_{nn}^2 + K_a u_{ij}^0 u_{ij}^0, \tag{47}$$

where K_b and K_a are the constant compressional and shear moduli, respectively, u_{nn} is the trace of the strain tensor u_{ij}, and u^0_{ij} is its traceless part, i.e., $u^0_{ij} = u_{ij} - u_{nn}\delta_{ij}/3$. This form of the free energy does not include the mechanical yield maximal shear stress that makes the system to become mechanically unstable. Moreover, it is quadratic in u_{ij}, providing a stable solution regardless of the stress and strain values (Jiang and Liu 2003). This is a major drawback because yield in granular matter is an essential phenomenon that leads to phase transition (Corwin et al. 2005). Therefore, under linear elasticity the system is unable to experience an unstable configuration for given values of the stress and strain. Another deficiency of the linear model lies on the stress-strain relation

$$\sigma_{ij} = -K_b u_{nn}\delta_{ij} - 2K_a u^0_{ij}, \tag{48}$$

which can be obtained from Eq. (47) by applying the thermodynamic relation $\sigma_{ij} = -(\partial f_{el}/\partial u_{ij})$ (Landau and Lifshitz 1970). The above expression implies that for pure shear stresses $\sigma_{ij} = 2K_a u^0_{ij}$ (Jiang and Liu 2007c). This result obviates the possibility of volume dilatancy (a physical effect observed in compact granular media, which induces an increase of their overall volume under shear stresses (Reynolds 1885). While this theory is adequate for describing ordinary solids (Jiang and Liu 2007c), it gives a poor description of granular matter. Also, Eqs. (47) and (48) do not take into account the interaction between granular particles, which may deform as a result of the contact with one another. This is known in the theory of elasticity as the contact problem, which was first solved by Hertz (Hertz 1882; Landau and Lifshitz 1970). A mathematical solution to this problem is given in Landau and Lifshitz (1970), where the free energy is proportional to $h^{5/2}$ and the applied force to $h^{3/2}$, with h being the relative change in height. Here h corresponds to u_{nn} so that K_b must be proportional to $u_{nn}^{1/2}$ in order to achieve Hertz's solution. This relation between the compressional modulus and the deformation experienced by the body is known as the Hertz's Contact Model and it is widely used to simulate nonlinear elastic contacts between grains (Corwin et al. 2005). In this approach, the grains are modelled as spheres and the applied forces are assumed to be sufficiently small (de Gennes 1999). In 1873, Boussinesq proposed a modified theory of elasticity, where Eq. (48) remains the same except for the fact that the stress-dependent elastic moduli K_b and K_a are both assumed to be approximately equal to $-u_{nn}^{1/2}$ (Boussinesq 1873). With this provision, Eq. (48) takes the form

$$\sigma_{ij} \sim u_{nn}^{1/2}\left(-u_{nn}\delta_{ij} - \frac{(3-6\nu)}{(1+\nu)}u^0_{ij}\right), \tag{49}$$

where ν is the constant Poisson ratio and $(3-6\nu)/(1+\nu) = 2K_a/K_b$ (Jiang and Liu 2007c). This nonlinear relation is consistent with a quasi-elastic model. However, it still possesses some deficiencies. First, it does not predict the effects of dilatancy; second, the yield condition is absent, and third, a free energy equation cannot be formulated because $\partial\sigma_{ij}/\partial u_{kl} \neq \partial\sigma_{kl}/\partial u_{ij}$ (Jiang and Liu 2007c).

This last point can be better understood if we return to the general thermodynamic expression: $\sigma_{ij} = \partial f_{el}/\partial u_{ij}$ (Landau and Lifshitz 1970), which implies that $\partial\sigma_{ij}/\partial u_{kl} = \partial^2 f_{el}/\partial u_{kl}\partial u_{ij}$. Since the free energy is an exact differential, we must have that $\partial^2 f_{el}/\partial u_{kl}\partial u_{ij} = \partial^2 f_{el}/\partial u_{ij}\partial u_{kl}$ so that the equality $\partial\sigma_{ij}/\partial u_{kl} = \partial\sigma_{kl}/\partial u_{ij}$ holds. Considering that Eq. (49) fails to satisfy this last equality, it follows that no free energy can be formulated.

5.1 The Jiang-Liu Model

Boussinesq's elastic theory represents an improvement on linear elasticity because it incorporates the interaction between bodies through the use of stress-dependent elastic moduli. This explains why Eq. (49) has been employed to deal with granular compression and sound wave propagation as shown by Jiang and Liu (2007c). However, a self-contained elastic theory must include the effects of mechanical yield and volume dilatancy in order to correctly describe the dynamics of granular media. Such a theory was recently developed by Jiang and Liu (2003, 2007a; 2007b). In particular, their model relies on a formulation of the free energy that accounts for the occurrence of instabilities as they are observed in granular media. They start by considering an arbitrary equilibrium state with thermal temperature T and packing density ρ_c, in the absence of external forces and with an associated free energy density equal to $f_1\rho/m$, where f_1 is the free energy and m is the mass per grain (Jiang and Liu 2003). In the presence of external forces and under the effects of gravity, the granular packing deforms leading to a change in the density given by $\delta = 1 - \rho_c/\rho = -u_{nn}$. This relation corresponds to a pure elastic change and it is valid only when the forces are small enough and applied slowly. Since the contact areas between grains are finite, the elastic moduli are also finite. Therefore, as long as the particles lose contact with one another, their contact areas are set to zero. As mentioned above, the formulation of the free energy that contains this observation as well as unstable configurations of the system is given by Eq. (14). A stress-strain relation can be derived by using Eqs. (14) and (15) into the energy and momentum conservation laws of Jiang and Liu (2003) to yield

$$\sigma_{ij} = -K_b u_{nn}\delta_{ij} + 2K_a u^0_{ij} - \frac{1}{\delta}\left(\frac{1}{2}bK_b u^2_{nn} + aK_a u^0_{kl}u^0_{kl}\right), \tag{50}$$

where the dominant non-linear terms are maintained. The above equation contains the stress elements of both the linear and Boussinesq elasticity models as we may see from inspection of the first two terms on the right-hand side of Eq. (50). In this way, the stress is completely defined once we specify a and b together with the stress-dependent elastic moduli, which account for the desired granular behaviour. If we apply Hertz contacts, which are appropriate for spheres subjected to small forces (Tykhoniuk et al. 2007), Eq. (50) becomes

$$\sigma_{ij} = Y\sqrt{\delta}\left(B\delta\delta_{ij} - 2Au_{ij}^0 + \frac{1}{2\delta}Au_s^2\delta_{ij}\right), \tag{51}$$

where $Y = 9K_bK_a/(3K_b + K_a)$, $B = \frac{5}{4}\widetilde{K}_b/Y$, $A = \widetilde{K}_a/Y$, and $u_s^2 = u_{ij}^0 u_{ij}^0$. We may see that Eq. (51) is similar to Boussinesq's stress formulation, given by Eq. (49), except for an additional term. This difference arises because the Jiang-Liu model is derived from an energy expression, whereas Eq. (50) does not (Bonneau et al. 2007). The free-energy expression that is consistent with Eq. (51) reads

$$f = Y\sqrt{\delta}\left(\frac{2}{5}Bu_{nn}\delta^2 + Au_s^2\right). \tag{52}$$

This expression is stable only in the range of strain values that keeps it convex. Therefore, Eq. (52) naturally accounts for unstable configurations of the system, as the yield, which appears as a phase transition on a potential-strain diagram. The stability condition that f is convex only for $\sigma_s/P \leq \sqrt{2/\xi}$ is reported in by Jiang and Liu (2007a), where σ_s is the pure shear stress, $P = \frac{1}{3}\sigma_{ii}$ is the hydrostatic pressure, and $\xi = B/A$. So, yield can be determined.

The Jiang-Liu model also includes the effects of volume dilatancy. These effects are represented by the first term between parentheses on the right-hand side of Eq. (51), where the pure shear stress is proportional to the shear strain and the volumetric deformation δ. In addition, the last term represents a component of pure shear deformation under hydrostatic pressure which also includes shear-induced anisotropies (Jiang and Liu 2007c). Jiang and Liu (2008) present a detailed account of how anisotropy arises from the shear stress. Here we shall only mention briefly that compared to linear elastic theory, where the compliance tensor which links the increments of stress and strain is both isotropic and constant (Landau and Lifshitz 1970; Jiang and Liu 2007c), in the Jiang-Liu formulation it is stress-dependent. If only pure compression is considered, the Jiang-Liu formulation of the compliance tensor takes the form appropriate for linear elastic theory, whereas in the presence of shear extra terms appear which are linear and quadratic in σ_{ij}. These extra terms are anisotropic and arise due to the effects of shear. This is precisely what is meant by shear-induced anisotropy in the Jiang-Liu formulation.

In summary, linear elasticity cannot be used to describe granular matter because it does not allow the system to achieve an unstable configuration. Consequently, mechanical yield will never be present. Moreover, it does not predict the effects of volume dilatancy and its constant elastic moduli do not account for interactions between the grains, making it only accurate for modelling ordinary solids. On the other hand, in the Boussinesq's elastic theory the interactions between particles are incorporated by means of a Hertz contact model. However, it does not include yield and volume dilatancy, and more importantly, it is not the result of a free-energy formulation. Conversely, the Jiang-Liu model is based on an energy expression which allows for unstable configurations so that yield is viable. The effects of dilatancy and shear-induced anisotropies are also included. In this model, the elastic moduli can be strain-dependent in a proportion that is determined by the type of contact. A drawback of this formulation is that it does not allow for irreversible grain deformation and that it works only when the forces involved are sufficiently small and slowly applied. In spite of this, the Jiang-Liu formulation provides a self-contained theory that allows for a fairly accurate description of granular elasticity.

6 From Discrete Particles to Continuous Matter Field: The Micro–Macro Problem

Exact continuum forms of matter fields, such as mass, momentum, and energy can be established as relations between averages of microscopic quantities calculated for macroscopic scales within a range of coarse-graining scales where the fields are resolution independent. Another problem is the calculation of realistic constitutive equations for continuous models. Here we shall briefly discuss the problem of the micro-macro transition and microscopic constitutive laws.

6.1 Weighted Averages of Microscopic Quantities

The rigorous passage from a microscopic to a macroscopic (continuum) mechanical description of granular and heterogeneous materials, including the mesoscopic disorder, has been recently addressed by Goldenberg and Goldhirsch (2002a, b, 2004, 2005, 2008); Goldenberg et al. (2006); Serero et al. (2008) and by Barrat and coworkers for the case of amorphous glasses (Tanguy et al. 2002; Leonforte et al. 2004, 2005, 2006; Goldenberg et al. 2007).

In particular, Goldenberg and Goldhirsch established continuum fields in terms of local spatially-weighted sums that average the particle motion in a way similar to Smoothed Particle Hydrodynamics (SPH) methods (Monaghan 1992). In SPH, the continuum fields are represented by a finite set of particles through the use of a smoothing procedure in which a quantity F (such as the density) is approximated by the summation interpolant

$$F(\mathbf{r}) = \sum_i F_i(\mathbf{r}) W(\mathbf{r} - \mathbf{r}', h), \tag{53}$$

where $W(\mathbf{r} - \mathbf{r}', h)$ is a smooth (differentiable) function, commonly referred to as the interpolating kernel, and h is the smoothing length, which determines the spatial resolution and defines the involved scales for the spatial averages (Murdoch and Bedeaux 1994; Goldenberg and Goldhirsch 2002b, 2004; Glasser and Goldhirsch 2001). Exact continuum forms of the balance equations (for mass, momentum, and energy) can be established as relations between weighted space (and time) averages (Murdoch and Bedeaux 1994; Glasser and Goldhirsch 2001;

Goldenberg and Goldhirsch 2002b, 2004). In this framework, it is possible to derive an exact expression for the stress field, providing a way for a systematic study of the constitutive relations. Goldenberg and Goldhirsch (2005, 2002a, b, 2008) have shown that this coarse-grained stress field reproduces the experimentally observed response of granular packings to localized perturbations in 2D (Geng et al. 2001, 2003) and 3D (Reydellet and Clément 2001; Serero et al. 2001). On the other hand, in applications to amorphous glasses (Leonforte et al. 2004), a striking outcome was that for a smoothing length $h \sim d$, where d is the particle diameter, the stress field is such that the force chains have no effects on scales above the spatial average, implying that they need not be seen as a fundamental structural feature for the force response at a macroscopic level (Goldenberg and Goldhirsch 2005, 2002a, b, 2008). The short correlation length of the contact forces suggests that both the macroscopic fields and their fluxes can be defined as scale invariant. For 2D polydisperse packings, with average size $\langle d \rangle$, the stress field is independent of the range of coarse-graining scales for $h \geq \langle d \rangle$ (Goldenberg et al. 2006). The range of scale independence can be as small as 3–5 particle diameters (Goldenberg et al. 2006). These findings justify the applicability of a continuum description for granular materials. On the other hand, further work is required to understand the correspondence between the spatially-weighted summation technique and the SPH computational method. This study opens the possibility to introduce a new computational method which should be very useful for simulations of granular materials in physics and engineering.

6.2 Non-Local Fluidity Relation

For continuous theories the link between the macroscopic fields and the microscopic properties of the material is established through the constitutive relations. In Sect. 4.1, we have illustrated how the kinetic theory of dense gases can be used to derive the transport constitutive relations for a continuum fluid dynamics description of granular flows. However, under this approach it is impossible to describe simultaneously the coexistence of fast and slow flows observed in experiments. The description of the re-arrangement of a granular packing induced by monotonic deformations and local spatial variations demands incorporating relaxation rates D in the constitutive equations that relate the time rates of the stress tensor σ_{ij} to D. Static states often reveal spatial fluctuations which are stronger at state limits. They change in jumps during monotonic deformations, i.e., as local plastic events occurring above a microscopic yield stress. On the other hand, non-local elastic reorganization occurs as a consequence of the microscopic fluctuations.

Recently, the rheology of dense granular flows have been analyzed using non-local constitutive laws for the flow of dense disordered materials (Picard et al. 2005; Bocquet et al. 2009). In this approach the dynamics is based on a mesoscopic treatment between the microscopic and macroscopic scales: local plastic events associated with a microscopic yield stress and the non-local elastic release of the stress over the system. This idea was extended by Kamrin and Koval (2012), postulating a non-local fluidity relation for flowing granular materials which captures several known finite-size effects observed in steady flow. The model expresses a continuum field equation for the flow law in terms of a fluidity ratio and an extra Laplacian term that is scaled by the grain size. This approach not only predicted areas of fast-flowing grains, but also the sites where grains would be slowly moving, at the very edges of each.

7 Conclusions

In this chapter, we have presented an introductory discussion of different theoretical scenarios for the mathematical analysis of physical properties observed in granular materials. The physical attributes and constraints in granular materials are so complex that the construction of a unified theoretical framework seems unreachable. However, such difficulties should be an appealing challenge for theoretical physicists and applied mathematicians. Three main foundational theoretical issues were addressed in this work: (i) *Granular thermodynamics*; (ii) *Granular hydrodynamics*; and (iii) *Granular elasticity*.

Concerning thermodynamics we would like to emphasize that although thermal energy is not the relevant parameter, one can construct thermodynamical-like state functions or potentials using other relevant variables for granular materials, as for example, volume, energy dissipation, the number of particles, or a modified elastic free energy. Nevertheless, it is important to explore possible connections between these formulations. The incorporation of a thermodynamical framework should be beneficial to state the postulates of a theory for granular matter. This framework has to be related with statistical average. Let us quote an interpretation of thermodynamics advanced by Callen (1974): *Thermodynamics is the study of those properties of macroscopic matter that follow from the symmetry properties of physical laws, mediated through the statistics of large systems.* Grounded on this interpretation, two underlying principles consistently emerged: The bases of macroscopic thermodynamic laws lie in the symmetries of microscopic physical laws; and the irrelevant details of specific systems are blurred by the properties of large statistical collections. To illustrate this, Callen presented a model in which the energy plays no relevant role in order to emphasize the lack of uniqueness of the energy in thermodynamics. Therefore, granular matter represents a quite good physical model to explore the ideas presented by Callen (1974), in congruence with the alternative *granular thermodynamics* reviewed in Sect. 3. This exercise should pave the way towards the collection of foundational blocks.

Granular hydrodynamics and kinetic theory are perhaps the major developed theoretical frameworks for granular matter. Due to the fact that the grains are macroscopic and one can often measure the dynamics of each and every grain, hydrodynamic theories are often expected to describe scales that they cannot model or resolve. There are many examples in which granular hydrodynamics provides good description for granular flows. Some remarkable analogies with different fluid-mechanical phenomena are the granular parallels of convection, buoyancy, the Leidenfrost effect, and the hydraulic jump. However, there are some experiments and simulations where granular hydrodynamics does not work. On the other hand, due to the lack of strong scale separation one needs to go beyond the Navier-Stokes level of description and this rises certain problems that require further study.

We have shown that for the *solid state* of granular matter it is possible to derive an elastic theory which includes important signatures characterizing particulate systems. A fundamental ingredient is that the Jiang-Liu granular elasticity takes into account a general form of the type of contacts between the grains, with the Hertz's contact model as a particular case. On the other hand, the Jiang-Liu formulation directly extends the Boussinesq modified theory of elasticity and removes their thermodynamic inconsistency via a correct formulation of the strain free energy functional. In addition, under certain limits the Jiang-Liu theory includes the well-known linear elasticity of isotropic and homogeneous elastic solids. The theory also emphasizes the role of intrinsic key features of granular dynamics such as volume dilatancy, mechanical yield, and anisotropies in the stress distribution. However, this formulation is not complete and must be extended to include the effects of grain rotation into the constitutive relations for the non-linear elastic theory. An open question is: How should the spatial distribution of elastic constants be in heterogeneous granular packings? An answer to this question is very important for the analysis of force networks which form the skeleton that carries most of the load in a static granular medium.

The theoretical ideas illustrated in this paper represent a starting point towards the construction of a general theory for granular matter. A necessary goal is to explore how granular thermodynamics, hydrodynamics, and elasticity work together to provide interpretations of observations and how this partnership can be made more efficient. Providing a formal justification for continuum matter fields, a promising program towards a unified theory could be advanced from the new recent ideas introduced by Yiang and Liu for the construction of a *granular solid hydrodynamics* (Jiang and Liu 2009). They proposed a continuum mechanical theory which includes the energy currents and the entropy production. Their principal postulate is that granular media are *elastic* when at rest, but turn *transiently elastic* when the grains are agitated.

Finally, let us remark the necessity to account for random effects to determine the response of granular matter and their heterogeneous nature.

References

Alam M, Willits JT, Arnarson BÖ, Luding S (2002) Kinetic theory of a binary mixture of neraly elastic disks with size and mass disparity. Phys Fluids 14:4085–4087

Aranson IS, Tsimring LS (2006) Patterns and collective behavior in granular media: theoretical concepts. Rev Mod Phys 78:641–692

Arnarson BÖ, Willits JT (1998) Thermal diffusion in binary mixtures of smooth, nearly elastic spheres with and without gravity. Phys Fluids 10:1324–1328

Arnarson BÖ, Jenkins JT (2004) Binary mixtures of inelastic spheres: simplified constitutive theory. Phys Fluids 16:4543–4550

Aumaître S, Fauve S, McNamara S, Poggi P (2001) Power injected in dissipative systems and the fluctuation theorem. Eur Phys J B 19:449–460

Baldassarri A, Barrat A, D'Anna G, Loreto V, Mayor P, Puglisi A (2005) What is the temperature of a granular medium? J Phys: Condens Matter 17:S2405

Barrat A, Trizac E (2002) Lack of energy equipartition in homogeneous heated binary granular mixtures. Granular Matter 4:57–63

Barrat A, Loreto V, Puglisi A (2004) Temperature probes in binary granular gases. Phys A 334: 513–523

Baxter GW, Olafsen JS (2007) The temperature of a vibrated granular gas. Granular Matter 9:135–139

Bocquet L, Colin A, Ajdari A (2009) Kinetic theory of plastic flow in soft glassy materials. Phys Rev Lett 103:036001

Bonneau L, Andreotti B, Clément E (2007) Surface waves in granular media under gravity and their relation to booming avalanches. Phys Rev E 75:016602

Borderies N, Goldreich P, Tremaine S (1985) A granular flow model for dense planetary rings. ICARUS 63:406–420

Boussinesq J (1873) Essai théorique sur l'équilibre d'élasticité des massifs pulvérulents et sur la poussée des terres sans cohésion. Comptes Rendus des Séances de l'Académie des Sciences 77:1521–1525

Brey JJ, Ruiz-Montero MJ, Moreno F (2000) Boundary conditions and normal state for a vibrated granular fluid. Phys Rev E 62:5339

Brilliantov NV, Pöschel T (2004) Kinetic theory of granular gases. Oxford University Press, Oxford

Callen H (1974) Thermodynamics as a science of symmetry. Found Phys 4:423–443

Chapman S, Cowling TG (1970) The mathematical theory of nonuniform gases. Cambridge University Press, Cambridge

Corwin EI, Jaeger HM, Nagel SR (2005) Structural signature of jamming in granular media. Nature 435:1075–1078

Cugliandolo L, Kurchan J, Peliti L (1997) Energy flow, partial equilibration, and effective temperature in systems with slow dynamics. Phys Rev E 55:3898–3914

de Gennes PG (1999) Granular matter: a tentative view. Rev Mod Phys 71:374–382

Edwards SF, Oakeshott RBS (1989) Theory of powders. Phys A 157:1080–1090

Esposito LW (2010) Composition, structure, dynamics, and evolution of Saturn's rings. Annu Rev Earth Planet Sci 38:383–410

Feitosa K, Menon N (2002) Breakdown of energy equipartition in a 2D binary vibrated granular gas. Phys Rev Lett 88:198301

García-Colín LS, Velasco RM, Uribe FJ (2008) Beyond the Navier-Stokes equations: burnett hydrodynamics. Phys Rep 465:149–189

Garzó V, Dufty JW, Hrenya CM (2007a) Enskog theory for polydisperse granular mixtures. I. navier-stokes order transport. Phys Rev E 76:031303

Garzó V, Dufty JW, Hrenya CM (2007b) Enskog theory for polydisperse granular mixtures. II. sonine polynomial approximation. Phys Rev E 76:031304

Garzó V, Vega-Reyes F, Montanero JM (2009) Modified Sonine approximation for granular binary mixtures. J Fluid Mech 623:387–411

Geng J, Howell D, Longhi E, Behringer RP, Reydellet G, Vanel L, Clément E, Luding S (2001) Footprints in sand: the response of a granular material to local perturbations. Phys Rev Lett 87:035506

Geng J, Rydellet G, Clément E, Behringer RP (2003) Green function measurements of force transmission in 2D granular materials. Phys D 182:274–303

de Gennes PG (1998) Reflections on the mechanics of granular matter. Phys A 261:267–293

Glasser BJ, Goldhirsch I (2001) Scale dependence, correlations, and fluctuations of stresses in rapid granular flows. Phys Fluids 13:407–420

Goldenberg C, Goldhirsch I (2002a) Force chains, microelasticity, and macroelasticity. Phys Rev Lett 89:084302
Goldenberg C, Goldhirsch I (2002b) On the microscopic foundations of elasticity. Eur Phys J E 9:245–251
Goldenberg C, Goldhirsch I (2004) Small and large scale granular statics. Granular Matter 6:87–96
Goldenberg C, Goldhirsch I (2005) Friction enhances elasticity in granular solids. Nature 435:188–191
Goldenberg C, Atman APF, Claudin P, Combe G, Goldhirsch I (2006) Scale separation in granular packings: stress plateaus and fluctuations. Phys Rev Lett 96:168001
Goldenberg C, Tanguy A, Barrat J-L (2007) Particle displacements in the elastic deformation of amorphous materials: local fluctuations vs. non-affine field. Europhys Lett 80:16003
Goldenberg C, Goldhirsch I (2008) Effects of friction and disorder on the quasistatic response of granular solids to a localized force. Phys Rev E 77:041303
Goldhirsch I (2008) Introduction to granular temperature. Powder Technol 182:130–136
Guyon E, Roux S, Hansen A, Bideau D, Troadec J-P, Crapo H (1990) Non-local and non-linear problems in the mechanics of disordered systems: application to granular media and rigidity problems. Rep Prog Phys 53:373–419
Haff PK (1983) Grain flows as a fluid-mechanical phenomenon. J Fluid Mech 134:401–430
Herrmann HJ (1993) On the thermodynamics of granular media. Journal de Physique II (France) 3:427–433
Herrmann HJ (2002) Granular matter. Phys A 313:188–210
Hertz H (1882) Über die Berührung fester elastischer körpe. J. Reine Angewandte Mathematik 92:156–171
Hong DC, Hayakawa H (1997) Thermodynamic theory of weakly excited granular systems. Phys Rev Lett 78:2764–2767
Jaeger HM, Nagel SR, Behringer RP (1996) Granular solids, liquids, and gases. Rev Mod Phys 68:1259–1273
Jenkins JT, Mancini F (1987) Balance laws and constitutive relations for plane flows of a dense, binary mixture of smooth, nearly elastic, circular disks. J Appl Mech 54:27–34
Jenkins JT, Mancini F (1989) Kinetic theory for binary mixtures of smooth, nearly elastic spheres. Phys Fluids A 1:2050–2057
Jia X (2004) Codalike multiple scattering of elastic waves in dense granular media. Phys Rev Lett 93:154303
Jiang Y, Liu M (2003) Granular elasticity without the coulomb condition. Phys Rev Lett 91:144301
Jiang Y, Liu M (2007a) From elasticity to hypoplasticity: dynamics of granular solids. Phys Rev Lett 99:105501
Jiang Y, Liu M (2007b) A brief review of "granular elasticity", why and how far is sand elastic? Eur Phys J E 22:255–260
Jiang Y, Liu M (2008) Incremental stress-strain relation from granular elasticity: comparison to experiments. Phys Rev E 77:021306
Jiang L, Liu M (2009) Granular solid hydrodynamics. Granular Matter 11:139–156
Jiang Y, Liu M (2007c) Hydrodynamic theory of granular solids: Permanent, transient, and granular elasticity. arXiv: 0706.1352v1
Kadanoff LP (1999) Built upon sand: theoretical ideas inspired by granular flows. Rev Mod Phys 71:435–444
Kamrin K, Koval G (2012) Nonlocal constitutive relation for steady granular flow. Phys Rev Lett 108:178301
Kudrolli A (2004) Size separation in vibrated granular matter. Rep Prog Phys 67:209–247
Landau LD, Lifshitz EM (1970) Theory of elasticity. Pergamon Press, New York
Leonforte F, Tanguy A, Wittmer JP, Barrat J-L (2004) Continuum limit of amorphous elastic bodies II: linear response to a point force. Phys Rev E 70:014203
Leonforte F, Boissière R, Tanguy A, Wittmer JP, Barrat J-L (2005) Continuum limit of amorphous elastic bodies III: three dimensional systems. Phys Rev E 72:224206

Leonforte F, Tanguy A, Wittmer JP, Barrat J-L (2006) Inhomogeneous elastic response of silica glass. Phys Rev Lett 97:055501

López de Haro M, Cohen EGD, Kincaid JM (1983) The Enskog theory for multicomponent mixtures. I. linear transport theory. J Chem Phys 78:2746–2759

Mehta A, Barker GC (1994) The dynamics of sand. Rep Prog Phys 157:384–416

Monaghan JJ (1992) Smoothed particle hydrodynamics. Annu Rev Astron Astrophys 30:543–574

Murdoch AI, Bedeaux D (1994) Continuum equations of balance via weighted averages of microscopic quantities. Proc R Soc Lond Ser A 445:157–179

Nagel SR (1992) Instabilities in a sandpile. Rev Mod Phys 64:321–325

Olafsen JS, Urbach JS (1998) Clustering, order, and collapse in a driven granular monolayer. Phys Rev Lett 81:4369–4372

Olafsen JS, Urbach JS (2005) Two-dimensional melting far from equilibrium in a granular monolayer. Phys Rev Lett 95:098002

Ottino JM, Khakhar DV (2001) Fundamental research in heaping, mixing, and segregation of granular materials: challenges and perspectives. Powder Technol 121:117–122

Picard G, Ajdari A, Lequeux F, Bocquet B (2005) Slow flows of yield stress fluids: complex spatiotemporal behavior within a simple elastoplastic model. Phys Rev E 71:010501

Prevost A, Melby P, Egolf DA, Urbach JS (2004) Nonequilibrium two-phase coexistence in a confined granular layer. Phys Rev E 70:050301

Rajchenbach J (2000) Granular flows. Adv Phys 49:229–256

Reydellet G, Clément E (2001) Green function probe of a static granular piling. Phys Rev Lett 86:3308–3311

Reynolds O (1885) On the dilatancy of media composed of rigid particles in contact, with experimental illustrations. Phil Mag 20:469–481

Serero D, Rydellet G, Claudin P, Clément E, Levine D (2001) Stress response function of a granular layer: quantitative comparison between experiments and isotropic elasticity. Eur Phys J E 6:169–179

Serero D, Noskowicz SH, Goldhirsch I (2007) Exact results versus mean field solutions for binary granular gas mixtures. Granular Matter 10:37–46

Serero D, Goldhirsch I, Noskowicz SH, Tan M-L (2008) Hydrodynamics of granular gases and granular mixtures. J Fluid Mech 554:237–258

Serero D, Goldenberg C, Noskowicz SH, Goldhirsch I (2008) The classical granular temperature and slightly beyond. Powder Technol 182:257–271

Da Silva M, Rajchenbach J (2000) Stress transmission through a model system of cohesionless elastic grains. Nature 406:708–710

Tanguy A, Wittmer JP, Leonforte F, Barrat J-L (2002) Continuum limit of amorphous elastic bodies: a finite-size study of low-frequency harmonic vibrations. Phys Rev E 66:174205

Tykhoniuk R, Tomas J, Luding S, Kappl M, Heim L, Butt H-J (2007) Ultrafine cohesive powders: from interparticle contacts to continuum behaviour. Chem Eng Sci 62:2843–2864

Willits JT, Arnarson BÖ (1999) Kinetic theory of a binary mixture of nearly elastic disks. Phys Fluids 11:3116–3122

Shock Structure and Acoustic Waves in a Supersonic Jet

Catalina Stern Forgach and José Manuel Alvarado Reyes

Abstract Even though supersonic flows have been studied for a long time, many questions remain unanswered about their behavior. The understanding of jet noise goes in parallel with the understanding of jet turbulence. It has been speculated that different kinds of vortex interactions in the near field, can produce sound. Also, that the interaction between the flow and the shock structure produces noise. It is now known that noise, in supersonic and subsonic jets, is made up of two basic components; one from the large turbulence structures and instability waves, and the other from the fine-scale turbulence. Measurements inside a supersonic jet are difficult. Hot wires are easily broken and homogeneous seeding for Laser Doppler and Particle Image Velocimetries is complicated. We have developed a non-intrusive technique that uses the heterodyne detection of Rayleigh scattering. The laser light scattered elastically by the molecules of the flow at a particular angle, has information about density fluctuations of a particular size. It can be shown that the signal that comes out of a quadratic photo detector is proportional to the spatial Fourier transform as a function of time, of the density fluctuations for a wave vector given by the scattering angle. The spectral analysis of the data has allowed us to identify fluctuations of different origins; entropic and acoustic. We have taken data at many points inside and outside the flow. The technique is sensitive to the wave vector so we can study fluctuations that propagate in different directions. Fluctuations in the direction of the flow are shifted in frequency with respect to fluctuations perpendicular to the flow at the same location. The frequency shift allows us to measure the local speed of the flow. Outside the flow, only acoustic fluctuations are detected. We have been able to determine the far field acoustic radiation pattern for a given wave vector. Inside the jet, the analysis is much more complicated because the acoustic and the

C. S. Forgach (✉) · J. M. A. Reyes
Departamento de Física, Facultad de Ciencias, Universidad Nacional Autónoma de México, UNAM, Circuito Exterior S/N, Ciudad Universitaria, CO 04510, México, D. F., Mexico
e-mail: catalinastern@gmail.com

J. M. A. Reyes
e-mail: mar_ney2003@yahoo.com.mx

L. Di G. Sigalotti et al. (eds.), *Computational and Experimental Fluid Mechanics with Applications to Physics, Engineering and the Environment*, Environmental Science and Engineering, DOI:10.1007/978-3-319-00191-3_9,

entropic peaks overlap when we use simple Fourier transforms. However, with the use of parametric periodgrams we have been able to identify each type of fluctuation. Moreover, we found a third peak at a much lower frequency that appears and disappears as we move along the centerline of the jet. This peak appears also in other positions outside the centerline. We have used Rayleigh scattering and Schlieren to visualize the shock structure. We can then associate each spectrum with a position in the jet relative to the shock structure. The slow peak appears always at a shock, probably due to the interaction between the flow and the shock structure. We are now working on the visualization of the flow, and hope that the combination of all the techniques will give us further insight into the global behavior of the flow, especially in the interfaces between the flow and the shocks and between the mixing layer and the stationary fluid.

1 Introduction

Supersonic noise is of great importance in several industrial applications (Nichols et al. 2011). The objective of this work is to understand the production and propagation of acoustic waves in a supersonic jet. On one hand, there are several hypotheses about acoustic production in supersonic jets, related to vortex interactions in the mixing layer, the interaction between the flow and the shock structure, and to small scale turbulence. Each kind of aerodynamic events produces waves of different frequency (Bodonyy and Lele 2006). The traditional methods to study these flows are intrusive: hot wire anemometry, laser Doppler, or particle image velocimetries and sometimes involve complicated correlations of signals from three dimensional arrangements of microphones in the far acoustic field outside the flow with flow measurements in the near field. We hope to develop simpler methods to relate the acoustic field to aerodynamic events in the flow.

In the late seventies, in the École Polytechnique in France, a non-intrusive optical technique was developed to detect density fluctuations inside and outside of the flow (Stern and Grésillon 1983). This technique takes advantage of the heterodyne detection of laser light elastically scattered by the molecules of a transparent gas in motion (Rayleigh scattering). It can also be used to measure the mean velocity in the scattering volume. In the Hydrodynamic and Turbulence Laboratory in the Physics Department of the School of Science in UNAM, the technique has been complemented with signal processing methods that increase the resolution in frequency space and with different forms of visualization of the shock structure (Aguilar 2003).

The acoustic emission pattern for some frequencies has been measured with the technique mentioned. Also, with the help of visualizations, it has been possible to relate measurements of density fluctuations with the shock structure.

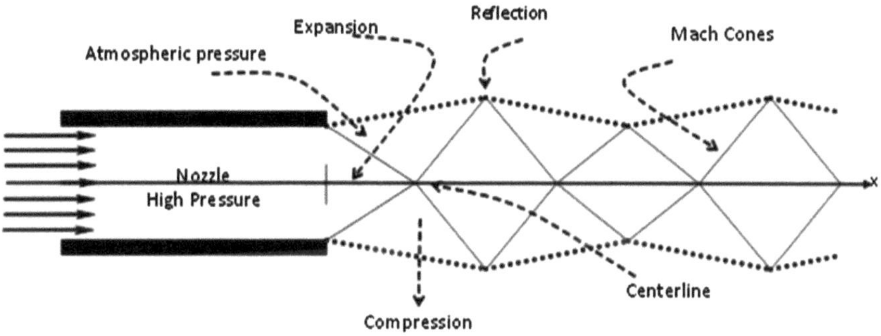

Fig. 1 The discontinuity at the end of the nozzle creates a perturbation that gives rise to a stationary shock pattern, when the flow is supersonic

2 Theoretical Background

In this section, some relevant concepts of supersonic flows will be reviewed first and then, some concepts from the electromagnetic theory behind the measuring and visualization techniques.

2.1 Supersonic Jets, Shock Structure, Density Fluctuations, and Sound Production

2.1.1 Supersonic Jets and Shock Structure

When a stream of fluid comes out of a nozzle, and mixes with the surrounding nozzle, a jet is formed. The two dimensionless parameters that characterize these flows are the Reynolds Number $R_e = vD/\nu$ and the Mach Number $M = v/c_s$, where v is the speed of the flow, D the exit diameter of the nozzle, ν the kinematic viscosity and c_s the local speed of sound. When M is close to or larger than 0.3, compressibility effects, that is density fluctuations, become important. The jet is supersonic for $M \geq 1$ at the exit.

Figure 1 shows the structure of a supersonic jet. The discontinuity at the edge of the nozzle produces a perturbation that propagates at the speed of sound. Each new perturbation catches on the previous one. The addition of these perturbations creates a conic region of very high density called shock. Starting with an expansion, a stationary pattern of shocks is formed in the supersonic region of the jet (Goldstein 1976).

As the speed decays the flow becomes subsonic and the shocks disappear.

2.1.2 Density Fluctuations

The equations that describe a compressible flow are more complicated than for the incompressible case. However, if small oscillations about a point of equilibrium are considered, the equations can be linearized. Monin and Yaglom (1987) have shown that if the equations of motion are written in terms of the vorticity Ω, the divergence D of the velocity, the entropy S, and the pressure P, all possible motions can be described by three non-interacting modes:

$$\begin{aligned}
&\frac{d\Omega(t)}{dt} = 0 \\
&\frac{dS(t)}{dt} = 0 \\
&\frac{d^2D(t)}{dt^2} + a_0^2k^2D(t) = 0 \\
&\frac{d^2P(t)}{dt^2} + a_0^2k^2P(t) = 0.
\end{aligned} \tag{1}$$

The incompressible vorticity (Ω) mode and the entropy (S) mode are stationary or move at constant speed. The acoustic (P) or potential (D) mode is related to pressure fluctuations that propagate at the speed of sound as can be seen from the wave equation. The entropic and acoustic modes, related to the compressible part of the flow, can be studied by Rayleigh scattering.

2.1.3 Acoustic Emission

There are several theories that try to explain acoustic emission by a jet based on the near field aerodynamics, either vortex interactions in the shear layer like pairing, tearing, or merging, or small scale motions due to instabilities generated at the nozzle. In general, it is now accepted that noise, in supersonic and subsonic jets, is made up of two basic components: one from the large turbulence structures and instability waves, the other from the fine-scale turbulence (Tam 1992, 1998, 2012; Veltin 2008). There are other possible sources of emission like the interaction of the flow with the shock waves and the feedback of acoustic waves that re-enter the flow.

Traditionally, experimental studies on acoustic waves are done by placing many microphones in the far field and correlating these measurements with events measured inside the flow. Most of the local techniques are intrusive, like hot wires or LDA and PIV that require homogeneous seeding (Goldstein 1983). Not only there is a problem trying to determine uniquely sources from far field measurements, but also certain phenomena like the diffraction of the acoustic waves by the mixing layer are usually not taken into account.

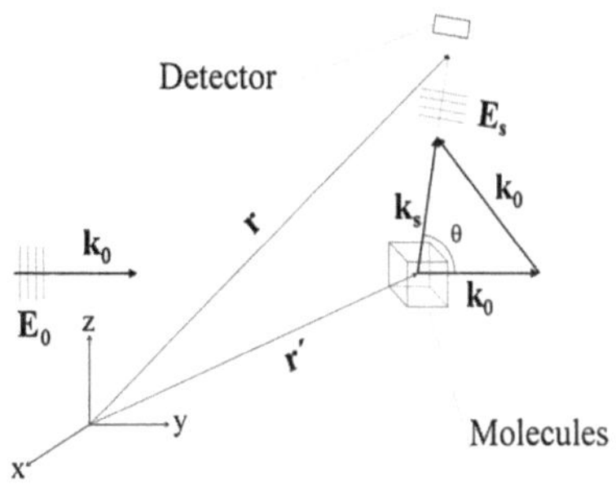

Fig. 2 Molecules scatter light in all directions. By selecting the orientation of the detector, we select the scattering angle and thus the size of the fluctuations to be studied

2.1.4 Rayleigh Scattering

The elastic scattering of an electromagnetic wave of wavelength λ_o by a neutral particle of dimensions smaller than the wavelength is known as Rayleigh scattering. In a static transparent gas, light is scattered homogeneously. If the gas is in motion or with strong density variations, the characteristics of the scattered light reflect the characteristics of the structure and motion of the gas (Jackson 1962). Figure 2 shows the wave vectors of the incident and scattered light.

The total scattered field can be obtained from the integral

$$\overrightarrow{E_{ST}} = \overrightarrow{E_S}(\vec{r}, t) \int d^3r'n\left(\overrightarrow{r'}, t\right) \exp\left(-i\overrightarrow{k_\Delta} \cdot \overrightarrow{r'}\right) = \overrightarrow{E_S}(\vec{r}, t)\, n(\overrightarrow{k_\Delta}, t), \quad (2)$$

where $\overrightarrow{E_S}(\vec{r}, t)$ is the field scattered by one molecule, $n(\vec{r}, t)$ is the distribution of molecules in the scattering volume Vs and $n\left(\vec{k_\Delta}, t\right)$ is the spatial Fourier transform of the density fluctuation. The scattered field has information about the motion of the molecules in the scattering volume through the spatial Fourier transform of the density.

2.2 Experimental Techniques

Rayleigh scattering has been used in two different ways: to visualize the flow and to obtain the spatial Fourier transform of the density fluctuations as a function of time. Shadowgraphs have been used also to visualize.

2.2.1 Visualization Methods; Rayleigh and Shadowgraphs

The shock structure can be visualized if the light scattered at small angles is captured with a lens and sent into a screen. The part of the laser bean that is not scattered is blocked (Azpeitia 2004) (see Fig. 3).

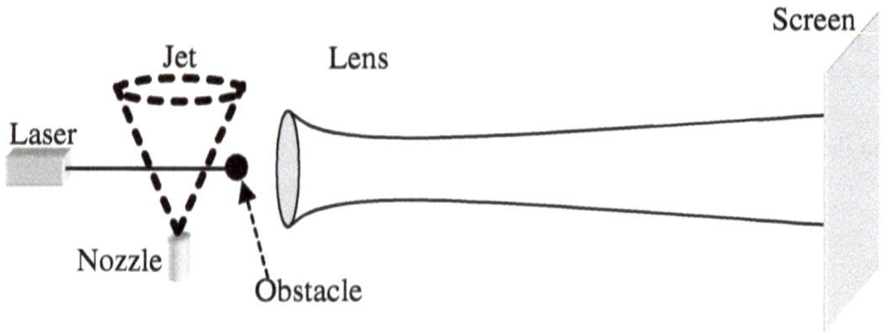

Fig. 3 Set-up for Rayleigh scattering

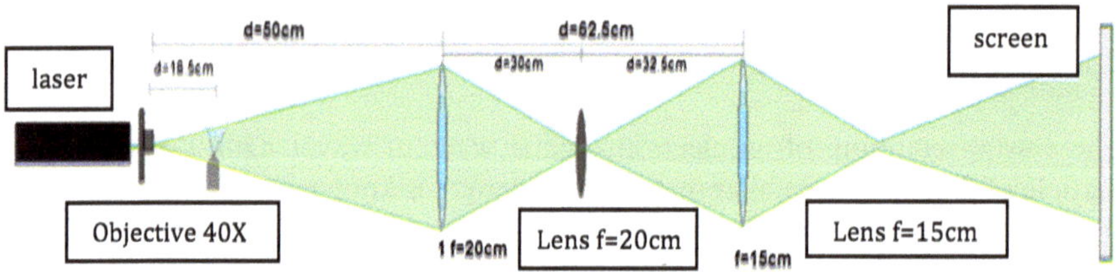

Fig. 4 Set-up for shadowgraphs (without attenuator in the middle) and Schlieren (with attenuator) (Salazar 2012)

If a cylindrical lens is placed in front of the laser, a sheet of light is formed and a larger section of the jet can be observed on the screen.

Shadowgraphs and Sclieren methods are used to detect changes in the index of refraction that cannot be seen by the naked eye (Settles 2001; Salazar 2012) (see Fig. 4).

2.2.2 Heterodyne Detection of Rayleigh Scattering

The amplitude of the light scattered by the molecules is extremely small, and it cannot be measured by a common diode. To solve this problem we mix it, on the surface of the photodetector, with a reference beam of light called the local oscillator. The local oscillator is displaced in frequency with respect to the incident beam by 110 MHz. This technique is known as heterodyne detection. Figure 5 shows the experimental set-up.

The beam that comes out of the laser is sent into an acoustic modulator. The modulator acts as a Bragg cell and several orders of diffraction come out. The order zero goes through without being deviated; we refer to this beam as incident or primary. The order one is diffracted at a certain angle, is less intense, and is displaced in frequency by 110 MHz. We will refer to this beam as the local oscillator. Both beams are manipulated so that they cross at an angle θ in the volume to be studied. The local oscillator is sent directly to a photodetector, the main beam is blocked just after the

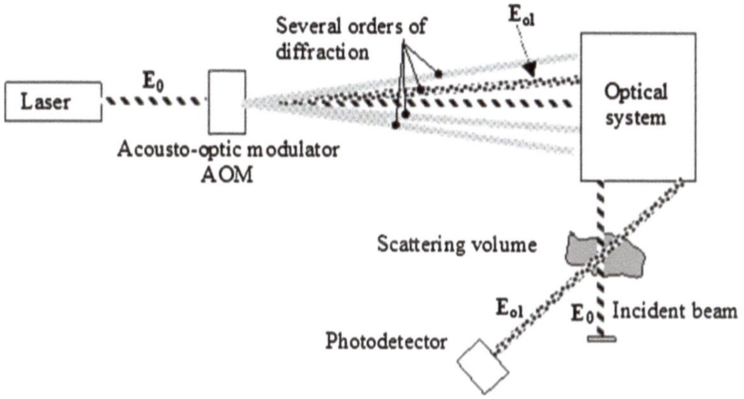

Fig. 5 The light from the laser is sent to an acousto-optic modulator to obtain several orders of diffraction displaced in frequency. One of these beams is used as the local oscillator to be mixed with the scattered light at the surface of the detector

scattering. On the photodetector then arrives the part of the incident field scattered at the angle θ and the local oscillator. The scattering angle determines the wavenumber of the fluctuations that are studied through the equation

$$k_{\Delta} = 2k_o \sin\frac{\theta}{2}, \tag{3}$$

where $\overrightarrow{k_o}$ is the wavevector of the incident field, $\overrightarrow{k_{\Delta}}$ is the wavevector of the density fluctuations, $k_o = \left|\overrightarrow{k_o}\right|$, and $k_{\Delta} = \left|\overrightarrow{k_{\Delta}}\right|$.

The photodetector is sensitive to the intensity of the incident light, so the current it produces is proportional to the square of the electric field incident on its surface. The current of the photodiode is then proportional to

$$\left(\overrightarrow{E_{ST}} + \overrightarrow{E_{OL}}\right)^2 = \left|\overrightarrow{E_{ST}}\right|^2 + \left|\overrightarrow{E_{OL}}\right|^2 + 2\vec{E}_{ST} \cdot \vec{E}_{OL}. \tag{4}$$

The first two terms are constant and give a constant voltage. The first is too small to be extracted from the total value, and the second is of no interest. The third term gives a time dependent current that oscillates at the frequency difference of the frequencies of both electric fields, contains the information we are interested in, and is modulated by the amplitude of the local oscillator. The current proportional to this term is known as the heterodyne current (Yariv 1976).

It can be shown that the spectral density of the heterodyne current I(ω) produced by all the scatterers is of the form

$$I(\omega) = \frac{1}{8\pi k_0^2}\left(\frac{\eta e}{\hbar\omega_0}\right)^2 n_0 \left(r_0^R\right)^2 \frac{\varepsilon_0}{\mu_0}\left(\overrightarrow{E_0}\cdot\overrightarrow{E_D}\right)^2 \int d^3k \left|W(\overrightarrow{k_\Delta} - \vec{k})\right|^2 \left[S(\vec{k}, \omega - \omega_\Delta) + S(\vec{k}, \omega - \omega_\Delta)\right] \quad (5)$$

where η is the efficiency of the detector, n_o the mean density, W is related to the Gaussian profiles of the beams and $S(\vec{k}, \omega)$ is the form factor defined by

$$S(\overrightarrow{k_\Delta}, \omega) = \frac{\left|n(\overrightarrow{k_\Delta}, \omega)\right|^2}{n_0 V} \quad (6)$$

Density fluctuations have been studied with Rayleigh scattering using other techniques (Panda and Seasholtz 1998).

3 Experimental Results

3.1 Visualization

When the near region is illuminated, the first shock created by the discontinuity at the nozzle can be observed (Fig. 6). The exiting flow expands as it comes out of the nozzle, but the stationary gas compresses it. The result is a high density region in form of a V.

When the flow is illuminated by a sheet of light, various shocks can be observed (Fig. 7).

In the shadowgraphs and Schlieren images the shock structure is better defined than in those shown above. Part of the internal structure can be observed. Figure 8 shows three first shocks for flows with different exit velocities. When the speed is very high, instead of a sharp cross, a flat region appears known as Mach's disk.

3.2 Heterodyne Detection and Spectral Densities

To obtain I(ω), the current that comes out of the photodetector i(ω,t) can be sent directly to a spectrum analyzer or acquired and processed with a computer. Due to the fact that the technique is sensitive to wave vector $\overrightarrow{k_\Delta}$ determined by the optical set-up, we can observe fluctuations propagating in different directions. To apply the previous equations, it is necessary to make sure that the beams remain Gaussian along the optical path. We have checked this in different points, and measured precisely the scattering volume.

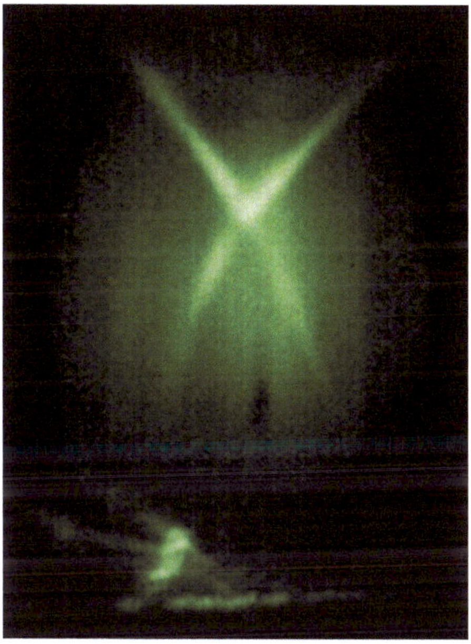

Fig. 6 The flow goes upward. The first shock composed of an expansion and a compression is observed

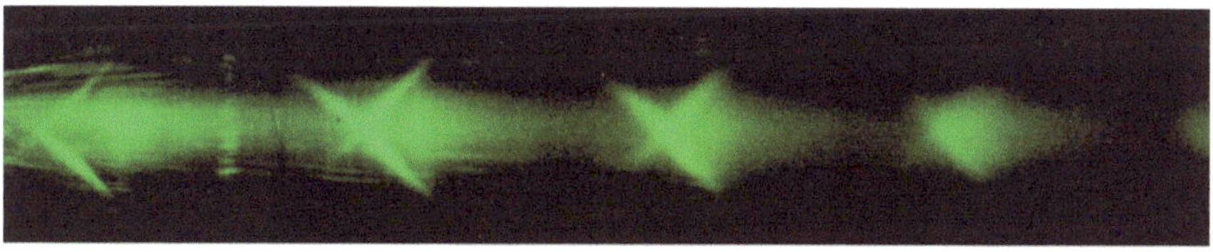

Fig. 7 The flow goes from left to right. A series of shocks can be observed

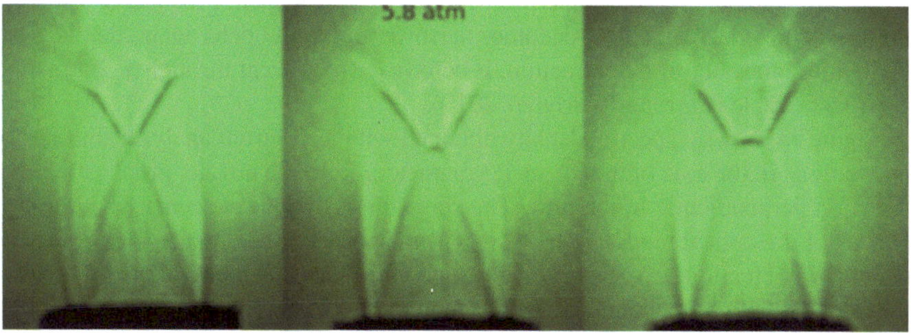

Fig. 8 First shock for three different exit velocities. The crossover from expansion to compression flattens as the speed of the flow increases

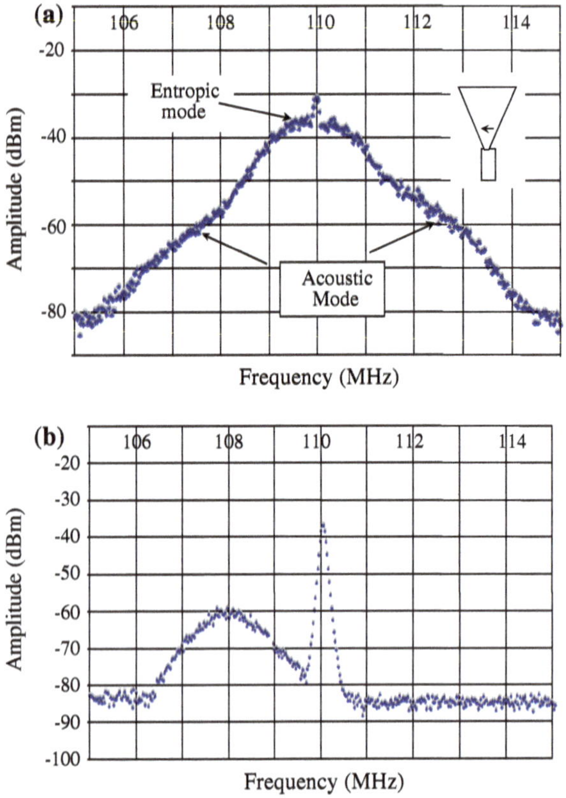

Fig. 9 Spectral densities for fluctuations traveling perpendicular to the flow. **a** centerline at 1.3 diameters from the nozzle, **b** outside the jet

We are also able to determine the width of the jet at each x location. When the flow exits the nozzle, the flow becomes narrower before it opens again; that is, even though it comes out of a contraction, it behaves as if it went through a contracting divergent nozzle (Carreño Rodríguez 2010).

Figure 9a shows a spectral density obtained with a spectrum analyzer for density fluctuations in the axis of the jet, at 1.3 diameters from the nozzle, propagating in a direction perpendicular to the flow. The spectrum is symmetric centered at the frequency of the optical modulator. The spectrum appears to be broad and corresponds mainly to fluctuations of entropic origin due to the turbulent nature of the flow. However, a small bump on either side of the center, at a frequency that can be identified with an acoustic fluctuation can be observed.

Figure 9b shows density fluctuations outside the flow. The center of the peak corresponds to an acoustic wave traveling in a direction opposite to the wave vector defined by the optics.

Figure 10a shows the spectral density obtained at the same point than Fig. 9a but for fluctuations traveling parallel to the flow. In this case, the fluctuations are convected with the flow and the entropic peak is displaced to the right because of the Doppler shift. The local mean velocity can be measured from the frequency shift. In this figure the acoustic peak cannot be observed on the right, probably because it is submerged under the entropic peak. The peak on the left comes from outside of the jet due to the length of the scattering volume.

Figure 10b is similar to Fig. 10a but at a different location along the centerline. It is important to notice a low frequency peak.

From Fig. 10 two important observations can be made. First, if the acoustic peak exists, it could be hidden under the broad entropic peak; or its frequency could be so close to the entropic frequency that they form together a broad peak and the two frequencies cannot be identified separately. The second observation is that in certain locations, always at the centerline of the jet, a peak appears at a much lower frequency than the entropic peak.

3.3 Signal Processing and Parametric Periodgrams

From the conclusions above, it became necessary to search for an appropriate signal processing that would increase the spectral resolution. The first thing to be considered, besides Nyquist theorem, is that a very high sampling frequency would indeed give a very faithful reconstruction of the signal in time but extremely poor in the frequency domain. The spectral densities obtained with the spectrum analyzer showed the need for much better frequency resolution. The second thing is the method to obtain a spectral density with less noise.

This seems a simple matter but most oscilloscopes and spectrum analyzers do not allow the user to choose freely the sampling frequency.

After analyzing several techniques, we decided to use Burg parametric periodgrams. If the acquisition is done right, there is a wide range within which the spectral density is independent of the number of parameters (Alvarado Reyes 2004, 2010).

3.4 Slow Mode and Shock Structure

Figure 11 shows a series of spectra for fluctuations traveling perpendicular to the flow, obtained with periodgrams at different locations along the centerline.

Three peaks are visible. The acoustic peak is always at the same location. The entropic peak changes with the local speed of the flow. As expected, the speed of the flow changes very little in the supersonic region. The slow peak appears and disappears along the centerline and changes slightly its frequency. It is interesting to note that when the new peak has its highest amplitude, the acoustic peak disappears and vice versa. We are searching for an explanation for this effect.

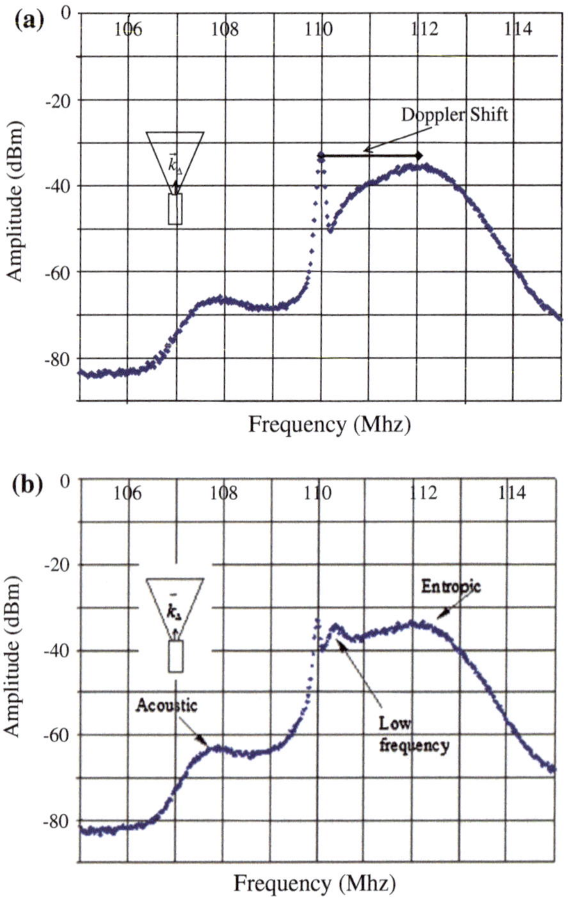

Fig. 10 Spectral densities for fluctuations traveling parallel to the flow. **a** at 1.3 diameters from the nozzle, **b** at another location along the centerline. A low frequency peak appears at certain positions. The acoustic peak on the left comes from outside of the jet. For small angles the scattering volume is very long in the direction perpendicular to the wave vector

If we compare Figs. 7 and 11, we obtain Fig. 12 where we can determine that the regions of maximum amplitude of the low frequency peak, correspond to the crossover between expansion and compression in the shocks. The x coordinate (along the centerline of the jet) is given in multiples of the nozzle diameter.

This low frequency can be related to the interaction between the shock structure and the flow. The shock structure appears very stable in all the images. However, there is a possibility that it oscillates at a frequency that cannot be detected by the eye. This will be determined with further tests with a high speed camera. In both cases, this slow mode would be maximum close to the shock structure.

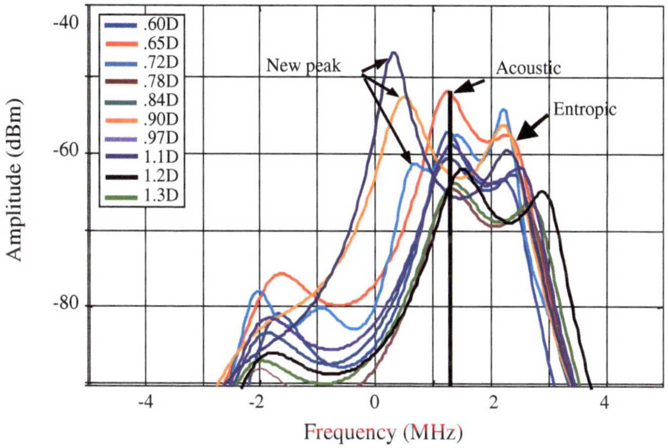

Fig. 11 Spectral densities obtained with Burg's parametric periodgrams at various positions along the centerline of the jet

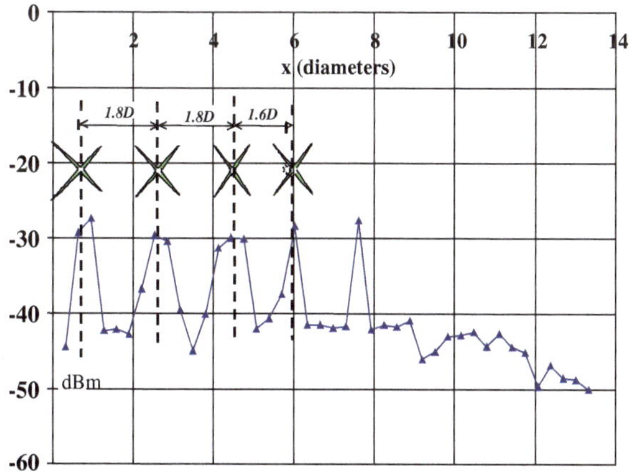

Fig. 12 Comparison between the locations at which the low frequency peak is maximum and the shock position along the centerline

3.5 Emission Pattern

We have taken data at many positions inside and outside the jet, and processed the signals with Burg's periodgrams, for different directions of the wave vector. At each point, we determined the direction for which the amplitude of the acoustic fluctuation is the largest. Figure 13 shows the spectra at the same point for different directions of the wave vector.

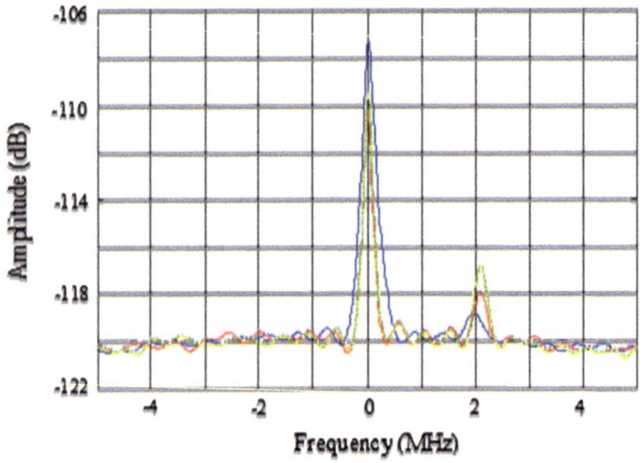

Fig. 13 Periodgrams obtained at the same point for different directions of the wave vector outside the jet. The maximum amplitude corresponds to the direction of propagation of the fluctuation

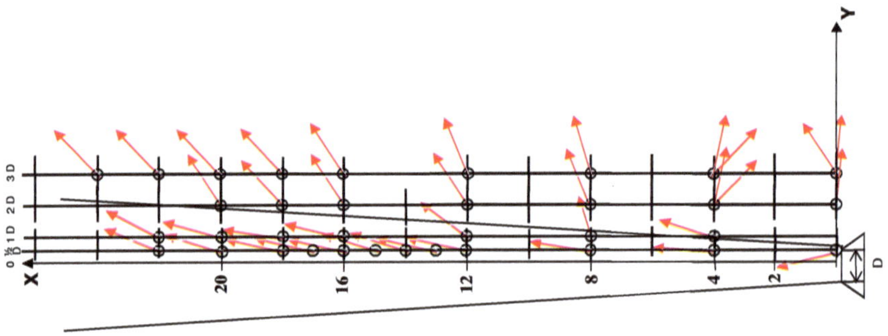

Fig. 14 Acoustic radiation pattern for a supersonic jet

We define the direction of maximum amplitude as the direction of propagation of the acoustic fluctuations. As shown in Fig. 14, we were able to construct an acoustic emission pattern for a jet.

It must be pointed out that with our technique, we can only detect ultrasound waves. To reduce the frequencies we can detect for acoustic waves a larger wavelength for the incident laser is required.

3.6 Particle Image Velocimetry (PIV)

So far, we have been able to determine the speed of the flow either by the Doppler shift of the spectrum, or by the angle of the shock wave. The two measurements give different results, probably due to the lack of precision in the measurement of the angle. To have a better measurement to compare with, we are in the process of

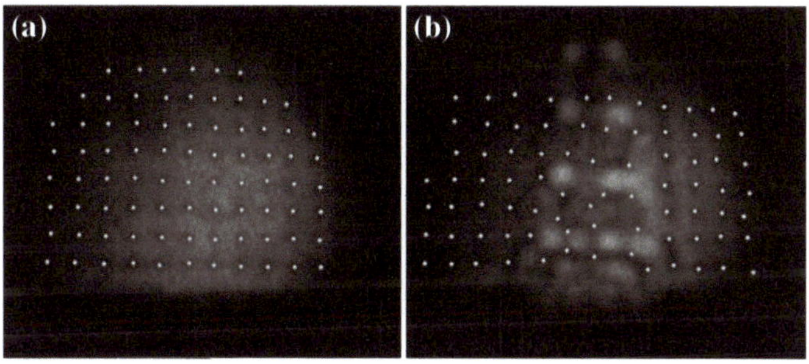

Fig. 15 Variation of the index of refraction. **a** Original dot net, **b** dot net modified by the jet flow

doing Particle Image Velocimetry on the flow. It is very difficult to seed high speed gas flows because most commercial particles are too heavy to follow the flow. The way seeds are introduced must not affect the flow speed and the distribution should be homogeneous at least for some time. It is well known that turbulence and large vertical motions redistribute the particles (Echeverría Arjonilla 2013).

In our experiments, we have detected that the seeds modify the shock structure. The first shock occurs closer to the nozzle, the angle becomes larger, and thus the local speed decreases. We are not sure yet whether it is the presence of the titanium dioxide particles or the system of injection that produces this modification.

3.7 Quantification of the Index of Refraction

The images shown in Figs. 6, 7, and 8 have been used so far to provide only qualitative information, except for the determination of the angle of the shock wave and thus the local speed.

Schlieren images and shadowgraphs can be used, with the help of a net of dots, to quantify the variation of the index of refraction. To do so, an image of a net of points is placed behind the location where the flow will be. An image without flow is taken as shown in Fig. 15a and b (Porta 2013)

The change in the position of each dot is related to the change in the index of refraction and thus to the density. Figure 16 shows the field of displacements of the dots. We are on the process of relating this field with the size of the shocks.

4 Conclusions and Future Work

Rayleigh scattering has proven to be a very useful tool because it is non-intrusive, does not require seeding, and measures directly the motion of the fluid.

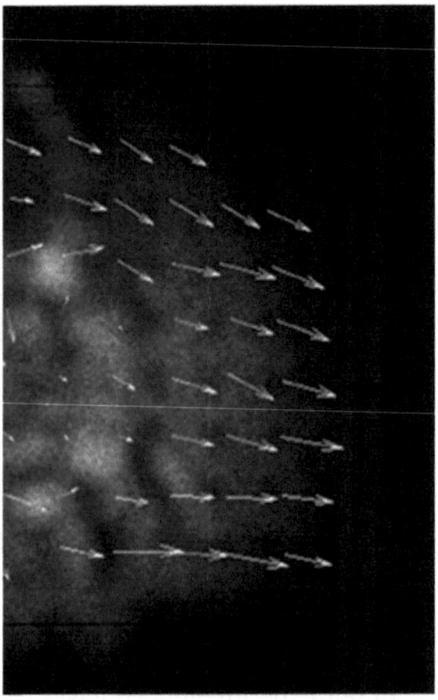

Fig. 16 Displacement field which is proportional to the gradient of the index of refraction

Heterodyne detection of Rayleigh scattering can differentiate among three different density fluctuations: acoustic waves propagating at the speed of sound, entropy fluctuations convected by the flow velocity, and a low speed fluctuation that might be related to interaction between the flow and the shock structure. It is still necessary to determine if the shock structures fluctuate with time.

With this technique we have observed that, contrary to what most textbooks prove, a contraction can produce a supersonic flow and a stable shock structure. We have shown that the flow gets narrower just outside of the nozzle and then reopens again (Carreño Rodríguez 2010). This happens in a length shorter than a diameter.

This technique can also be used to measure the local speed of flow in any direction. The precise measurement of the frequency, that can be done when the spectral density is calculated with periodgrams, gives a precise measurement of the speed.

The system is sensitive to the direction of propagation of the fluctuations, so it can be used to determine the acoustic radiation pattern of the flow for a wave number determined by the optics. It is quite easy to change the scattering angle, and thus the wave number. Its limitation consists that only fluctuations with the wave vector determined by the optical set up can be detected.

The spectral density is related to the energy, so it can be seen how the energy is distributed among the different kinds of fluctuations for each wave number.

The supersonic structure can be visualized with Rayleigh scattering, with Schlieren images and with shadowgraphs. Each technique gives different information about the shock structure and about the changes in the index of refraction.

The work that is presently being done in PIV and the quantification of the index of refraction will complement the information we already have. The visualization of the flow simultaneously to the shock structure is still a big challenge, because the shocks scatter an enormous amount of light that hides all other motions.

The biggest problem of the heterodyne detection of Rayleigh scattering is the spatial resolution of the scattering volume. The scattering volume is formed by the intersection of the local oscillator and the primary beam. Very small angles (about 2 mrad) have to be used to detect lower frequencies of acoustic interest. As a consequence, the scattering volume is very thin but very long, so the spatial resolution in the direction perpendicular to the direction of propagation is very bad. Presently, a system with two local oscillators crossing the primary beam at the same angle in the same location is being developed. Correlations between the two signals should help reduce the problem.

In textbooks, shocks are lines representing sudden changes in the density. We know however, that the change in density is continuous. Our next interest is to determine experimentally the changes in the density across the shocks, and give an operational way to measure the interface. Also Schlieren images and shadowgraphs will show if there is any internal structure.

Acknowledgments We acknowledge support from UNAM through DGAPA projects IN107599, IN104102, IN116206 and IN117712. Also the participation of several undergraduate students: Cesar Aguilar, Carlos Azpeitia, Alejandro Carreño, Yadira Salazar, Carlos Echeverría, and David Porta.

References

Aguilar C (2003) Detection of acoustic waves in a supersonic jet using Rayleigh scattering. BS thesis, Department of Physics, School of Science, UNAM, Mexico City, Mexico

Alvarado Reyes JM (2004) Spectral analysis of signals from Rayleigh scattering experiment. M.E, Thesis, School of Engineering, UNAM, Mexico City, Mexico

Alvarado Reyes JM (2010) Técnicas modernas para el tratamiento de señales turbulentas, PhD thesis, Universidad Nacional Autónoma de México

Azpeitia C (2004) Use of Rayleigh scattering to localize acoustic sources in a supersonic jet. BS thesis, Department of Physics, School of Science, UNAM, Mexico City, Mexico

Bodonyy DJ, Lele SK (2006) Low frequency sound sources in high-speed turbulent jets. Annual Research Briefs, Center for Turbulence Research, Stanford University, Palo Alto. Ca, USA

Carreño Rodríguez AS (2010) Reconstruction of gaussian beams to increase the spatial resolution in a Rayleigh scattering experiment. BS thesis, Department of Physics, School of Science, UNAM, Mexico City, Mexico

Chapman CJ (2000) High speed flow. Cambridge University Press, Cambridge

Echeverría Arjonilla C (2013) PIV measurements in a supersonic flow. BS Thesis, Department of Physics, School of Science, UNAM, Mexico City, Mexico

Goldstein ME (1976) Aeroacoustics, 1st edn. McGraw-Hill, USA

Goldstein RJ (1983) Fluid Mechanics Measurements, 1st edn. Edit, Hemisphere, USA

Jackson JD (1962) Classical electrodynamics. John Wiley & Sons Ltd, New York

Monin AS, Yaglom AM (1987) Statistical fluid mechanics. MIT Press, Cambridge

Nichols JW, Ham FE, Lele SK, Monin P, Ham FE, Lele SK, Moin P (2011) Prediction of supersonic jet noise from complex nozzles. Annual Research Briefs, Center for Turbulence Research, Stanford University, Palo Alto. Ca, USA

Panda J, Seasholtz R (1998) Density measurement in underexpanded supersonic jets using Rayleigh scattering. 36th AIAA Aerospace Sciences Meeting and Exhibit, 1998,10.2514/6.1998-281
Porta D (2013) Interfaces in a supersonic jet using shadowgraphs. BS Thesis, Department of Physics, School of Science, UNAM, Mexico City, Mexico
Salazar Romero MY (2011) Advanzed optical techniques applied to fluid dynamics. BS Thesis, Department of Physics, School of Science, UNAM, Mexico City, Mexico
Settles GS (2001) Schlieren and shadowgraph techniques. Visualizing Phenomena in Transparent Media, Edit, Springer, USA
Stern C, Grésillon D (1983) Fluctuations de Densité dans la Turbulence d'un Jet. Observation par Diffusion Rayleigh et Détection Heterodyne. J Phys 44:1325–1335
Tam C (2012) Computational aeroacoustics: A wave number approach. Cambridge University Press, Cambridge
Tam C (1998) Jet Noise: Since 1952. In: Theoretical and computational fluid dynamics. Springer, New York
Tam C (1992) Broadband shock associated noise from supersonic jets measured by a ground observer. 30th Aerospace Sciences Meeting and Exhibit, 1992,10.2514/6.1992-502
Veltin J (2008) On the characterization of noise sources in supersonic shock containing jets. PhD Thesis, Pennsylvania State University, Google Books
Yariv A (1976) Introduction to optical electronics. Holt, Rinehart and Winston, New York

Complex Fluids, Soft Matter and the Jamming Transition Problem

Alberto A. Díaz and Leonardo Trujillo

Abstract We present an introductory view of the jamming transition problem, starting from Soft Matter, passing through Granular Matter and ending up with Jamming. Various properties of Soft Matter are discussed, because almost all the systems included in this category can be jammed. Then, we discuss fundamental and intrinsic aspects of Soft Matter systems. Although they look like a hodgepodge of things, they share some common features. Here, we propose that Granular Matter could provide a framework to understand essential aspects of Soft Matter. Granular materials can mimic glassy, liquid, solid, and gas-like behaviours and one can use them to understand the other members of Soft Matter. Finally, we present an overview of the jamming transition problem and outline a program towards a unified theory of Soft Matter.

1 What is Soft Matter?

Before entering into our discussion on the jamming transition problem, we have to address some questions and definitions briefly. We start with Soft Matter[1] (SM), because jamming transition appears in systems included under this common denomination, so *What is Soft Matter?*

[1] fr. Matière Molle.

A. A. Díaz · L. Trujillo
Centro de Física, Instituto Venezolano de Investigaciones Científicas, IVIC, Apartado Postal 20632, Caracas 1020-A, Venezuela
e-mail: aadv777@gmail.com

L. Trujillo (✉)
The Abdus Salam International Centre for Theoretical Physics, ICTP, Strada Costiera 11, 34014 Trieste, Italy
e-mail: leonardo.trujillo@gmail.com

L. Di G. Sigalotti et al. (eds.), *Computational and Experimental Fluid Mechanics with Applications to Physics, Engineering and the Environment*, Environmental Science and Engineering, DOI:10.1007/978-3-319-00191-3_10,

The definition of SM has been controversial owing to the existence of a plethora of features characterizing each one of the different members of this rare family. Kleman and Lavrentovich (2002) wrote that what we call SM covers a large variety of systems; from polymers to colloids, from liquid crystals to surfactants, and from soap bubbles to solutions of macromolecules. Regarding the above intuitive conception of SM, Jones (2002) expressed that many such materials are familiar from everyday life: for example, glues, paints, and soaps, while others are important in industrial processes, such as polymer melts which are moulded and extruded to form plastics; colloids which are used in pharmaceutical applications; surfactants which may act as detergents, wetting agents, emulsifiers, foaming agents, and dispersants; and liquid crystals in LCDs and glasses. Much of the food we eat and indeed much of the stuffs of life are part of this wide family, and all of them share similar qualities of mutability and responsiveness to its surrounding. Biological systems are of great importance, in fact, Poon et al. (2002) established that soft condensed matter is the place where physics meets biology and Jones (2002) quoted a William Burroughs'(1961) apt phrase—*we are ourselves soft machines*—.

The above sketch brings to us an idea of the vast field of SM physics. It looks like a hodgepodge of weird characters (and granular materials (GM) have not been introduced yet. We consider that GM systems play a key role in our approach of unification of SM; they could have a sort of "messianic presence" because they will appear as unifiers). So, how can we define SM?, what kind of materials can be classified as SM?, and what do these apparently disparate materials have in common? In regard to a definition of SM many questions may still arise.

In his Nobel lecture in 1991 (de Gennes 1991), the French physicist de Gennes said that the name *complex fluids* to talk about SM is an ugly name, which tends to discourage young students. However, Piazza (2011) shows us that complex fluids could be a convenient subclass of SM. de Gennes (2005) defines SM as all physio-chemical systems which have a large response function.[2] As a realization of this, in the same report, four examples are illustrated. They are related to vulcanization of rubber, nematic liquid crystals, ferroelectric smectics C^*, and single domain nematic rubbers. In the first of those examples, a very diluted additive is enough to switch a system of polymer chains (the latex of hevea tree) from liquid to solid state: a rubber.

A further definition of SM was given by Möhwald (2005) as materials that are held together by non-covalent interactions. This interactions are typically weak, often on the order of $k_B T$ (where k_B is the Boltzmann's constant and T is the temperature), and thus comparable to entropic forces. Möhwald gave this definition by analogy with supramolecular chemistry. In particular, he considered that SM is not easily

[2] When experiments are performed on thermodynamic systems, the quantities which are easiest to measure are the response functions. Generally, we change one parameter in the system and see how other parameters respond to that change under highly controlled conditions. They also provide a measure of the size fluctuations in a thermodynamic system (Reichl 1998). Response functions are the usual method for characterizing the macroscopic behaviour of a system. They are experimentally measured from changes in thermodynamic coordinates with external probes.

defined in terms of elastic moduli. For instance, polymer science is part of SM but there exist polymers as hard as steel.

Piazza (2011) established a broad classification of SM in contrast to Jones' definition. The second dubbed SM as a convenient term for materials in states of matter that are neither simple liquids nor crystalline solids of the type studied in other fields of solid state physics (Jones 2002). This definition is more related to stiffness deformability, i.e., a gentle external influence has a big effect. The first wrote that there is not a single name to label all those systems, and although scientists dubbed them SM, this term is suitable for a group of materials, like those similar to ricotta, whipped cream, toothpaste, etc. However, others are really too soft, to the point that they are not even solids, but rather liquids. In these cases, we should speak more precisely of complex fluids to distinguish them from simple fluids such as air or pure water. On the other hand, opals and concrete do not fit neither soft nor fluid, and so they should be dubbed supramolecular materials.[3]

Kleman and Lavrentovich (2002) wrote that every system that falls under the name of SM belongs, with very few exceptions, to organic chemistry because the building blocks of SM are organic molecules with often complicated architectures, anisometric in shape, and bounded by weak interactions. In fact, they considered colloids, biological matter (proteins, membranes, DNA and their associates, like viruses or microtubules[4]), liquid crystals, and polymers as SM. In contrast, in Piazza's book (Piazza 2011) we find that colloids, micelles, vesicles, and emulsions; colloid crystals, gels, and glasses; liquid crystals, granular matter, membranes, biopolymers, and biological machines can be classified as SM. However, this classification will make sense if we would be able to highlight common features for all of these systems. Here we are concerned with this labour but not following the same pathway.

A very interesting definition of SM was given by Witten (1999)—SM is a fluid in which large groups of the elementary molecules have been constrained so that the permutation freedom within the group is lost—. This definition is related to the following facts: i) the mobile molecules of a simple fluid may freely exchange positions, so that their new positions are permutations of their old ones, ii) in contrast, the molecules of an ideal solid have fixed positions and may not readily permute in this way.

Hamley (2007) wrote the following about SM: *The term soft matter originates from macroscopic mechanical properties. We mean here, materials such as colloids, surfactants, liquid crystals, certain biomaterials, and polymers in the melt or solution. Many soft materials can be induced to flow under certain conditions (this could be dubbed in some cases as an unjammed state). This weak ordering results from the lack of three dimensional atomic-range order as found in a crystalline solid. Nevertheless, there is always a degree of local order at least as great as that in a liquid. From the viewpoint of kinetic energy, a crude distinction between soft materials and*

[3] Supramolecular structures are large molecules formed by bonding smaller molecules together.

[4] They are the third component of the cytoskeleton and are rigid hollow rods approximately 25 nm in diameter. They are dynamic structures that undergo continual assembly and disassembly within the cell, and are composed of a single type of globular protein, called *tubulin*.

hard is on the basis that the molecular kinetic energy for the former is close to $k_B T$, whereas for the latter it is much less than $k_B T$ (when temperature is near ambient).

We can make two observations on Hamley's consideration for SM: (i) it is in agreement with Mohwald's definition and (ii) Hamley establishes that SM can be, under certain conditions, in congruence with another key issue: the unjammed state in SM.

Let us summarize that SM is a convenient classification for materials in physical states that are neither simple liquids nor crystalline solids of the type studied in other branches of condensed matter physics. Physicists are able to describe with good approximation fluids and ordered solids, but what do they do with SM? It does not exist a theory to do it, and there is a need to describe them. Nowadays, the interest is centred on: colloidal dispersions, where sub-micrometric particles of solid or liquid are dispersed in another liquid, for instance: blood and milk; polymer melts or solutions in which the size and connectivity of the molecules lead to striking new properties, such as viscoelasticity[5]; liquid crystals, where anisotropic molecular shapes lead to states with an intermediate order between a crystalline solid and a liquid; and GM. The latter are broadly defined as the class of materials composed of a large number of discrete solid particles or grains, where the grains are large enough that they are free of thermal fluctuations. They are very common, but their description is one of the hardest problems in modern science and engineering, with several implications in technological and industrial processes. We could certainly ask why? Some reasons are: they are complex heterogeneous materials, which are out of equilibrium, interacting through dissipative dynamics. Their sizes range from very tiny particles such as sand grains to very big objects such as rocks, stellar dust, and asteroids. Moreover, GM are neither microscopic nor macroscopic enough to try studying them with "classical" statistical mechanics, condensed matter physics, or quantum mechanics.

[5] Viscoelasticity is referred to as the phenomenon in which the stress and strain of some materials depends on time. Viscoelasticity is the combination of viscous and elastic response of a material subjected to constant strain, constant stress, or oscillatory stress and strain. Let recall that elasticity deals with the mechanical properties of elastic solids, which obey Hooke's law: stress (σ) is proportional to strain (γ), i.e., $\sigma = G\gamma$, where G is the shear modulus which is independent of the applied strain at low values. On the other hand, viscosity deals with the properties of liquids in the classical theory of hydrodynamics according to Newton's law: $\sigma = \eta\dot{\gamma}$, where η is the viscosity which is independent of the applied shear rate at low values. Whether a material behaves as an elastic solid or a viscous liquid depends on the length time over which an experiment will be done. Shear modulus is defined as the ratio of shear stress to the shear strain, and it is useful for measuring the stiffness of materials.

1.1 *What do These Apparently Disparate Materials Have in Common?*

Jones (2002) establishes a group of common features for SM listed as follows[6]:

- Length scales intermediate between atomic sizes and macroscopic scales:
  1. Colloid particles are typically less than a micrometre in size.
  2. Polymeric chains have overall dimensions of tens of nanometres.
  3. The self-assembled structures formed by amphiphilic[7] molecules have dimensions in a similar range.
  4. The size for grains in GM are roughly 1 μ. On the upper size limit, the physics of granular materials may be applied to ice floes, where the individual grains are icebergs, and to asteroids belts of the solar system with individual grains being just the asteroids.

- Fluctuations and Brownian motion: Although typical structures in SM are larger than atomic sizes, they are small enough for Brownian motion—the fluctuations that take place in any thermal system—. For Brownian motion to be important, the typical energies associated with the bounds between structures and with the distortions of those structures must be comparable in size to the thermal energies. SM systems should be visualized as being in a constant state of random motion: polymer chains in solution are continually writhing and turning, and the membranes formed by sheets of self-assembled amphiphilic molecules are not rigid plates but are continually buckling and flexing under the influence of Brownian fluctuations. Returning to GM, we face a wall: *they do not undergo thermal fluctuations*, but the above explanation contradicts this statement, so what is the problem here? The answer could be that we should expand the definition of SM to include non-thermal systems as well.
- The propensity of SM to self-assemble[8]: Related to the importance of Brownian motion is the fact that most SM systems are able to move towards equilibrium. But the equilibrium state of lowest free energy in a SM system is often not a state of dull uniformity; the balance of energy and entropy in SM systems yield rich phase behaviour in which complex structures arise spontaneously. This self-assembly can take place at the level of molecules, but even more complexity occurs when the ordering takes place hierarchically, with molecules coming together to

[6] We have introduced features for granular matter trying to make a broad classification of all of them.

[7] Amphiphilic molecules possess both hydrophilic (polar) and lipophilic (fat loving) parts. They are related to molecules having a polar, water-soluble group attached to a nonpolar, water-insoluble hydrocarbon chain.

[8] What does a self-assembly mean? It is a type of process in which a disordered system of pre-existing components forms an organized structure of patterns as a consequence of specific, local interactions among the components themselves, without external direction.

form supramolecular structures (such as micelles[9]), which self-organize with a high level of order. In this way, structures with high complexity are put together without external intervention, driven solely by the second law of thermodynamics (a spontaneous process is the time evolution of a system in which it releases free energy, usually as heat, and moves to a lower thermodynamically stable state). In contrast, GM are driven by dissipative forces. In fact, one of the most important properties of GM is the irreversible energy dissipation in the course of interaction between particles. We can then ask: Should we speak about self-assembly in GM? What is the meaning of the Second Law of thermodynamics here? These are two important questions that we can take from the plethora of possibilities, and we must think about them.

2 What is the Basic Aim of Condensed Matter Physics?

Physicists have created many theories to explain the Universe, one of them is condensed matter physics. It is one of the most important, which has allowed us to develop almost all the current technology: transistors, AFM, computers, cell phones, an so forth. However, it fails when it tries to explain SM. One example of this failure is related to the presence of the boson peak (BP) in glasses. Remarkable differences between crystalline solids and amorphous solids, like glasses, are observed mainly in the thermodynamics and transport properties. At low temperature, glasses reveal an excess of specific heat and a plateau in the thermal conductivity. This anomalies originate from the anomaly in the frequency spectra of atomic vibrations. These phenomena are related to the occurrence of the BP, which is universally observed for all glasses in measurements of specific heat, light, X-ray, and neutron scattering. However, the nature of the BP is a theoretical open problem (Chumakov et al. 2011). It is important to recall here that, in crystals, the low frequency DOS follows the Debye model (Zorn 2011). In spite of apparent deficiencies in condensed matter theory, we have learnt something important. As in it, we should tackle the problem in a similar way: *understanding the collective properties of large assemblies of atoms and molecules in terms of the interactions between their components.*

In regard to the aim of SM physics, we would like to talk about the style of SM research following de Gennes' arguments (de Gennes 1991). One major feature is the possibility to do very simple experiments. Second, some outstanding theoretical analogies between SM and other fields sometimes emerge, for instance:

- Edwards (1964) showed a correspondence between the conformations of a flexible chain and the trajectories of a non relativistic particle; the statistical weight of the chain corresponding to the propagator of the particle. In the presence of external potentials, both systems are ruled by exactly the same Schrödinger equation.

---

[9] A micelle is an aggregate of surfactant molecules dispersed in a liquid colloid. A surfactant is a substance which exhibits some superficial o interfacial activity.

- A wonderful analogy was found by McMillan and by de Gennes and coworkers. It consisted in relating smectic A phase[10] in liquid crystals to superconductors. They knew that the problem of type-II superconductors in a magnetic field is exactly equivalent to that of a vortex state in rotating superfluid helium. The analogy here is a smectic A phase inside which we add chiral solutes playing the role of the field. In some favorable cases, as predicted in 1988 by Lubensky (1988), this may generate a smectic phase drilled by screw dislocations—the so called A* phase.
- Another interesting example is the protein folding. Prasanth and Ioan (2012) investigated similarities between protein folding and granular jamming. They demonstrated that folded proteins exhibit signatures common to both glassiness and jamming by using temperature—and force—unfolding molecular dynamics simulations. They generalized the phase diagram proposed by Liu and Nagel (1998) to include folded-unfolded proteins. Below certain values of the external force, temperature, and environmental parameter concentration proteins fold and undergo changes in mechanical properties similar to jamming of grains.[11] As Cates et al. (1998) proposed, fragile materials may jam or unjam depending on the direction of the applied forces. Based on this consideration Prasanth and Ioan advanced that proteins seem to share such unusual mechanical properties in the sense that an individual protein exhibits different resilience to unfolding under external pulling with atomic force microscopy or single-molecule tweezers, depending on the direction of the pull (see Wilcox et al. 2000; Brockwell et al. 2003).

3 Granular Materials: Are they Solid or Liquid?

Granular materials are collections of many macroscopic solid grains large enough (with typical sizes $> 1\,\mu$m) so that thermal fluctuations are negligible (Aranson and Tsimring 2009), i.e., Brownian motion is unimportant (de Gennes 1999). They are highly dissipative, and this is one of the reasons for the uselessness of thermodynamic concepts, such as energy equipartition, which become generally inapplicable. To illustrate the importance of the energy scales in GM, or the unimportance of thermal fluctuations, Aranson and Tsimring (2006) established that the energy of a 1 mm grain moving with a typical velocity of 1 cm/s exceeds its thermal energy by at least ten orders of magnitude. Compared to String Theory, Particle Physics, or Quantum

[10] In the smectic-A mesophase, the director is perpendicular to the smectic plane and there is no particular positional order in the layer. Similarly, the smectic-B mesophase is faced with the director perpendicular to the smectic plane, but the molecules are arranged into a network of hexagons within the layer. In the smectic-C mesophase, molecules are arranged as in the smectic-A mesophase, but the director is at a constant tilt angle measured normally to the smectic plane.

[11] They should have considered the work done by Ciamarra et al. (2010), where they made certain modifications to the Liu-Nagel's phase diagram taking into account the role of *infinite* relaxation times and the convex shape of the jamming region.

Information the physics of granular matter looks "mundane". However, GM have striking features and despite their humble appearance, their bulk properties are often different from conventional solids, liquids, and gases due to the dissipative nature of forces acting on interacting grains, such as inelastic collisions, dry friction, or viscous drag (similar to fluid dynamics when Stokes' law is invoked). On the other hand, tangential friction forces play an important role in the dynamics of granular matter, especially in dense systems, and we should have in mind that these forces are hysteretic and history dependent (de Gennes 1999).

Piazza (2011) gives a definition of GM that fits well with our approach towards unification. Nevertheless, it is different to our strategy. One statement is that GM are often regarded as colloidal suspensions that have lost their suspending agent, but this is not rigorously true since air could be considered as a solvent and it is always there, exactly as in aerosols. GM could be considered as an additional state of matter in its own right. Their pressure is not uniform as in a liquid and they press more on the sidewalls than on the bottom of a container. For granular systems to remain active they have to gain energy from external sources. This is an important factor in the study of hopper collapse (Janssen and Vereins 1895; Forterre and Pouliquen 2009). In addition to shear and vibrations, external volume forces such as gravity, electric, and magnetic fields as well as the flow of interstitial fluids may also activate grains. When subjected to a large enough driving force, a granular system may exhibit a transition from a granular solid-like state to a liquid or flowing state, and various ordered patterns of grains may develop (Aranson and Tsimring 2009). In particular, dissipation is responsible for the fact that most static regimes of GM are metastable and so the system does not typically evolve towards a certain free energy minimum. In fact, dissipation (or dissipative interactions) among grains leads to different dynamics that cannot be described by standard equilibrium statistical thermodynamics.

A very old and famous problem in GM is related to sphere packings in three-dimensional Euclidean space. This is known as Kepler's problem. This problem can be stated as follows: What is the arrangement of (many) identical spheres that has a maximum density packing? (Weaire 1999). Kepler formulated this question around 1611. In relation to this problem, in 1960, Bernal (1960) published a paper in Nature, where he went far beyond by studying the coordination of randomly packed spheres. This problem is connected to the foundations of the geometry of the Jamming Transition Problem. It is curious that yet, we fall so frequently in the realm of GM-problems and its puzzles, or in the words of Corwin (2012): *"while the horizons of physics have stretched to encompass the extremes of the universe, it is remarkable that there are still fundamental puzzles awaiting us at bottom of a bucket of sand"*.

GM could be considered as a *tool* to *simulate* other SM systems or complex fluids. An example of this is the work done by Weeks (2012): *Melting Colloidal Crystals from the Inside out*, where it is established that a collection of hard spheres, that interact only through short-range repulsive forces, are useful for simulations of melting in colloidal crystals. Another beautiful example is the work of D'Anna et al. (2003), where they report an experimental investigation of an archetypal non-equilibrium system. They used a sensitive torsion oscillator immersed in a granular system of millimetre-size grains that were fluidized by strong external vibrations. The vibro-fluidized granular medium is a driven environment, with continuous injection and dissipation of energy, and the immersed oscillator can be seen as analogous to an elastically bound Brownian particle. By measuring the noise and the susceptibility, they showed that the experiment can be treated with an equilibrium formalism. They proposed that the vibro-fluidized granular system behaves as a thermal bath satisfying a fluctuation-dissipation relation, thus giving access to an experimental estimation of the granular viscosity and effective temperature. Nevertheless, these quantities are anisotropic and inhomogeneous.

As was mentioned before, in SM the interactions are typically weak and on the order of k_BT, but not zero. Nonetheless, in GM, $k_BT \sim 0$ implies that entropy considerations can easily be outweighed by dynamical effects, that now become of a paramount importance. A key role of the temperature is that it allows a system to explore phase space. In a granular material, $k_BT \sim 0$, precludes such exploration. Unless perturbed by external disturbances, each metastable configuration of the material will last indefinitely, and so no thermal averaging over nearby configurations will take place (Jaeger and Nagel 1996). Another role of temperature in ordinary gases or liquids is to provide a microscopic velocity scale. Again in GM this role is completely suppressed, because the only velocity scale is imposed by the macroscopic flow itself. It is possible to formulate an effective granular temperature in terms of velocity fluctuations around the mean flow velocity (Owaga 1978; Savage 1984; Walton and Braun 1986; Haff 1985; Campbell 1990; Ippolito et al. 1995; Warr and Huntley 1995; Warr et al. 1995). However, such approaches do not always recover thermodynamics or hydrodynamics due to the inelastic nature of collisions between grains.

Regarding to phases in GM, Jiang and Liu (2007) made a classification of them in terms of energy. Independently of the grain ratio of deformation and kinetic energy, these phases may loosely be referred to as gaseous, liquid, or solid. The first phase is relatively well understood: moving fast and being free most of the time, the grains in the gaseous phase have much kinetic energy, which is dissipated by collisions between the particles (Jiang and Liu 2007). In the denser liquid phase, there is less kinetic energy, more deformation, and a complex rheology that has been scrutinized recently (Jiang and Liu 2007). In the static regime, with the grains deformed in a compressed packing, the energy is almost elastic. Up to now, there is no theory capable of accounting for GM statics and dynamics simultaneously. However, Aranson and Tsimring (2009) proposed that some combination of fluid dynamics with phase-field modelling, based on an order-parameter description of the state of a granular material, could be one possible candidate for a unified theory. Kadanoff (1999) expresses that the study of collisions and flow in these materials requires new theoretical ideas beyond those in the standard statistical mechanics, hydrodynamics, or traditional solid mechanics. These materials show complex flow patterns similar to those of ordinary liquids, but also freezing, plasticity, and hysteresis. GM can mimic: glass, liquid, plastic flow, glassy behaviour, folding of proteins, and so on.

We employ points and straight lines to understand the dynamics of rigid solids, particles, and systems of them. This allowed us to construct buildings, bridges, rockets, vehicles, etc. We use points to understand some properties of matter and interactions so that all the complexity is reduced to "tiny" non-physical objects. So, when GM is considered as a set of spheres, we could use them to understand the dynamics of gels, colloids, polymers, etc. We join them to form chains and other structures. GM could be considered as the next step in abstraction to model and understand heterogeneous materials. Perhaps, if we go far beyond, GM can be used as a cornerstone to construct a unified theory for SM. The jamming transition problem in granular matter could help us to understand and develop such a theory for SM.

4 What is Jamming?

In this section, we review some significative scientific papers about the jamming transition problem. We present a brief discussion and quotations about some phenomenological, experimental, computational, and theoretical facts.

Let us start with a brief history about jamming transition (JT). The study of jammed systems began as a culinary curiosity in 1727, when the reverend Stephen Hales studied how peas pack when compressed in an iron pot. If we fill a pot with peas, we are then able to pass our hands through the bulk of peas because they flow out of the way like a liquid. However, as the pressure and the density, are increased, we will find a critical point, above which the peas *jam* into a stable amorphous solid (Corwin 2012). When the packing fraction is increased sufficiently, loose particles jam to form a rigid solid in which the constituents are no longer free to move (Zhang et al. 2009).

Above this threshold for stress (the critical point in "the flowing of peas"), a concentrated colloidal dispersion of hard particles under shear may jam. According to Ball and coworkers (Ball and Melrose 1995; Melroseb and Ball 1995; Farr et al. 1997), the jamming apparently occurs because the particles form *force chains*[12] along the compressional direction. Indeed, the shape of the distribution of normal forces is a manifestation of the jamming/unjamming transition (Corwin et al. 2005). As was argued by Cates et al. (1998), once the contacts arise, an array or network of force chains can support the shear stress indefinitely, and hence the material is a solid.

Liu and Nagel (1998) sketched a speculative phase diagram for jamming. This phase diagram depends on temperature, load, and density. According to this picture, jamming can occur only when the density is high enough. One can then unjam the system either by raising the temperature or by applying a stress. The ideas introduced by Liu and Nagel were more general than those of Cates et al. (1998) (CWBC) work in the sense that the former pointed out that the class of jammed materials

[12] An important feature of granular materials is that the internal forces are not carried uniformly by the material, but instead through long chain-like structures whose density and orientation depend on the state and history of the sample. This feature allows us to study some problems of information propagation in GM using percolation theory.

may actually be broader than suggested by the latter, because the CWBC work considered jamming only in systems with no attractive interactions, where the particle dynamics is constrained through an applied stress, and where the individual particles are large enough that there is no thermal motion. These two constraints may not be essential. Liu and Nagel conjectured that it is possible to correlate jamming and glass transitions. The phase diagram, proposed by them, was an attempt to answer the following question: *Might the concept of jamming and fragility include microscopic systems with attractive interactions, which unjam as one raises the temperature, as well as stressed macroscopic systems with repulsive interactions, which unjam as one applies an incompatible stress?* (Liu and Nagel 1998)

The off-equilibrium solidification in colloids, characterized by the sudden arrest of the dynamics of their constituent particles, shows kinetic heterogeneities near the onset of solidification. The particles' mobilities become heterogeneous in space and intermittent in time. This general observations led to Liu and Nagel to speculate that different transitions due to the kinetic arrest can be unified by a jamming phase diagram (Kumar and Wu 2004).

The term jamming has been reserved to describe situations in which both the flowing phase and the jammed phase are disordered (Ellenbroek 2007). In this sense, we think that it could be possible to apply the concept given by Witten in relation to SM, that is, to define jamming using the concept of permutation freedom.

Trappe et al. (2001) provided a link between the glass transition, gelation, and aggregation. Their results support the concept of a jamming phase diagram for attractive colloidal particles. They plotted the full 3D jamming phase diagram for attractive colloidal systems, using $\phi_c = \phi_0 e^{-U_c/\alpha k_B T}$, $\sigma_y = \sigma_\phi(\phi - \phi_c)^{\mu_\phi}$, and $\sigma_y = \sigma_U(U - U_c)^{\mu_U}$ to determine the phase boundaries. In this equations ϕ_c, ϕ_0, α, U, U_c, k_B, σ_y, and T are, respectively, the critical value of the volume fraction, the initial volume fraction, the parameter depending on the relative range of the depletion interaction ($\alpha \sim 1$), the interparticle attractive energy and its critical value, the Boltzmann's constant, the yield stress,[13] and a temperature scale fixed by U. This is a compendium of contributions from different physical systems and which correctly captures the overall shape and behaviour. According to Trappe et al. (2001), the shape of the experimental phase diagram differs significantly from that proposed by Liu and Nagel (1998). In the new phase diagram the curvature is contrary and diverging at each vertex. They wrote that: *These divergences reflect the particular details of attractive colloidal particles and correspond to irreversible aggregation, where* ${\phi_c}^{-1}$ *is large; to the limit of hard spheres, where* T/U *is large; and to high volume fractions of strongly attracting particles that form, for example sintered solids, where* σ_y *is large.*

On the other hand, O'Hern et al. (2003) explained that jamming occurs when a system develops a yield stress in a disordered state and that the appearance of a yield

[13] Following H. A. Barnes (1999), yield stress is defined for liquids and solids. In the first case, yield stress is a point at which, when decreasing the applied stress, solid-like behaviour is first seen, i.e., no continual deformation. In the latter case, it is essentially the point at which, when increasing the applied stress, the solid first shows liquid-like behaviour, i.e. continual deformation.

stress is equivalent to an infinite stress relaxation time. They presented similarities between the transitions in all the systems able to undergo JT. First, the increase of the stress relaxation time tends to be super-Arrhenius[14] as a function of the control parameter (Pusey and van Megan 1987; Kivelson and Tarjus 1998; D'Anna and Gremaud 2001). The other property is that all the systems show kinetic heterogeneities near the onset of jamming (i.e., where particle mobilities became heterogeneous in space and intermittent in time) (Langer and Liu 1997; Kob et al. 1997; Doliwa and Heuer 1998; Ediger 2000; Weeks et al. 2000).

There exists the possibility to unify different non-equilibrium transitions in colloids and other SM systems under the concept of jamming. In this sense, Kumar and Wu (2004) offered a convincing proof towards this unification. Phases such as gelation, coagulation, kinetic arrest, dynamic slow-down, and ergodic-nonergodic transitions have been used to loosely describe the solidification phenomena. Their proof is based on simulations showing that near the jamming transition, the shear viscosity diverges following a critical-like scaling law as observed for realistic colloids.

Let us address the following question: Is there a structural signature of jamming in SM or in GM, that could distinguish it from its flowing counterpart? Corwing, Jaeger, and Nagel (2005) found evidence for such a signature by measuring the contact force distribution between particles during shearing. Since the forces are sensitive to small variations of the particle positions, the distribution of forces can serve as a probe to observe correlations in the position of nearest neighbours. Their result demonstrated a clear signature of the jamming/unjamming transition, which manifests itself in the shape of the distribution of normal forces $\rho(f)$. Jammed packings are characterized by a distribution that decays exponentially at large forces. This shape reflects the non-equilibrium character of the jammed state. By contrast, $\rho(f)$ in the flowing regime is well described by a model that assumes that the system is in equilibrium.

Moreover, Biroli (2007) presented a brief *status quo* of the jamming transition as well as its relation with glass-transition, suggesting that jamming is a new kind of phase transition. On the other hand, Keys et al. (2007) showed that, when approaching the jamming transition, these systems display growing length and timescales, thus strengthening the idea that the jamming transition is indeed a critical phenomenon.[15] Biroli proposed some questions that must be addressed in deep:

- What is the physical mechanism behind the formation of amorphous solids?
- Is the abrupt dynamical arrest merely a crossover with little or no universality or is there a true underlying phase transition of a new kind?
- Could it be that explanations of the glass transition work for the jamming transition?

[14] For some super-cooled liquids, the temperature dependence on relaxation times or on transport properties, such as the diffusion constant, is stronger than predicted by Arrhenius law. Arrhenius law refers to the fact that in some viscous liquids $\log \eta$ (η is the viscosity) is linear in T^{-1} (T is the temperature).

[15] Critical phenomena are phenomena observed near a critical point and this is precisely a point in the phase diagram where a continuous phase transition takes place.

- What is the nature of the underlying phase transition? (if any).
- Could the emergence of rigidity be the driving mechanism of the transition, as has been proposed?

In connection to jamming and glass transition, Biroli wrote that *Keys et al. fitted the growth of the time and length scales using laws that were introduced for glass forming liquids, the Vogel-Fulcher-Tammann (VFT)*[16] *law and power laws. Surprisingly, the VFT fit gave a transition value,* $\phi_c \approxeq 0.84$, *which is very close to the one independently determined for the JT.*

When jammed, the disordered system is caught in a small region of phase space without possibility to scape. A wide class of materials, ranging from thermal systems such as structural glasses, pastes, gels, and colloids, to the systems of particle whose thermalization is negligible, such as granular materials, are characterized by a JT from a flowing liquid-like state to a disordered solid-like state with mechanical strength. Control parameters of the jamming transition include the temperature, the density and the applied shear stress, and the flowing phase occurring at high temperature, low density, and high applied shear stress (Ciamarra et al. 2010).

Mills et al. (2008) proposed a mechanism to explain the JT in GM. They suggest that shear stress of non-cohesive GM in the vecinity of the JT is supposed to be connected to the formation of transient clusters of particles. They considered that *dense granular flows are made of transient rigid clusters of caged particles immersed in free particles. When approaching the quasistatic regime, the growth of these clusters is responsible for jamming when they invade all the flow.* Near the jamming transition, GM fluctuates between two jam-flow states which generates, on average, a frictional force proportional to the mean velocity relative to the jammed state.

There are important results related to jamming, which must be reproduced in a complete theory if they are correct at all. One of them was reported by Song, Wang, and Makse (2008). They presented a statistical description of jammed states in which the random close packing can be interpreted as the ground state of the ensemble of jammed matter. They demonstrated that random packings of hard spheres in three dimensions cannot exceed a density limit of ~63.4 %. On the other hand, Lu, Brodsky, and Kavenpoor (2008) proposed a thermodynamic unification of jamming. They quantified jamming using a thermodynamic approach by accounting for the structural aging and the shear-induced compressibility of dry sand. Specifically, the jamming threshold is defined using a non-thermal temperature. An equation of state was derived, which governs the mechanism of shear banding and the associated features of shear softening and thickness invariance. A controversial problem then emerged in relation to the definition of temperature because it is defined in thermal equilibrium.

Jamming was suggested as a viable mean to unify all fragile systems, and this was thought using three independent variables: pressure, packing fraction, and an effective temperature. We shall go further trying to unify granular matter with SM, including liquid crystals. We think that it is possible to construct this unification.

[16] $\xi = \xi_0 e^{\frac{A}{T-T_0}}$.

However, we need a confirmation of jamming in liquid crystals using the arguments discussed above in this article. Perhaps, more than a unification this could be interpreted as a link between GM and SM so that some results applicable to one system may also apply to other systems, e.g., emulsions with granular matter, liquid crystals, and foams. While this could be accomplished using thermodynamics or some other theoretical approach, it is perhaps more suitable to use wave propagation in random media to determine structural signatures of jamming or associated symmetries. Following the thermodynamic approach Song et al. (2008) introduced an aging temperature and a external pressure depending on the shear-rate and free volume to construct an equation of state that unifies different behaviours of fragile matter. On the other hand, Jacquin et al. (2011) elaborated a microscopic mean-field theory for the JT. In this theory, they describe athermal packings and observed geometric phase transitions by using equilibrium statistical mechanics. They developed a fully microscopic, mean-field theory of the JT for soft repulsive spherical particles. Then, they derived some of the scaling laws characterizing the JT.

We would like to address a couple of facts: first, Liu and Nagel (1998) speculated about the existence of a more profound analogy between temperature and shear stress and this was depicted in their phase diagram. On the other hand, O'Hern et al. (2003) investigated the jamming transition at zero temperature and zero applied stress. In this limit there is no particle motion and so the JT is identified by the appearance of mechanical strength. They found that the jamming point (J-point) has an extremely interesting mixed first-order/second-order character.[17] On approaching the JT from the flowing state, the relaxation time τ grows rapidly as measured, for instance, from the decay of density auto-correlation functions $G(\tau) = \langle \rho(\tau)\rho(t+\tau)\rangle$. It is therefore difficult to locate the precise points of the jamming phase diagram where the relaxation time diverges. For this reason, it could be convenient to consider a *jamming surface* of constant relaxation time τ_g. This time is the timescale accessible to a typical experiment or numerical simulation. In this spirit Zhang et al. (2009) introduced a new speculative jamming phase diagram. The jamming surface is a surface of constant relaxation time τ_g and the depicted surface does not touch the axes. A conclusion of this work, is that the shape of the jamming surface at a infinite relaxation time is questioned and that the jamming transition (infinite relaxation time) does not occur at any finite value of the temperature, while it does occur in the inverse-volume-fraction shear plane. The new hypothesis is that *at zero temperature and zero applied stress, the JT may occur in a whole density range, not exclusively at the J-point.*

Experimentally, it is extremely difficult to access the complex dynamics of systems very near the jamming transition. Instead, computational simulations allow to explore the dynamics close to jamming. At this point, it is important to highlight the work of Dagois-Bohy et al. (2012). They demonstrated that properly jammed systems must be stable not only to compression but also to shear. However, widely used

---

[17] Owing to the co-existence of quantities that vary continuously at the transition, such as the pressure and the shear modulus, and of quantities that change discontinuously, such as the mean contact number per particle (van Hecke 2010).

numerical methods, in which particles are compressed together, can produce packings that are unstable to shear. On the other hand, Goodrich et al. (2012) carried out an analysis of finite-size effects in jammed packings of N soft, frictionless spheres at zero temperature. They showed that the scaling behaviour of finite-size systems jammed under varying constrains can provide strong evidence that the JT should be considered as a phase transition. Their results are consistent with jamming being either a random first-order or a mean-field second-order transition (Corwin 2012). Ciamarra and Sollich (2012) introduced another point of view. They suggested that the JT could be a first-order crossover of high-order jamming crossovers and density anomalies. For athermal systems, these crossovers induce an anomalous behaviour of the bulk modulus, which varies non-monotonically with the density. This transition occurs when particles (interacting via repulsive contact forces) start to be closer to each other to the point of contact (*percolation transition*). Otherwise, at finite temperature they induce density anomalies consisting in an increased diffusivity upon isothermal compression and in a negative thermal expansion coefficient.

5 What Kind of Materials can Undergo a Jamming Transition?

Some polymer-like magnetic fluids are capable to undergo a JT: if the magnetic particles are large enough, they will, in the presence of a strong field, form long chains bridging the whole system (this is the JT after which the system is truly elastic) (e.g., Jiang and Liu 2004).

It is also possible to observe JT in magnetofluidized beds (MFBs). They consist of magnetic particles suspended in a vertical gas flow and anchored to a magnetic field (Valverde et al. 2010). The JT in this case is a jamming of athermal magnetic particles driven by the magnetic stress induced by the externally applied magnetic field. The size of particles in these systems usually range from tens to hundreds of microns. Thus, van der Waals forces and Brownian motion are fully negligible. The field strength needed for jamming is presented as a function of a dimensionless number, defined as the ratio of the solid kinetic energy per unit volume to the magnetic energy. An open problem in this field is related to the lack of an equation to predict the strength of the field necessary to stabilize MFBs. The growth of particle chains as the gas velocity is decreased drives the system to a JT at a critical gas velocity.

Other systems capable to undergo JT are: concentrated colloidal dispersions (Cates et al. 1998), star polymers (Loppinet et al. 2001), gels (Ovarlez et al. 2010), glasses (Liu and Nagel 1998; Trappe et al. 2001), granular polymers (Lopatina et al. 2011), GM, and molecular systems. In particular, Trappe et al. (2001) reports results that support the concept of a jamming phase diagram for attractive colloidal particles, providing a unifying link between the glass transition, gelation, and aggregation. This work could be considered as a strong support to our leading hypothesis because it is related to JT as a unifying framework between aggregation and formation of amorphous solids in soft matter physics.

Jamming occurs when a system develops a yield stress in a disordered state (O'Hern et al. 2003). Following Barnes (1999), it is possible to find yield stress in: clay, oil paint, toothpaste, drilling mud, molten chocolate, creams, ketchups, and other culinary sauces, molten filled rubbers and printing inks, ceramic pastes, electro-viscous fluids, thixotropic paints, heavy-duty washing liquids, surface-scouring liquids, mayonnaise, yoghurts, purees, liquid pesticides, bio-mass broths, blood, water-coal mixtures, molten liquid-crystalline polymers, plastic explosives, foams, rocket propellant pastes, and fire retardants among others.

It is possible to measure a yield stress in liquid crystals. Watanabe et al. (2006) studied electro-rheological properties in smectic A phase of a liquid crystal, using a parallel-plate type rheometer. They found that the yield stress was depending on the current type of the electric field. Moreover, the yield stress only depends on the electric field condition when the smectic structure grows and the yield stress, i.e., the structure generated, is hardly changed by variations in the electric field. Narumi et al. (2008) found that there were partially collapsed structures at the first stage of the yield process generating a large scale deformation. These studies have led us to propose that it is possible to measure a jamming transition in liquid crystals.

A further definition is that jamming occurs when a system develops a stress relaxation time that exceeds the experimental time scales in a disordered state (O'Hern et al. 2003). According to this definition, many systems can jam. For example, granular materials (Jaeger et al. 1996; Majmudar et al. 2007; van Hecke 2010) and colloidal suspensions. The latter are fluids, but jam when the pressure or packing fraction raises (Kumar and Wu 2004; Ballesta et al. 2008; van Hecke 2010). Foams and emulsions are able to undergo JT too. Let us remind that they are concentrated suspensions of deformable bubbles or droplets. They flow when a large shear stress is applied, but jam when the shear stress is lowered below the yield stress (Zhang and Makse 2009; Denkov et al. 2009; van Hecke 2010; Katgert and van Hecke 2010; Berthier et al. 2011) In foams and emulsions, jamming at low shear rates is explained by the thinning dynamics of the transient films formed between neighboring bubbles and drops (Denkov et al. 2009). Foam is a good example of a system that is jammed, but it is not fragile. Shaving foam, for example, is jammed because the bubbles are tightly packed together under atmospheric pressure (Liu and Nagel 1998).

In molecular liquids, composed of molecules that do not form a covalent network, but interact only through weak van der Waals forces or through transient hydrogen bonds, temperature is important. They can jam if crystallization does not intervene first. This process is called glass transition (Ediger et al. 1996; Debenedetti and Stillinger 2001). Loppinet et al. (2001) demonstrated that solutions of star polymers above the overlap concentration undergo a reversible thermal gelation. This phenomenon was attributed to the formation of clusters causing a partial dynamic arrest of the swollen interpenetrating spheres at high temperatures. They proposed a generalized jamming phase diagram for regular and irregular star polymers in which gelation is an alternative route to jamming and thus show analogies to glass formation.

An interesting paper by Kumar and Wu (2004) reports a three-dimensional jamming phase diagram by considering inter-colloidal forces represented by a Derjaguin-

Landau-Verwey-Overbeek potential.[18] This potential describes the forces between charged surfaces interacting in a liquid medium. It includes van der Waals forces and electrostatic double layer effects. The jamming threshold is uniquely defined in terms of the critical volume fraction, the critical temperature, and the critical yield stress. The results of the simulations indicate that near the jamming transition the shear viscosity diverges, following a critical-like scaling law as observed for realistic colloids. These results offer a convincing proof for unifying different non-equilibrium transitions in colloids under the concept of jamming. Some of this transitions are: gelation, coagulation, kinetic arrest, dynamic slow-down, and ergodic-nonergodic transition. Concerning this work, it is important to quote the following fact: the shear viscosity diverges as the transition temperature is approached with a power law behaviour of the form: $\eta/\eta_s = (T_c^{-1} - T^{-1})^{-0.155}$. Kumar and Wu observed that as in the experiments, the shape of the surface separating the jammed and unjammed states is concave everywhere in contrast to the original proposal of jamming phase diagram by Liu and Nagel.

6 What is the Deal with Soft Matter and Jamming Transition?

As it was shown by Cates et al. (1998), Liu and Nagel (1998), and other authors, JT can be used to unify fragile matter. Examples of fragile matter are: GM, foams, and emulsions. They are athermal systems with no-minimum free energy state because they are marginally stable. However, jamming is not only restricted to fragile matter, as was suggested by Trappe et al., because JT can be used to unify different transitions in fragile matter and SM. For attractive colloidal particles, we have: glass transition, gelation, and aggregation. Let us recall that SM has two main common features: (i) large scales between atomic size and macroscopic scales; (ii) fluctuations, Brownian motion, and self-assemblage. In consequence, SM can be considered to be thermal systems. But, as we argued, SM is able to undergo JT. A digression here concerned with liquid crystals: they have two main ingredients to undergo JT, namely disorder-disorder transition and the possibility to measure yield stress. Nonetheless, it is not so simple, jammed state in these systems have to show a slow response to external perturbations and the onset of structural heterogeneities.

As we quoted, magneto-rheological fluids are able to undergo JT. This suggests that the phase diagram proposed by Liu and Nagel, Trappe et al., Ciamarra, and others, should be extended to include external action-at-distance fields effects because in MFBs the magnetic field plays an important role since the JT occurs at a certain value of this field. It could be possible that for liquid crystals the phase diagram should be expanded or even depicted another route to jamming in some transition in liquid crystals, like gelation (Loppinet et al. 2001).

[18] $V(r_{i,j}) = \frac{A}{r_{i,j}^{36}} + B\frac{e^{-kr_{i,j}}}{r_{i,j}} - \frac{C}{r_{i,j}^{6}}$.

Witten defined SM *as a fluid in which large groups of the elementary molecules have been constrained so that the permutation freedom within the group is lost.* If we can associate a symmetry to this freedom, we could describe jamming in terms of some non-continous-symmetry breaking in disordered systems. One candidate is a kind of relabelling symmetry. In Hamiltonian fluid dynamics the relabelling symmetry is referred to as the symmetry under the exchange of labels between fluid elements. Permutation freedom is related with spatial exchange. Therefore, in a first attempt, we can define JT as the process *in which permutation freedom is globally lost in a disordered system, giving rise to a yield stress.* Another important point to be considered is that jamming is related to a percolation process. This was pointed out in Ciamarra and Sollich (2012): *at the JT a percolating network of contact forces appears and the system acquires mechanical rigidity.* This statement is presented in gelation processes too (Jones 2002). In gels, if we reduce the number of bonds towards the critical fraction E_c, we would expect the modulus of the gel (the shear or the Young modulus) to vanish. Nonetheless, the process is not so simple because we have to take into account the dangling ends. This ends are part of the infinite network, but since they are only connected to the network by one end, they cannot contribute to the transfer of stress across the sample. This is similar to the case of wave propagation in GM. Regarding to gelation processes, a more realistic approach was developed by Suarez et al. (2009). They considered dangling bond effects and a soft-particle potential that stabilizes the structure, starting with a fractal structure generated with diffusion limited aggregation (DLA).

The observation presented with gelation and geometry of paths (or graphs) led us to a question: Is it possible to identify critical scaling laws joining geometry and elastic or rheological properties in SM or fragile matter systems? Ciamarra and Sollich (2012) stated the problem in the following sentence: *Simple arguments to explain mechanical properties in terms of geometric ones away from jamming should therefore be developed; we speculate that they would need to take account of the heterogeneity of the force network as a key feature of highly compressed particle systems.* The situation calls for findings related to the following equation: $\Xi \sim (E - E_c)^{\alpha}$, where Ξ is an elastic property and E is a geometrical property of the system. The exponent α should be peculiar of the system considered and their interactions. For example, it would characterize granular matter, gels, polymers, emulsions, foams, liquid crystals, and so on.

References

Anderson PW (1995) Science 267:1615

Araki T, Tanaka H (2006) Colloidal aggregation in a nematic liquid crystal: topological arrest of particles by a single-stroke disclination line. Phys Rev Lett 97:127801.1–127801.4

Aranson IS, Tsimring LS (2009) Granular patterns. Oxford University Press, Oxford

Aranson IS, Tsimring LS (2006) Patterns and collective behavior in granular media: theoretical concepts. Rev Mod Phys 76:641–692

Bak P (1997) How nature works; the science of self-organized critically. Oxford University Press, Oxford
Ball RC, Melrose JR (1995) Lubrication breakdown in hydrodynamic simulations of concentrated colloids. Adv Colloid Interface 59:19–30
Ballesta P, Duri A, Cipelletti L (2008) Unexpected drop of dynamical heterogeneities in colloidal suspensions approaching the jamming transition. Nature 4:550–554
Barnes HA (1999) The yield stress - a review or $\pi\alpha\nu\tau\alpha\ \rho\epsilon\iota$ - everything flows? J Non-Newtonian Fluid Mech 81:133–178
Berghmans M, This S, Cornette M, Bergbans H, De Schryver FC, Moldenaers P, Mewis J (1994) Thermoreversible gelation of solutions of syndiotactic poly(methy1 methacrylate) in toluene: a two-step mechanism. Macromolecules 27:7669
Bernal JD (1960) Geometry and the structure of monatomic liquids. Nature 185:6870
Berthier L, Biroli G (2011) Theoretical perspective on the glass transition and amorphous materials. Rev Mod Phys 83:587–645
Berthier L, Jacquin H, Zamponi F (2011) Can the jamming transition be described using equilibrium statistical mechanics? J Stat Mech P01004
Bi D, Zhang J, Chakraborty B, Behringer RP (2011) Jamming by shear. Nature 480:355–358
Biroli G (2009) Glass and Jamming Transition. Séminaire Poincaré XIII, 37 67.
Biroli G (2007) Jamming: a new kind of phase transition? Nature 3:222–223
Brockwell DJ, Paci E, Zinober RC, Beddard GS, Olmsted PD, Smith DA, Perham RN, Radford SE (2003) Pulling geometry defines the mechanical resistence of a beta-sheet protein. Nat Struct Biol 10:731–737
Campbell CS (1990) Rapid granular flows. Annu Rev Fluid Mech 22:57–92
Cang H, Li J, Novikov VN, Fayer MD (2003) Dynamical signature of two ideal glass transition in nematic liquid crystals. J Chem Phys 119:10421–10427
Cates ME, Wittmer JP, Bouchaud JP, Claudin P (1998) Jamming, force chains, and fragile matter. Phys Rev Lett 81:1841–1844
Chumakov AI, Moncaco G, Crichton WA, Bosak A, Rüffer R, Meyer A, Kargl F, Comez L, Fioretto D, Giefers H, Roitsch S, Wortmann G, Manghnani MH, Hushur A, Williams Q, Balogh J, Parlinski K, Jochym P, Piekarz P (2011) Equivalence of the boson peak in glasses to the transverse acoustic van hove singularity in crystals. Phys Rev Lett 106:225501
Ciamarra MP, Sollich P (2012) High-order jamming crossovers and density anomalies. arXiv:1209.3334
Ciamarra MP, Sollich P (2013) The first jamming crossover: geometric and mechanical features. J Chem Phys 138: 12A529.
Ciamarra MP, Nicodemi N, Coniglio A (2010) Recent results on the jamming phase diagram. Soft Matter 6:2871–2874
Corwin EI, Jaeger HM, Nagel SR (2005) Structural signature of jamming in granular media. Nature 435:1075–1078
Corwin E (2012) Viewpoint: getting into a proper jam. Physics 5:97
Dagois-Bohy S, Tighe BP, Simon J, Henkes S, van Hecke M (2012) Soft-sphere packings at finite pressure but unstable to shear. Phys Rev Lett 109:095703
D'Anna G, Gremaud G (2001) The jamming route to the glass state in weakly perturbed granular media. Nature 413:407–409
D'Anna G, Mayor P, Barrat A, Loreto V, Nori F (2003) Observing brownian motion in vibration-fluidized granular matter. Nature 424:909–912
Dawson K, Foffi G, McCullagh GD, Scortino F, Tartaglia P, Zaccarelli E (2002) Ideal glass in attractive systems with different potentials. J Phys Condens Matter 14: 2223
de Gennes PG (1991) Nobel lecture: soft matter http://www.nobelprize.org/nobelprizes/physics/laureates/1991/gennes-lecture.pdf
Debenedetti PG, Stillinger FH (2001) Supercooled liquids and the glass transition. Nature 410:259–267

Denkov ND, Tcholakova S, Golemanov K, Lips A (2009) Jamming in sheared foams and emulsions, explained by critical instability of the films between neighboring bubbles and drops. Phys Rev Lett 103:118302
Doliwa B, Heuer A (1998) Cage Effect, Local anisotropies, and dynamic heterogeneities at the glass transition: a computer study of hard spheres. Phys Rev Lett 80:4915–4918
Durian DJ, Weitz DA (1994) Foams. In: Kirk-Othmer, Kroschwitz JL (eds) Encyclopedia of chemical technology. Wiley, New York
Ediger MD, Angell AC, Nagel SR (1996) Supercooled liquids and glasses. J Phys Chem 100:13200–13212
Ediger MD (2000) Spatially heterogeneous dynamics in supercooled liquids. Annu Rev Phys Chem 51:99–128
Edwards SF (1964) The statistical mechanics of polymers with excluded volume. Proc Phys Soc 85:613–624
Ellenbroek WG (2007) Response of granular media near the jamming transition. Ph.D. Thesis (Delft-Leiden, Casimir PhD Series)
Ellenbroek WG, van Hecke M, van Saarloos W (2009) Jammed frictionless disks: connecting local and global response. Phys Rev Lett E 80:061307
Farr RS, Melrose JK, Ball RC (1997) Kinetic theory of jamming in hard-sphere startup flows. Phys Rev E 55:7203–7211
Forterre Y, Pouliquen O (2009) Granular flows. Séminaire Poincaré XIII http://iusti.polytech.univ-mrs.fr/pouliquen/publiperso/bourbaphyspouliquen2light.pdf
de Gennes PG (1999) Granular matter: a tentative view. Rev Mod Phys 71:S374–S382
de Gennes PG (2005) Soft matter: more than words. Soft Matter 1:16
Goodrich CP, Liu AJ, Nagel SR (2012) Finite-size scaling at the jamming transition. Phys Rev Lett 109:095704
Guenet JM (1999) Physical gels from PVC: molecular structure of pregels and gels to chain microstructure. In: Morishima Y, Norisuye T, Tashiro K (eds) Molecular interactions and time-space organization in macromolecular systems. Springer, Berlin
Haff PK (1985) Physical picture of kinetic granular fluids. J Rheology 30:931–948
Hamley IW (2007) Soft matter: synthetic and biological self-assembling materials. Wiley, West Sussex
Ippolito I, Annie C, Lemaitre J, Oger L, Bideau D (1995) Granular temperature: experimental analysis. Phys Rev E 52:2072–2075
Jacob X, Aleshin V, Tournat V, Leclaire P, Lauriks W, Gusev VE (2008) Acoustic probing of the jamming transition in a unconsolidated granular material. Phys Rev Lett 100:158003
Jacquin H, Berthier L, Zamponi F (2011) Microscopic mean-field theory of the jamming transition. Phys Rev Lett 106:135702
Jaeger HM, Nagel SR (1996) Granular solids, liquids and gases. Rev Mod Phys 68:1259–1273
Jaeger HM, Nagel SR, Behringer RP (1996) Granular solids, liquids, and gases. Rev Mod Phys 68:1259–1273
Janssen HA, Vereins Z (1895) Versuche ber getreidedruck in silozellen. Dtsch Eng 39:1045
Jiang YM, Liu M (2004) Energy instability unjams sand and suspension. Phys Rev Lett 93:148001
Jiang Y, Liu M (2007) From elasticity to hypoplasticity: dynamics of granular solids. Phys Rev Lett 99:105501
Jones RAL (2002) Soft condensed matter. Oxford University Press, Oxford
Kadanoff LP (1999) Built upon sand: theoretical ideas inspired by granular flows. Rev Mod Phys 71:435–444
Katgert G, van Hecke M (2010) Jamming and geometry of two-dimensional foams. EPL 92:34002
Keys AS, Abate AR, Glotzer SC, Durian DJ (2007) Measurment of growing dynamical length scales and prediction of the jamming transition in a granular material. Nature 3:260–264
Kivelson D, Tarjus G (1998) SuperArrhenius character of supercooled glass-forming liquids. J Non-Cryst Solids 86:235–237

Kleman M, Laverntovich OD (2002) Soft matter physics: an introduction. Springer, Berlin
Kob W, Donati C, Plimpton SJ, Poole PH, Glotzer SC (1997) Dynamical heterogeneity in a supercooled Lennard-Jones mixture. Phys Rev Lett 79:2827–2830
Kumar A, Wu J (2004) Jamming phase diagram of colloidal dispersions by molecular dynamics simulations. Appl Phys Lett 84:4565–4567
Langer SA, Liu AJ (1997) Effect of random packing on stress relaxation in foam. J Phys Chem B 101:8667–8671
Liu AJ, Nagel SR (2001) Jamming and rheology. Taylor and Francis, New York
Liu AJ, Nagel SR, van Saarloos W, Wyart M (2011) The jamming scenario-an introduction and outlook. Oxford University Press, Oxford
Liu AJ, Nagel SR (1998) Jamming is not just cool anymore. Nature 396:21–22
Lopatina LM, Olso Reichhardt CJ (2011) Jamming in granular polymers. Phys Rev E 84:011303
Loppinet B, Stiakakis E, Vlassopoulos D, Fytas G, Roovers J (2001) Reversible thermal gelation in star polymers: an alternative route to jamming of soft matter. Macromolecules 34:8216–8223
Lu K, Brodsky EE, Kavehpour HP (2008) A thermodynamic unification of jamming. Nature 4:404–407
Majmudar TS, Sperl M, Luding S, Behringer RP (2007) Jamming transition in granular systems. Phys Rev Lett 98:058001
Melrose JR, Ball RC (1995) The pathological behavior of sheared hard-spheres with hydrodynamic interactions. Europhys Lett 32:535–540
Mills P, Rognon PG, Chevoir F (2008) Rheology and structure of granular materials near the jamming transition. EPL 81:64005
Mills P, Chevoir F (2009) Rheology of granular materials and sound emission near the jamming transition, powder and grains. In Nakagawa N, Luding S (eds) Proceedings of the 6th international conference on micromechanics of granular media
Miskin MZ, Jaeger HM (2013) Adapting granular materials through artificial evolution. Nat Mater (published online)
Möhwald H (2005) Not a soft science. Soft Matter 1:328
Müller O, Hahn D, Liu M (2006) Non-Newtonian behaviour in ferrofluids and magnetization relaxation. J Phys Condens Matter 18:s2623–s2632
Narumi T, Uematsu H, Hasegawa T (2008) Solid-like properties of liquid crystal in smectic phase controlled with electric field applied. In: Proceeding of AIP conference, vol 1027, pp 484–486
O'Hern CS, Silbert LE, Liu AJ, Nagel SR (2003) Jamming at zero temperature and zero applied stress: the epitome of disorder. Phys Rev E 68:011306
Olsson P (2010) Diffusion and velocity autocorrelation at the jamming transition. Phys Rev E 81: 040301(R)
Ovarlez G, Barral Q, Coussot P (2010) Three-dimensional jamming and flows of soft glassy materials. Nature 9:115–119
Owaga S (1978) In: Cowin SC, Satake M (eds) Proceedings of US-Japan seminaron continuum-mechanical and statistical approaches in the mechanics of granular materials, Gakujutsu Bunker Fukyukai, Tokyo, Japan
Parisi G, Zamponi F (2010) Mean-field theory of hard sphere glasses and jamming. Rev Mod Phys 82:789–845
Piazza R (2011) Soft matter: the stuff that dreams are made of. Springer, Berlin
Pontoni D, Finet S, Narayanan T, Rennie AR (2003) Interactions and kinetic arrest in an adhesive hard-sphere colloidal system. J Chem Phys 119:6157
Poon W, McLeish T, Donald A (2002) Soft condensed matter: where physics meets biology. Phys Educ 37:25–33
Prasanth J, Ioan A (2012) Similarities between protein folding and granular jamming. Nature commun 3:2177
Pusey PN, van Megan W (1987) Observation of a glass transition in suspensions of spherical colloidal particles. Phys Rev Lett 59:2083–2086

Pusey PN (1991) Les Houches Summer School Proceedings. In: Hansen PJ, Levesque D, Zinn-Justin J (eds) Liquids, freezing and the glass transition Part II. Elsevier, Amsterdam
Reichl LE (1998) A modern course in statistical physics. Wiley interscience publication, New York
Renn SR, Lubensky TC (1988) Abrikosov dislocation lattice in a model of the cholesterictosmectic-A transition. Phys Rev A 38(4):2132–2148
Saitoh K, Magnanimo V, Luding S (2012) Slow dynamics near jamming. In: Proceeding of 28th international symposium on rarefied gas dynamics AIP conference, vol 1501, pp 1038–1043
Sander LM (2000) Diffusion-limited aggregation: a kinetic critical phenomenon? Cont Phys 41:203–218
Savage SB (1984) The mechanics of rapid granular flows. Adv Appl Mech 24:289
Segre PN, Prasad V, Schofield AB, Weitz DA (2001) Glasslike kinetic arrest at the colloidal-gelation transition. Phys Rev Lett 86:6042
Siemens AON, van Hecke M (2010) Jamming: a simple introduction. Physica A 389:4255–4264
Silbert LE, Liu AJ, Nagel SR (2005) Vibrations and diverging length scales near the unjamming transition. Phys Rev Lett 95:098301
Song Ch, Wang P, Makse H (2008) A phase diagram for jammed matter. Nature 453:629–632
Stiakakis E, Vlassopoulos D, Loppinet B, Roovers J, Meier G (2002) Kinetic arrest of crowded soft spheres in solvents of varying quality. Phys Rev E 66:051804
Suarez MA, Kern N, Pitard E, Kob W (2009) Out-of-equilibrium dynamics of a fractal model gel. J Chem Phys 130:194904
Tighe BP (2011) Relaxation and rheology near jamming. Phys Rev Lett 107:158303
Trappe V, Prassad V, Cipelletti L, Segre PN, Weitz DA (2001) Jamming phase diagram for attractive particles. Nature 411:772–775
Valverde JM, Espin MJ, Quintanilla MAS, Castellanos A (2010) Fluid to solid transition in magnetofluidized beds of fine powders. J Appl Phys 108:054903
van Hecke M (2010) Jamming of soft particles: geometry, mechanics, scaling and isostaticity. J Phys Condens Matter 22: 033101
Vitelli V (2010) Attenuation of shear sound waves in jammed solids. Soft Matter 6:3007–3012
Vollmayr-Lee K, Kob W, Binder K, Zippelius A (2002) Dynamical heterogeneities below the glass transition. J Chem Phys 116:5158
Walton OR, Braun RL (1986) Viscosity, granular-temperature and stress calculations for shearing assemblies of inelastic, frictional disks. J Rheology 30:949
Warr S, Huntley JM (1995) Energy input and scaling laws for a single particle vibrating in one dimension. Phys Rev E 52:5596–5601
Warr S, Huntley JM, Jaques GTH (1995) Fluidization of a two-dimensional granular systems: experimental study and scaling behaviour. Phys Rev E 52:5583–5595
Watanabe K, Narumi T, Watanabe H, Hasegawa T (2006) Influence of several conditions on yield stress of smectic liquid crystal. In: Proceeding of JSME fluid engineering conference, pp 06–21, CD
Weaire D (1999) A short history of packing problems. Forma 14:279285
Weeks ER, Crocker JC, Levitt AC, Schofield A, Weitz DA (2000) Three-dimensional direct imaging of structural relaxation near the colloidal glass transition. Science 287:627–631
Weeks ER (2012) Melting colloidal crystals from the inside out. Science 338:55–56
Wilcox AJ, Choy J, Bustamante C, Matouschek A (2000) Effect of the protein structure on mitochondrial import. Proc Natl Acad Sci 9:1399–1401
Witten TA (1999) Insights from soft condensed matter. Rev. Mod. Phys. 71:S367–S373
Witten TA (2005) How Soft Matter correlates: three examples. J Phys Condens Matter 17:S1651–S1658
Xu N (2011) Mechanical, vibrational, and dynamical properties of amorphous systems near jamming. Front Phys 6:109–123
Zhang Z, Xu N, Chen DTN, Yunker P, Alsayed AM, Aptowicz KB, Habdas P, Liu AJ, Nagel SR, Yodh AG (2009) Thermal vestige of the zero-temperature jamming transition. Nature 459:230–233

Zhang HP, Makse HA (2009) Jamming transition in emulsions and granular materials. Phys Rev E 72:011301
Zhao C, Tian K, Xu N (2011) New jamming scenario: from marginal jamming to deep jamming. Phys Rev Lett 106:125503
Zorn R (2011) The boson peak demystified? Physics 4:44

A Multiphase Approach to Model Blood Flow in Micro-tubes

T. M. Mubita, L. R. Rojas-Solórzano and J. B. Moreno

Abstract The development of micro-fluidic devices to support the systemic circulation of blood has been used either as a temporary bridge or as a recovery method to treat different heart diseases. Blood flow through these artificial micro-channels is a major challenge because blood at scales from tens to hundreds of microns behaves as a multiphase suspension of deformable particles. A homogeneous model of blood is not adequate if the effect of cell segregation through these devices is of interest to evaluate blood cell damage (e.g., hemolysis or thrombosis). To determine the flow field and model the occurrence of segregation, an Eulerian frame of reference is employed. The simulations are performed in a tube of internal diameter of 217 μm. We find that the results contribute to improve the understanding of the fluid dynamics of blood as a multi-component medium. Our simulations are based on an alternative methodology for blood modelling at a lower computational cost compared to DNS.

1 Introduction

The use of micro-fluidic components has become important in developing equipment for blood transport. During the design process, it is critical to predict the location of regions with large damage generation and potential deposition of blood aggregates in order to minimize these conditions. This fact is particularly relevant to devices

T. M. Mubita · J. B. Moreno
Department of Thermodynamics and Transport Phenomena, Universidad Simón Bolívar, USB, Apartado Postal 89000, Caracas, Venezuela
e-mail: 08-86884@usb.ve

J. B. Moreno
e-mail: morenoj@usb.ve

L. R. Rojas-Solórzano (✉)
School of Engineering, Nazarbayev University, Astana 010000, Republic of Kazakhstan
e-mail: luis.rojas@nu.edu.kz

L. Di G. Sigalotti et al. (eds.), *Computational and Experimental Fluid Mechanics with Applications to Physics, Engineering and the Environment*, Environmental Science and Engineering,
DOI:10.1007/978-3-319-00191-3_11,

containing small channels, junctions, or gaps, in which near wall shear forces and contact with a foreign surface can cause cell activation or damage.

To accurately predict the rate of damage in these "hot spot" zones, it is very important to take into account that blood can be regarded as a homogeneous fluid at a macroscopic level, while at the microscopic level, blood must be considered as a tissue comprising various types of cells (i.e., red blood cells, RBCs; white blood cells, WBCs; and platelets, PLTs) and a liquid intercellular material (i.e., the plasma). This corpuscular nature of blood makes the blood cells have different flow patterns under microcirculation, which are distributed in different regions around the conduct where they flow through.

It has been well recognized that in the flow of blood, the mutual interactions of the RBCs with each other and with platelets lead to an organization of these individual blood components, such that RBCs tend to accumulate at the centreline of the blood vessels (the Fahraeus–Lindquist effect), while platelets are observed to be displaced laterally towards the walls of the vessels in a non-diffusive manner (Almomani et al. 2008).

It has also been reported that due to the presence of RBCs, an increased platelet concentration near the wall (platelet margination) can be observed. In dilute-rich plasma, suspension platelets behave similarly to rigid particles and accumulate at a radial location near 60 % of the tube radius (Aarts et al. 1988). Goldsmith et al. (1995) studied whether red cells exert a physical or chemical effect that promotes aggregation of platelets. Zhao et al. (2008) studied a suspension of RBCs and platelet-sized fluorescent polystyrene particles at controlled flow rates (6–30 mL h^{-1}) through a micro-channel containing a 100–200 μm expansion. Zhao et al.'s work focused on determining the effect of hematocrit on the enhanced platelet concentration in sudden expansions.

Numerical simulations and mathematical formulations have been conducted by a number of research groups to model the spatial distributions of blood elements. Approaches like kinetic theory model, employed by Gidaspow and Huang (2009) and mixture theory model proposed by Massoudi and Antaki (2008) only assume blood as a mixture consisting of RBCs suspended in the plasma, while ignoring the presence of other cells. A two-blood-cell approach was used by Yeh et al. (1994), who characterized the platelet lateral migration (i.e., towards the wall) using a convective diffusive transport equation for platelets, and incorporating an empirical drift function, based on the flow visualization of RBCs and platelet-sized latex beads solutions in a micro-tube. Jung and Hassanein (2008) used the averaging technique to simulate leukocyte migration due to flow-dependent interactions with RBCs within disturbed flow in a sudden expansion and a carotid artery by using a multiphase non-Newtonian model. Jung and Hassanein's research constitutes a good approach in which the multiphase nature of the blood is taken into account despite there is not enough information about how they validated the model.

The aim of this research is to propose and validate a numerical model of blood flow that uses a multiphase computational fluid dynamics approach, to investigate the blood flow behaviour at the microscopic level in order to gain accurate insight into the detailed migration of blood cells within a straight micro-tube.

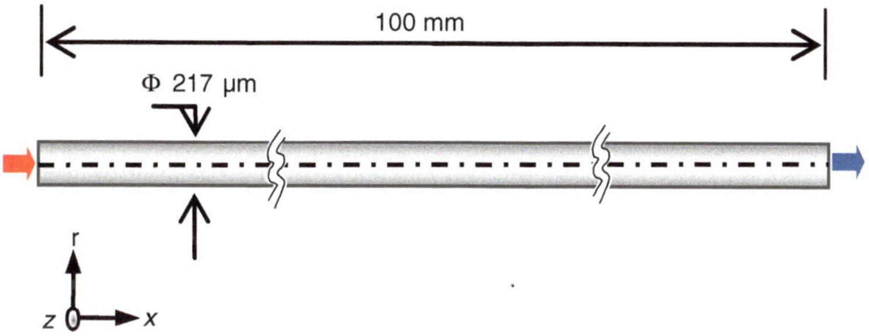

Fig. 1 Computational domain

2 Physical Properties and Numerical Set-up

2.1 Computational Domain and Boundary Conditions

The numerical simulations are performed using an Eulerian–Eulerian multiphase model. The blood is considered as a fluid that consists of plasma, RBCs, and PLTs. The plasma is modelled as a continuous Newtonian fluid with a viscosity of 0.0012 Pa-s and a density of 1,025 kg m^{-3}. RBCs and PLTs are taken as a spherical particle-like fluid dispersed with a density of 1,100 and 1,040 kg m^{-3}, respectively. Sauter mean diameters of 6 μm for RBCs and 2.5 μm for PLTs are assumed. In this study, the viscosity of the disperse phases is considered to be constant through the computational domain, with a value of 0.002 Pa s.

Figure 1 illustrates a schematic of the computational domain showing the tube with a constant cross-section. The internal diameter of the tube is 217 μm (equivalent to ~35 times the mean Sauter diameter of RBCs) and the geometry is depicted by assuming axis-symmetry. At the beginning of the simulation, the whole tube was full of a uniform mixture of plasma, RBCs, and PLTs with volume fractions (α) of 58, 40, and 2 %, respectively. The inlet flow is adjusted such that the axial velocities within the tube lead to similar values to the equivalent Poiseuille average wall shear rate of 555 s^{-1}used by Yeh et al. (1994). Initially, the radial velocities of each phase were set to zero. A free-stress boundary condition is assumed at the exit of the domain, while a zero velocity/no-slip condition is assumed at the walls. Because of the low Reynolds number (<5), turbulence effects were not considered.

2.2 Computational Methodology

The numerical simulations of the 3D, steady, laminar, multiphase flow are performed using ANSYS-CFX ® v12. The numerical model proposed is based on the Eulerian-Eulerian approach. The conservation equations for multiphase flows are presented below. The continuity equation for each phase (k = plasma, RBCs, or PLTs) is given by:

$$\frac{\partial \alpha_k \rho_k}{\partial t} + \nabla.(\alpha_k \rho_k \mathbf{u}_k) = 0 \tag{1}$$

where t is time, ρ_k ,α_k , and u_k are, respectively, the density, volume fraction, and velocity vector of phase k. In addition, the volume fraction of each phase must satisfy the following constraint,

$$\sum_{k=1}^{n} \alpha_k = 1, \tag{2}$$

with $n=3$ being the total number of phases. The volume fraction occupied by one phase cannot be occupied by other phases.

The momentum equation for each phase is given:

$$\begin{aligned} \frac{\partial}{\partial t}(\alpha_k \rho_k \mathbf{u}_k) + \nabla.(\alpha_k \rho_k \mathbf{u}_k \mathbf{u}_k) \\ = -\alpha_k \nabla_p + \nabla \cdot \tau_k + \alpha_k \rho_k \mathbf{g} \\ + \sum_{P=1}^{n} R_{\mathrm{kl}}(\mathbf{u}_{\mathrm{k}} - \mathbf{u}_{\mathrm{l}}) + \boldsymbol{F}, \end{aligned} \tag{3}$$

where p is the pressure shared by all phases, $\mathbf{g}$ is the gravity, $R_k l$ is the interaction force coefficient between the continuum and disperse phases, which depends on the geometry of the interface and local slip Reynolds number, $\boldsymbol{F}$ is a force term containing the lift force, virtual mass, and drag, and τ_k is the stress tensor of phase k for Newtonian fluids, given by:

$$\tau_k = \alpha_k \mu_k (\nabla \mathbf{u}_k + \nabla \mathbf{u}_k^T) + \alpha_k \left(\lambda_k - \frac{2}{3}\mu_k\right)(\nabla \cdot \mathbf{u}_k)\mathbf{I}, \tag{4}$$

where μ_k and λ_k are, respectively, the dynamic and bulk viscosity of phase k, and $\mathbf{I}$ is the unit tensor.

2.3 Verification of Grid Independence

A second-order discretization scheme is used to calculate the advection terms in the discrete finite volume equations. The standard calculation is case-dependent and takes a CPU time from 8 to 36 h on a processor AMD Turion64 with operative systems MS Windows 7, SP2.

In order to corroborate that the solution is independent of the grid resolution, a study of grid convergence is undertaken. With this study, the truncation error is reduced and the best degree of grid resolution is defined. The method consists of iterative and parameter convergence studies using multiple solutions with sys-

Table 1 Uniform structured grids with total number of cells

Mesh	Elements
Coarse	61,740
Medium	152,920
Fine	387,266

tematic parameter refinement to estimate numerical errors and uncertainties. The adopted method for discretization error estimation is the Grid Convergence Index (GCI) based on the Richardson Extrapolation (RE) method, which involves comparisons between three different grid sizes (Roache 1994). The GCI method has shown good performance for numerous and different CFD applications. In order to quantify the discretization error, the systematic procedure recommended by Celik (2008) is followed. The method considers the situation for three solutions corresponding to fine ϕ_1, medium ϕ_2, and coarse ϕ_3 grids with hexahedral elements; three convergence conditions are possible: (i) monotonic convergence; (ii) oscillatory convergence; and (iii) divergence.

The finite element meshes carefully modelled the near-wall region. Table 1, gives the number of computational cells for each one of the meshes generated. The refinement factors, $r_{21} = h_{medium}/h_{fine}$ and $r_{32} = h_{coarse}/h_{medium}$, which are based on a representative cell, mesh, or grid size h, were 1.3 in all cases. The grid refinement should be done systematically; the use of a constant parameter refinement ratio is not required but simplifies the analysis. Celik (2008) suggested that a factor greater than 1.3 would be desirable, based on experience and not on formal derivation. As illustrated in Fig. 2a, there is a reduction of the PLTs volume fraction as the grid is refined, indicating that the dependence of the numerical simulation on the cell size is reduced. Further refinement of the grid will give oscillatory convergence in most of the measurements points, which is an indication of the small difference between the results of the grids.

The three grids used with the GCI method showed that the apparent order of accuracy ranges from 2.64 to 13.96, with an average of 7.15 in the measurement line located at 70 mm from the inlet. This averaged apparent order of accuracy is used to assess the GCI index values for individual grids, which is plotted in the form of error bars, as shown in Fig. 2b. The maximum discretization uncertainty is 6.52 % with an average value of 1.39 %; the highest differences obtained occur at distances from 5.9 to 6.8 µm from the tube wall.

3 Results and Discussions

The blood cells are modelled as liquid dispersed droplets in a three-dimensional Poiseuille flow with gravitational acceleration perpendicular to the tube axis, where the Froude number is 0.33. The section of the micro-tube subjected to analysis corresponds to a region of hydrodynamically developed flow far downstream from the inlet.

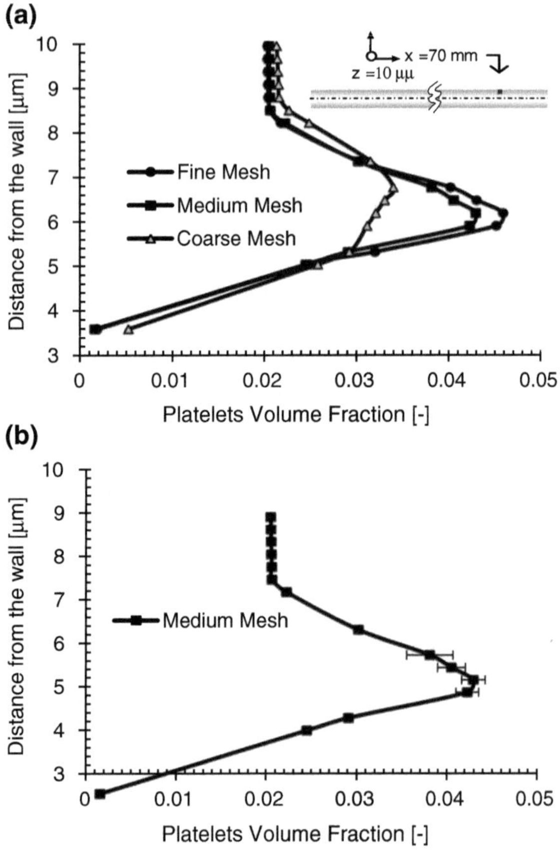

Fig. 2 **a** Platelets concentration profile in a section of the tube obtained for three different meshes, **b** Medium-grid solution, with discretization error bars

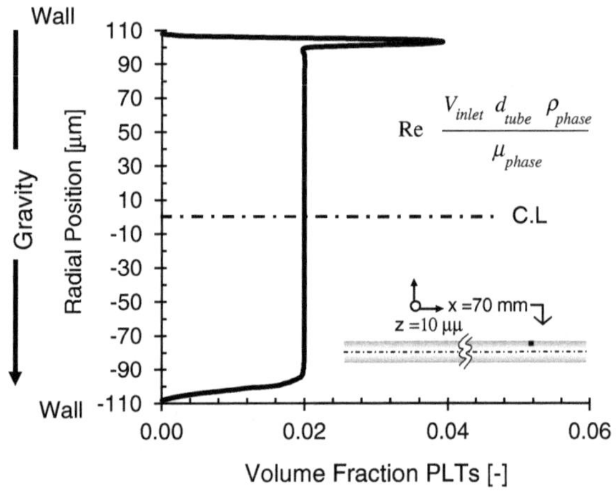

Fig. 3 Platelets' volume fraction. The Reynolds number as calculated at the inlet of the channel was $Re_{PLTs} = 1.62$

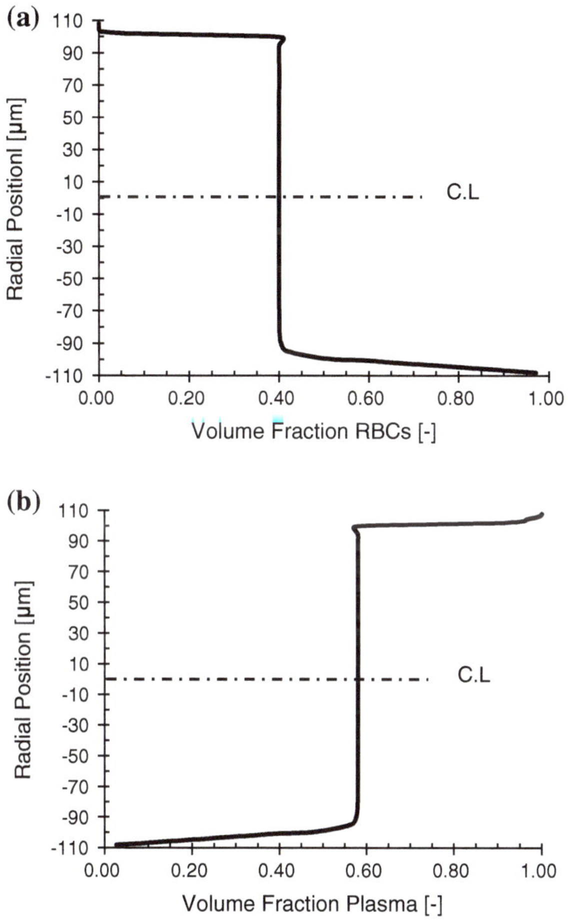

Fig. 4 **a** RBCs concentration profile, Reynolds number, calculated at the inlet of the channel, $Re_{RBCs} = 1.71$, **b** Plasma concentration profile, $Re_{Plasma} = 2.62$

The volume fraction of the disperse phases and plasma are plotted as a function of the radial location in Figs. 3 and 4.

Because of the low Froude number, RBCs tend to migrate towards the bottom of the pipe, increasing their concentration in this region. As they are settled down, a displacement of plasma and platelets occurs towards the upper side of the pipe. The maximum concentration of plasma and RBCs are in the upper and bottom wall, respectively. However, in the case of PLTs, the highest concentration is located very near the upper wall. This behaviour has been reported by several authors for neutrally buoyant solid spherical particles (Won and Yul 2008), buoyant particles (Hogg 1994), drops (Nourbakhsh and Mortazavi 2010), and in Eckstein et al. (1988)'s work, where they found that this behaviour is also characteristic of the flow of platelets

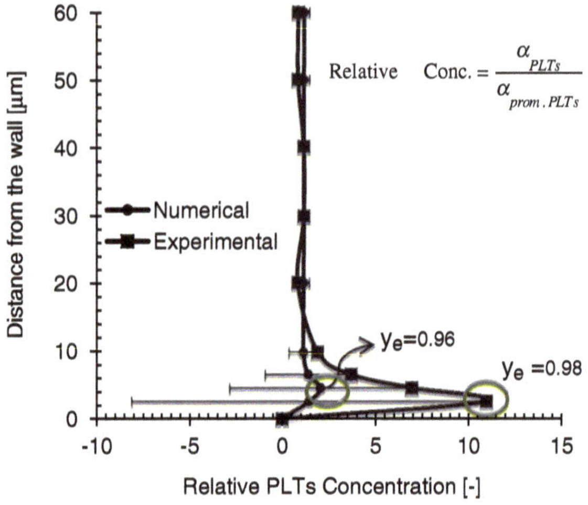

Fig. 5 Comparison between the platelets' concentration profile as obtained experimentally and numerically. The error bars represent the relative errors of computed results

in microcirculation. In general, this behaviour is related to multiphase flow, when the relationship between the channel size and the particle size is relatively large and the Reynolds number (Re) is low.

The peak in the platelets concentration profile is attributed to a lateral migration of particles that tend to cluster at a distance from the wall known as the *equilibrium position* y_e, which is defined as the lateral distance from the centreline of the channel to the peak position of the particle concentration distribution, (r/R). According to Ho and Leal (1974) and the first studies of Segré and Silberberg (1962), using spherical particles with neutral buoyancy, it is observed that in the absence of any slip velocity, the particles tend to migrate away from the wall and the centreline canals, accumulating in an equilibrium position equal to 0.6. However, if we compare with the results reported by Yeh et al. (1994) for a geometry equal to the one used in this chapter (see Fig. 5), the equilibrium position for their experimental data is 0.98, while in our simulation is 0.96, yielding approximately a relative error of 2 %. In addition, we may see that our model is able to predict the platelet concentration at the centre of the pipe; though as we approach to the wall, the concentration numerically obtained is approximately five times lower than that reported experimentally.

To analyze the reason for this deviation, it is important to note that Tilles and Eckstein (1987), Aarts et al. (1988), and Zhao et al. (2007) among others, have reported that the margination and accumulation of platelets in a region near the wall is influenced mainly by the hydrodynamic interaction of these cells with the red blood ones, thus increasing its concentration as the hematocrit increases. In principle, the deviation can be explained by the fact that our computational model does not take into account the interactions among the disperse phases, but only those between the disperse and continuum phases, under the assumption that each phase is present in each control volume and has a volume fraction equal to the fraction of the control volume occupied by that phase.

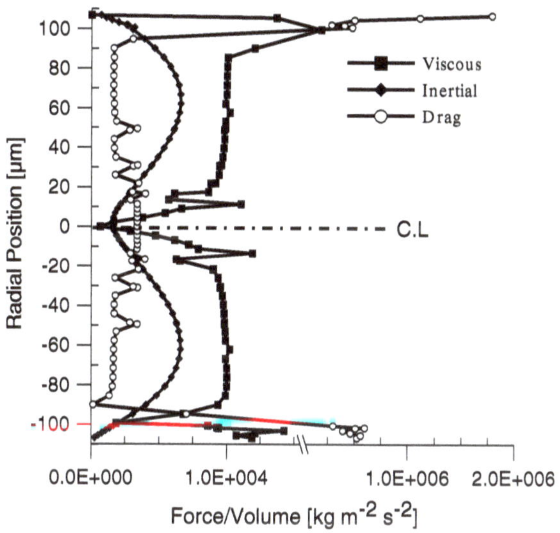

Fig. 6 Radial profiles of the forces that affect the movement of the phases through the tube

Since in this case the Froude number is very small, PLTs lateral migration is controlled by gravity. RBCs, which are heavier than the fluid, migrated to the bottom of the tube, while all PLTs eventually tended to migrate to the top. Therefore, in a given section of the pipe there is a single equilibrium position. A region of dense solids can be supported by a less concentrated (and therefore, less-dense) layer owing to the interplay of sedimentation and shear-induced diffusion (Matas et al. 2004).

In addition, the interplay of viscous, drag, and inertial forces gives rise to a variety of possible fluid phase distributions. Ranking the importance of different forces helps predicting multiphase flow behaviour. In Fig. 6, the different forces are plotted in the radial position. The drag force, due primarily to the viscous nature of the carrier-continuum fluid phase, is dominant at the walls and is responsible for entraining particles along the flow main stream. The lateral dispersion, responsible for the transverse migration of particles across the flow streamlines, is likely to be dominated by the wall inertial forces.

It is clear that the drag force increases as the distance between the wall and the particles decreases. The increase is due to viscous effects derived from the wall presence; when the particle Reynolds number, $Re_p = d_p u_s/\nu$, is small, the inertial term in the Navier-Stokes equations is small compared to the viscous term at distances of the order of the particle radius, $d_p/2$, from the particle centre. Here u_s is the dimensional magnitude of the slip velocity and ν is the kinematic viscosity of the continuous phase.

The buoyant weight of the particles defines a slip velocity that can be either upward or downward, depending of the disperse phase concentration ratio (see Fig. 7). Although, in our case the difference is not so remarkable, this small variation is the effect of the particle migration close to the tube wall which generates a wall-induced lift force, driving the PLTs away from the wall.

Fig. 7 Platelets slip velocity obtained by varying the volume fraction of phases

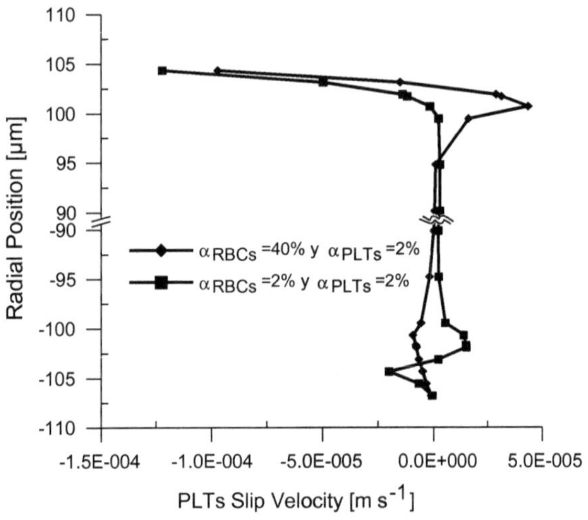

It is important to note that the lateral migration is only observed in the PLTs, while for RBCs only sedimentation occurs. This implies that the phenomenon depends on the size of the particles among other factors. This can be seen in Fig. 8, where the results are compared when PLTs and RBCs are modelled by assuming they have the same volume fraction (2 %) or when they have volume fractions of 2 and 40 %, respectively.

The peak of highest concentration is still observed for the platelet phase at the same equilibrium position. However, it is placed to the other side of the tube. For RBCs, the concentration profile does not change even though the volume fraction varies from 40 to 2 %.

Since the density difference between the plasma and the platelets is small and the platelet diameters are smaller than the RBC ones, the magnitude of the gravitational force is higher for RBCs and makes the major contribution to the motion of these particles. So, if the particle is only slightly buoyant, it stabilizes at a position close to the wall, and depending on the concentration of the other disperse phases, this will be at the top or at the bottom of the tube. In this case, the particles have sufficient buoyancy to dominate the far field of the velocity field, but insufficient to affect the region where there is a balance of the viscous forces with the shear advection term. Then, platelets cannot go all the way onto the wall because of the repulsion and so it will stabilize at a stand-off distance from one of the walls.

For larger buoyancy, the migration of the phase settling under gravity resembles that caused by a sedimenting particle, and therefore the effect of shear and wall repulsion is negligible.

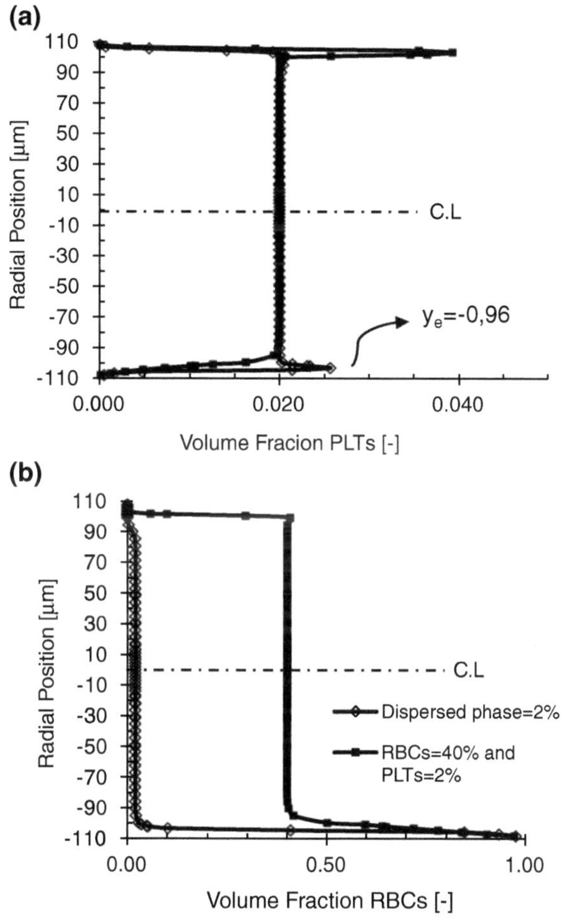

Fig. 8 Concentration profile when varying the volume fraction of the dispersed phase at the tube inlet: **a** platelets, **b** red blood cells

4 Concluding Remarks

Numerical simulations of the motion of two disperse phases, of different densities and concentrations, have been carried out in order to model the migration of blood cells through a micro-tube.

In summary, although the physical mechanisms of the lateral migration are complex and are still under active research, we have presented a simple and accurate approach to model the movement of blood cells through micro-conducts. The results obtained in this investigation show that the underlying dynamics is due to the competition of three effects: one linked to wall-induced shear stress, producing a core-ward drift; one associated to the shear and the curvature of the Poiseuille flow, producing a motion towards the wall (Ho and Leal 1974); and the other one related to the concentration and diameter of the disperse phases, which become a key factor when the phases are not neutrally buoyant.

It should be pointed out that although the scenario of platelet cell migration in microcirculation is qualitatively described by our numerical simulation, we were unable to quantitatively account for the effects of the RBCs migration. In this sense, we may state that the numerical model is able to describe with a good approximation the physical phenomenon of segregation, but further fine tune-up is still required to account for the effects of RBCs migration on the PLTs segregation.

Acknowledgments The authors gratefully acknowledge the financial support from Universidad Simón Bolívar's Deanship of Research and Development under a Research Assistantship grant. Thanks also to the Laboratory of Fluid Mechanics at the Universidad Simón Bolívar, whose PC cluster hosted a large number of the simulations performed in this investigation, and supplied the ANSYS CFX licenses for the study. We would also like to thank Dr. Samuel Hund and Dr. James Antaki for their insightful guidelines and suggestions on benchmark databases to compare with our numerical model.

References

Aarts P, van den Broek S, Prins G, Kuiken G, Sixma J, Heethaar R (1988) Blood platelets are concentrated near the wall and red blood cells, in the center in flowing blood. Arterioscler Thromb Vasc Biol 8:819–824

Almomani T, Udaykumar H, Marshall J, Chandran K (2008) Micro-scale dynamic simulation of erythrocyte-platelet interaction in blood flow. Ann Biomed Eng 36:905–920

Celik I (2008) Procedure for estimation and reporting of discretization error in CFD applications. J Fluid Eng 130:1–4

Eckstein E, Tilles A (1988) Conditions for the ocurrence of large near wall excess of small particles during blood flow. Microvasc Res 36:31–39

Gidaspow D, Huang J (2009) Kinetic theory based model for blood flow and its viscosity. Ann Biomed Eng 37:1534–1545

Goldsmith H, Bell D, Braovac S, Steinberg A, McIntosh F (1995) Physical and chemical effects of red cells in the shear-induced aggregation of human platelets. Biophys J 69:1584–1595

Ho B, Leal L (1974) Inertial migration of rigid spheres in two-dimensional unidirectional flows. J Fluid Mech 65:365–400

Hogg A (1994) The inertial migration of non-neutrally buoyant spherical particles in two-dimensional shera flows. J Fluid Mech 272:285–318

Jung J, Hassanein A (2008) Three-phase CFD analytical modeling of blood flow. Med Eng Phys 30:91–103

Massoudi M, Antaki J (2008) An anisotropic constitutive equation for the stress tensor of blood based on mixture theory. Math Prob Eng 2008:1–30

Matas J, Morris J, Guazzelli E (2004) Lateral forces on a sphere. Oil Gas Sci Technol Rev IFP 50:59–70

Nourbakhsh A, Mortazavi S (2010) A three dimensional study of the motion of a drop in plane Poiseuille flow at finite Reynolds numbers. Iranian J Sci Technol Trans B Eng 34:179–196

Roache P (1994) Perspective: a method for uniform reporting of grid refinement studies. J Fluid Eng 116:405–413

Segré G, Silberberg A (1962) Behaviour of macroscopic rigid spheres in Poiseuille flow: part 2. Experimental results and interpretation. J Fluid Mech 14:136–157

Tilles A, Eckstein E (1987) The near wall excess of platelets-sized particles in blood flow: Its dependance on hematocrit and wall shear rate. Microvasc. Res. 33:211–223
Won Y, Yul J (2008) The lateral migration of neutrally-buoyant spheres transported through square microchannels. J Micromech Microeng 18:065015
Yeh C, Calvez A, Eckstein E (1994) An estimated shape function for drift in a platelet-transport model. Biophys J 67:1252–1259
Zhao R, Kameneva M, Antaki J (2007) Investigation of platelet margination phenomena at elevated shear stress. Biorheology 44:161–177
Zhao R, Marhefka J, Shu F, Hund S, Kameneva M, Antaki J (2008) Micro-Flow visualization of red blood cell-enhanced platelet concentration at sudden expansion. Ann Biomed Eng 36:1130–1141

Perspective: The Breakup of Liquid Jets and the Formation of Droplets

José R. Castrejón-Pita and Ian M. Hutchings

Abstract The breakup of liquid surfaces is a topic of great relevance to industry that often presents complications for both experimental and theoretical physicists. Although they have been widely studied since the end of the eighteenth century, many of the phenomena involved in the processes of the breakup of liquids and the formation of new surfaces and droplets are still not fully understood. In this chapter we discuss some of the current issues faced by researchers working in the field of droplet dynamics.

1 Introduction

The creation and behaviour of droplets might appear superficially to be perfectly predictable phenomena and thoroughly explored topics of research. After all, several numerical methods are available both commercially and non-commercially that appear to model not only the breakup of streams or jets of Newtonian liquids from dripping taps (faucets), spray nozzles, and inkjet printers, but also the breakup of viscoelastic and granular flows. In reality, however, only a few of these simulations have ever been rigorously experimentally validated, and most of those that have were assessed under conditions outside the range covered by real industrial processes. As a consequence, the use of modelling methods is not extensive in industry and, in the few cases in which modelling is used, it is merely employed as a first-approach tool. More often, the development and optimization of products and manufacturing processes rely on empirical trial and error testing. In industries dealing with liquid

J. R. Castrejón-Pita (✉) · I. M. Hutchings
Department of Engineering, University of Cambridge, 17 Charles Babbage Road, Cambridge, CB3 0FS, United Kingdom
e-mail: jrc64@cam.ac.uk

I. M. Hutchings
e-mail: imh2@cam.ac.uk

L. Di G. Sigalotti et al. (eds.), *Computational and Experimental Fluid Mechanics with Applications to Physics, Engineering and the Environment*, Environmental Science and Engineering,
DOI:10.1007/978-3-319-00191-3_12,

delivery, these optimization processes involve not only adjustment of the properties of the fluids, but also modification of the geometry of the system (e.g. a nozzle and/or the liquid conduits) and its drivers (e.g. pumps, heating elements, or piezoelectric transducers). It is clear that, given the maturity of some of the technologies, most current practical methods for the production of droplets, aerosols, or liquid jets (*e.g.* inkjet and sprays) are very complex and as such are very difficult to characterize or model. In general terms, the modelling of commercially available systems is not easy as most are 'black boxes' in which none of the fluid dynamical conditions are known accurately. Studies of inkjet print-heads present unique difficulties because they contain very small internal channels, operate at very high repetition frequencies, tend to lack optical access to the print-head interiors, and the droplets produced are very small and travel at high speeds. This discussion is therefore focused on inkjet systems and, in particular, on the processes of droplet generation and deposition. Although there are many fields of research into applications of droplets, this chapter discusses only some of the topics of academic and scientific interest presently faced by inkjet technology.

2 Droplet Production

It is generally accepted that there is still much to understand about the breakup of liquids into droplets (Eggers 1997; Basaran 2002). Some remarkably complete and early studies were performed by Lord Rayleigh during the nineteenth century, providing much of the basic understanding that is still currently used to predict jet behaviour and drop formation. In recent years, the study of the formation and behaviour of jets and drops has gained additional importance, motivated by their application in inkjet printing technology. Inkjet printing has been identified by many as a technology that can potentially change the way some products are delivered by manufacturing processes. This is because inkjet offers several advantages over traditional techniques. Inkjet printing is very efficient, robust, versatile, digital, and non-contact. It is efficient because it utilises only the material required, producing little or no waste; it is robust because it is a mature technique widely used, and it is versatile because several techniques exist for the printing of different inks or liquids. In addition, it offers the advantage of being digital so that no master template is required and thus printed patterns can be readily modified or changed on demand. Another advantage is that the print-head is never in physical contact with the substrate. All these characteristics have made inkjet a technology with enormous potential, with applications being actively explored in novel areas of biological deposition and the printing of electronic devices. Traditional inkjet processes are also being increasingly employed to increase versatility in commercial printing. Current investigations of inkjet systems are focused on fundamental studies to describe the physics of liquid jetting and on empirical investigations aimed at adapting current technologies to the printing of non-conventional materials (Hutchings and Martin 2013). Examples of these are investigations of the thinning and breakup of liquids, the delivery of active

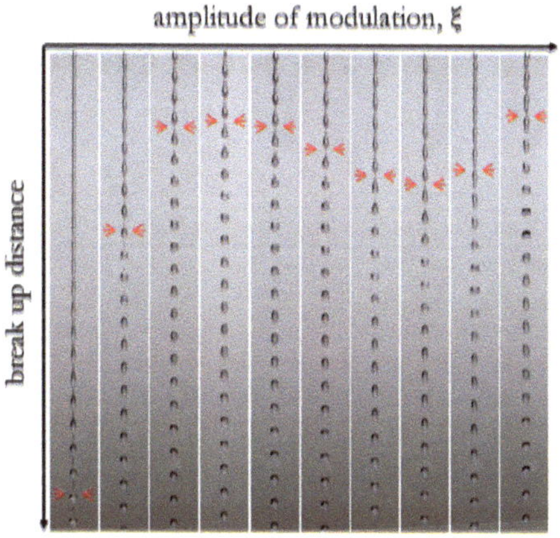

Fig. 1 Images showing the breakup of a modulated continuous liquid jet, travelling downwards from a nozzle at the *top* of the image, as the modulation amplitude is increased. Theory predicts a monotonic decrease in the breakup length (i.e. the distance from the nozzle to the breakup point) with the modulation amplitude. In most commercial examples of continuous inkjet printing, the breakup length exhibits a reversal as shown here, i.e., above a certain amplitude of modulation the breakup distance increases with the amplitude

components in pharmaceutical applications, and printing processes for conductors and semi-conductors for printed electronics, displays, sensors, and photovoltaic cells.

Two basic inkjet processes are commonly used by industry: *Drop on Demand* (DoD) and *Continuous InkJet* (CIJ). In the first method, droplets are produced by the action of a heater or piezoelectric transducer in response to an electric signal (called the *waveform*) (Castrejón-Pita et al. 2011a). In contrast, in CIJ mode a continuous jet of ink is harmonically perturbed to produce a train of equally spaced droplets (Castrejón-Pita et al. 2011b). Most CIJ systems rely on piezoelectric elements to produce the perturbation that initiates the breakup into droplets. Once the droplets are formed, these are usually electrically charged and then directed by electric fields to form the desired pattern on the substrate. In many commercial and industrial systems, the pressure or velocity responses to the electrical driving signal, the waveform in DoD and the harmonic signal in CIJ, are very often unknown, which presents a major complication for modelling (Castrejón-Pita et al. 2011a).

The surface perturbation (as illustrated in Fig. 1) that leads to breakup of modulated continuous jets is described by the Weber-Rayleigh equation:

$$r = a + \xi_0 e^{ikz+(\alpha+i2\pi f)z/v}, \tag{1}$$

where v and a are the speed and the initial radius of the jet, ξ_0 and k are the initial amplitude and the wave number of the harmonic perturbation ($k = 2\pi/\lambda$), and λ is its wavelength. The growth rate α depends on the fluid properties as follows:

Fig. 2 Different stages of the thinning and breakup of a drop of water (a near-inviscid Newtonian liquid) emerging from a nozzle. The whole process shown in the pictures takes approximately 33 ms. Visualised by using high speed shadowgraph imaging, details in Castrejón-Pita et al. (2012)

$$\alpha^2 \frac{ka I_0(ka)}{2I_1(ka)} + \alpha \frac{k^2\mu}{\rho}\left[2ka\frac{I_0(ka)}{I_1(ka)}\frac{l^2}{l^2-k^2} - 1 - 2la\frac{I_0(la)}{I_1(la)}\frac{k^2}{l^2-k^2}\right] = \frac{\sigma k^2}{2\rho a}(1-k^2a^2), \tag{2}$$

where $l = k^2 + \rho/\mu$. Here ρ, σ, and μ are the density, surface tension, and viscosity of the fluid (or ink), respectively. Detailed derivations of these equations are available elsewhere (Strutt 1896; Brenn et al. 2000). Equation (1) can be used to calculate fluid properties from observed jet profiles (Castrejón-García et al. 2011). In commercial CIJ systems, the precise point of breakup which defines the breakup distance (*i.e.*, $r = 0$), is important as it needs to be in the correct position to ensure that charge is correctly induced on the drops so they can be deflected in a controlled way by an electrostatic field. It has been demonstrated that Eq. (1) can be used very accurately to predict fluid properties and breakup distances for low-amplitude modulations (González and García 2009) yet it fails to predict the behaviour observed experimentally for high amplitude oscillations. Figure 1 shows the point of breakup for different amplitudes of modulation, from which it is clear that the breakup length does not decrease monotonically with the amplitude of modulation. This phenomenon is often seen in industry and called *breakup reversal*; it is not yet understood. Various hypotheses have been proposed to explain this effect but none has yet been successfully verified experimentally. This is an important commercial issue as industry relies on empirical experimentation to set up the jetting conditions, which is a complex procedure as it depends on the fluid properties and the nozzle design.

Other studies of drop formation aim to understand and predict the breakup behaviour of droplets and liquid filaments. This is a fundamental step towards the complete modelling of inkjet systems. The different behaviours observed for liquids of different viscosities are very familiar, as water dripping from a tap shows very different behaviour from syrup flowing from a spoon, as illustrated in Figs. 2 and 3. The breakup of viscous liquids has been extensively studied theoretically but little experimental work exists, Brenner et al. (1996); Day et al. (1998); Basaran (2002); Chen et al. (2002). During the filament thinning process that eventually leads to breakup, several regimes have been identified. These regimes depend on the internal dynamics

Fig. 3 Imaging of the thinning and pinch-off process of a droplet of glycerol and water mixture with a viscosity of 365 mPa s. The images cover a period of 120 ms and were captured with the same system as in Fig. 2

of the liquid and its properties, but the transitions from one regime to another are processes that still require more experimental study. Experimental studies of thinning liquid filaments are complicated, because the processes occur over a wide range of timescales. Various regimes can be identifed which are dominated by the action of different forces: in dripping the first forces to act are gravity and surface tension, while later in time viscous forces come into play, then inertia and finally the viscosity of the surrounding (outer) medium (Basaran 2002). This presents practical complications for high speed imaging as the resolution of conventional setups is compromised by the frame speed. It is relatively easy to record images of high resolution at a slow rate and to record low resolution images quickly, but not to record high resolution images at high rates. This conflict has so far limited the observation of regime transitions in drop formation. The gap between theory and experimental observation must be bridged in order to understand the whole process of drop formation.

There are many industrial scenarios in which there is a need to understand the mechanisms involved in drop formation. Current industrial challenges include studies to improve the reliability, quality, and speed of inkjet printing. Other efforts are focused on the development and testing of methods to characterize dynamic liquid properties such as viscosity, other rheological properties, and surface tension at appropriate time and length scales found in industrial applications, which are believed to play a decisive role in determining the dynamics of the droplet in both CIJ and DoD systems.

3 Droplet Deposition

Another topic of industrial and academic interest is the impact and coalescence of droplets on both solid and liquid substrates. On solid substrates the dynamics of the contact line and the hysteresis of the dynamic contact angle are themes that have

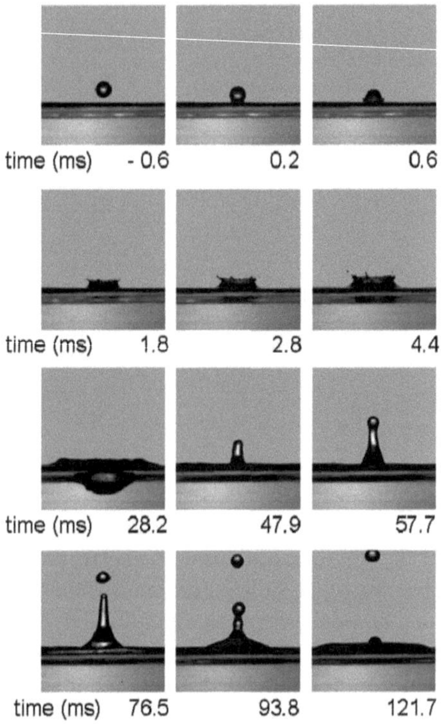

Fig. 4 The impact of a 2 mm diameter drop of a glycerol and water mixture with a viscosity of 10 mPa s onto a pool of the same liquid; the impact speed was 2 m/s. Observed by high speed shadowgraphy, as described in Castrejón-Pita et al. (2012)

been widely studied but still remain to be completely understood. Liquid-on-liquid deposition is a rich field of research that relates to common events such as rain droplets impacting onto rivers or the sea, or milk being poured into a cup of tea. Single droplets impacting on a stationary pool of water are the simplest example that can be studied and yet offer a very complex dynamics, as seen in the example in Fig. 4. This type of impact is currently being investigated using state of the art technologies such as high speed X-ray photography and ultra high speed imaging. By using these methods, recent experiments have uncovered previously unknown phenomena happening at the very first instants after the impact and coalescence of fluids. One such previously unstudied phenomenon is the ejection of very thin and fast sheets of liquids whose appearance can only be recorded by ultra high speed imaging. This liquid sheet has been termed the *fluid ejecta* and precedes the spreading of the droplet and the formation of the well-known splashing crown (Thoroddsen 2002, 2012).

Studies on the behaviour of the fluid ejecta still have much to offer and this is an active area of research (Thoraval et al. 2012). Current studies on this topic are focused on numerical modelling and on experimental work aiming to understand the dynamics of the ejecta in terms of the fluid properties and the speed of impact. It has been observed that the fate of the ejecta is highly dependent on the viscosity of

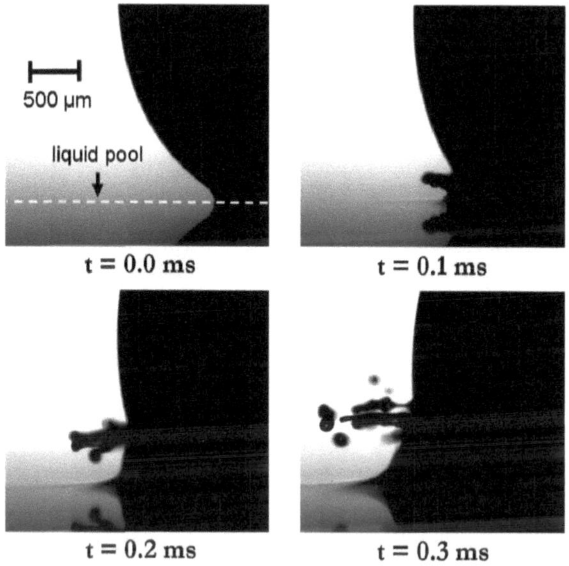

Fig. 5 Close-up of the interface formed between an impacting water droplet and a pool of the same liquid at different times. An *ejecta sheet* is clearly visible 100 μs after impact. The sheet rapidly becomes unstable due to surface tension and breaks up into micro-drops

the fluid as it can break up into micro-drops or remain as a liquid sheet (see Figs. 5 and 6). Recent work has suggested that the ejecta sheet formed by rain droplets impacting the sea may be responsible for the formation of very small droplets that form salt crystals that eventually nucleate clouds (Thoroddsen 2012). Furthermore, the formation of such ejecta is a process that could be exploited in commercial applications to produce very small droplets. Generally speaking, conventional DoD droplet generators produce drops with diameters similar to the size of the nozzle used to jet them (Chen and Basaran 2002). As the ejecta can breakup into droplets that are much smaller than the size of the droplet, this is a mechanism that might be used to produce micro-drops, although in industrial applications such as inkjet printing the production of such tiny additional droplets is already known and is undesirable.

Novel applications for droplet-based technologies are currently being explored in diverse manufacturing processes such as rapid prototyping, the printing of electronics and bio-materials, 3D printing and coatings. The commercial success of many of these ventures really depends on the capabilities of current technologies to handle a large range of fluid viscosities, liquids with complex viscosities, fluids with high solid contents, and granular material. In all these areas there is still much to be learned both theoretically and experimentally. In particular, there are topics such as non-Newtonian fluids and granular media, where great advances have been made theoretically but technological limitations have prevented associated experimental work. Novel applications are not only attracting the attention of industry but driving science into previously unexplored regions.

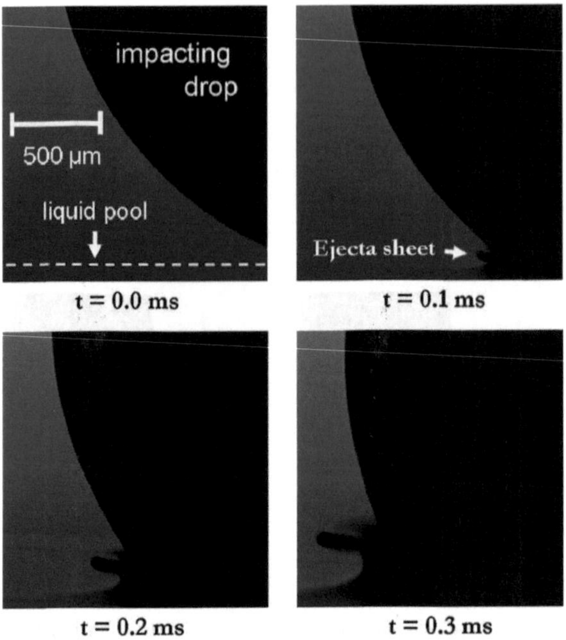

Fig. 6 Close-up of the interface formed between an impacting droplet of a glycerol and water mixture with a viscosity of 10 mPa and a pool of the same liquid. In this case, in the *ejecta sheet* the breakup of the liquid surface is prevented by the action of viscosity and as a consequence micro-drops droplets are not formed

4 Conclusions

Many current studies in fluid mechanics are focused on real industrial problems, such as the modelling of inkjet technologies. These studies are necessary in order for industry to reduce the costs of development by transferring effort from empirical testing to modelling. For that to happen, several limitations need to be overcome. The validation of theoretical and numerical methods will benefit inkjet technologies and enable conditions to be investigated that are not cost-effective to explore experimentally. The realization of fully-validated algorithms can lead to the production of systems with enhanced capabilities to jet and deposit micrometre-scale droplets of complex liquids, increase the speed of printing, achieve high quality single-pass printing, the sequential deposition of different fluid compositions, better feature definition and the ability to deposit a much wider range of materials or liquids with high solid content.

Acknowledgments We are grateful for the support of the UK Engineering and Physical Sciences Research Council (Grant EP/H018913/1: Innovation in Industrial Inkjet Technology).

References

Basaran OA (2002) Small-scale free surface flows with breakup: drop formation and emerging applications. AIChE J 48:1842–1848
Brenn G, Liu Z, Durst F (2000) Linear analysis of the temporal instability of axisymmetrical non-Newtonian liquid jets. Int J Multiph Flow 26:1621–1644
Brenner MP, Lister JR, Stone HA (1996) Pinching threads, singularities and the number 0.0304. Phys Fluids 8:2827–2836
Castrejón-García R, Castrejón-Pita JR, Martin GD, Hutchings IM (2011) The shadowgraph imaging technique and its modern application to fluid jets and drops. Rev Mex Fis 57:266–275
Castrejón-Pita JR, Morrison NF, Harlen OG, Martin GD, Hutchings IM (2011a) Experiments and Lagrangian simulations on the formation of droplets in drop-on-demand mode. Phys Rev E 83:036306
Castrejón-Pita JR, Morrison NF, Harlen OG, Martin GD, Hutchings IM (2011b) Experiments and Lagrangian simulations on the formation of droplets in continuous mode. Phys Rev E 83:016301
Castrejón-Pita JR, Castrejón-Pita AA, Hinch EJ, Lister JR, Hutchings IM (2012) Self-similar breakup of near-inviscid liquids. Phys Rev E 86:015301
Castrejón-Pita AA, Castrejón-Pita JR, Hutchings IM (2012) Experimental observation of von Kármán vortices during drop impact. Phys Rev E 86:045301
Chen AU, Notz PK, Basaran OA (2002) Computational and experimental analysis of pinch-off and scaling. Phys Rev Lett 88:174501
Chen AU, Basaran OA (2002) A new method for significantly reducing drop radius without reducing nozzle radius in drop-on-demand drop production. Phys Fluids 14:L1–L4
Day RF, Hinch EJ, Lister JR (1998) Self-similar capillary pinchoff of an inviscid fluid. Phys Rev Lett 80:704–707
Eggers J (1997) Nonlinear dynamics and breakup of free-surface flows. Rev Mod Phys 69:865–929
González H, García FJ (2009) The measurement of growth rates in capillary jets. J Fluid Mech 619:179–212
Hutchings IM, Martin GD (2013) Inkjet technology for digital fabrication. Wiley, Chichester
Strutt (Lord Rayleigh) JW (1896) The theory of sound, vol II. Macmillan, New York
Thoraval M-J, Takehara K, Etoh TG, Popinet S, Ray P, et al. (2012) von Kármán vortex street within an impacting drop. Phys Rev Lett 108:264506
Thoroddsen ST (2002) The ejecta sheet generated by the impact of a drop. J Fluid Mech 451:373–381
Thoroddsen ST (2012) The making of a splash. J Fluid Mech 690:1–4

Experimental Investigation of Thermal Diffusion in Binary Fluid Mixtures

Humberto Cabrera

Abstract The mass transport of chemical species in response to a temperature gradient, referred to as the Soret effect or thermal diffusion, leads under certain conditions to a separation of the chemical constituents. The Soret coefficient is the ratio of the thermodiffusion coefficient to the molecular diffusion coefficient. This effect along with molecular diffusion occurs in many natural phenomena and engineering systems. One early application of this effect was the separation of isotopes. Understanding the Soret effect is also important for exploring the mechanics of crude oil extraction and its reservoir characterization, as well as in the research of the global circulation of see water. It has also been used for polymer characterization by thermal field flow fractionation. Moreover, recent studies on the Soret effect of bio-systems, like protein and DNA solutions, indicate that it might help revealing the mechanisms behind the mysterious phenomenon of life. Many experimental techniques have been developed for investigation of the Soret effect: thermogravitational columns, thermal lens, diffusion cells, thermal diffusion forced Rayleigh scattering, thermal field flow fractionation, and microfluidic fluorescence. In this chapter, we focus on the investigation of thermal diffusion behaviour in simple liquid mixtures by a thermal lens method. The big advantage of the thermal lens method is that it is fast, simple, and the experimental set-up is much cheaper compared to other methods. In particular, a calibrated two-beam mode-mismatched thermal lens experiment is used for determining the Soret coefficient for isopropanol/water and ethanol/water mixtures.

H. Cabrera (✉)
Centro Multidisciplinario de Ciencias, Laboratorio de Óptica Aplicada,
Instituto Venezolano de Investigaciones Científicas, IVIC,
Loma de Los Guamos, Vía Jají, Mérida, Estado Mérida, Venezuela
e-mail: cabrera_25@yahoo.com

H. Cabrera
The Abdus Salam International Centre for Theoretical Physics, ICTP,
Strada Costiera 11, 34014 Trieste, Italy
e-mail: hcabrera@ictp.it

L. Di G. Sigalotti et al. (eds.), *Computational and Experimental Fluid Mechanics with Applications to Physics, Engineering and the Environment*, Environmental Science and Engineering, DOI:10.1007/978-3-319-00191-3_13,

The fitting curves show a very good agreement between the theoretical model and the experimental data. The experimental results have also shown good agreement with available thermodiffusion coefficient data.

1 Introduction

Thermal diffusion, known also as the Ludwig–Soret effect, plays an important role in our understanding of the properties of liquid mixtures (Sitzber 1856; Soret 1879). It characterizes the flux of matter in response to a temperature gradient, which leads to the formation of a concentration gradient (Bierlein 1955). This stationary concentration gradient is given by

$$\nabla c = -S_T c_0 (1 - c_0) \nabla T, \tag{1}$$

where $S_T = D_T / D$ is the Soret coefficient, D the mass-diffusion coefficient, D_T the thermal diffusion coefficient, T the temperature, c the molar fraction of component 1 (the heaviest component), and c_0 its equilibrium value (Tyrell 1961).

The thermal lens method (TL) is a powerful method that can be used to study the Soret effect in transparent liquid mixtures. The TL measures the amount of heat deposited in a medium after the absorption of light. Immediately after the absorption of photons, the spatial heat distribution resembles the intensity distribution of the beam spot. Later on, the thermal diffusion spreads the heat over distances more than 1 order of magnitude larger than the excitation beam spot radius. Since the refractive index depends on temperature, a spatial distribution of the refractive index of a similar extent is generated in the absorbing medium. In addition to the temperature-dependent refractive index gradient, the Soret effect can produce refractive index changes and influence the total signal (Giglio and Vendramini 1974). For CW excitation the generation of a stationary TL can take milliseconds to several seconds, depending on the beam spot radius and the thermal diffusivity of the sample. The TL alters the propagation of the probe beam through the medium by generating a phase shift on the beam wavefronts. Since the first report on the thermal lens effect, the sensitivity of the technique has been improved by changing the experimental configuration. Early experimental arrangements used a single-beam configuration, which employed a laser beam to excite the TL and probe it (Whinnery 1974). The use of a second probe beam for testing the TL has improved the versatility and sensitivity of the technique (Long et al. 1976; Shen et al. 1992; Marcano et al. 2006; Cabrera et al. 2009a). The dual-beam configuration provides the possibility to use signal processing devices, e.g. lock-in amplifiers, to improve the signal-to-noise ratio and hence the sensitivity of the TL measurement. In addition, detection optics and detectors can be optimized for a single, convenient probe laser wavelength rather than requiring detection in what may be a more difficult spectral region, which is a convenient choice for generating continuous scanning TL spectra (Shen et al. 1992).

An advantage of the thermal lens method compared to the diffusion cell is the short duration of a typical experiment (short equilibration times) due to small distances on the order of the focal beam width. Furthermore, in thermal lens experiments typical temperature changes are on the order of $10^{-2} - 10^{-5}\,°C$ (Gordon et al. 1965; Whinnery 1974). This allows to work with extremely small temperature and concentration gradients, so that the addition of a dye can be avoided by using the natural absorption of the molecules. Such advantages make the Soret coefficient independent of the effects of convection. In the same way, the sensitivity to convection may be neglected for fast diffusing systems such as, for example, water/organic solvent mixtures.

In this work, a calibrated two-beam mode-mismatched thermal lens experiment, where the pump beam is tightly focused and the probe beam is collimated, was used for determining the Soret coefficient of alcohol-water systems. This particular set-up provides the maximum possible response in terms of the amplitude of the signal in a CW photothermal experiment (Marcano et al. 2006). In addition, the alignment procedure, the experiment itself as well as its calibration, and the interpretation of the experimental results are simpler than working with the mode-matched configuration (Long et al. 1976).

2 Theoretical Model

In the mode-mismatched dual-beam thermal lens experiment when an infinite medium is illuminated, at time $t = 0$ along the z-axis, with a light beam of Gaussian intensity $I(r, z) = [2P_e/\pi\omega_e^2(z)]\exp[-2r^2/\omega_e^2(z)]$, in the limit of small absorption values, the temperature distribution is given by Shen et al. (1992)

$$\Delta T(r, z, t) = \frac{2P_e\alpha}{\pi\rho C_p\omega_e^2}\int_0^t \frac{t_c}{t_c + 2t'}\exp\left[-\frac{2t_c r^2}{\omega_e^2(t_c + 2t')}\right]\mathrm{d}t', \tag{2}$$

where P_e is the excitation power, and α, ρ, and C_p are, respectively, the absorption coefficient, the density, and the specific heat of the medium. The quantity $\omega_e(z) = \omega_{0e}[1 + (z - a_e)^2/z_e^2]^{1/2}$ is the excitation beam radius at the sample cell, where a_e, ω_{0e}, $z_e = \pi\omega_{0e}^2/\lambda_e$, and λ_e are the waist position, the radius at the waist, the Rayleigh parameter, and the excitation wavelength, respectively. The coordinate z is the sample position, r is the coordinate radius, $t_c(z) = \omega_e^2(z)/4D_{th}$ is the characteristic thermal time constant, with $D_{th} = k/\rho C_p$ and k being the thermal diffusivity and thermal conductivity, respectively.

If the refractive index n of the liquid mixture satisfies the condition $\Delta n/n << 1$ (Gordon et al. 1965), its change with temperature and concentration is given by (Bierlein 1955)

$$\Delta n(r,z,t) = \frac{\partial n}{\partial T}\Delta T(r,z,t) + \frac{\partial n}{\partial c}\Delta c(r,z,t), \tag{3}$$

while the concentration change is (Bierlein 1955)

$$\Delta c(r,z,t) = -S_T c_0(1-c_0)\Delta T(r,z,t)\Gamma(z,t), \tag{4}$$

with

$$\Gamma(z,t) = 1 - \sum_{i=1}^{\infty} \frac{4}{(2i-1)\pi} \sin\left[\frac{(2i-1)\pi}{2}\right] \exp\left[-(2i-1)^2 \frac{t}{t_D(z)}\right], \tag{5}$$

where $t_D(z) = \omega_e^2(z)/4D$ is the mass-diffusion time (Arnaud and Georges 2001).

The expansion in Eq. (5) converges rapidly. In order to obtain an accuracy better than 1 % it is sufficient to retain terms up to $i = 4$ in the summation. When $t >> t_D$, the exponential term vanishes and $\Gamma = 1$, and so Eq. (4) reduces to Eq. (1).

The change of the refractive index acts as an optical element producing a phase shift Φ on the wavefronts of the probe beam. Inserting Eq. (4) into Eq. (3) and then using Eq. (3), we obtain for the phase shift (Marcano et al. 2002; Cabrera et al. 2009a)

$$\begin{aligned}\Phi(r,z,t) &= \frac{2\pi}{\lambda_p} l[\Delta n(r,z,t) - \Delta n(0,z,t)] \\ &= \frac{2\pi}{\lambda_p} l\left[\frac{\partial n}{\partial T} - \frac{\partial n}{\partial c} S_T c_0(1-c_0)\Gamma(z,t)\right] \\ &\quad \times [\Delta T(r,z,t) - \Delta T(0,z,t)],\end{aligned} \tag{6}$$

where l is the sample cell length and λ_p is the wavelength of the probe field. Substituting Eq.(2) into Eq.(6), Φ can be written as (Marcano et al. 2002; Cabrera et al. 2009a)

$$\begin{aligned}\Phi(g,z,t) &= \frac{\phi_s \Gamma(z,t) - \phi_{th}}{2} \\ &\quad \times \int_{(1+2t'/t_c)^{-1}}^{1} \frac{[1-\exp(-2m(z)g\tau)]}{\tau} d\tau,\end{aligned} \tag{7}$$

where $\phi_s = (\partial n/\partial c)[P_e \alpha l S_T c_0(1-c_0)/k\lambda_p]$ is the induced concentration phase shift amplitude and $\phi_{\text{th}} = (\partial n/\partial T)(P_e \alpha l/k\lambda_p)$ is the induced thermal lens phase shift amplitude. In the above equation, the ratio $m(z) = [\omega_p(z)/\omega_e(z)]^2$ accounts for the level of mode-matching between the beams, $g = r/\omega_p(z)$ is a dimensionless parameter, and $\omega_p(z) = \omega_{0p}[1 + (z-a_e)^2/z_p^2]^{1/2}$ is the probe beam radius at the sample cell, with a_p, ω_{0p}, and $z_p = \pi\omega_{0p}^2/\lambda_p$ being the waist position, the radius at the waist, and the Rayleigh parameter.

The phase shift is added to the wavefront of the propagating probe light. At the exit of the sample the amplitude of the probe light is (Marcano et al. 2002)

$$E_p(r,z,t) = \frac{\sqrt{2P_p/\pi}}{\omega_p(z)} \exp\left[-\frac{r^2}{\omega_p(z)} - ik_p z - i\frac{k_p r^2}{2R(z)} + i\arctan\left(\frac{z-a_p}{z_p} - i\Phi\right)\right], \tag{8}$$

where P_p is the total power of the probe light beam, $k_p = 2\pi/\lambda_p$ is the probe light wave number, and $R(z) = [(z-a_p)^2 + z_p^2]/z$ is the radius of curvature of the probe beam at the sample position z.

Behind the sample cell the probe beam propagates freely up to the position of the aperture. The next step is to calculate the probe beam amplitude at the position of the detector. The Fresnel diffraction approximation provides the solution of this problem. It is given by the convolution of the probe field amplitude $E_p(r,z,t)$ at the exit of the sample with the impulse response of free space propagation (Marcano et al. 2002).

Finally, the time-dependent total signal can be calculated using the definition (Shen et al. 1992):

$$S(z,t) = \frac{I(z,t) - I_0}{I_0}, \tag{9}$$

where $I(z,t) = 2\pi \int_a |E_p(r,z,t)|^2 r dr$ is the transmission of the aperture in the presence of the excitation beam, I_0 is its transmittance in the absence of the excitation beam, and a represents the aperture surface. If we consider a small phase shift ($\Phi << 1$) and aperture dimensions much smaller than the probe beam spot, a simple solution can be obtained by means of Eq. (8) and the Fresnel diffraction approximation. Considering the Soret effect in liquid mixtures, the time-dependent total signal which is the sum of the pure thermal lens plus the Soret concentration lens can be expressed as follows (Marcano et al. 2002; Cabrera et al. 2009a):

$$S_{\text{total}}(z,t) = S_{th} - S_s = \frac{P_e \alpha l K(z,t)}{k\lambda_p}\left[\frac{\partial n}{\partial T} - \frac{\partial n}{\partial c} S_T c_0(1-c_0)\Gamma(z,t)\right], \tag{10}$$

with

$$K(z,t) = \arctan\left\{\frac{4m\nu t/t_c}{\nu^2 + [1+2m]^2 + [1+2m+\nu^2]2t/t_c}\right\}, \tag{11}$$

$$\Gamma(z,t) = 1 - \sum_{i=1}^{\infty} \frac{4}{(2i-1)\pi} \sin\left[\frac{(2i-1)\pi}{2}\right] \times \exp\left[-(2i-1)^2 \frac{t}{t_D(z)}\right], \quad (12)$$

where $\nu(z) = (z - a_p)/z_p + (z_p/L - z)[1 + (z - a_p)^2/z_p^2]$ is the geometrical parameter of the probe beam and L denotes the position of the plane detector.

The mode-mismatched scheme optimizes the value of the stationary total signal when $K(z,t)$ in Eq. (10) reaches its maximum value of $\pi/2$. In the stationary situation ($t \to \infty$), this value is reached at $z = 0$ for $z_p >> L >> z_e$. This corresponds to a situation with a collimated probe beam and a tightly focused pump beam. Under these conditions $K(z,t) = \pi/2$, $\Gamma = 1$, and Eq. (10) can be re-written as (Marcano et al. 2006; Cabrera et al. 2009a)

$$S_{total\infty} = S_{th} - S_s = \phi_{th}\frac{\pi}{2} - \phi_s\frac{\pi}{2} = \frac{P_e \alpha l \pi}{k \lambda_p 2}\left[\frac{\partial n}{\partial T} - \frac{\partial n}{\partial c} S_T c_0 (1 - c_0)\right]. \quad (13)$$

Equation (13) provides a relation between the total signal $S_{total\infty}$ and the Soret coefficient S_T. Measuring the total signal and calculating the thermal lens signal, the Soret signal can be obtained, which is then used to determine the Soret coefficient whenever the rest of the parameters are known. However, the ratio of the Soret signal and the pure thermal lens signal provides the more compact expression (Polyakov and Wiegand 2009; Cabrera et al. 2013):

$$S_T = \frac{\frac{\partial n}{\partial T}}{\frac{\partial n}{\partial c} c_0 (1 - c_0)} \frac{S_s}{S_{th}}. \quad (14)$$

In this relation there are three unknown parameters: the Soret coefficient S_T, the thermal lens signal S_{th}, and the Soret signal S_s. With the use of Eq. (14) we do not need to determine the absorption coefficient, and so avoid the addition of dye to increase the absoption. This particular property reduces the uncertainty in the determination of the Soret coefficient.

We can obtain the Soret signal S_s as the difference between the final steady-state total signal $S_{total\infty}$ and the value of the pure thermal lens signal S_{th}, which is obtained by extrapolating the fit of the thermal lens contribution from Eq. (10) in the 0–500 ms range. Finally, the Soret coefficient S_T is calculated using Eq. (14).

3 Experimental Method

The dual-beam thermal lens experimental set-up (see Fig. 1) has recently been used for the measurement of the Soret effect (Cabrera et al. 2009a,b, 2013). The system is composed of two blocks as follows:

1. *Probe beam and recording block.* The He–Ne laser ($\lambda_p = 632.8$ nm, $P_p = 10$ mW, Spectra Physics) generates the probe beam. This beam passes through a 1 cm L_1 and 30 cm L_2 focal-length lenses, resulting in a 6 mm width near-collimated probe beam. This beam is reflected by mirror E_1, passes through the dichroic beam combiner DivD, the sample M (contained in a 1 cm quartz cuvette), and the interference filter F_1 (central wavelength $\lambda = 632.8$ nm, Thorlabs). The dichroic mirror E_2 then reflects the beam towards the 0.3 mm pinhole A and the photodetector D (model DET 110/M, Thorlabs). The output signal of the photodetector is transformed by the current pre-amplifier Amp (model SR570, Stanford Research Systems) and introduced into the digital oscilloscope Osc (model TDS 3052, Tektronix), where the signal is digitized, saved, and processed statistically. The radiant flux of the probe beam at the sample surface is 0.2 mW.
2. *Excitation beam block.* The second harmonic of a diode pumped neodymium yttrium aluminum garnet (Nd:Yag) CW laser (model GSF32–200,$\lambda_e = 532.8$ nm, $P_e = 200$ mW, Intelite) delivers the excitation beam through the shutter Sh (model 846, Newport Corporation), the beam splitter Div, and the 180 mm focal length lens L_3 to the beam combiner DivD, which reflects it to the sample M. The excitation beam passes from the left to the right through the sample cell. The interference filter F and the dichroic mirror E_2 block the excitation beam preventing it from reaching the pinhole A and the photodetector. The beam reflected from the beam splitter Div is sensed by the photodetector Ref (model DET 110/M, Thorlabs), which transforms the incident radiant flux into a synchronization signal for the oscilloscope.

Measurements were taken according to the following procedure: the shutter Sh modulates the beam with a period of 10 s, and for each sample the signal was obtained from the average of 128 recordings at the digital oscilloscope. In order to measure the value of the final steady-state total signal $S_{total\infty}$, we have recorded the relative change of the transmission of the probe light through the aperture according to the definition given by Eq. (9), which allowed us to determine S_{th}, S_s, and S_T.

The water/isopropanol and water/ethanol mixtures were prepared using distilled and de-ionized water (with a purity of better than 99 %) and Fisher Scientific chemical organic components (with a purity of 99.8 %).

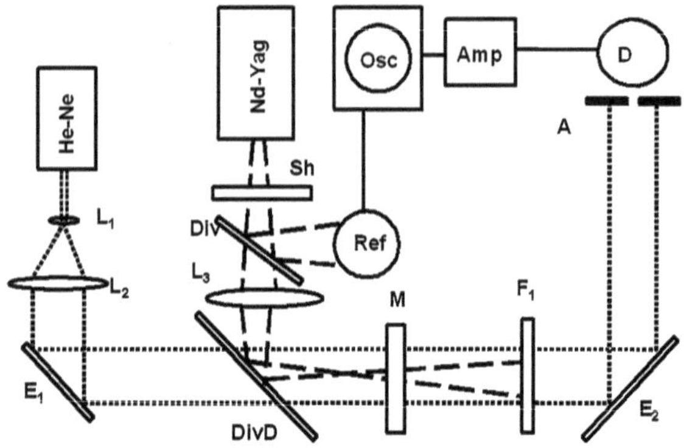

Fig. 1 Experimental set-up. Lenses L_1 and L_2: form the beam expander, E_1: mirror, DivD: dichroic beam combiner, M: sample, F: interference filter, E_2: dichroic mirror, A: pinhole, D: photodetector, Amp: current pre-amplifier, Osc: digital oscilloscope, Sh: shutter, Div: beam splitter, L_3: lens, and Ref: reference photodetector

4 Results and Discussion

The evolution of the total experimental signal with time is shown in Fig. 2 for a water/isopropanol mixture with an initial mass fraction of isopropanol $c_0 = 0.5$ (at 298 K). Using Eq. (10), the best fit to the experimental data gives $S_T = -5.2 \times 10^{-3}\,K^{-1}$, when the fitting parameters are: $\lambda_e = 532.8$ nm, $\lambda_p = 632.8$ nm, $z_p =$ 100, 000 cm, $z_e = 0.0001$ cm, $a_e = 0$, $a_p = 0$, $L = 50$ cm, $m = 10,000$, $\partial n/\partial T = -3.474 \times 10^{-4}\,K^{-1}$ (Mialdun et al. 2012), $P_e = 73\,mW$, $\alpha = 2 \times 10^{-4}\,cm^{-1}$, $l = 1$ cm, $k = 3 \times 10^{-3}\ Wcm^{-1}K^{-1}$, $\partial n/\partial c = -0.0364$ (Mialdun et al. 2012), $c_0 = 0.5$, and $t_D = 1$ s. If we cut the fit at $t_c = 500$ ms, only the pure thermal lens signal takes place (first term of Eq. (10), represented by the red line).

As we may see from this figure, the signal first shows a rapid reduction due to the pure thermal lens effect, followed by a further, much slower, decrease due to the build-up of the Soret concentration gradient. Both processes can be easily separated because the Soret component builds up with a time constant which is much greater ($t_d = 1$ s) than the characteristic time constant of the thermal lens ($t_c = 500$ ms) (Arnaud and Georges 2001; Cabrera et al. 2009a). The separation of the time scales allows for an analysis of the temporal build-up of the pure thermal lens, which is shown in expanded scale in the inset box of Fig. 2. The fit was limited to a short period, typically 500 ms, within which mass diffusion is inoperative and the signal is only governed by the temperature-dependent refractive index gradient. The fit allowed the determination of the transient steady-state thermal lens signal ($S_{\mathrm{th}} = -0.036$), from which the Soret signal ($S_s = 0.005$) was taken as the difference between the final steady-state total signal ($S_{total\infty} = -0,041$) and the steady-state pure thermal lens signal extrapolated from the fit using only the first term of Eq. (10). Then we have determined the Soret coefficient ($S_T = -5.34 \times 10^{-3}\,K^{-1}$) by means of Eq. (14) with $c_0 = 0.5$, $\partial n/\partial T = -3.474 \times 10^{-4}\,K^{-1}$ (Mialdun et al. 2012), and

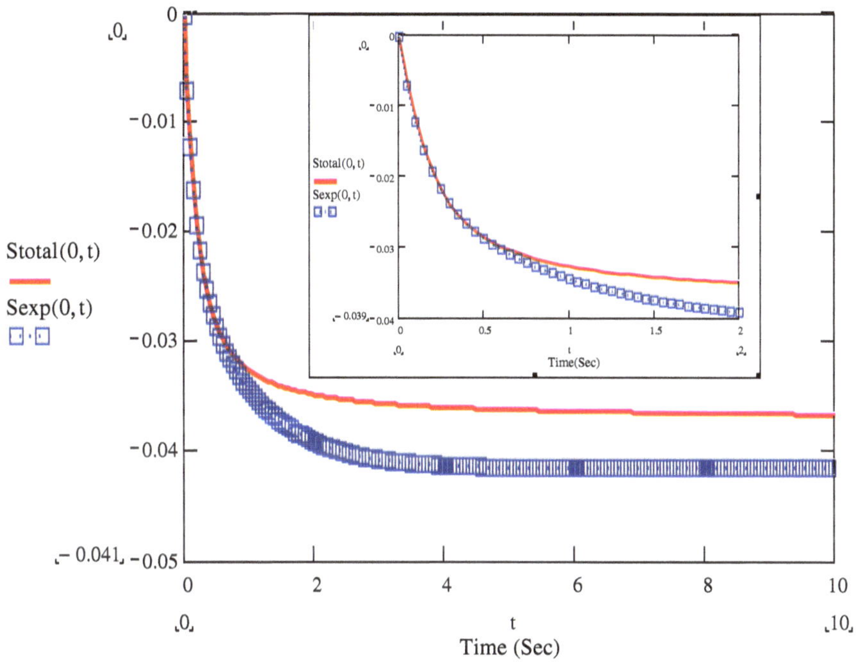

Fig. 2 Time evolution of the total experimental signal ($S_{total\infty} = -0,041$) for a water/isopropanol mixture with $c_0 = 0.5$ at $T = 298$ K. The *red line* is the best fit of the first term of Eq. (10) to the experimental data and represents the pure thermal lens signal. The *inset box* shows the build-up of the initial thermal lens effect in a 2 s period with $t_c = 500$ ms

$\partial n/\partial c = -0.0364$ (Mialdun et al. 2012). The Soret coefficient obtained from the fitting procedure ($S_T = -5.2 \times 10^{-3}\ K^{-1}$) does coincide with the calculated one using Eq.(14) ($S_T = -5.34 \times 10^{-3}\ K^{-1}$). This result demonstrates that the model predictions are in good agreement with the experimental data.

We have applied a similar procedure to determine the Soret coefficients for different concentrations of water/isopropanol mixtures. The results are summarized in Fig. 3 (Cabrera et al. 2013). For comparison, the data for the Soret coefficients from recent measurements by Mialdun et al. (2012) and Mialdun and Shetsova (2008), using three different instrumental techniques, and from early measurements by Poty et al. (1974), using flowing cell methods, are also displayed in Fig. 3. These results show a reasonably good agreement with ours in the region $0.2 < c_0 < 0.8$, except for the Soret coefficient reported by Poty et al. (1974) for $c_0 = 0.7$, which being too high, deviates significantly from all other data. In the region with low water content ($c_0 < 0.2$), our results also disagree with those reported by Poty et al. (1974). In addition, at low water concentrations the contrast factors are very low, which means that the concentration variations become invisible to the optical techniques and so the measurements of the transport properties are subject to large errors. The results of the measurements in this region are not shown in Fig. 3 because they were con-

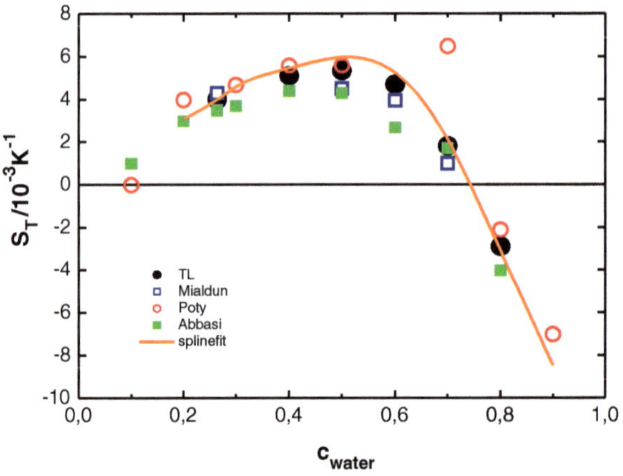

Fig. 3 Soret coefficient of water/isopropanol mixtures as a function of the mass fraction of water (*black circles*: present work). The *open blue squares* represent the experimental data reported by Mialdun and Shetsova (2008), while *open red circles* correspond to data from Poty et al. (1974). The *green squares* are the predicted values of the viscous energy model developed by Abbasi et al. (2009). The *continuous orange line* is a spline fit to all available data (Mialdun et al. 2012)

sidered to be unreliable. Note that the Soret coefficient changes sign at low and high isopropanol concentrations. In Fig. 3 we also compare our experimental results with the predictions of the viscous energy model developed by Abbasi et al. (2009). A reasonably good agreement is also observed.

We have also applied our technique to mixtures of ethanol and water at 298 K. The solid red circles in Fig. 4 show the measured Soret coefficients S_T as a function of the mass fraction of ethanol. Positive Soret coefficients are observed for low ethanol contents. By increasing the ethanol content, the Soret coefficient decays and changes sign at $c \sim 0.3$. In a mixture with high ethanol content, S_T is negative (because the ethanol molecules migrate to the hot side), whereas at low ethanol content, S_T is positive (in this case, the ethanol molecules migrate to the cold side). For comparison, Fig. 4 also depicts the experimental results reported by Kolodner et al. (1988), Zhang et al. (1996), and Kita et al. (2004). In addition, we also compare the results with the predictions of Abbasi et al. (2009). We see that the agreement between these three data sets, the viscous energy model, and our results is excellent.

We notice that for both mixtures the change of sign in the Soret coefficient occurred at an alcohol mass fraction of $c \sim 0.3$. Regarding this fact, a model for the calculation of thermodiffusion in associating mixtures was proposed. In this model a new approach for calculating the viscous energy and the ratio of the evaporation energy to the viscous energy was developed (Abbasi et al. 2009). Thus, the variation of the viscous energy of the organic component is considered to control the sign change in the Soret coefficient. In these two mixtures, the viscous energy of the organic component decreases sharply with its concentration, which leads to a sign change in the Soret coefficient. It then increases gradually with decreasing water con-

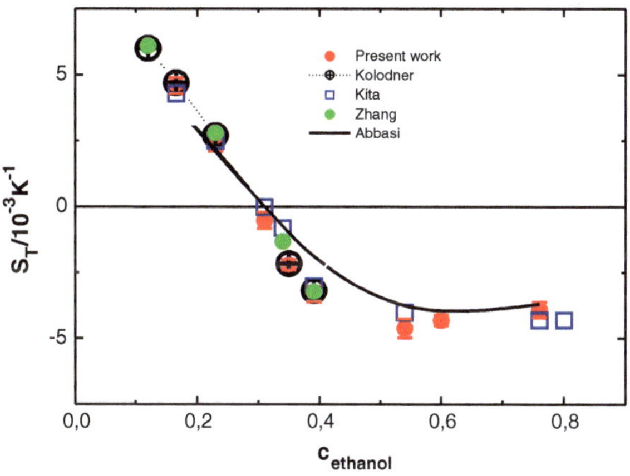

Fig. 4 Soret coefficient as a function of the mass fraction of ethanol. *Solid red circles* refer to the experimental data measured in the present work. *Cross black circles* represent previous results from Kolodner et al. (1988), *green solid circles* from Zhang et al. (1996), and *open blue squares* from Kita et al. (2004). The *continuous black line* draws the predictions of the viscous energy model developed by Abbasi et al. (2009)

centration in the systems. For the isopropanol-water mixture, the isopropanol viscous energy increases to the point where a second change of sign occurs. However, the viscous energy of ethanol does not increase enough to cause a second change of sign in the thermodiffusion factor (Abbasi et al. 2009).

5 Conclusions

We have used an optimized thermal lens experiment that allowed the precise determination of the values of the Soret coefficient for isopropanol/water and ethanol/water mixtures. Our experimental results were seen to compare well with existing data in the literature. We observed that the Soret coefficient of alcohol/water mixtures depends strongly on the concentration, with a change of sign at alcohol mass fractions of $c \sim 0.3$. In addition, for the isopropanol/water mixture, the isopropanol viscous energy increased to the point where a second change of sign occurred for the Soret coefficient. These results suggest a relation with the variation of the activation energy of the organic component. While there is not a unique technique for measuring the Soret coefficient, the results obtained here shows that a calibrated two-beam mode-mismatched thermal lens experiment represents an additional independent method, which provides new reliable benchmark data that agree quite well with previously reported measurements in the literature.

References

Abbasi A, Saghir MZ, Kawaji M (2009) A new approach to evaluate the thermodiffusion factor for associating mixtures. J Chem Phys 130:064506

Arnaud N, Georges J (2001) Investigation of the thermal lens effect in water-ethanol mixtures: composition dependence on the refractive index gradient, the enhancement factor and the Soret effect. Spectrochim Acta A 57:1295–1301

Bierlein SB (1955) A phenomenological theory of the Soret diffusion. J Chem Phys 23:10–15

Cabrera H, Marcano A, Castellanos Y (2009a) Absorption coefficient of nearly transparent liquids measured using thermal lens spectrometry. Cond Matt Phys 9:385–389

Cabrera H, Sira E, Rahn K, García-Sucre M (2009b) A thermal lens model including the Soret effect. Appl Phys Lett 94:051103

Cabrera H, Marti-López L, Sira E, Rahn K, García-Sucre M (2009) Thermal lens measurement of the Soret coefficient in acetone/water mixtures. J Chem Phys 131:031106

Cabrera H, Cordido F, Velasquez A, Moreno P, Sira E, López-Rivera SA (2013) Measurement of the Soret coefficients in organic/water mixtures by thermal lens spectrometry. Comptes Rendus Mécanique 341:372–377

Giglio M, Vendramini A (1974) Thermal lens effect in a binary liquid mixture: a new effect. Appl Phys Lett 25:555–557

Gordon J, Leite R, Moore R, Porto S, Whinnery J (1965) Long transient effects in lasers with lasers with inserted liquid samples. J Appl Phys 36:3–8

Kita R, Wiegand S, Strathmann JL (2004) Sign change of the Soret coefficient of poly(ethylene oxide) in water/ethanol mixtures observed by thermal diffusion forced Rayleigh scattering. J Chem Phys 121:3874–3885

Kolodner P, Williams H, Moe C (1988) Optical measurement of the Soret coefficient of ethanol water solutions. J Chem Phys 88:6512–6524

Long M, Swofford R, Albrecht A (1976) Thermal lens technique: a new method of absorption spectroscopy. Science 191:183–185

Marcano A, Loper C, Melikechi N (2002) Pump-probe mode-mismatched thermal-lens Z scan. J Opt Soc Am B 19:119–124

Marcano A, Cabrera H, Guerra M, Cruz RA, Jacinto C, Catunda T (2006) Optimizing and calibrating a mode-mismatched thermal lens experiment for low absorption measurement. J Opt Soc Am B 23:1408–1413

Mialdun A, Shetsova VM (2008) Development of optical digital interferometry technique for measurement of thermodiffusion coefficients. Int J Heat Mass Transf 51:3164–3178

Mialdun A, Yasnou V, Shetsova VM, Königer A, Köhler W, Bou-Ali MM (2012) A comprehensive study of diffusion, thermodiffusion, and Soret coefficients of water-isopropanol mixtures. J Chem Phys 136:244512

Polyakov P, Wiegand S (2009) Investigation of the Soret effect in aqueous and non-aqueous mixtures by the thermal lens technique. Phys Chem Chem Phys 11:864–871

Poty P, Legros JC, Thomaes G (1974) Thermal diusion in some binary liquid mixtures by the flowing cell methods. Z Naturforsch 29A:1915–1916

Shen J, Lowe R, Snook R (1992) A model for cw laser-induced mode-mismatched dual-beam thermal lens spectrometry. Chem Phys 165:385–396

Sitzber L (1856) Diffusion zwischen ungleich erwärmten Orten gleich Zusammengesetz Lösungen. Akad Wiss Wien Math- Naturwiss Kl 20:539

Soret C (1879) Sur l'état d'équilibre que prend, du point de vue de sa concentration, une dissolution saline primitivement homogène, dont deux parties sont portées à des températures différentes. Arch Sci Phys Nat 2:48–61

Tyrell H (1961) Diffusion and heat flow in liquids. Butterworths, London

Whinnery J (1974) Laser measurement of optical absorption in liquids. Acc Chem Res 7:225–231

Zhang KJ, Briggs ME, Gammon RW, Sengers JV (1996) Optical measurement of the Soret coefficient and the diffusion coefficient of liquid mixtures. J Chem Phys 104:6881–6892

Stellar Mass Accretion Rates from Fragmentation of a Rotating Core

Jaime Klapp, Leonardo Di G. Sigalotti and Miguel Zavala

Abstract We investigate the details of protostellar mass accretion, $\dot{M}$, during the collapse of isolated, initially uniformly rotating, low-mass cores, using hydrodynamic models of star formation. The assumption of rigid rotation is supported by recent observations that there is no apparent correlation between the level of turbulence and fragmentation in dense cores, suggesting that turbulence works mainly before gravitationally bound pre-stellar cores form and that their inner parts are likely to be velocity coherent. We perform high-resolution calculations using the Smoothed Particle Hydrodynamics (SPH) code GADGET-2, modified by the inclusion of sink particles. We compare our results with theoretical models of star formation based on gravoturbulent fragmentation and with observational data. We find that on the small scales of low-mass, dense cores the details of mass accretion and the statistical properties of the resulting stellar ensembles bear little dependence on whether the contracting gas is turbulent or rotating as a whole.

J. Klapp (✉) · M. Zavala
Departamento de Física, Instituto Nacional de Investigaciones Nucleares, ININ, Km. 36.5, Carretera México-Toluca, 52750 La Marquesa, Estado de México, Mexico
e-mail: jaime.klapp@inin.gob.mx; jaime.klapp@hotmail.com

M. Zavala
e-mail: zvl_2006@yahoo.com.mx

J. Klapp
Departamento de Matemáticas, Cinvestav del Instituto Politécnico Nacional (I.P.N.), 07360 México D. F., Mexico

L. Di G. Sigalotti
Centro de Física, Instituto Venezolano de Investigaciones Científicas, IVIC, Apartado Postal 20632, Caracas 1020-A, Venezuela
e-mail: leonardo.sigalotti@gmail.com

L. Di G. Sigalotti et al. (eds.), *Computational and Experimental Fluid Mechanics with Applications to Physics, Engineering and the Environment*, Environmental Science and Engineering, DOI:10.1007/978-3-319-00191-3_14,

1 Introduction

As far as we know, all stars form by gravitationally-driven mass accretion from the protostellar envelope of fragmenting molecular cloud cores. Although considerable progress has been made in observing and understanding pre-main sequence evolution, the earliest phases of star formation, i.e., the events which lead to Class I young stellar objects (YSOs), still remain highly enigmatic. The Class 0 YSOs—the protostars that have yet to accrete the bulk of their masses (André et al. 1993)—represent the most crucial phase in the star formation process, when the physics that determines the mass of the final star is imprinted on the system.

After a central protostellar core develops, theory predicts that material with higher angular momentum continues to infall from the circumstellar envelope along streamlines which eventually miss the central protostar and is deposited in a circumstellar disk. The disk quickly becomes rotationally unstable and transport mass onto the protostar while redistributing angular momentum outwards. This process lasts for a few times 10^5 yr when infall onto the protostar-plus-disk system terminates, signaling the end of the Class I phase (Shu et al. 1987). To build up a star over these timescales requires time-averaged accretion rates of $\sim 2 \times 10^{-6}$–$10^{-5} M_\odot$ yr^{-1}, which are typically used in evolutionary calculations of protostars at the end of accretion (Stahler 1988; Hartmann et al. 1997). In a first approximation, the mass accretion rate can be defined as the ratio of the local Jeans mass, M_J, over the free-fall time, t_ff: $\dot{M} \approx M_\mathrm{J}/t_\mathrm{ff} = 5.4 c_\mathrm{s}^3/G$ (Schmeja and Klessen 2004), where c_s is the isothermal sound speed and G is the gravitational constant. This gives a higher value than the constant accretion rate of $0.975 c_\mathrm{s}^3/G$ for the collapse of the singular isothermal sphere (Shu 1977). Numerical simulations of star cluster formation predicts that stars and brown dwarves may be formed in burst over about 2×10^4 yr, implying infall rates of $\sim 10^{-4}$–$10^{-5} M_\odot$ yr^{-1} (Bate et al. 2003). However, these infall rates imply much higher luminosities than the values typically observed in protostars (Kenyon et al. 1994). One way to solve this problem is by invoking a highly time-dependent accretion. For instance, FU Orionis stars provide direct evidence for short episodes of rapid accretion in early stages of stellar evolution, with accretion rates of $10^{-4} M_\odot$ yr^{-1} or even more (Hartmann and Kenyon 1996), which are from one to two orders of magnitude larger than typical values for low-mass objects (Furlan et al. 2008).

An alternative approach to the standard theory of star formation is to consider the process of star formation itself as being controlled by the interplay between gravity and supersonic turbulence (Mac Low and Klessen 2004). Numerical models of star formation based on gravoturbulent fragmentation have predicted highly time-variant accretion rates, with a sharp peak shortly after the formation of the protostellar core in the range between 3 and $50 c_\mathrm{s}^3/G$, or equivalently, between 5×10^{-6} and $10^{-4} M_\odot$ yr^{-1} (Schmeja and Klessen 2004). This range is in good agreement with the observed one of $\sim 10^{-5}$–$10^{-4} M_\odot$ yr^{-1} for Class 0 YSOs (André et al. 1999; Ceccarelli et al. 2000; Maret et al. 2002). The transition phase between Class 0 and Class I sources takes place when about half of the final stellar mass has been accreted by the protostar (André et al. 2000). This borderline typically takes place during or at the end of the peak accretion phase, thereby determining the lifetime of Class 0 objects. The accretion rates of Class I sources are believed to be between $\sim 10^{-7} M_{\odot}$ yr^{-1}, for the most evolved objects, and $\sim 5 \times 10^{-6} M_{\odot}$ yr^{-1}, for the youngest ones (Greene and Lada 2002; Boogert et al. 2002; Young et al. 2003). An analysis of the outbursting system FU Ori based on Spitzer IRS data predicted that the disk accretion of low-mass protostars will generally be unsteady for typical infall rates (Zhu et al. 2007). A recent study has shown that magnetorotational instabilities in the inner disk and gravitational instabilities in the outer disk are likely to combine to produce outbursts of rapid accretion at a few AU from the central protostar (Zhu et al. 2009), consistent with a highly time-dependent accretion scenario.

The mass accretion rates of protostellar cores in a dense cluster are different from those of isolated protostars due to the dynamical interaction and competition between the cores. For instance, in the first stage a protostar accretes local gas from its immediate vicinity. Once the local reservoir is depleted, the protostar may accrete gas streaming in from farther away or by encounters with non-collapsed clumps (Klessen and Burkert 2000). This may well result in secondary accretion peaks owing to the protostar first accreting only about half of its final mass from its direct environment (first peak) and then the rest from later accretion events (secondary peaks). Numerical simulations of small cluster formation based on gravoturbulent fragmentation have shown the formation of secondary peaks in the mass accretion rates of most protostars in the cluster (Schmeja and Klessen 2004). These models have also predicted a strong correlation between the maximum mean accretion rate and the final stellar mass. In addition, the maximum accretion rate was shown to be one order of magnitude higher than the constant rate produced by the collapse of a classical singular isothermal sphere (Shu 1977), while the accretion rates are seen to decline exponentially from Class 0 to Class I phase, matching the observational findings (Bontemps et al. 1996; Myers et al. 1998).

In this chapter, we present numerical simulations of the fragmentation of an isolated, initially uniformly rotating, pre-stellar core, using the three-dimensional, parallelized Smoothed Particle Hydrodynamics (SPH) code GADGET-2 (Springel 2005), which has been modified to include "sink" particles to increase the computational efficiency when small regions of high density contrast form (Bate et al. 1995; Federrath et al. 2010). Except for turbulence, the initial conditions are the same used in previous simulations of gravoturbulent fragmentation (Goodwin et al. 2004a,b; Attwood et al. 2009). The assumption of initial uniform rotation is based on recent evidence that the distribution of turbulent to gravitational energy versus core size in detected protostellar cores is highly scattered, implying that there is no correlation between the level of turbulence and fragmentation in these cores (Chen 2008). This suggests that turbulence works mainly before gravitationally bound pre-stellar cores form and that on such small scales the cores, at least in their inner parts, are velocity coherent and essentially free of turbulence (Goodman et al. 1998). We address the simulations to analyze the time evolution of the mass accretion process and values of the peak accretion rates from pure rotational fragmentation and compare the results with observations and previous models based on gravoturbulent fragmentation.

2 Initial Model and Numerical Methods

The pre-stellar core is modelled as an isothermal sphere of molecular gas of temperature $T = 10$ K and radius $R \approx 0.24$ pc with initial uniform rotation about the z-axis. Since the density structure of pre-stellar cores appears to consist of a central kernel of approximately uniform density, surrounded by an outer envelope in which the density falls off radially, we adopt a Plummer-like initial density profile,

$$\rho(r) = \rho_0 \left[\frac{R_0}{(R_0^2 + r^2)} \right]^4 , \tag{1}$$

where $\rho_0 = 3 \times 10^{-18}$ g cm^{-3} is the central density and $R_0 \approx 0.024$ pc is the radius of the central kernel. The total mass of the core is $M_0 = 5.4 M_\odot$, while the mass inside the central kernel is about $2 M_\odot$. With the above parameters the ratio of thermal to gravitational energy is $\alpha = 0.3$ and the free-fall time is $t_{\rm ff} \approx 38.5$ kyr.

In the simulations presented here, we consider initial ratios of the rotational to gravitational energy characterized by $\beta_0 = 0.05$, 0.10, and 0.25, corresponding to specific angular momenta of ≈ 1.6–3.5×10^{21} cm^2 s^{-1}, which are a bit lower than the observationally estimated value of $\sim 5 \times 10^{21}$ cm^2 s^{-1} for the high-density nucleus of the pre-stellar core L1544 in Taurus (Crapsi et al. 2007). The thermodynamics of the system is described by a barotropic equation of state such that the sound speed is given by

$$c_{\rm s}^2 = \frac{p}{\rho} = c_0^2 \left[1 + \left(\frac{\rho}{\rho_{\rm cr}} \right)^{2/3} \right], \tag{2}$$

where p is the gas pressure, $c_0 \approx 0.19$ km s^{-1} is the isothermal sound speed, and $\rho_{\rm cr} = 10^{-13}$ g cm^{-3}. In the low-density gas ($\rho \ll \rho_{\rm cr}$), with composition of 70 % molecular hydrogen and 28 % atomic helium at 10 K, $c_{\rm s} \approx c_0$. At high densities, i.e., when $\rho > \rho_{\rm cr}$, the gas becomes opaque to its own radiation and heats up adiabatically with an effective adiabatic index $\gamma \sim 5/3$. In this case, the sound speed rises as $\sim \rho^{1/3}$ and the Jeans mass increases more rapidly as $M_{\rm J} \sim \rho^{1/2}$.

The simulations are performed using a variant of the GADGET-2 code (Springel 2005), which includes sink particles (Bate et al. 1995). The code solves the three-dimensional equations of hydrodynamics using a standard SPH discretization (Monaghan 2005) and a TreePM method for calculating the gravitational forces. In contrast to other standard SPH formulations, GADGET-2 employs an entropy conserving scheme which replaces the usual evolution equation for the internal energy. The smoothing length, h, is allowed to vary adaptively in such a way that the mass within the kernel volume remains a constant during the evolution. For equal-mass particles this implies that the number of neighbours, $N_{\rm neigh}$, is always the same. This keeps numerical dissipation and diffusion at very low rates, thereby increasing the reliability of the SPH calculations (Attwood et al. 2007). In the present simulations the particles have unequal masses and so $N_{\rm neigh}$ is not exactly the same at each timestep.

Therefore, in order to keep numerical dissipation and diffusion at acceptable low levels we use large numbers of particles, $N_{\rm p}$, so that the ratio $N_{\rm p}/N_{\rm neigh}$ is large enough to make the timescale of numerical dissipation to become close to the evolution time. For the runs of this chapter we use a spherically symmetric M_4 interpolation kernel (Monaghan and Lattanzio 1985), vary $N_{\rm p}$ between 5×10^5 and 2×10^6, and set $N_{\rm neigh} = 50 \pm 3$. In order to avoid spurious fragmentation, the gravitational softening, ϵ, is set equal to the minimum smoothing length, $h_{\rm min}$, (Whitworth 1998) and gravitational forces are spline-softened using the same M_4 kernel function. Finally, the code uses a leapfrog integration scheme to advance the hydrodynamical variables from one timestep to another.

A Cartesian square box of sides equal to $2R$ is chosen as the computational volume. The box is subdivided into cubic cells and the spherical cloud core is built up by placing particles at the centres of cells at distances $\leq R$ from the box centre. A small random perturbation is applied to the position of particles so that they will be slightly shifted from the centres of cells, solid-body rotation is enforced in a counterclockwise sense by assigning to each particle an initial velocity $\boldsymbol{v} = (\Omega_0 x, -\Omega_0 y, 0)$, where Ω_0 is the initial core angular velocity, and the mass of particles is modified by adding a small-amplitude ($a = 0.1$), bar-mode perturbation. Moreover, artificial fragmentation is suppressed as long as the minimum resolvable mass satisfies the following condition (Hubber et al. 2006):

$$M_{\rm min} = N_{\rm neigh} m \leq M_{\rm J} = \frac{\pi^{5/2} c_{\rm s}^3}{6 G^{3/2} \rho^{1/2}}. \tag{3}$$

If we assume that all particles have approximately the same mass m, then $m = M_0/N_{\rm p}$, and from Eq. (3) it follows that

$$N_{\rm p} \geq N_{\rm neigh} \left(\frac{M_0}{M_{\rm J}} \right), \tag{4}$$

which defines a lower limit to the number of particles required to guarantee genuine fragmentation in SPH simulations. At $\rho = \rho_{\rm cr} = 10^{-13}$ g cm^{-3}, the Jeans mass is $\approx 0.005 M_\odot$ and from Eq. (3) it follows that $m \leq m_{\rm max} \approx 0.0001 M_\odot$, which represents an upper limit on the particle mass m. For unequal particle masses, this inequality can be re-written as

$$\frac{1}{N_{\rm neigh}} \sum_{i=1}^{N_{\rm neigh}} m_i \leq m_{\rm max} \approx 0.0001 M_\odot, \tag{5}$$

which implies that the mean mass of particles within the kernel volume must be less than about $0.0001 M_\odot$ in order to satisfy the Jeans condition. In our simulations, we always use $N_{\rm p} \geq 5 \times 10^5$ SPH particles which is quite above the lower limit of $\sim 8 \times 10^4$ SPH particles required to fulfill the Jeans condition when $\rho = \rho_{\rm cr}$, according to Eq. (4). Furthermore, a collapsing fragment region is replaced by a sink particle when the density of at least one SPH particle within that region exceeds the threshold value of $\rho_s = 10^{-11}$ g cm^{-3}. When this happens a control (spherical) volume of radius $r_{acc} \approx 4$ AU centred at the position of the particle of maximum density is temporarily created from the gas, which must pass a number of checks before becoming a permanent sink (Federrath et al. 2010). In this way, the runaway collapse of the fragments is effectively controlled and can be followed together with the global evolution of the core.

3 Rotation Versus Turbulence

Extensive N_2H^+ (1–0) and NH_3 mapping surveys of dense cloud cores indicate the presence of local velocity gradients in the range between 0.5 and 6 km s^{-1} pc^{-1}, in both starless cores and cores with embedded YSOs (Caselli et al. 2002). If these gradients represent rotation, the ratio β of kinetic rotational energy to gravitational energy ranges between $\sim 10^{-4}$ and 0.07, implying an average value of $\beta \sim 0.03$. Such low values suggest that rotation is not significant in the support of most cores. However, even relatively small rotational energies may strongly influence protostellar formation during core collapse (Bodenheimer et al. 2000). For instance, angular momentum evolution due to contraction may well induce disk formation accompanied by fragmentation of the disk into multiple protostars (Hennebelle et al. 2004; Walch et al. 2009). While detected local velocity gradients may be intrinsically related to complex internal motions that deviate strongly from a simple model of coherent rotation of the whole core, more recent observations have however interpreted these velocity gradients as a signature of core rotation (Di Francesco et al. 2007). In addition, high angular resolution observations indicate rotational energies in the range of $\beta \sim 0.01$–0.1 for a large sample of both starless and dense cores with embedded single or binary protostars in the NH_3 and N_2H^+ emission lines (Chen 2008).

As an alternative to rotationally induced fragmentation, the supersonic turbulence observed in the interstellar gas, on scales larger than the molecular gas clouds themselves, has been invoked as a mechanism capable to induce strong local fluctuations in the velocity and density that may favour gravitational collapse on scales of ~0.05 pc (Larson 2003; Klessen 2004; Mac Low and Klessen 2004), where turbulence turns from supersonic to subsonic (Ballesteros-Paredes et al. 2007). For instance, turbulence can carry enough energy to support typical molecular gas clouds from collapsing and, at the same time, to generate a complex network of interacting shocks, where filaments of enhanced density form at the stagnation points of convergent flows. Eventually, gravity dominates in the densest and more massive parts of a filament, inducing localized collapse and the birth of stellar masses. In this picture, the main source of angular momentum on global scales lies in the differential rotation of the galactic disk, while on intermediate to small scales, comparable to the sizes of dense cloud clumps and cores, angular momentum may be derived from the high degree of vorticity inherent to turbulent flows (Jappsen and Klessen 2004).

Although angular momentum is channeled mostly into vorticity as turbulence distributes its energy among smaller scales, it is unknown whether turbulence can also explain the low levels of rotation detected in dense cloud cores. In particular, random Gaussian velocity fields with power spectra $P(k) \propto k^{-3}$ to k^{-4} have been found to reproduce both the observed line width-size relationship and the observed projected rotational properties of molecular cloud cores (Burkert and Bodenheimer 2000). Since the fluctuation spectrum is dominated by the large-scale eddies, the shape of the angular momentum distribution as inferred from the line-of-sight velocity gradients is, at least, on a statistical basis, in good agreement with that of the intrinsic angular momentum distribution of turbulent cores. In other words, even if turbulent motions are random and chaotic, dominant large-wavelength modes can lead to velocity gradients that look like ordered rotation. Although cores that are described by the same power spectrum show a large spread in their rotational properties, which is in qualitative agreement with the large spread in observed binary periods, the median angular momentum of the cores was an order of magnitude larger than in the observations. While these results apply to cores that are mildly subsonic, it has to be clarified whether uniform rotation may represent a reasonable condition in shock-generated clumps and cores.

Observations in several molecular emission lines of the velocity structure of dense cores suggest that turbulence is still present down to subcore scales of $\sim$2400 AU, implying that cores are not quiescent, homogeneous structures (Volgenau et al. 2006). However, recent high angular resolution observations of embedded binary protostars indicate that the distribution of the ratio of turbulent energy to gravitational energy versus core size is highly scattered (Chen 2008). This implies that there is no correlation between the level of turbulence and fragmentation of the cores, suggesting that turbulence works mainly before gravitationally bound pre-stellar cores form and that down to subcore scales they may well be velocity coherent and essentially free of turbulence. The verification or refutation of this trend must certainly await for future higher sensitive observations of the velocity fields and structure of dense cloud cores. However, it would be interesting to see whether initial uniform rotation may reproduce many of the statistical properties predicted by gravoturbulent fragmentation models. Here we will focus primarily on the details of the time-dependent mass accretion rates in initially uniformly rotating cores that fragment in multiple protostars.

4 Results

In this section, we describe the results of five model calculations with varying levels of initial uniform rotation (i.e., $\beta_0 = 0.05, 0.10$, and 0.25, corresponding to angular velocities of $\Omega_0 = 3.68 \times 10^{-14}$, 5.21×10^{-14}, and 8.23×10^{-14} s^{-1}, respectively) and varying spatial resolution. The models are labelled as Aβ_0Nn, where $n = N_{\rm p}/10^5 = 5$, 10, and 20 denotes the total number of SPH particles filling the computational volume. The numerical models were all evolved up to $6t_{\rm ff}$ ($\approx$231 kyr), i.e., quite beyond the theoretically predicted Class 0/I borderline ($\approx$115 kyr, when 50 % of the total mass has reached the central protostar) for a non-rotating ($\beta_0 = 0$) core starting to collapse with the same initial parameters as ours (Whitworth and Ward-Thompson 2001).

4.1 Pre-Stellar Contraction

All core models collapsed initially to form an over-dense, central disk owing to spin angular momentum. By about $1t_{\rm ff}$, the central density field approaches an $r^{-3/2}$ profile, while the density of the outer core regions maintains its initial r^{-4} dependence. With the exception of the central disk flattening owing to rotation, this phase is qualitatively similar to the semi-analytic model of Whitworth and Ward-Thompson (2001), who assumed negligible internal pressure and hence free-fall, and the three-dimensional numerical calculations of non-turbulent cores of Goodwin et al. (2004a), where only one central protostar formed after $\approx 1t_{\rm ff}$, when the central density reached the threshold value of 10^{-11} g cm^{-3} for sink formation. After this time, the accretion rate became non-negligible, marking the end of the initial pre-stellar contraction stage and the beginning of the Class 0 phase. During this latter phase, the accretion rose to values as high as $\sim 10^{-4} M_\odot$ yr^{-1} before declining steadily.

In contrast, the formation of a central point-mass (hereafter referred to as the primary protostar) is delayed in our simulations until a substantial disk is built up due to rotational motion of the core as a whole. For example, in the rapidly rotating core model A0.25N5 the low angular momentum gas converging to the centre of the disk condenses to form the first protostar on about 50 kyr ($\approx 1.35t_{\rm ff}$). At this time, the mass interior to 70 AU was just around the primary protostar and beyond this the disk extended to $\sim$1500 AU, with density variations of $r^{-5/2}$. Although more rapidly rotating cores are expected to reach high densities after longer times, resulting in less dense and more extended disks, than slowly rotating cores, the primary protostar was however seen to form at approximately the same time for all models independently of β_0. In all cases, the disk forms before the primary protostar and grows faster than it. This is consistent with previous SPH collapse simulations of initially rigidly rotating cores, including the effects of molecular line cooling (Walch et al. 2009), where for $\beta > 0.03$ the disks always formed before the primary protostar, while for lower β the primary formed first. This latter pattern has been also seen in collapse models of turbulent dense cores (Goodwin et al. 2004a; Attwood et al. 2009), where the incipient angular momentum produced by the turbulent flow led first to the formation of a primary and then to a circumprimary disk.

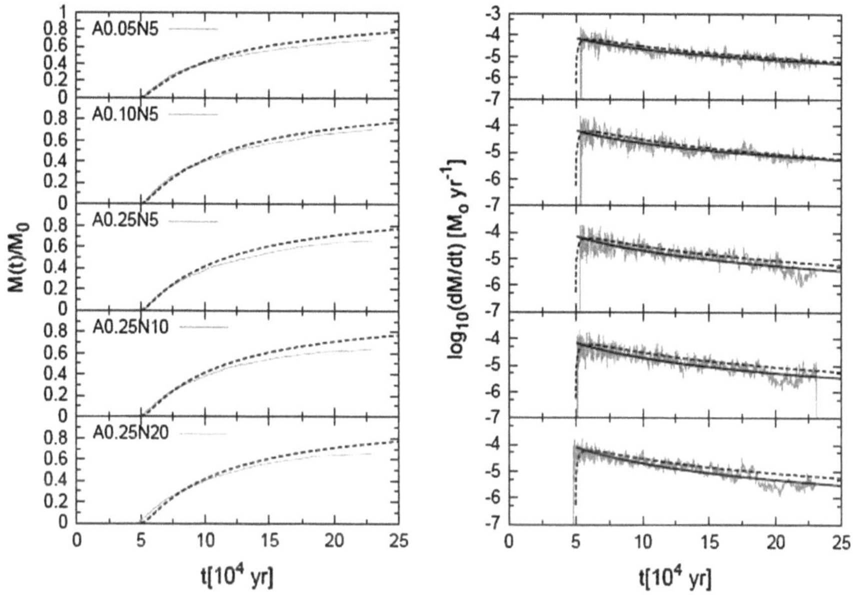

Fig. 1 Mean accreted mass (*left panel*) and mean accretion rate (*right panel*) for all fragmentation models (*thin solid lines*). The numerical data are compared with the semi-analytic solution of Whitworth and Ward-Thompson (2001) for a pre-stellar core collapse starting with the same initial conditions, but with negligible pressure and $\beta_0 = 0$ (*dashed lines*). The *thick solid lines* on the plots of the *right panel* are the applied fits to the actual mean accretion rates, which scale with time as $\sim t^{-2}$ for all models independently of the level of rotation and spatial resolution

4.2 Protostellar Accretion Phase

The protostellar evolution phase will critically depend on the rotational energy. Cores with initially higher β_0 produce disks that favours fragmentation into larger numbers of secondary protostars. After about 77 kyr ($\approx 2t_{\rm ff}$) in our reference model A0.25N5, the gas interior to $\approx$80 AU is undergoing adiabatic collapse with $\rho > \rho_{\rm cr}$, whereas the gas contained within the first 60 AU from the core centre has already achieved peak densities above the threshold value for sink formation. At this time 19 newly born protostars have been counted around the primary, which have accreted $\sim$24 % of the total core mass. The delay between the primary formation and fragmentation of the disk into the first secondary is typically less than 1 kyr. During growth of the primary the disk has already accumulated sufficient mass from the outer core that it quickly fragments into several secondaries in the space of a few kyr.

The actual mean accreted mass and mean accretion rates for all runs are depicted in the left and right panels of Fig. 1, respectively. The mean values correspond to averages of $M_{\rm s}$ and $\dot{M}_{\rm s}$, i.e., the masses and accretion rates of the individual sinks, over the protostellar mass range at any time. The numerical data is compared with the

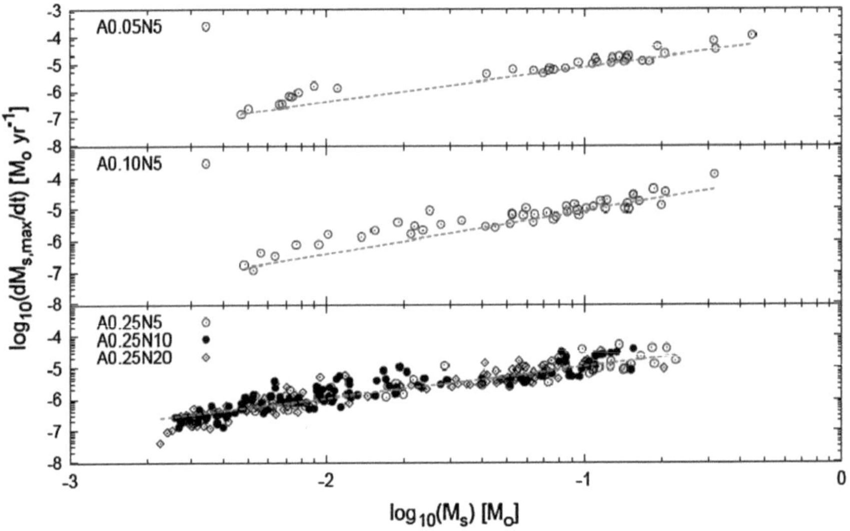

Fig. 2 Peak accretion rates against final stellar masses for all runs. The *straight dashed lines* on each plot are linear fits to the numerical data, suggesting a positive linear correlation between the maximum accretion rates and the final masses of the stars

semi-analytic solution of Whitworth and Ward-Thompson (2001) for $\beta_0 = 0$. The thick solid lines on the plots of the right panel are the applied fits to the actual mean accretion rates. For all models, the numerical data fits the power law $\dot{M}_{s,av} \propto t^{-2}$ independently of β_0 and the spatial resolution. For comparison, the semi-analytic solution of Whitworth and Ward-Thompson (2001) predicts a decline of the accretion rate that varies with time as $\sim t^{-1.67}$ in the long evolution.

The dependence of the maximum accretion rates on the final stellar masses is shown in Fig. 2. The straight line in each plot is a linear fit to the numerical data. Evidently, there is a positive correlation between the peak accretion rate and the final mass of the protostar in which the peak accretion rate increases with the stellar mass. Similar positive correlations were found in gravoturbulent fragmentation models of massive ($120 M_\odot$) cores (Schmeja and Klessen 2004), implying that this feature is independent of the initial core parameters and whether the collapsing medium is turbulent or not.

4.3 Accretion History of Primary Protostars

The accretion rates of the primaries that formed in our simulations are plotted in Fig. 3 as a function of time (left) and the parameter $m = (M_s - M_i)/(M_f - M_i)$ (right), where M_i and M_f denote the initial and final mass of the primary, respectively. The horizontal dashed lines on the plots of the left panel confine the observed range of

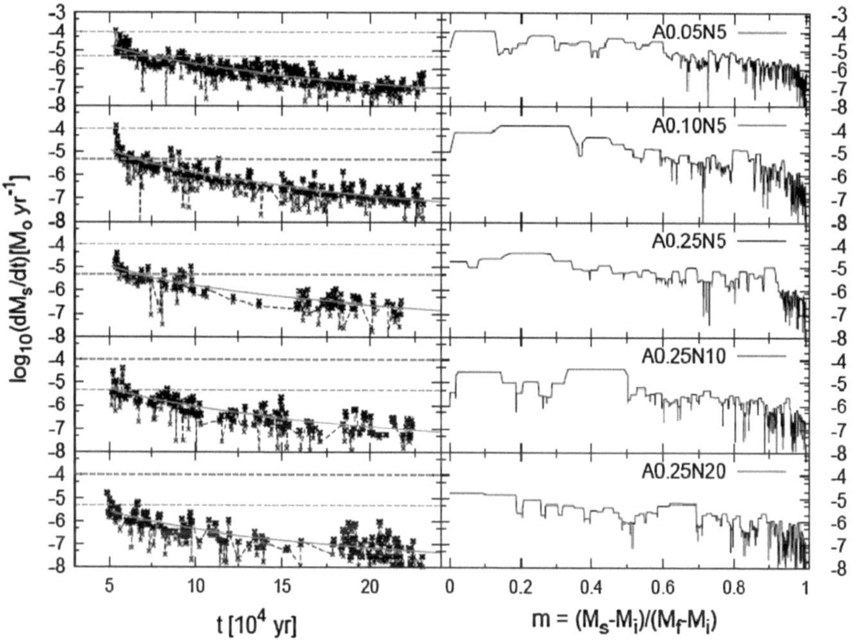

Fig. 3 Mass accretion rates of primary protostars as functions of time (*left panel*) and the mass parameter, $m = (M_s - M_i)/(M_f - M_i)$ (right panel), where M_i and M_f are the initial and final mass of the primary, respectively. The *horizontal dashed lines* on the plots of the *left panel* confine the observed range of peak accretion rates for Class 0 objects (André et al. 2000), while the *thick solid lines* are applied fits to the actual data, showing the t^{-2} decline of the accretion rate with time

maximum accretion rates between $\sim 5 \times 10^{-6}$ and $10^{-4} M_{\odot}$ yr^{-1} for Class 0 YSOs (André et al. 2000). The solid lines are applied fits to the actual accretion rates, which, as for the mean accretion rates, closely follows a $\sim t^{-2}$ variation in the long-term evolution.

In general, the primaries exhibit peak accretion rates that are towards the upper end of the observed range for Class 0 protostars. The decline is often interrupted by the occurrence of secondary peaks, suggesting that primaries might have been accreting gas from farther away or by encounters with non-collapsed clumps in their long-term evolution. Towards the end of the calculations, when $m \approx 1$, the accretion rates decline to values around $10^{-7} M_{\odot}$ yr^{-1} for all models, which match the typical accretion rates for evolved Class I objects (Greene and Lada 2002; Young et al. 2003).

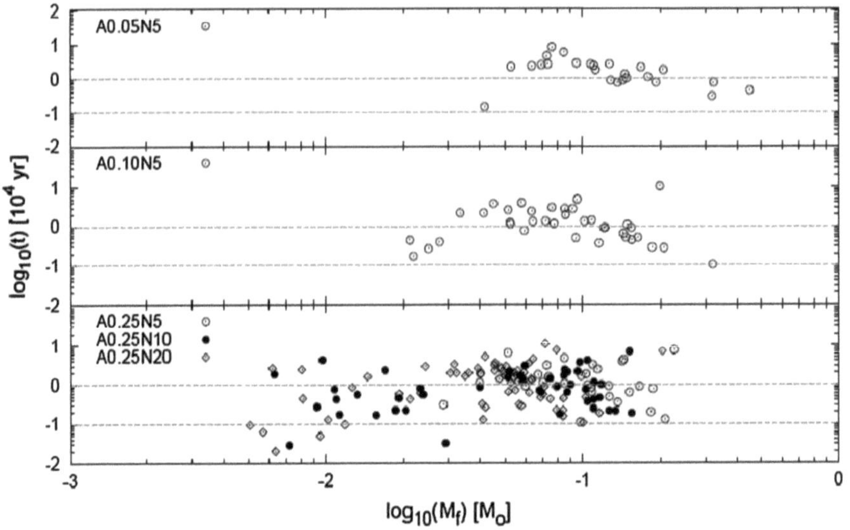

Fig. 4 Duration of the Class 0 stage as a function of the final stellar mass for all runs. The plotted data correspond to the precise time when the protostars have accreted half of their final masses. The *horizontal lines* in each plot confine the observationally detected range of Class 0 lifetimes (André et al. 2000)

4.4 Class 0 Lifetimes

The time at which about 50 % of the core mass has been converted into stars provide a fair estimate for the Class 0 lifetime. This time is slightly shorter for the slowly rotating cores ($\beta_0 \leq 0.10$) than for the $\beta_0 = 0.25$ core. However, in this latter model the Class 0/I borderline also shifts towards shorter times when the initial number of particles is increased. This occurs because the lower is the rotational energy and the larger is the number of particles, the faster is the overall evolution. On average, the models show that 50 % of the core mass has been converted into stars after $\sim 3.5 t_{\rm ff}$ ($\approx$130 kyr) so that the Class 0 stage lasts for $\sim$80 kyr in fairly good correspondence with the observationally inferred Class 0 lifetimes (André et al. 2000).

We may also estimate the duration of the Class 0 phase for each individual protostar by assuming that the transition time from Class 0 to Class I objects takes place when the protostar has accreted half of its final mass. These times are shown in Fig. 4 against the final stellar mass. The horizontal lines confine the observed range of Class 0 lifetimes (André et al. 2000). We see that the number of stars with masses $<0.1 M_{\odot}$, including brown dwarves, drastically increases with increasing β_0. Most stars in the $\beta \leq 0.10$ cores had Class 0 lifetimes that are beyond the upper limit indicated by the observations. A similar trend was also observed in gravoturbulent fragmentation models, where most final stars spent more time in the Class 0 stage than suggested by observations. Only for the most rapidly rotating cores ($\beta_0 = 0.25$), a larger number of stars was seen to fall within the observed range. Although a tendency exists for most stars to clump around $\approx 0.1 M_{\odot}$ (a tenth of the total core mass), there is no a clear correlation between the Class 0 lifetimes and the final stellar masses.

5 Number and Mass of Protostars

The temporal evolution of all stellar masses is displayed in Fig. 5. The birth of Class 0 YSOs is marked with plus signs, which here correspond to the precise instant when a permanent sink particle is activated. The two horizontal lines indicate the mean mass of the stellar ensembles (upper line) and the hydrogen-burning limit of $0.08 M_{\odot}$ (lower line), separating low-mass hydrogen-burning stars from brown dwarves. After $6t_{\mathrm{ff}}$, models A0.05N5, A0.10N5, A0.25N5, A0.25N10, and A0.25N20 formed a total number of 34, 48, 56, 112, and 115 sources, respectively. We see that more brown dwarves are formed in the higher β_0 cores and significantly more when the number of particles is increased. Previous fragmentation calculations of massive clumps ($120 M_{\odot}$) with driven supersonic turbulence have resulted in similar numbers of protostars (Schmeja and Klessen 2004). However, in these models the protostars are formed through filament fragmentation rather than disk fragmentation. On the other hand, our simulations form too many stars per core compared to turbulent fragmentation models of low-mass cores with similar initial conditions (Goodwin et al. 2004a, 2006; Attwood et al. 2009). A plausible explanation for this discrepancy can be given in terms of (a) these latter models being calculated with a much smaller number of particles, i.e., 2.5×10^4 against 5×10^5 particles for our lower resolution runs, which in our case may favour the growth of density fluctuations due to a much lower rate of numerical diffusion, and (b) the ability of initially rotating cores to build up more massive and extended disks early in the collapse. Thus, increasing both the kinetic rotational energy and the number of particles results in an increased number of stellar masses. A similar trend was also observed in gravoturbulent fragmentation models, where increasing the level of turbulence results in more forming stars (Attwood et al. 2009).

As β_0 is increased, the mean stellar mass decreases from $\approx 0.11 M_{\odot}$ for the $\beta_0 = 0.05$ core to $\approx 0.06 M_{\odot}$ for $\beta_0 = 0.25$. Note that increasing the number of particles to one and two million in the $\beta_0 = 0.25$ core lowered the mean stellar mass to $\approx 0.03 M_{\odot}$. In the low-β_0 (≤ 0.10) cores stars with masses $< 0.08 M_{\odot}$ form $\sim$5 kyr after the primary. This gap shortens when both β_0 and the number of particles are increased. Moreover, in our simulations all stars with masses $\leq M_{\mathrm{s,av}}$ are always formed after about 100 kyr from the beginning of collapse independently of β_0 and the number of particles. In general, the first stars to form are the more massive ones. This is particularly true for the $\beta_0 \leq 0.10$ cores, where the primary and the next stars that quickly form around it decline their accretion rates just before the disk becomes Toomre unstable. This behaviour is less evident in the more rapidly rotating cores, where many secondaries end up with masses comparable to or even higher than the primary. Many other stars forming with masses $< 0.08 M_{\odot}$ remain in the disk and accrete sufficient mass to become hydrogen-burning stars. However, since most of

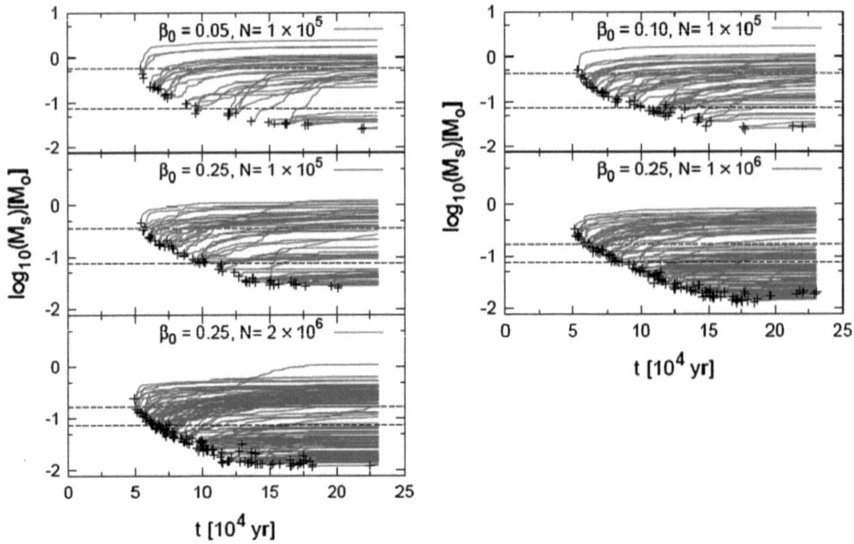

Fig. 5 Evolution of the protostellar masses for all simulations. The plus signs mark the birth point of each protostar in the ensemble. The *upper* and *lower horizontal lines* indicate the mean final stellar mass and the limit of $0.08 M_{\odot}$, separating hydrogen-burning stars from brown dwarves, respectively

these stars form in the outer parts of the disk, many of them undergo little accretion, ending up essentially with the same mass with which they formed. They become likely candidates to be ejected from the ensemble due to gravitational interactions with other more massive stars.

Our simulations produced final primaries of mass ≈ 0.45, 0.32, and $0.20 M_{\odot}$ in models A0.05N5, A0.10N5, and A0.25N5, respectively. In contrast, models A0.25N10 and A0.25N20, with much higher number of particles, produced final primaries of lower mass ($\sim 0.16 M_{\odot}$). In these cores the primary star is not necessarily the most massive one. The gravoturbulent fragmentation models of Attwood et al. (2009) predicted primary masses that were always $< 1 M_{\odot}$, with a clear delay between the formation time of the primary and fragmentation of the disk into secondary companions. The delay observed before any secondaries form is substantially reduced in our models. This occurs because accretion of high angular momentum gas from the core envelope creates dense and extended disks that favours fragmentation into secondaries soon thereafter the birth of the primary. Conversely, in the turbulent models the angular momentum required to create a circumprimary disk is produced by the vorticity inherent to the turbulent flow in the vicinity of the primary and its instability will depend on the ability of the inhomogeneous flow to deliver material onto it. As a consequence, the circumprimary disk needs more time to become over-dense and extended to fragment into secondary protostars. For instance, the β-distribution resulting from the collapse of highly turbulent, massive cloud clumps has on average $\beta < 0.05$ (Jappsen and Klessen 2004).

At the termination of the calculations ($t = 6t_{\rm ff}$), the efficiency of star formation, that is, the mean fraction of the total core mass that ends up into stars, is fairly the same for all models, with an average of ~65 %. This is a bit higher than the ~60 % reported in gravoturbulent fragmentation models of low-mass cores (Attwood et al. 2009).

6 Conclusions

We have presented a set of SPH collapse calculations of pre-stellar, low-mass cloud cores starting with uniform rotation and same initial parameters to previous gravoturbulent fragmentation models (Goodwin et al. 2004a,b, 2006; Attwood et al. 2009) in order to contrast the details of the accretion phase and the statistical properties of the forming stars.

The models are consistent with an intermediate mode of star formation where groups of 10–100 stars form from a single core (Adams and Myers 2001). They reproduce many of the statistical properties predicted by gravoturbulent fragmentation models, including the positive correlation between the peak accretion rates and the final stellar masses, the values of the peak accretion rates, the lifetimes of Class 0 protostars, and the efficiency of star formation. In particular, the primary protostars condense out from a central protostellar disk after ~50 kyr from the beginning of collapse. For similar initial conditions, the primaries were seen to form after 50 to 70 kyr in the turbulent core models (Attwood et al. 2009), with the primary forming before the disk similarly to the case of slowly rotating cores with $\beta < 0.03$ (Walch et al. 2009).

Our simulations predict peak accretion rates between $\sim 10^{-5}$ and $10^{-4} M_\odot$ yr^{-1} in close correspondence with the observed range for Class 0 protostars (André et al. 2000). After ≈231 kyr, the accretion rates are seen to decline to values around $\sim 10^{-7} M_\odot$ yr^{-1}, which matches the observed values for the most evolved Class I objects. The models also predict a positive linear correlation between the peak accretion rates and the final stellar masses in as much as the same way as predicted by highly turbulent collapse models of massive ($120 M_\odot$) clumps (Schmeja and Klessen 2004). In addition, the efficiency of star formation does not seem to change appreciably with varying levels of the kinetic rotational energy and number of particles, with about 65 % of the core mass being converted into stars. Similar high levels (~60 %) were also found in turbulent fragmentation simulations of low-mass cores (Attwood et al. 2009).

We anticipate that most results of subsonically turbulent models at scales of ~0.05 pc are reproducible from the collapse of initially slowly rotating ($\beta < 0.03$) cores. In a further paper, we plan to perform new high-resolution fragmentation calculations of both slowly rotating and weakly turbulent core models, but with the same net angular momentum, in order to provide a better comparison for both types of initial conditions.

Acknowledgments This work has been partially supported by the Consejo Nacional de Ciencia y Tecnología of Mexico (CONACyT) under the project CONACyT-EDOMEX-2011-C01-165873 and the Fondo Nacional de Ciencia, Tecnología e Innovación of Venezuela (FONACIT) under grant PC 201204710 (contract 112-1077).

References

Adams FC, Myers PC (2001) Modes of multiple star formation. Astrophys J 553:744–753

André P, Ward-Thompson D, Barsony M (1993) Submillimeter continuum observations of ρ Ophiuchi A: the candidate protostar VLA 1623 and prestellar clumps. Astrophys J 406:122–141

André P, Motte F, Bacmann A (1999) Discovery of an extremely young accreting protostar in Taurus. Astrophys J 513:L57–L60

André P, Ward-Thompson D, Barsony M (2000) From prestellar cores to protostars: the initial conditions of star formation. In: Mannings V, Boss AP, Russell SS (eds) Prostostars and Planets IV. University of Arizona Press, Tucson, pp 59–96

Attwood RE, Goodwin SP, Whitworth AP (2007) Adaptive smoothing lengths in SPH. Astron Astrophys 464:447–450

Attwood RE, Goodwin SP, Stamatellos D, Whitworth AP (2009) Simulating star formation in molecular cloud cores. IV. The role of turbulence and thermodynamics. Astron Astrophys 495:201–215

Ballesteros-Paredes J, Klessen RS, Vázquez-Semadeni E (2007) Molecular cloud turbulence and star formation. In: Reipurth VB, Jewitt D, Keil K (eds) Protostars and Planets V. University of Arizona Press, Tucson, pp 63–80

Bate MR, Bonnell IA, Price NM (1995) Modelling accretion in protobinary systems. Mon Not R Astron Soc 277:362–376

Bate MR, Bonnell IA, Bromm V (2003) The formation of a star cluster: predicting the properties of stars and brown dwarfs. Mon Not R Astron Soc 339:577–599

Bodenheimer P, Burkert A, Klein RI, Boss AP (2000) Multiple fragmentation of protostars. In: Mannings V, Boss AP, Russell SS (eds) Protostars and Planets V. University of Arizona Press, Tucson, pp 675–701

Bontemps S, André P, Terebey S, Cabrit S (1996) Evolution of outflow activity around low-mass embedded young stellar objects. Astron Astrophys 311:858–872

Boogert ACA, Hogerheijde MR, Blake GA (2002) High-resolution 4.7 micron Keck/NIRSPEC spectra of protostars. I. Ices and infalling gas in the disk of L1489 IRS. Astrophys J 568:761–770

Burkert A, Bodenheimer P (2000) Turbulent molecular cloud cores: rotational properties. Astrophys J 543:822–830

Caselli P, Benson PJ, Myers PC, Tafalla M (2002) Dense cores in dark clouds. XIV. N_2H^+ (1–0) maps of dense cloud cores. Astrophys J 572:238–263

Ceccarelli C, Castets A, Caux E, Hollenbach D, Loinard L, Molinari S, Tielens AGGM (2000) The structure of the collapsing envelope around the low-mass protostar IRAS 16293–2422. Astron Astrophys 355:1129–1137

Chen X (2008) High angular resolution observations of binary protostars. PhD thesis, The Ruperto-Carola University of Heidelberg, Germany

Crapsi A, Caselli P, Walmsley MC, Tafalla M (2007) Observing the gas temperature drop in the high-density nucleus of L1544. Astron Astrophys 470:221–230

Federrath C, Banerjee R, Clark PC, Klessen RS (2010) Modeling collapse and accretion in turbulent gas clouds: implementation and comparison of sink particles in AMR and SPH. Astrophys J 713:269–290

Di Francesco J, Evans NJ II, Caselli P, Myers PC, Shirley Y, Aikawa Y, Tafalla M (2007) An observational perspective of low-mass dense cores I: internal physical and chemical properties.

In: Reipurth VB, Jewitt D, Keil K (eds) Protostars and Planets V. University of Arizona Press, Tucson, pp 17–32

Furlan E, McClure M, Calvet N, Hartmann L, D'Alessio P, Forrest WJ, Watson DM, Uchida KI, Sargent B, Green JD, Heter TL (2008) Spitzer IRS spectra and envelope models of class I protostars in Taurus. Astrophys J Suppl Ser 176:184–215

Goodman AA, Barranco JA, Wilner DJ, Heyer MH (1998) Velocity coherence in dense cores. Astrophys Lett Commun 37:109–111

Goodwin SP, Whitworth AP, Ward-Thompson D (2004a) Simulating star formation in molecular cloud cores. I. The influence of low levels of turbulence on fragmentation and multiplicity. Astron Astrophys 414:633–650

Goodwin SP, Whitworth AP, Ward-Thompson D (2004b) Simulating star formation in molecular cloud cores. II. The effects of different levels of turbulence. Astron Astrophys 423:169–182

Goodwin SP, Whitworth AP, Ward-Thompson D (2006) Star formation in molecular cores. III. The effect of the turbulent power spectrum. Astron Astrophys 452:487–492

Greene TP, Lada CJ (2002) Spectroscopic detection of a stellar-like photosphere in an accreting protostar. Astron J 124:2185–2193

Hartmann L, Kenyon SJ (1996) The FU Orionis phenomenon. Ann Rev Astron Astrophys 34:207–240

Hartmann L, Cassen P, Kenyon SJ (1997) Disk accretion and the stellar birthline. Astrophys J 475:770–785

Hennebelle P, Whitworth AP, Cha SH, Goodwin SP (2004) Protostellar collapse induced by compression II. Rotation and fragmentation. Mon Not R Astron Soc 348:687–701

Hubber DA, Goodwin SP, Whitworth AP (2006) Resolution requirements for simulating gravitational fragmentation using SPH. Astron Astrophys 450:881–886

Jappsen AK, Klessen RS (2004) Protostellar angular momentum evolution during gravoturbulent fragmentation. Astron Astrophys 423:1–12

Kenyon SJ, Gomez M, Marzke RO, Hartmann L (1994) New pre-main-sequence stars in the Taurus-Auriga molecular cloud. Astron J 108:251–261

Klessen RS, Burkert A (2000) The formation of stellar clusters: Gaussian cloud conditions. I. Astrophys J Suppl Ser 128:287–319

Klessen RS (2004) Comments of gravoturbulent star formation. Astrophys Space Sci 292:215–223

Larson RB (2003) The physics of star formation. Rep Prog Phys 66:1651–1697

Mac Low MM, Klessen RS (2004) Control of star formation by supersonic turbulence. Rev Mod Phys 76:125–194

Maret S, Ceccarelli C, Caux E, Tielens AGGM, Castets A (2002) Water emission in NGC 1333-IRAS 4. The physical structure of the envelope. Astron Astrophys 395:573–585

Monaghan JJ, Lattanzio JC (1985) A refined particle method for astrophysical problems. Astron Astrophys 149:135–143

Monaghan JJ (2005) Smoothed particle hydrodynamics. Rep Prog Phys 68:1703–1759

Myers PC, Adams FC, Chen H, Schaff E (1998) Evolution of the bolometric temperature and luminosity of young stellar objects. Astrophys J 492:703–726

Schmeja S, Klessen RS (2004) Protostellar mass accretion rates from gravoturbulent fragmentation. Astron Astrophys 419:405–417

Shu FH (1977) Self-similar collapse of isothermal spheres and star formation. Astrophys J 214: 488–497

Shu FH, Adams FC, Lizano S (1987) Star formation in molecular clouds: observation and theory. Ann Rev Astron Astrophys 25:23–81

Springel V (2005) The cosmological simulation code GADGET-2. Mon Not R Astron Soc 364:1105–1134

Stahler SW (1988) Deuterium and the stellar birthline. Astrophys J 332:804–825

Volgenau NH, Mundy LG, Looney LW, Welch WJ (2006) Dense cores with multiple protostars: the velocity fields of L1448 IRS 3, NGC 1333 IRAS 2, and NGC 1333 IRAS 4. Astrophys J 651:301–320

Walch S, Burkert A, Whitworth AP, Naab T, Gritschneder M (2009) Protostellar disks formed from rigidly rotating cores. Mon Not R Astron Soc 400:13–25

Whitworth AP (1998) The Jeans instability in smoothed particle hydrodynamics. Mon Not R Astron Soc 296:442–444

Whitworth AP, Ward-Thompson D (2001) An empirical model for protostellar collapse. Astrophys J 547:317–322

Young CH, Shirley YL, Evans NJ II, Rawlings JMC (2003) Tracing the mass during low-mass star formation. IV. Observations and modeling of the submillimeter continuum emission from class I protostars. Astrophys J Suppl Ser 145:111–145

Zhu Z, Hartmann L, Calvet N, Hernandez J, Muzerolle J, Tannirkulam AK (2007) The hot inner disk of FU Orionis. Astrophys J 669:483–492

Zhu Z, Hartmann L, Gammie C (2009) Nonsteady accretion in protostars. Astrophys J 694: 1045–1055

Biocompatible Treatment of Extra Heavy Oil Produced in Venezuela

Ledys Y. Sánchez, Efrén D. J. Andrades, Erick A. Pacheco, Hilda C. Grassi, Carlos R. Vera-Lagos and Victor J. Andrades-Grassi

Abstract In this chapter, we investigate the behaviour of biocompatible mixtures in the treatment of Venezuelan extra heavy oil, using the non-ionic surfactant Polysorbate 80 (Tween 80) and low molecular weight linear n-alcohols with even and odd numbers of carbon atoms. Venezuelan extra heavy oil was recovered from mixtures that contained water, NaCl, polysorbate 80, and n-alcohols ranging from 1 to 8 carbon atoms. Water retained (in enhanced oil), density, conductivity, viscosity, drop weight, and retained oil (in the borosilicate glass tube) were measured and compared for the different n-alcohols in the mixture. We find that the mean (10.99 mS) of the conductivities of the aqueous phase from mixtures with C-2 – C-5 alcohols is statistically different and higher than the mean (4.91 mS) of the conductivities of the aqueous phase from mixtures with C-6 – C-8 alcohols. Among the proper-

L. Y. Sánchez
Laboratorio de Cinética y Catálisis, Departamento de Química, Facultad de Ciencias, Universidad de Los Andes, Mérida 5101, Estado Mérida, Venezuela
e-mail: ledys@yahoo.com

E. D. J. Andrades (✉) · E. A. Pacheco · H. C. Grassi · V. J. Andrades-Grassi
UBIP Project, Sección de Biotecnología, Instituto de Investigaciones, Escuela de Farmacia, Facultad de Farmacia y Bioanálisis, Universidad de Los Andes, Mérida 5101, Estado Mérida, Venezuela
e-mail: andradesefren@cantv.net

E. A. Pacheco
e-mail: alejandropc@gmail.com

H. C. Grassi
e-mail: cristinagrassi@gmail.com

V. J. Andrades-Grassi
e-mail: vicandrades@gmail.com

C. R. Vera-Lagos
Marketing Department, Laboratorios Servier S.A., Av. Sanatorio del Ávila, Centro Empresarial Ciudad Center, Torre F, Piso 3, Boleíta Norte 1071, Caracas, Venezuela
e-mail: carlos.vera@ve.netgrs.com

L. Di G. Sigalotti et al. (eds.), *Computational and Experimental Fluid Mechanics with Applications to Physics, Engineering and the Environment*, Environmental Science and Engineering, DOI:10.1007/978-3-319-00191-3_15,

ties of the recovered oil we find that there is a direct and oscillating correlation of viscosity and water retained in the crude oil fraction, and an inverse correlation of both with drop weight, indicating that the viscoelastic properties of recovered crude oil after treatment are a function of the n-alcohol in the mixture. Oil retained in the borosilicate glass tube as a function of the carbon number of the n-alcohol is directly proportional to toxicity of the alcohol (expressed as A) and ovality of the alcohol (expressed as molecular volume, θ3D), and inversely proportional to acentric factor of the alcohol (expressed as ω). Moreover, the polarity, shape, and size associated with the number of carbons in the n-alcohol may be responsible for the high conductivity (10.99 mS) in the aqueous phase released after treatment with the C-2 – C-5 alcohols and the low conductivity (4.91 mS) in the aqueous phase released after treatment with the C-6–C-8 alcohols.

1 Introduction

The use of biocompatible molecules in the treatment of extra heavy oil, bitumens, asphalts, and other hydrocarbons has been limited. Due to physical and chemical conditions of oil deposits, the well is scarcely compatible with biological processes, reducing the possibilities of oil enhancement *in situ*. Oil treatment *ex situ* faces the challenges presented by complex hydrocarbon mixtures in their limitations as microbial substrate, metabolic access, and toxicity of inorganic components such as heavy metals (Leahy and Colwell 1990; Chase et al. 1991). Thus, any attempt to take advantage of biological processes in enhanced oil recovery, requires the improvement of the physico-chemical properties that focus on this purpose (Chang et al. 1992; Torsvik et al. 1995; Kowalewski et al. 2006).

In any case, it is always desirable to treat the oil with the least toxic components in order to approach biocompatibility. Also, the selection, adaptation, and genetic modification of microorganisms is a necessary consideration in order to achieve microbial growth in biocompatible formulations (Van Hamme et al. 2003). These are the main objectives of this work, which is based on a formulation previously proposed for the treatment of light and heavy oil that supports microbial growth (Grassi 2001), its application to extra heavy oil with an optimal formulation, the incorporation of other alcohols in addition to ethanol, and the evaluation of other properties in addition to density, viscosity, and microbial metabolism of hydrocarbons previously evaluated (Grassi 2001).

2 Materials and Methods

10 ml of extra heavy oil from Morichal, Estado Monagas, Venezuela, with 9 °API gravity and density 1.01 g/mL were mixed with 5 mL of distilled water containing 17,800 ppm NaCl and left at 61 °C for 24 hours. 100 μL of polysorbate 80 (Tween 80

pharmaceutical grade) and 10 μL of the test alcohol (analytical grade) were added (optimal formulation 10:5:0.1:0.01 oil:brine-water:non-ionic-surfactant:alcohol).

In some experiments, Tween 80 was substituted by I200 (from Orimulsion®), which was kindly supplied by the UBIP Project. This was mixed at 200 rpm at 61 °C for 90 min and then left at room temperature without mixing, allowing for the separation of the respective aqueous phases from the recovered oil phases. The water retained in the oil phase was calculated by the difference of volumes between the separated aqueous phase and the added water (5 mL); the density (in units of mg/mL) was measured by weighing several times a fixed volume of the recovered oil phase; and the conductivity (mS) of the aqueous phase was measured in a TDSTestr 20 (OAKTON Instruments).

Video recordings (Tirtaatmadja et al. 2006) were taken for viscosity and drop diameter. The viscosity (in units of cP) was calculated using the formula given by Grassi (2001), after taking the time required (hours) for the oil phase to travel through a calibrated cylindrical borosilicate glass tube (Corning Low Expansion Glass Pyrex 7740, error <2 %), while the drop weight (in grams, g) was calculated from the drop volume (using the mean of the drops' diameters) and density. The retained oil (g) is the weight of the oil that remains in the glass tube, which is not referred to as oil wettability because we cannot exclude the possibility of a catalytic reaction such as those already described (Hayashi and Uozumi 1992). The significance of the conductivity values obtained with alcohols C-2–C-5 and C-6–C-8 was calculated by the Student's t-test.

3 Results and Discussion

Table 1 and Fig. 1 show the density of the crude oil recovered after treatment with Polysorbate 80 (Tween 80), brine water, and each of the linear n-alcohols containing from one to eight saturated carbons. Alcohols with an even number of carbon atoms (2, 4, 6, 8) yield lower density values than odd alcohols (1, 3, 5, 7).

The conductivity values of the aqueous phase that was separated from the mixtures are also reported. High values were obtained for alcohols with 2–5 carbons, while low values were obtained for alcohols with 6–8 carbons. We apply the Student's t-test on the results of conductivity for both groups of data (see Table 2), i.e., the aqueous phase recovered from mixtures containing C-2–C-5 alcohols and from mixtures that contain C-6–C-8 alcohols. This includes all the alcohols that are shown in Table 1, except methanol and sec-butanol that though they behave as expected, the values contribute to a distortion in the standard deviation. The null hypothesis, stating that the means of the conductivities of the two groups are equal, is rejected. Therefore, the alternative hypothesis indicating that the mean (10.99 mS) of the conductivities of the aqueous phase from mixtures with C-2–C-5 alcohols is proven to be statistically different and higher than the mean (4.91 mS) of the conductivities of the aqueous phase from mixtures with C-6–C-8 alcohols.

Table 1 Properties of crude oil and aqueous phase after treatment and recovery from mixtures

Alcohol	Number of carbons	Density g/mL	Viscosity cP	Conductivity mS	Drop weight g	Retained oil g	Retained water mL
Methanol	1	0.95	15831.5	926.8	0.062176	1.8474	2.0
Ethanol	2	0.77	6492.0	10.3	0.087082	1.6986	0
n-propanol	3	0.84	4573.7	11.2	0.150855	1.7666	0
n-butanol	4	0.74	6587.0	12.2	0.048431	1.179	0.5
n-pentanol	5	0.82	657.95	10.7	0.312988	1.5883	0.2
n-hexanol	6	0.61	2672.2	6.2	0.039923	1.7000	0.6
n-heptanol	7	0.89	3621.2	4.9	0.029823	2.2000	1.0
n-octanol	8	0.78	389.8	5.5	0.140079	2.979	0.1
iso-propanol	3	0.78	10754.7	11.2	0.088213	1.2232	0.4
Sec-butanol	4	0.89	2220.2	19.0	0.238594	2.0335	0.5
Iso-butanol	4	0.94	599.0	10.6	0.655097	1.8666	0
Tert-butanol	4	0.91	787.2	11.1	0.102916	2.7474	0.5
2-hexanol	6	0.82	1326.7	4.7	0.219828	3.2137	0.5
Benzyl alcohol	7	0.92	3824.8	4.1	0.104046	1.6137	0.5
Ethanol+ I200	2	0.96	855.3	10.6	0.10857	2.3474	1.8
Benzyl alcohol+ I200	7	0.66	7750.4	4.1	0.043196	4.1666	1.0

Figure 2 shows some of the properties of the crude oil recovered after treatment with polysorbate 80, brine water, and each of the n-alcohols. As expected, the dependence of the viscosity on the number of carbon atoms in the n-alcohol follows the same trend as the retained water. Conversely, the drop weight exhibits an opposite trend. In previous studies (Tirtaatmadja et al. 2006) of the relationship among elasticity, dynamic surface tension, and viscosity in polyethylene oxide solutions containing 0.4 % 2-butanol, elasticity was shown to be independent of dynamic surface tension and viscosity.

The addition of 2-butanol had the effect of decreasing the dynamic surface tension of the solutions at short times (10 ms) to the equilibrium value (>1 s). Therefore, the authors suggest that the differences in the dynamic surface tension of the polyethylene oxide solutions are not the cause of the observed differences in the evolution of elasticity (Tirtaatmadja et al. 2006). In addition, we observe that the use of different n-alcohols in the mixtures produces a profile for the drop weight that is opposite to the trends for the viscosity and water retained (as functions of the number of carbon

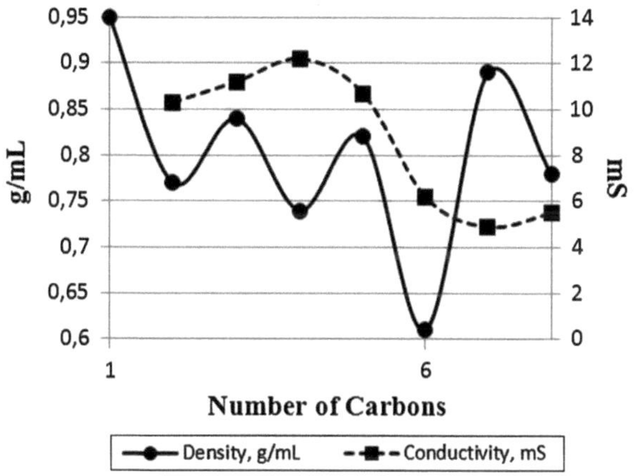

Fig. 1 Density (in units of g/mL) of the recovered crude oil (*solid line*) and conductivity (in units of mS; *dashed line*) of the recovered aqueous phase after treatment, as a function of the number of carbon atoms in the n-alcohol used in the mixture (methanol was omitted due to its high value)

Table 2 Student's t-test of conductivity of the aqueous phase recovered from mixtures containing C-2–C-5 alcohols and mixtures that contain C-6–C-8 alcohols

Variable	Conductivity
T	−19.2739
Df	5
P-value	6.934e-06
Lower confidence interval	−7.597823
Upper confidence interval	−4.968844
Mean of the differences	−6.283333
Interpretation	$P < 0.001$ The alternative hypothesis is accepted

atoms). The polysorbate 80 in its amphiphilic nature could behave as a compatibilizing agent (Koning et al. 1998) between the alcohols and the rest of the aqueous phase, on one hand, and the crude oil, on the other hand. Under the conditions of the optimal formulation used in this work, it is possible to reach a transient miscibility window (Guigley 2001) that allows for the improvement of extra heavy oil and the separation of an aqueous phase that causes a partial exclusion of ionic and other components from the crude oil.

We found that the retained oil (defined as the weight in grams of the recovered oil that remains in the borosilicate glass tube) is proportional to the molecular shape of butanol in the mixture (see Table 1): n-butanol (1.179 g), iso-butanol (1.8666 g), sec-butanol (2.0335 g) and tert-butanol (2.7474 g). This indicates that the more linear molecules promote less retention of the oil on the glass tube. It has been previously found that there exists a relationship between ovality of the alcohols and toxicity to *Tetrahymena pyriformis* (Vlaia et al. 2009). In order to ascertain the relationship

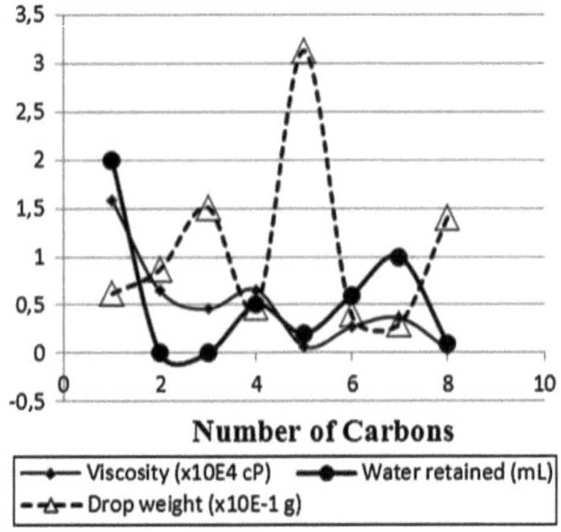

Fig. 2 Viscosity (in units of cP, thin *solid line*), drop weight (in units of g, *dashed line*), and water retained (in units of mL, *thick solid line*) in the recovered crude oil as a function of the number of carbon atoms in the n-alcohol used in the mixture

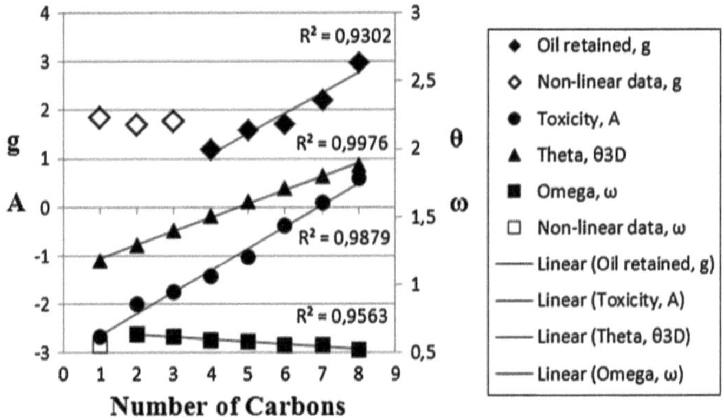

Fig. 3 Oil retained in the glass tube in grams as a function of the number of carbon atoms in the alcohol included in the mixture, compared with the acentric factor (omega value of the alcohol, ω) (Reid et al. 1977), the ovality of n-alcohols (measured as molecular volume and described as θ3D) (Vlaia et al. 2009), and the toxicity to *Tetrahymena pyriformis* (A) (Vlaia et al. 2009)

between the molecular shape of the alcohol and the toxicity, we compared the reported biocompatibility/toxicity of the n-alcohols used in the mixtures to the amount of oil retained in the calibrated borosilicate glass tube, assuming that the molecular shape of the alcohol is relevant for this relation.

Figure 3 shows a linear relation for the toxicity (A) (data taken from Vlaia et al. 2009) and the retained oil (data from Table 1) with the number of carbons of the n-alcohol; the former having a slope of 0.45 (R^2 = 0.9879) and the latter of 0.42 (R^2 = 0.9302). This analysis has excluded the values for oil retained using methanol, ethanol, and n-propanol (Fig. 3, non-linear data) because these are nearly constant.

A linear correlation is also found when comparing the data taken from Vlaia et al. (2009) for the ovality of n-alcohols, measured as molecular volume and described as θ3D, with a slope of 0.1 ($R^2 = 0.9976$).

Furthermore, the acentric factor omega (ω) was originally proposed to represent only the acentricity (or non-sphericity) of a molecule. However, it is currently widely used as a parameter which in some manner measures the complexity of a molecule with respect to both, geometry and polarity. In general, it increases with polarity. Using the data for the acentric factor omega from (Reid et al. 1977), and including it in Fig. 3, we found a linear relationship of ω as a function of the number of carbons of the n-alcohols, with a slope of −0.018 ($R^2 = 0.9563$). Linear n-alcohols with increasing number of carbon atoms become less polar thus influencing the amount of retained oil in the glass tube, which also correlates with ovality and toxicity. Moreover, the polarity, shape, and size associated with the number of carbons in the n-alcohol may be responsible for the high conductivity (10.99 mS) in the aqueous phase released after treatment with the C-2 – C-5 alcohols as well as the low conductivity (4.91 mS) in the aqueous phase released after treatment with the C-6–C-8 alcohols.

4 Conclusions

Attempts to take advantage of biological processes in enhanced oil recovery require the improvement of physico-chemical properties that focus on microbial substrate, metabolic access, and toxicity of inorganic components such as heavy metals. Moreover, it is desirable to treat oil with the least toxic components in order to approach biocompatibility. Treatment of extra heavy oil with polysorbate 80, brine water, and a linear n-alcohol results in the separation of enhanced petroleum from the aqueous phase. Among the properties of the recovered oil, we found that there is a direct and oscillating correlation of viscosity and water retained in the crude oil fraction, and an inverse correlation of both with drop weight. We conclude that the viscoelastic properties of recovered crude oil after treatment are a function of the n-alcohol in the mixture. On the other hand, conductivity of the released aqueous phase under the conditions of this work falls into two distinguishable categories: high conductivity for n-alcohols that are smaller than six carbons and low conductivity for n-alcohols that are C-6–C-8.

Previous studies have shown that the ovality and polarity of linear n-alcohols correlate with the toxicity and acentric factor of n-alcohols, respectively (Reid et al. 1977; Vlaia et al. 2009). In this work, we found that, after treatment of extra heavy oil with polysorbate 80, brine water, and each linear n-alcohol, the values of the retained oil in the calibrated borosilicate glass tube correlate linearly with the former parameters. Therefore, linear n-alcohols with increasing number of carbon atoms become less polar influencing the amount of retained oil in the borosilicate glass tube, which also correlates with ovality and toxicity. Moreover, the polarity associated with the number of carbons in the n-alcohol may be responsible for the high conductivity (10.99 mS) in the water released after treatment with the C-2–C-5 alcohols and the low conductivity (4.91 mS) in the water released after treatment with the C-6–C-8 alcohols. We conclude that it is possible to adapt the crude oil to biocompatible purposes, using the appropriate mixtures designed with optimal formulations of polysorbate 80, brine water, and low molecular weight linear n-alcohols.

Acknowledgments This work was supported in part by the UBIP Project.

References

Chang M-M, Bryant RS, Stepp AK, Bertus KM (1992) Modeling and laboratory investigations of microbial oil recovery mechanisms in porous media. NIPER-629 Bartlesville, National Institute for Petroleum and Energy Research

Chase KL, Bryant RS, Burchfield TE, Bertus KM, Stepp AK (1991) Ch. R-4 Investigations of microbial mechanisms for oil mobilization in porous media. In: Proceedings of the 1990 international conference on microbial enhancement of oil recovery, Dev Petr Sci 31:79–94

Grassi HC (2001) Patent Pending Code 224288. Producto y Proceso para Mejoramiento, Recuperación y Biotransformación de Crudo y sus Derivados. Request #2001-000538 www.patentesonline.com.ve.

Guigley KS (2001) Hydogen Bonded Polymer Blends. A Thesis in Materials Science and Engineering,The Graduate School, Department of Materials Science and Engineering, The Pennsylvania State University

Hayashi T, Uozumi Y (1992) Catalytic asymmetric synthesis of optically active alcohols via hydrosilylation of olefins with a chiral monophosphine-palladium catalyst. Pure Appl Chem 64(12):1911–1916

Koning C, Van Duin M, Pagnoulle C, Jerome R (1998) Strategies for compatibilization of polymer blends. Prog Polym Sci 23:707–757

Kowalewski E, Rueslåtten I, Steen KH, Bødtker G, Torsæter O (2006) Microbial improved oil recovery-bacterial induced wettability and interfacial tension effects on oil production. J Petrol Sci Eng 52(1–4):275–286

Leahy JG, Colwell RR (1990) Microbial degradation of hydrocarbons in the environment. Microbiol Rev 54(3):305–315

Reid RC, Plausnitz JM, Sherwood TK (1977) The properties of gases and liquids. McGraw-Hill Book Compay, New York

Tirtaatmadja V, McKinley GH, Cooper-White JJ (2006) Drop formation and breakup of low viscosity elastic fluids: effects of molecularweight and concentration. Hatsopoulos microfluids laboratory, Department of Mechanical Engineering, MIT, HML, Report Number 06-P-03.

Torsvik T, Gilje E, Sunde E (1995) Aerobic microbial enhanced oil recovery. Paper presented at the 5th international conference on microbial enhanced oil recovery, Dallas TX, USA, September 1995

Van Hamme JD, Singh A, Ward OP (2003) Recent advances in petroleum microbiology. Microbiol Mol Biol R 67(4):503–549

Vlaia V, Olariu T, Ciubotariu C, Medeleanu M, Vlaia L, Ciubotariu C (2009) Molecular descriptors for quantitative structure-toxicity relationship (QSTR). Rev. Chim-Bucharest 60(12):1357–1361

Dynamical Behaviour of As(V) and Se(IV) Adsorption in Biofilters: Analysis of Dimensions, Flux and Removal Percentage

Jaime Klapp, Carlos E. Alvarado-Rodríguez and Elizabeth Teresita Romero-Guzmán

Abstract In this work, we study the dynamical behaviour of As(V) and Se(IV) absorption in a biofilter through a mathematical simulation, comparing dimensions, flux, and removal percentage in packed columns. From the numerical simulation we obtained breakthrough curves. When comparing the experimental and numerical breakthrough curves, the best correlation gives $R^2 = 0.9825$. A set of columns with different dimensions and feed streams were simulated. "The higher and lower" calculated values in the adsorption of selenium (IV) were 3 and 17 %, respectively, whereas the corresponding values for arsenic (V) adsorption were 6 and 17 %, respectively.

1 Introduction

Arsenic is a commonly occurring toxic metal in natural systems and is the root cause of many diseases and disorders. Occurrence of arsenic contaminated water is reported from several countries all over the world (Giles et al. 2011). A great deal of research

J. Klapp (✉) · C. E. Alvarado-Rodríguez
Departamento de Física, Instituto Nacional de Investigaciones Nucleares, ININ, Km 36.5, Carretera México-Toluca, 52750 La Marquesa, Estado de México, Mexico
e-mail: jaime.klapp@hotmail.com

E. T. Romero-Guzmán
Departamento de Química, Instituto Nacional de Investigaciones Nucleares, ININ, Km 36.5, Carretera México-Toluca, 52750 La Marquesa, Estado de México, Mexico
e-mail: elizabeth.romero@inin.gob.mx

C. E . Alvarado-Rodríguez
Divisi6n de Ciencias Naturales y Exactas, Universidad de Guanajuato, Campus Guanajuato, Noria Alta S/N, Guanajuato, Mexico
e-mail: q_l_o@hotmail.com

J. Klapp
Departamento de Matemáticas, Cinvestav del Instituto Politécnico Nacional (I.P.N.), 07360 México D.F., Mexico

L. Di G. Sigalotti et al. (eds.), *Computational and Experimental Fluid Mechanics with Applications to Physics, Engineering and the Environment*, Environmental Science and Engineering, DOI:10.1007/978-3-319-00191-3_16,

over recent decades has been motivated by the requirement to lower the concentration of arsenic in drinking water and the need to develop low cost techniques which can be widely applied for arsenic removal from contaminated water (Giles et al. 2011).

For many years, selenium has been a largely unrecognized pollutant, particularly in developing countries, and has been overshadowed by issues involving contaminants such as industrial chemicals, heavy metals, pesticides, and air pollutants just to name a few. A continuous intake greater than 8 mg of selenium per day can produce harmful health effects. The EPA limits the amount of selenium allowed in drinking water supplies to 50 parts total selenium per billion parts of water (50 ppb) (Lemly 2004; WHO 2003). In Mexico, there have been reports of selenium contamination in water. For instance, in the state of Puebla, the denim laundries have polluted the waters, leaving residues of selenium (Hurtado and Gardea 2007).

Several vital characteristics are available and need to be listed to render the materials valuable enough as an industrial sorbent. Such characteristics are: high adsorption capacity, available in large quantities at one location, low economic value and less useful in alternative products, and attached metals can be easily recovered while biosorbent is reusable (Sisca et al. 2009).

If the biofilter is considered as a porous medium, a mathematical model can be constructed for representing arsenic and selenium sorption. Using the mathematical model and the simulation we can make an analysis of how the removal percentage changes with the dimensions and the flux.

2 The Mathematical Model

For the simulations we use equations that represent the flow and adsorption in each one of the different zones in the column.

For the unpacked zones the Navier–Stokes (1) and continuity (2) equations that are used for calculating the velocity field in the column are (Bird et al. 2003):

$$\rho\frac{\partial \boldsymbol{v}}{\partial t} + (\boldsymbol{v}\cdot\nabla)\boldsymbol{v} = -\nabla p + \mu\nabla^2\boldsymbol{v} + \rho\boldsymbol{g}, \tag{1}$$

$$\frac{\partial \rho}{\partial t} = -(\nabla\cdot\rho\boldsymbol{v}), \tag{2}$$

where ρ is the density, v is the velocity, t is the time, p is the pressure, μ is the viscosity, and $\boldsymbol{g}$ is the gravity.

For the packed zone the velocity field in the porous media is obtained by solving the Brinkman (3) and the modified continuity (4) equations that can be written in the form:

$$-\nabla p - \frac{\mu}{k}\boldsymbol{v}_0 + \mu\nabla^2 v_0 + \rho\boldsymbol{g} = 0, \tag{3}$$

$$\varepsilon \frac{\partial p}{\partial t} = -(\nabla \cdot \rho \boldsymbol{v}_0), \tag{4}$$

where k is the permeability, $\boldsymbol{v}_0$ is the average lineal velocity in the porous media, and ε is the porosity. The value of the porosity was determined in the laboratory. In a dynamical system, the transport velocity manages the effective sorption velocity in the inner surfaces, more than the sorption reactions (Walter et al. 1990).

Using the velocity field obtained from Eqs. (3) and (4), we then calculate the time rate of change of the concentration in the column through the equation

$$\varepsilon \frac{\partial c}{\partial t} + \rho_b \frac{\partial c_p}{\partial c} \frac{\partial c}{\partial t} + \nabla \cdot [-\varepsilon D_L \nabla_c + \boldsymbol{v} c] = 0. \tag{5}$$

In Eq. (5), the velocity v in the advective term is calculated using Eq. (1) for the unpacked zones and using Eq. (3) for the zone with porous media.

For the absorption we use a Freundlich (6) isothermal for arsenic and a linear isothermal (7) for selenium:

$$c_p = K_F c^N, \tag{6}$$

$$c_p = K_L c, \tag{7}$$

where c is the solute concentration in the solution, ρ_b is the porous medium density, c_p is the concentration of the solute sorbed in the biomass (the mass amount of solute sorbed per unit of biomass), D_L is the hydrodynamic dispersion tensor, and K_F and $K_L$are the isotherm constants. The parameter values for the solution of the model are listed in Table 1.

Equations (1, 2, 3 and 4) were solved with the following initial and boundary conditions: Every limit except the upper and lower limits are homogeneous:

Table 1 Model parameters

Symbol	Parameter	Value
ρ_L	Density of *lemna minor*[b]	109 kg/m^3
ρ_E	Density of *eichhornia crassipes*[b]	570 kg/m^3
E	Porosity[b]	15 %
K	Permeability[b]	1.4×10^{-3} m/s
N	Freundlich isotherm exponent[b]	0.7578
K_F	Constant of Freundlich isotherm[b]	1.1007 mg/gr
K_L	Constant of lineal isotherm[b]	0.1201 L/kg
α_L	Longitudinal dispersivity[a]	0.5 m
α_{TH}	Horizontal transversal dispersivity[a]	0.005 m
α_{TV}	Vertical transversal dispersivity[a]	0.005 m
μ	Water viscosity	0.001 Pas·s

([a]Bear 1972; [b]Marín 2010)

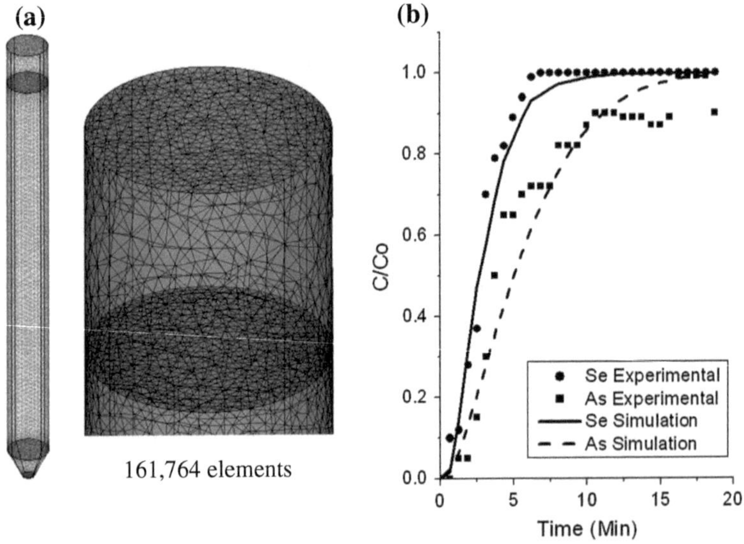

Fig. 1 **a** Computational mesh used in the numerical simulation. **b** Validation of the model by comparing the experimental and numerical breakthrough curves

Dirichlet condition ($\boldsymbol{v} = 0$). The upper limit is: *Dirichlet* condition $\boldsymbol{v} = -\boldsymbol{v}_0\boldsymbol{n}$, where $\boldsymbol{v}_0 = 1.48\times10^{-4}$ m/s for As(V) and $\boldsymbol{v}_0$=1.031×10^{-4} m/s for Se(IV). For the lower limit we use the pressure condition: $p = p_0$, where $p_0 = 0$. Equation (5) is solved using the following initial and boundary conditions: The upper and lower limits are *Dirichlet* conditions: $\boldsymbol{n}(-\theta_s D_L \nabla_c + \boldsymbol{v}c) = \boldsymbol{v}{*}c_{in}$ and $-\boldsymbol{n}(-\theta_s D_L \nabla_c + \boldsymbol{v}c) = \boldsymbol{v}{*}c$, respectively. The other boundaries are walls with the homogeneous *Dirichlet* condition: $\boldsymbol{n}(-\theta_s D_L \nabla_c + \boldsymbol{v}c) = 0$ (Fig. 2). $\boldsymbol{v}$ is the magnitude of velocity vector.

The solution of the model was obtained using the COMSOL Multiphysics software, version 3.5a. This software uses finite element methods to solve the relevant partial differential equations. The final mesh is composed of 161,764 elements (Fig. 1).

The model was validated through direct comparison between the numerical and experimental breakthrough curves. The correlation obtained between the experimental and numerical results gave $R^2 = 0.9825$ for selenium (IV) and $R^2 = 0.9441$ for arsenic (V). The experimental results for arsenic and selenium were taken from Marín (2010) and Rodríguez (2011), respectively. Similar correlations were obtained by Aksu and Gönen (2004) and Hasan et al. (2010) using different models.

3 Image Formation and Object Illumination

Using the validated mathematical model, we simulated the columns with the dimensions shown in Table 2 for analyzing the change in the removal percent due to a

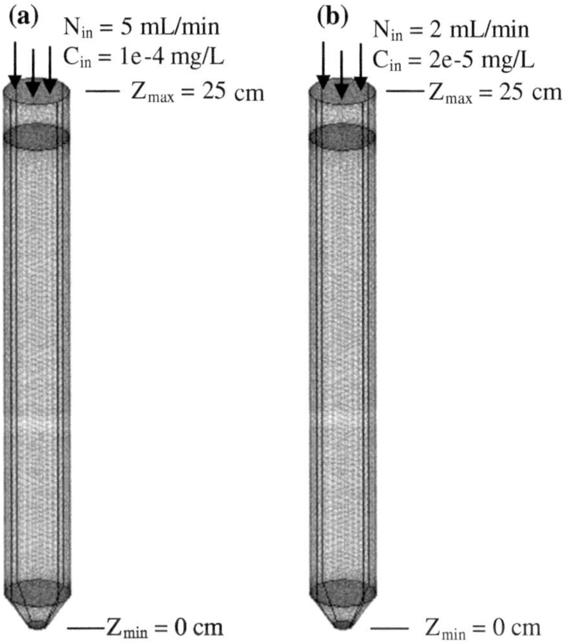

Fig. 2 **a** Feeding conditions for As(V). **b** Feeding conditions for Se(IV). Here N_{in} and C_{in} are the inlet flux and the inlet concentration, respectively

Table 2 Dimensions of columns

Column	Nominal diameter (in)	Inner diameter (cm)	Length (cm)
A	8	20.2	72
B	10	24.1	72
C	12	28.7	53
D	14	31.5	72
E	16	36.1	53

change in the flux of solution and dimensions of the column. The dimensions of the columns are similar to those used in standard commercial columns.

The columns in Table 2 were evaluated for arsenic (V) absorption in *Eicchornia crassipes* and for selenium (IV) absorption in *Lemna minor*. Each column was simulated using the parameters listed in Table 1 and a flux of 25, 50, 250 and 500 l/day.

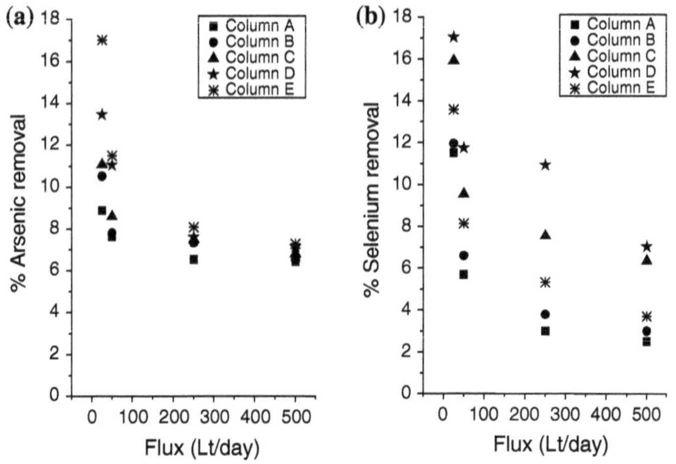

Fig. 3 **a** Arsenic (V) and **b** Selenium (IV) removal for different fluxes as listed in the columns of Table 2

4 Results

The higher percentage of arsenic (V) and selenium (IV) retention was obtained with the lower flux in all cases.

For arsenic (V), the retention increases when the dimensions of the column increase in both diameter and length. Column E in Table 2 had the higher removal percent of arsenic (V), 17 %. In this case, the minimum value of arsenic (V) removal was 6 % at 500 l/day. However, Fig. 3 shows that this is a convergence value for all columns with the same flux.

In the case of selenium (IV), column D in Table 2 had the higher percent removal, 17 %. In this case, however, the minimum value of percent removal changes in each column with a flux of 500 l/day (see Fig. 3).

The minimum percent removal of selenium (IV) was in column A, 2.5 %.

The arsenic (V) and selenium (IV) adsorption was simulated assuming different isotherms. The results indicate that the absorption process that can be represented with a linear isotherm is by far more affected by the dynamical conditions of the process than with a Freundlich isotherm.

5 Conclusions

The absorption of selenium (IV) and arsenic (V) in columns packed with biomasses of *Lemna minor* and *Eicchornia crassipes*, respectively, is affected significantly by the flux in the inlet of the columns.

The retention of arsenic (V) increases with increasing the length diameter of the columns. However, the retention of selenium (IV) is seen to increase when the length increases to a diameter of 14 in and decreases for a diameter of 16 in. This means that there exists an optimal dimension for retention of selenium (IV) in columns with diameters between 14 and 16 in.

Acknowledgments This work was partially supported by ABACUS and the Consejo Nacional de Ciencia y Tecnología (CONACyT) of Mexico through the project EDOMEX-2011-C01-165873.

References

Aksu Z, Gönen F (2004) Biosorption of phenol by immobilized activated sludge in a continuous packed bed: prediction of breakthrough curves. Proccess Biochem 39:599–613
Bear J (1972) Dynamics of fluids in porous media. Elsevier Scientific Publishing, Amsterdam
Bird RB, Stewart WE, Lightfoot EN (2003) Fenómenos de transporte. Barcelona, 4–29
Giles DE, Mohapatra M, Issa TB, Anand S, Singh P (2011) Iron and aluminium based adsorption strategies for removing arsenic from water. J Environ Manage 92:3011–3022
Hurtado JR, Gardea TJ (2007) Evaluación de la exposición a selenio en los altos de Jalisco, México Salud Pública de México 49(4):
Hasan SH, Ranjan D, Talat M (2010) Agro-industrial waste 'wheat bran' for the biosorptive remediation of selenium through continuous up-flow fixed-bed column. J Hazard Mater 181:1134–1142
Lemly AD (2004) Aquatic selenium pollution is a global environmental safety issue. Ecotoxicol Environ Saf 59:44–56
Marín AMJ (2010) Remoción de arsénico en solución empleando biomasas no vivas de malezas acuáticas. Universidad Autónoma del Estado de México, México, Tesina de Maestría
Rodríguez MCE (2011) Optimización de condiciones dinámicas para sorción y desorción de selenio en solución empleando biomasas no vivas de malezas acuáticas. Tesina de licenciatura, ININ-ITT, Toluca, Estado de México, México
Sisca OL, Novie F, Felycia ES, Jaka S, Suryadi I (2009) Studies of potential applications of biomass for the separation of heavy metals from water of wastewater. Biochem Eng J 44:19–41
Walter J, Wener PM Jr (1990) Sorption phenomena in subsurface systems: concepts, models and effects on contaminant fate and transport. Water Res 25:499–528
WHO (2003) Selenium in drinking-water. Background document for development of WHO Guidlines for Drinking-water Quality. World Health Organization Geneva (WHO/SDE/WSH/03.04/13), 14 pp

Part II
Drops, Particles and Waves

The Geometry of Drop-Formed Vortex Rings

Franklin Peña-Polo, Armando Blanco and Leonardo Di G. Sigalotti

Abstract Vortex rings are among the most important objects of fluid mechanics for their numerous technological applications. They are common in inviscid and low viscosity fluids and represent the simplest examples of organized structure formation. The coalescence of a drop with a liquid surface is one process that can result in a vortex ring. Although the phenomenon was observed more than a century ago, the appearance of vorticity and its organization into a toroidal vortex have not yet been completely understood. Here we use fast digital video imaging to study the geometry of drop-formed vortices and the dependence of the early drop inflow on the underlying dynamics above the liquid surface.

1 Introduction

Since its first investigation more than a century ago, it is well known that vortex rings, similar to ordinary smoke rings in air, can also be formed when a drop coalesces with the flat surface of another liquid (Thomson and Newall 1885). However, it was not until the last twenty years that the problem has received considerable attention. A key practical difficulty in resolving and predicting coalescence exists due to the small length and time scales associated with the transition in topology when the drop surface ruptures upon contact with the liquid surface.

F. Peña-Polo (✉) · L. Di G. Sigalotti
Centro de Física, Instituto Venezolano de Investigaciones Científicas, IVIC,
Apartado Postal 20632, Caracas 1020-A, Venezuela
e-mail: franklin.pena@gmail.com

L. Di G. Sigalotti
e-mail: leonardo.sigalotti@gmail.com

A. Blanco
Departamento de Mecánica, Universidad Simón Bolívar, USB, Caracas 1080, Venezuela
e-mail: ajblanco@usb.ve

L. Di G. Sigalotti et al. (eds.), *Computational and Experimental Fluid Mechanics with Applications to Physics, Engineering and the Environment*, Environmental Science and Engineering, DOI:10.1007/978-3-319-00191-3_17, © Springer International Publishing Switzerland 2014

A number of investigations of drop coalescence through planar liquid surfaces has focused on describing the underlying dynamics and morphology of merging just above the interface (Charles and Mason 1960; Thoroddsen and Takehara 2000; Aryafar and Kavehpour 2006; Blanchette and Bigioni 2006; Yue et al. 2006; Blanchette and Bigioni 2009; Ray et al. 2010). It was demonstrated on experimental grounds that coalescence of a free drop with a flat interface may either result in total merging of the drop or, alternatively, in partial coalescence as it eventually pinches off, defying surface tension and leaving behind a daughter drop (Charles and Mason 1960; Thoroddsen and Takehara 2000; Aryafar and Kavehpour 2006). Partial coalescence may occur several times in succession to form what has been termed a coalescence cascade (Thoroddsen and Takehara 2000). In particular, when the drop touches the interface, its surface ruptures and part of its liquid is drained into the pool through a rapidly expanding orifice due to the excess capillary pressure inside it, $\Delta p = 2\sigma/R$, where σ is the strength of the surface tension with the ambient fluid and R is the radius of the spherical drop. Depending on the drop size, the orifice widens on a time scale of 10–200 μs (Blanchette and Bigioni 2006). When the drop surface ruptures, the unbalanced surface tension forces initiate capillary waves that rise up the surface of the drop and deform it as they converge on its summit. As a consequence, the draining drop first deforms into a bell-shaped figure and then into a stretched cylindrical column due to the inward pull of surface tension. If enough momentum is carried by the capillary waves to delay the inward pull of surface tension at the drop's summit, the orifice collapses upon itself and the remnant drop liquid above the surface pinches off. In general, this occurs for drops with $Oh < (\sigma/\rho R)^3$, where

$$Oh = \nu \left(\frac{\rho}{\sigma R}\right)^{1/2}, \tag{1}$$

is the dimensionless Ohnesorge number relating the viscous forces to the inertial and surface tension forces, ν is the kinematic viscosity of the drop liquid, and ρ is its density. A critical value of $Oh \sim 0.026$, below which pinch-off is expected to occur, was derived for ethanol/water and ethanol/glycerin solutions (Blanchette and Bigioni 2006). We shall see that the same criterion also applies to other liquids as, for instance, pure water and water/glycerin mixtures.

The instability that gives rise to organization of the ensuing flow into a vortex ring beneath the interface follows a sequence of events that strongly depends on the dynamics of the draining drop. It was long pictured that the drop discharges its liquid through a thin microjet, which quickly rolls up giving rise to the vorticity (Glezer 1988; Anilkumar et al. 1991; Baumann et al. 1992; Peck and Sigurdson 1994; Shankar and Kumar 1995; Residori et al. 2000). In this view, the returning circulatory motion inside the nascent vortex due to viscous friction, even at low Reynolds numbers, impedes the expansion of the orifice, thereby maintaining the microjet. While this geometry is tantamount of that employed in laboratory experiments of piston-generated vortex rings (Pullin 1979; Lim 1997; Cater et al. 2004), it does not correctly describe the dynamics of early drop inflow.

Investigations of drop inflow to explain the origin of vorticity have also been carried out by a number of workers. Experimental observations by Cresswell and Morton (1995) indicate that vorticity is generated at the free surface due to the formation and separation of a boundary layer, presumably at the junction between the drop and the receiving liquid, and the requirement that the viscous stress be zero at the surface. A separate analysis by Rood (1994) and further experiments by Dooley et al. (1997) interpreted the vortex roll-up at the junction of the drop-pool interface as due to a surface-parallel acceleration of the fluid elements near the free surface, which causes the flux of vorticity into the bulk fluid. While almost all previous studies on drop inflow have concentrated on either describing the morphology of drop penetration or explaining the sources of vorticity, there is a complete lack of experiments describing the dependence of drop inflow on the dynamics above the interface. With this in mind we have conducted a sequence of experimental observations of partial drop coalescence for pure water and water/glycerin solutions.

2 Experimental Setup and Procedure

A Plexiglas container, with a base size of $3 \times 3\,\text{cm}^2$ and 10 cm high, was filled with a solvent of pure distilled water and mounted on a rigid metallic support. After pouring, the solvent was kept overnight to ensure complete quiescence. The drops, once formed at the tip of an electronic pipette, were deposited on the liquid surface by means of a micrometric translation stage with the aid of a Computer Numerical Controlled (CNC) apparatus (Peña-Polo et al. 2010).

The translation of the drops was obtained using consecutive discrete steps of length $\approx$5.6 μm and duration of 200 ms in order to minimize induced drop vibrations and produce sufficiently accurate contact with the liquid surface. The drops were made up of distilled water with water:glycerin volume concentrations of 100:0, 90:10, 80:20, 70:30, and 60:40 and seeded with a small amount of sodium fluorescein (10^{-8} moles/l) for visualization purposes. Drops with higher volume concentrations of glycerin were also tried in a separate set of experiments. Different drop sizes were employed, corresponding to volumes between 1 and 10 μl. Depending on the drop size, plastic capillary tips of varied diameters from 0.3 to 1.2 mm were used to produce pendant drops close to spherical. The size of the container was chosen large enough to avoid undesirable effects of the wall boundaries on the drop behaviour. All the experiments were carried out under isothermal conditions at a room temperature of 25 °C. Lateral photographs of the drop inflow were obtained using a PixeLINK PL-B774-PL-BL colour CCD camera, which allows a maximum resolution of 1280×1024 pixels and maximum recording rate of 4,400 frames per second (fps). In order to allow for a sufficiently long depth of field, a Schneider Kreuznach 1.4/8 lens is used with the CCD camera at a distance of 10 mm from a lateral wall of the container. Acceptable sharp images of the drops during coalescence are obtained by fine tuning the lens focus and adjusting the lens aperture along with the intensity of the incident light. Illumination of the container was provided with blue light-emitting diodes ($\lambda = 480$ nm) of very high brightness.

An image processing software was developed to digitize the images and convert the edge of the projected drop into a large number of closely space data points for analysis. Once the images are acquired, the data are transferred to a host computer where a specialized software is used to record the drop vibrations in real time during micrometric translation, perform accurate edge detection of the drop image, evaluate its deformation at contact with the interface and during penetration into the bulk fluid, and calculate other characteristics, including its centre of mass, surface area, volume, and infall velocity.

3 Experimental Results

Here we deal with experimental observations of the partial coalescence of pendant drops for which the interfacial tension force at the top of the capillary wall exceeds the buoyancy force. This way, the drop is initially steadily pinned at the capillary tip and macroscopic motion is minimal. This initial setup is different from most experiments documented in the literature, where coalescence is driven by gravity, i.e., the drop is freely suspended on a thin film of surrounding air above the liquid surface. The initial drop shape and degree of interface deformation depend on the Bond number

$$Bo = \frac{g \Delta\rho R^2}{\sigma}, \tag{2}$$

where g is the Earth's gravitational acceleration and $\Delta\rho$ is the density difference between the disperse (drop) and continuous phase (air). As the Bond number increases above one, the drop shape evolves from spherical to oblate. The drops employed in our experiments all have Bond numbers in the range $0.02 \le Bo \le 0.7$ and Ohnesorge numbers $Oh < 0.026$ so that pinch-off is always observed.

Figure 1 depicts the experimental drops in the Oh–Bo plane. The data markers define five different sequences of water drops with varying glycerin concentrations from 0 to 40 %. The solid line with crosses separates the region of total coalescence for highly viscous drops (with > 50 % glycerin) from that of partial coalescence for drops of lower viscosity. The two sequences of drops in region I produced a transient pair of vortex rings with subsequent pairing off through a single passage, while pairing off in the two sequences of region II occurred through a merging type interaction. The drops in region III all coalesced to form a single vortex ring downstream. Drops located above the solid line (not depicted here) fully merged with the lower liquid and experienced rapid vortex dissipation. We note that for pendant drops the behaviour is distinct from the "coalescence cascade" driven by inertial recoil (Charles and Mason 1960; Thoroddsen and Takehara 2000). For pendant drops no bouncing droplets are left behind because the portion of liquid that pinches off rapidly retracts into a smaller drop that remains steadily pinned at the capillary tip.

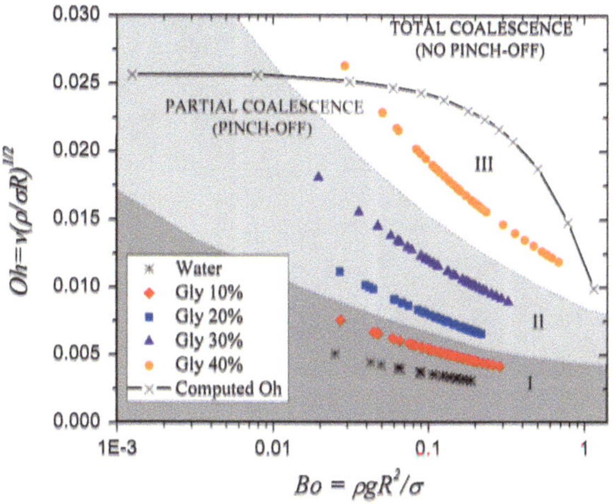

Fig. 1 Experimental drops in the Oh–Bo plane. The five sets of data markers from *bottom to top* represent water drops with increasing glycerin concentration. Each set contains drops with varied volumes from 1 to 10 μl. The *solid line with crosses* separates the region of total drop absorption from that of partial coalescence. The sequences in the *gray*(**I**) and *light*–*gray*(**II**) coloured regions produced a transient pair of vortex rings with subsequent pairing off through a single passage and through direct merging, respectively, while in region (**III**) all drops formed a single vortex ring downstream.

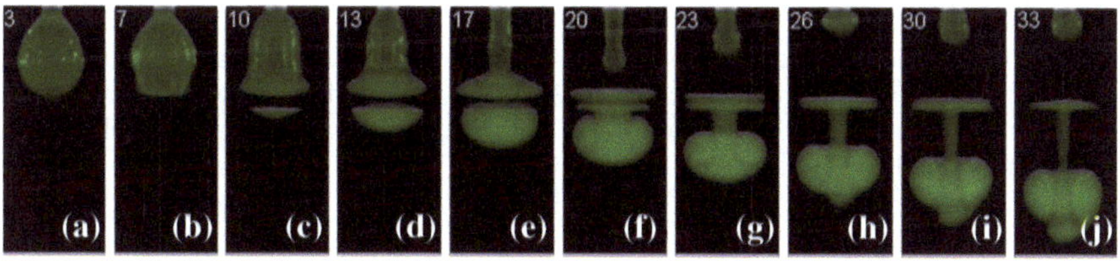

Fig. 2 Side-view images showing the partial coalescence of a pendant drop of distilled water and volume 10 μl ($Oh = 0.003$, $Bo = 0.187$, lower sequence of region I) with a reservoir of the same liquid. The numbers on the *left top* of each photograph give the time in milliseconds. The horizontal black fringe in photographs c-g, separating the *upper drop* image from the *lower one*, is an optical effect due to the refraction index of water and the inclination of the CCD camera with respect to a plane perpendicular to the *front wall* of the Plexiglas container.

Figure 2 shows time lapse images of the process of partial coalescence for a pure water drop on both sides of the interface. After contact with the surface, an orifice forms which rapidly expands (Figs. 2a and b). The drop drains liquid into the pool through the orifice as it is deformed by the running capillary waves into a bell-shaped body (Fig. 2c). By this time, the orifice reaches its maximum opening (about 60–70 % of the initial drop diameter). Owing to viscous friction, circulatory motion has already penetrated the bulk fluid in the form of a single-branched vortex sheet spiral (Figs. 2d and e). Meantime the draining drop stretches into a cylindrical column, while a significant portion of its liquid has already submerged. The spiraling,

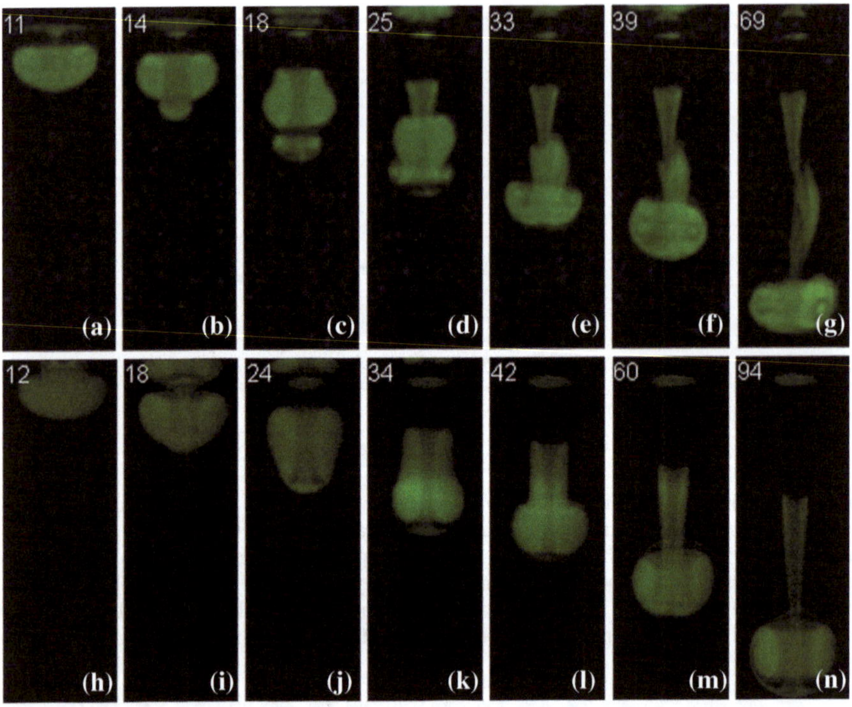

Fig. 3 Side-view images of vortex penetration for a water drop of 10 μl *(lower sequence of region I; top row)* and a drop of the same volume with 30 % glycerin composition *(upper sequence of region II; bottom row)* in a reservoir of distilled water. The numbers on the *left top* of each photograph give the time in milliseconds.

which carries with it an increasing volume of dyed fluid, rapidly organizes itself into a circular tube of vorticity. The base of the cylindrical column undergoes further stretching due to the inward pull of surface tension and pinches off (Fig. 2f). During this process, the drop discharges liquid into the pool through a thick laminar jet, which then contracts and decays into a thin thread as the vortex ring descends (Figs. 2g–j). The leading edge of the jet is unstable and rolls up into the nascent vortex, while only a small amount of mass at its centre flows down undisturbed, crossing the plane of the ring and bulging the lower cap of the descending drop.

The side-view images of Fig. 3 display the vortex evolution farther in time for a drop of pure water (top row) and one composed of 30 % glycerin (bottom row). For pure water, the lower bulge (Fig. 3b) detaches from the bulk of the drop and organizes itself into a new vortex ring (Fig. 3c). Meantime, the primary vortex weakens due to both entrainment of external liquid back into the moving region of vorticity and convection of diffused vorticity towards its upper part, where it is deposited into a trailing wake (Fig. 3d) (Maxworthy 1972). The two vortices interact coaxially, producing an unstable manifold for leapfrogging motion (Figs. 3d–e). Due to mutual induction, the rear ring descends faster and shrinks at the same time the front one widens and slows down. This typifies a kind of slip-through interaction with only one passage as it has been observed in piston-generated vortices at low Reynolds numbers ($<10^3$) (Yamada and Matsui 1978, 1979) and in numerical solutions of the Navier-Stokes equations (Riley and Stevens 1993). The fatter vortex suffers severe core distortion during its own passage due to the straining field of the leading one, whereas the trailing wake shrinks and elongates upstream (Figs. 3f and g). Drops with 10 % glycerin composition experienced a similar evolution independently of their size.

For heavier drops (30 % glycerin composition), the lower bulge never detaches from its parent drop due to increased viscous dissipation (Fig. 3i). However, a second vortex ring of much smaller core radius suddenly emerges at the circular top of the bulge (Fig. 3j). The new vortex strengthens and expands in radius, while the rear one stretches and merges with it (Fig. 3k). Only the lower part of the weakened vortical liquid is sucked by the pursuing vortex, while the upper part diffuses into a conical trailing wake whose axis of symmetry approximately coincides with that of the final vortex (Figs. 3l–n). When the glycerin concentration is raised to 40 %, the bulge no longer forms and no secondary vortex ring appears. In this case, the primary vortex ring evolves undisturbed into the pool. For even higher concentrations of glycerin there is no conclusive evidence of sustained vorticity. For highly viscous drops strong dissipation may lead to inhibition of the vorticity field.

4 Concluding Remarks

We have performed experimental observations of the process of partial coalescence of pendant drops with a quiescent liquid surface. After rupture of the drop interface upon contact with the surface, the drop discharges most of its liquid through an expanding orifice. The erupting liquid rolls up due to viscous friction with the host liquid. Just prior to pinch-off above the surface, the drop drains its last portion of liquid through a thick laminar jet, which feeds the growing vortex. At pinch-off, the orifice closes and the jet contracts, eventually diffusing away as the vortex continues to move downwards. Recent experiments of electrically driven partial coalescence of freely suspended charged drops have revealed a similar sequence of dynamical events (Hamlin et al. 2012).

Our results suggest that the details of the dynamics of vortex penetration depends on the relative density of the two liquids. For low density contrasts, a transient pair of vortices is produced followed by pairing-off through a slip passage. When the density contrast is increased, the two vortices merge with each other. For much higher density differences, the dynamics simplifies as a single vortex is produced. Highly viscous drops undergoing complete coalescence result in much weaker vortices, implying a close relationship between the amount of stretching, or equivalently, the strength of capillary wave damping and the strength of vorticity.

References

Anilkumar AV, Lee CP, Wang TG (1991) Surface-tension-induced mixing following coalescence of initially stationary drops. Phys Fluids A 3:2587–2591

Aryafar H, Kavehpour HP (2006) Drop coalescence through planar surfaces. Phys Fluids 18:072105

Baumann N, Mohr DDJ, Mohr P, Renardy Y (1992) Vortex rings of one fluid in another in free fall. Phys Fluids A 4:567–580

Blanchette F, Bigioni TP (2006) Partial coalescence of drops at liquid interfaces. Nat Phys 2:254–257

Blanchette F, Bigioni TP (2009) Dynamics of drop coalescence at fluid interfaces. J Fluid Mech 620:333–352

Cater JE, Soria J, Lim TT (2004) The interaction of the piston vortex with a piston-generated vortex ring. J Fluid Mech 499:327–343

Charles GE, Mason SG (1960) The coalescence of liquid drops with flat liquid/liquid interfaces. J Colloid Sci 15:236–267

Cresswell RW, Morton BR (1995) Drop-formed vortex rings - The generation of vorticity. Phys Fluids 7:1363–1370

Dooley BS, Warncke AE, Gharib M, Tryggvason G (1997) Vortex ring generation due to the coalescence of a water drop at a free surface. Exp Fluids 22:369–374

Glezer A (1988) The formation of vortex rings. Phys Fluids 31:3532–3542

Hamlin BS, Creasey JC, Ristenpart WD (2012) Electrically tunable partial coalescence of oppositely charged drops. Phys Rev Lett 109:094501

Lim TT (1997) A note on the leapfrogging between two coaxial vortex rings at low Reynolds numbers. Phys Fluids 9:239–241

Maxworthy T (1972) The structure and stability of vortex rings. J Fluid Mech 51:15–32

Peck B, Sigurdson L (1994) The three-dimensional vortex structure of an impacting water drop. Phys Fluids 6:564–576

Peña-Polo F, Trujillo L, Sigalotti L Di G (2010) A computer-controlled apparatus for micrometric drop deposition at liquid surfaces. Rev Sci Instrum 81:055107

Pullin DI (1979) Vortex ring formation at tube and orifice openings. Phys Fluids 22:401–403

Ray B, Biswas G, Sharma A (2010) Generation of secondary droplets in coalescence of a drop at a liquid-liquid interface. J Fluid Mech 655:72–104

Residori S, Pampaloni E, Buah-Bassuah PK, Arecchi FT (2000) Surface tension effects in the zero gravity inflow of a drop into a fluid. Eur Phys J B 15:331–334

Riley N, Stevens DP (1993) A note on leapfrogging vortex rings. Fluid Dyn Res 11:235–244

Rood EP (1994) Interpreting vortex interactions with a free surface. J Fluids Eng 116:91–94

Shankar PN, Kumar M (1995) Vortex rings generated by drops just coalescing with a pool. Phys Fluids 7:737–746

Thomson JJ, Newall HF (1885) On the formation of vortex rings by drops falling into liquids, and some allied phenomena. Proc R Soc Lond A 39:417–436

Thoroddsen ST, Takehara K (2000) The coalescence cascade of a drop. Phys Fluids 12:1265–1267

Yamada H, Matsui T (1978) Preliminary study of mutual slip-through of a pair of vortices. Phys Fluids 21:292–294

Yamada H, Matsui T (1979) Mutual slip-through of a pair of vortex rings. Phys Fluids 22:1245–1249

Yue P, Zhou C, Feng JJ (2006) A computational study of the coalescence between a drop and an interface in Newtonian and viscoelastic fluids. Phys Fluids 18:102102

Hydrodynamics of Multiple Coalescence Collisions of Liquid Drops: From the Modelling of the Coalescence Phenomenon to Flocculation of Drops in 3D Using the SPH Formalism

Alejandro Acevedo-Malavé

Abstract In this chapter, we present three-dimensional numerical calculations of the collision and coalescence of multiple water drops of equal size in a vacuum environment, using a Lagrangian mesh-free scheme based on the Smoothed Particle Hydrodynamics (SPH) formalism. The water drops are modelled using a general Mie-Grüneisen equation of state. Attention is focused for collision velocities from low to moderate so that shattering separation is excluded. Depending on the collision velocity three different possible outcomes are predicted by the calculations: permanent coalescence, coalescence accompanied by fragmentation into a few satellite droplets, and flocculation of the drops with no coalescence.

1 Introduction

Many studies have been proposed for the numerical simulation of the coalescence and break up of liquid drops (Foote 1974; Nobari et al. 1996; Eggers et al. 1999; Cristini et al. 2001; Mashayek et al. 2003; Narsimhan 2004; Roisman 2004; Pan and Suga 2005; Meleán and Sigalotti 2005; Jia et al. 2006; Decent et al. 2006; Azizi and Al Taweel 2010; Acevedo-Malavé and García-Sucre 2011a,b,c, 2012). In these studies, different numerical methods have been proposed to simulate the dynamics of liquid drops by numerical integration of the Navier-Stokes equations. While most of them have focused on describing the dynamics of drop coalescence, only a few has studied the details of the liquid bridge that arises when two drops collide. The effects of the Reynolds number, the impact velocity, the drop size ratio, and the internal circulation on coalescence have been investigated and different regimes for droplets' collisions have been simulated. For binary collisions, these calculations have predicted four

A. Acevedo-Malavé (✉)
Centro Multidisciplinario de Ciencias, Instituto Venezolano de Investigaciones Científicas, IVIC, Mérida 5101, Estado Mérida, Venezuela
e-mail: alaceved@ivic.gob.ve

L. Di G. Sigalotti et al. (eds.), *Computational and Experimental Fluid Mechanics with Applications to Physics, Engineering and the Environment*, Environmental Science and Engineering, DOI:10.1007/978-3-319-00191-3_18,

different regimes, namely bouncing, coalescence, reflexive separation, and stretching separation. These numerical simulations indicate that collisions leading to rebound between colliding drops can be described macroscopically.

The mechanism of formation of satellite drops at moderate collision velocities was also studied, confirming that the principal cause of the formation of satellite drops is the "end pinching", while the capillary wave instabilities are the dominant features in cases where a large value of the parameter impact is employed. Whereas almost all of these previous calculations were confined either to 2D or axi-symmetry, Acevedo-Malavé and García-Sucre applied the Smoothed Particle Hydrodynamics (SPH) method to model for the first time the hydrodynamic coalescence collision of two liquid drops in 3D. As a result of the collision process, the formation of a circular flat film is first observed. In these references three possible outcomes for the collision between the drops were found, which correspond to: permanent coalescence, fragmentation, and flocculation of drops. If the collision velocity is greater than the range of values for permanent coalescence, a fragmentation phenomenon is observed, and if the collision velocity is too low the surface tension forces are dominant and the drops interact only through their deformed surfaces. A detailed description of the SPH equations and the approach used in this chapter can be found in these previous references. In this work, the same SPH method is applied to simulate the 3D hydrodynamic collision of multiple liquid drops and the formation of drop clusters in a vacuum environment.

2 The Smoothed Particle Hydrodynamics Formalism

Smoothed Particle Hydrodynamics is a mesh-free, Lagrangian method for solving the equations of fluid dynamics (Monaghan 1985). In the SPH model, the fluid is represented by a discrete set of N particles. The position of particle i is denoted by the position vector $\mathbf{r}_i$, with $i = 1, \ldots, N$. In essence, the scheme is based on the idea that the smoothed representation $A_s(r)$ of a continuous function $A(r)$ can be represented by the convolution integral over the product of a smoothing function (or interpolating kernel) and the function itself.

The smoothing function W must satisfy the normalization condition

$$\int W(\mathbf{r} - \mathbf{r}', h) d\mathbf{r}' = 1, \tag{1}$$

where the integration is performed over all space and h is the smoothing length that determines the spatial resolution. In the limit when h tends to zero, the smoothing function W becomes a Dirac delta function and so the smoothed representation of $A_s(r)$, as stated above, tends to the exact function $A(r)$. In SPH the integrals over all space are replaced to second-order accuracy by summations over all neighbouring particles.

In this way, in SPH the mass density ρ_i at the location of particle i is given by the summation

$$\rho_i = \sum_j m_j W(\mathbf{r}_i - \mathbf{r}_j, h), \tag{2}$$

where m_j is the mass of the jth particle that is a neighbour of particle i.

The position $\mathbf{r}_i$ and velocity $\mathbf{v}_i$ of particle i are given by the well-known SPH discretization equations

$$\begin{aligned} \frac{d\mathbf{r}_i}{dt} &= \mathbf{v}_i, \\ \frac{d\mathbf{v}_i^{\alpha}}{dt} &= \sum_{j=1}^{N} m_j \left(\frac{\sigma_i^{\alpha\beta}}{\rho_i^2} + \frac{\sigma_j^{\alpha\beta}}{\rho_j^2} \right) \cdot \nabla W_{ij}^h, \end{aligned} \tag{3}$$

where σ is the total stress tensor.

A cubic B-spline kernel is used as the smoothing function (Monaghan 1985). We consider water drops and use a general Mie-Grüneisen equation of state with different analytic forms for the states of compression $(\rho/\rho_0 - 1) > 0$ and tension $(\rho/\rho_0 - 1) < 0$ (Acevedo-Malavé and García-Sucre 2011a).

3 Coalescence Collision of Multiple Drops and the Formation of Drop Clustering

Here we describe the results of a set of collision models in which five liquid drops are made to coalesce permanently for different choices of the value of the collision velocity. In a first model configuration, four drops with a diameter of 30 µm and 5512 particles each are made to collide with a central one with a collision velocity of 15.0 mm/ms directed towards the centre of mass of the system.

We may see in Fig. 1 that at $t = 3.08 \times 10^{-4}$ ms a flat circular section forms, which increases its diameter as the coalescence dynamics progresses. At $t = 4.77 \times 10^{-4}$ ms the coalescence process begins, and the fluid of the four converging drops starts to penetrate in the drop placed at the centre of the system. At $t = 7.38 \times 10^{-4}$ ms a little wave front arises in the (x = 0, y = 0)-plane, which then disappears at $t = 8.46 \times 10^{-4}$ ms. As a result of coalescence, a bigger drop forms, which tends to a spherical shape as the evolution proceeds ($t = 1.10 \times 10^{-3}$ ms). Figure 2 shows the velocity vector field inside the drops as well as in their regions of contact at $t = 3.08 \times 10^{-4}$ ms. It is important to note that inside the drops, the fluid tends to a velocity value close to the initial velocity of 15.00 mm/ms, while in the areas of contact of the drops the fluid velocity increases to about 17.00 mm/ms. This difference in the velocity is due to the non-uniform pressure field that sets in inside the drops when the coalescence process has begun.

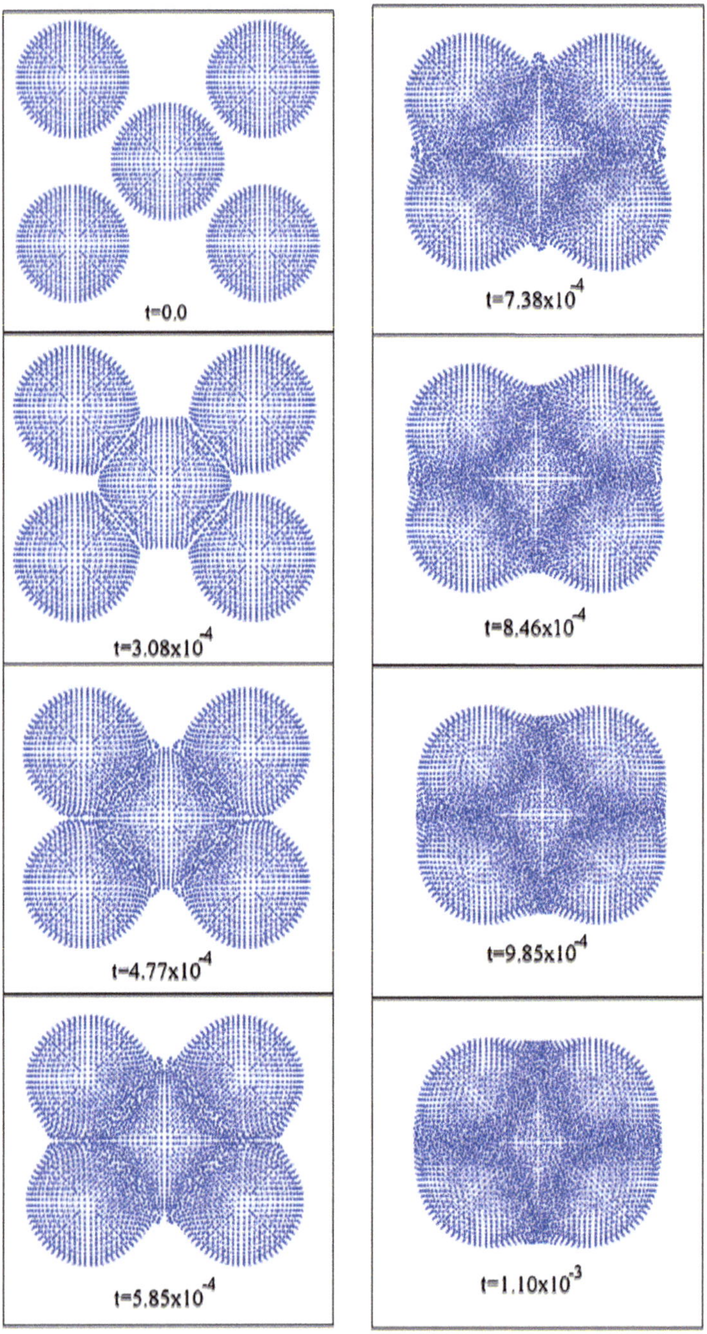

Fig. 1 Sequence of snapshots showing the coalescence collision of five drops (permanent coalescence) for $V_{col} = 15.0$ mm/ms. Time is given in milliseconds. The origin of the coordinate system coincides with the centre of each panel

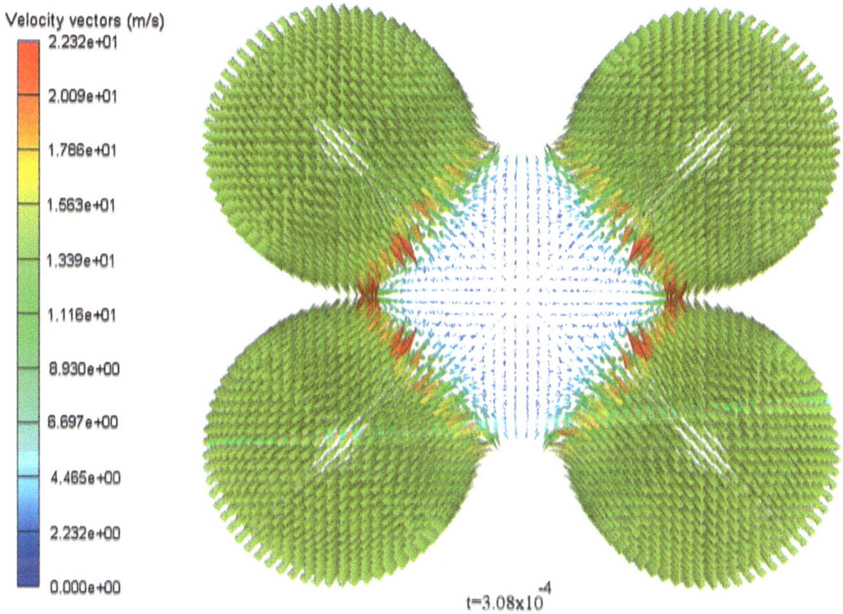

Fig. 2 Velocity vector field during the collision of five drops at $t = 3.08 \times 10^{-4}$ ms (permanent coalescence) for $V_{col} = 15.0$ mm/ms

In Fig. 3, we show the evolution when five drops of diameter 30 μm and 5512 particles each collide with a higher velocity $V_x = 30.0$ mm/ms. As before the configuration is such that four drops are made to collide with a central one. At the beginning of the calculation at $t = 1.51 \times 10^{-4}$ ms, a flat circular section between the drops forms again. This section disappears at $t = 2.42 \times 10^{-4}$ ms when coalescence begins. At $t = 4.20 \times 10^{-4}$ ms four wave fronts arise from the resulting mass of fluid and these wave fronts start to form little satellite droplets. Figure 4 shows the velocity vector field before fragmentation of the drops has taken place. Note that the fluid velocity at the centre of the drops is around the initial collision value of 30.00 mm/ms, while the fluid that is spread to the edges is accelerated reaching speeds of about 49.00 mm/ms.

In Fig. 5, we model the flocculation process for five drops having 30 μm of diameter and 5512 particles each, and a much lower collision velocity $V_x = 0.2$ mm/ms. As in the previous cases, the velocities of the four approaching drops are directed towards the central drop. At $t = 1.13 \times 10^{-3}$ ms, a flat circular section forms among the five droplets. As the simulation continues, the drops do not coalesce but rather form a cluster, remaining in contact only through their deformed (flat) surfaces.

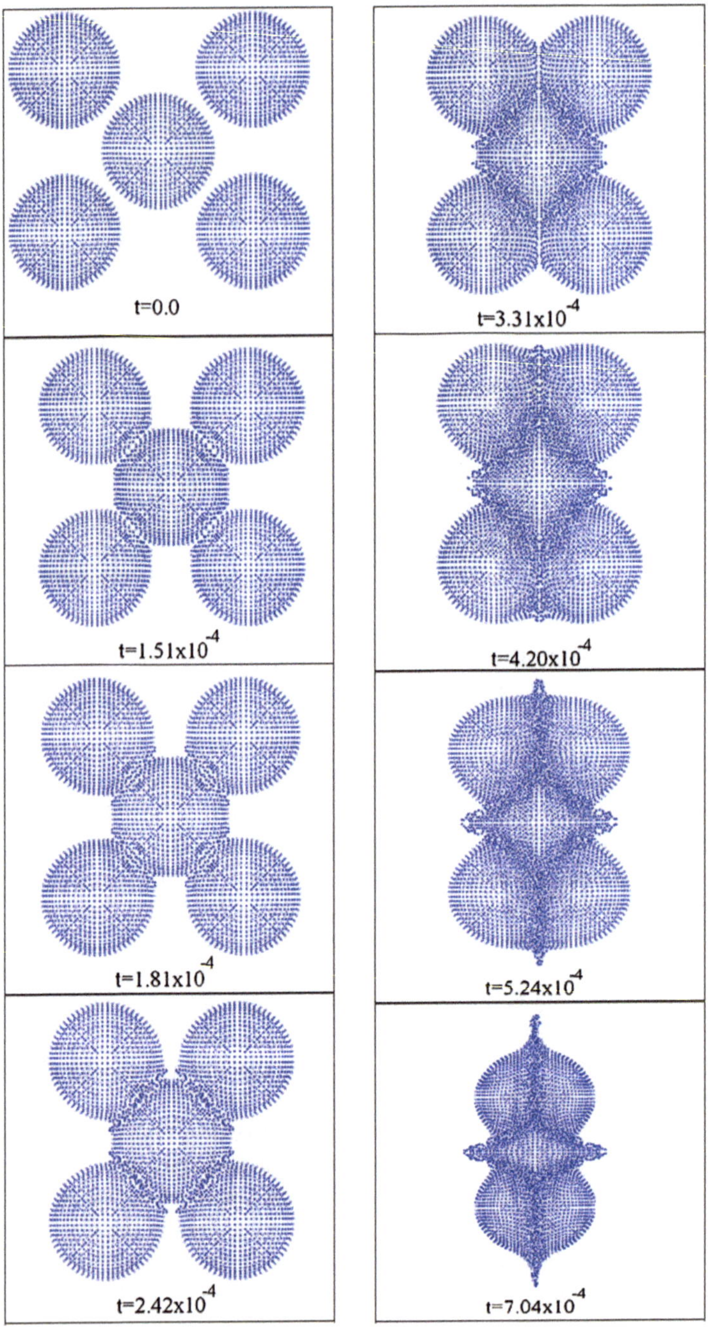

Fig. 3 Sequence of times showing the evolution of the collision of five drops with $V_{col} = 30.0$ mm/ms. Time is given in milliseconds. The origin of the coordinate system was made to coincide with the centre of each panel

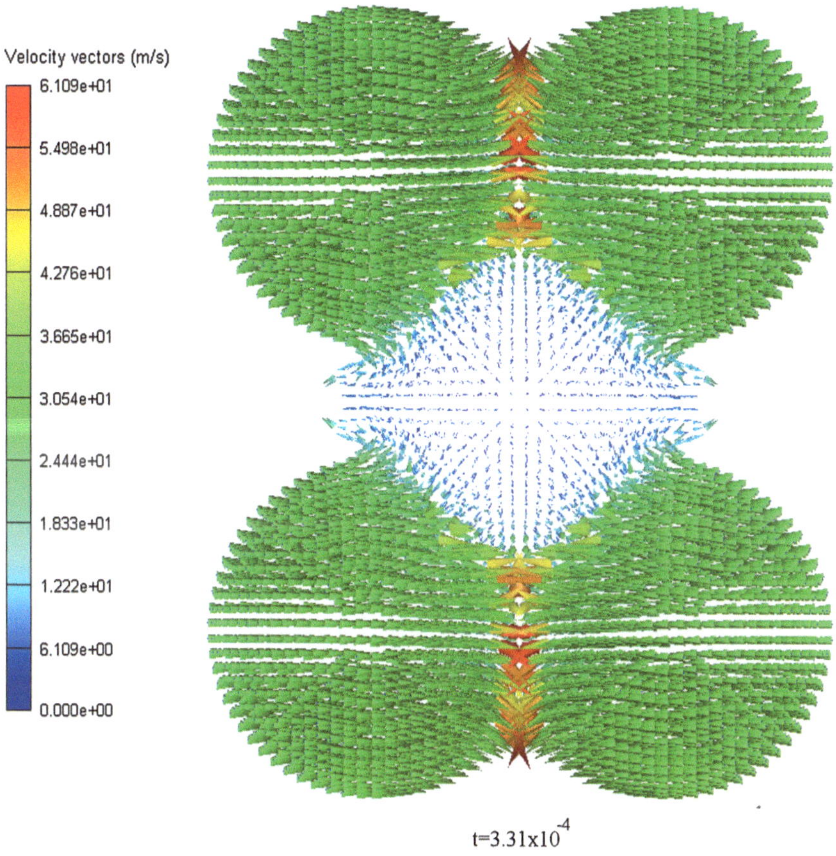

Fig. 4 Velocity vector field for the collision of five drops at $t = 3.31 \times 10^{-4}$ ms for $V_{col} = 30.0$ mm/ms

4 Conclusions

The SPH method was used to simulate the hydrodynamic collision of multiple drops and the formation of clusters of liquid drops in a vacuum environment in three-space dimensions. The characteristic behaviour of the collision process when multiple drops are involved was reproduced. A circular section between the drops appears as a consequence of the surface tension forces acting on the drop surfaces at contact. Depending on the collision velocity, three possible outcomes are predicted after the collision process: coalescence, coalescence accompanied by fragmentation into satellite droplets, and clustering of drops with no coalescence. The non-uniform pressure difference inside the drops tends to accelerate the fluid close to the zones of contact between them. This work represents a step ahead towards the modelling of drop coalescence in emulsion systems. In a future work, we plan to study the

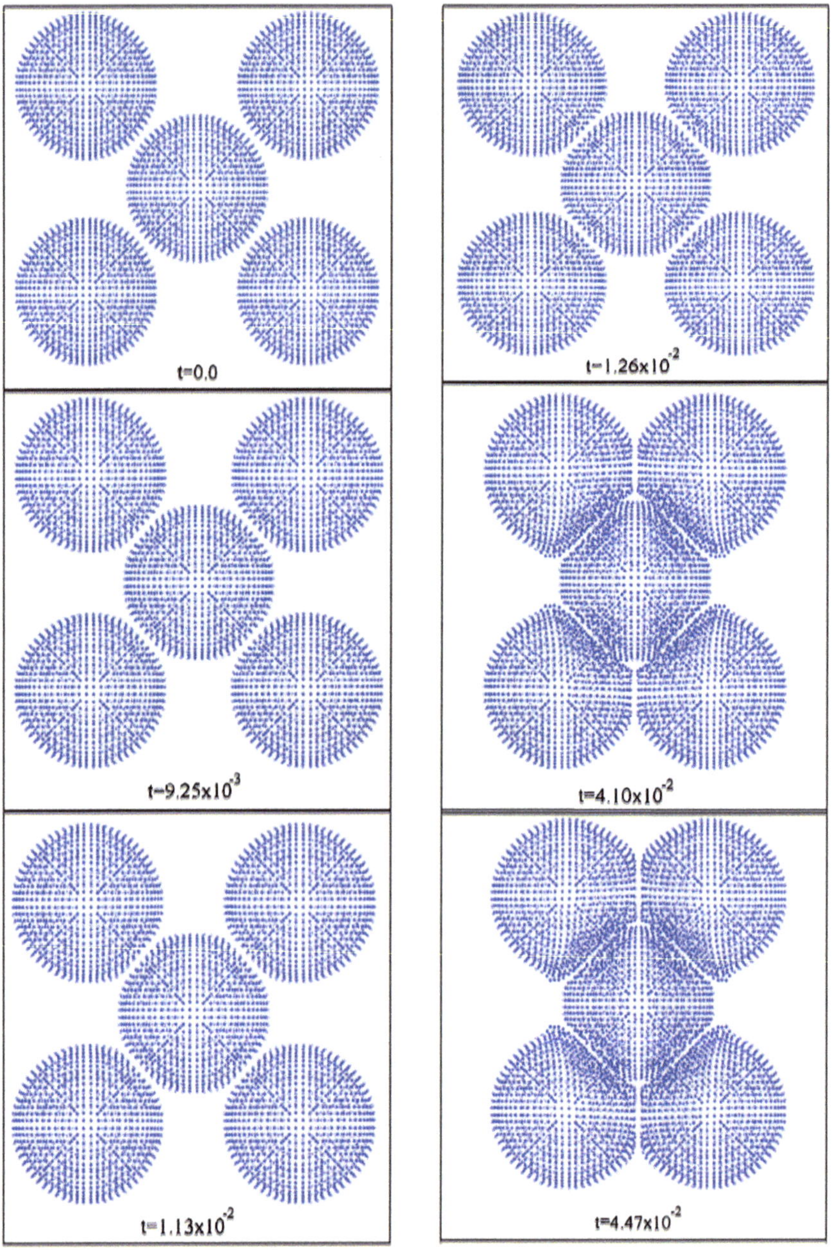

Fig. 5 Sequence of snapshots showing the evolution of the collision of five drops for V_{col} = 0.2 mm/ms. Time is given in milliseconds. The origin of the coordinate system is made to coincide with the centre of each panel

collapse of the interfacial film in emulsion systems during drop coalescence. For this purpose, a hybrid approach that combines SPH with molecular dynamics must be implemented in order to solve the interfacial film just before the collapse of the film occurs (100Å of thickness). In a future work this aspect of the problem will be explored.

References

Acevedo-Malavé A, García-Sucre M (2011a) 3D coalescence collision of liquid drops using smoothed particle hydrodynamics, vol 5. INTECH Publishers, pp 85–106
Acevedo-Malavé A, García-Sucre M (2011b) Coalescence collision of liquid drops. I. Off-center collisions of equal-size drops. AIP Adv 1:1–17
Acevedo-Malavé A, García-Sucre M (2011c) Coalescence collision of liquid drops. II. Off-center collisions of unequal-size drops. AIP Adv 1:1–12
Acevedo-Malavé A, García-Sucre M (2012) Head-on binary collisions of unequal size liquid drops with smoothed particle hydrodynamics. Transworld Res Netw 9:245–266
Azizi F, Al Taweel AM (2010) Algorithm for the accurate numerical solution of PBE for drop breakup and coalescence under high shear rates. Chem Eng Sci 65:6112–6127
Cristini V, Bawzdziewicz J, Loewenberg M (2001) An adaptive mesh algorithm for evolving surfaces: simulations of drop breakup and coalescence. J Comput Phys 168:445–463
Decent SP, Sharpe G, Shaw AJ, Suckling PM (2006) The formation of a liquid bridge during the coalescence of drops. Int J Multi-phase Flow 32:717–738
Eggers J, Lister JR, Stone HA (1999) Coalescence of liquid drops. J Fluid Mech 401:293–310
Foote GB (1974) The water drop rebound problem: dynamics of collision. J Atmos Sci 32:390–401
Jia X, McLaughlin JB, Kontomaris K (2006) Lattice Boltzmann simulations of drop coalescence and chemical mixing. Phys. A 362:62–67
Mashayek F, Ashgriz N, Minkowycz WJ, Shotorban B (2003) Coalescence collision of liquid drops. Int J Heat Mass Transf 46:77–89
Meleán Y, Di Sigalotti LG (2005) Coalescence of colliding van der Waals liquid drops. Int J Heat Mass Transf 48:4041–4061
Monaghan JJ (1985) Extrapolating B splines for interpolation. J Comput Phys 60:253–262
Narsimhan G (2004) Model for drop coalescence in a locally isotropic turbulent flow field. J Coll Interf Sci 272:197–209
Nobari MR, Jan YJ, Tryggvason G (1996) Head-on collision of drops—a numerical investigation. Phys Fluids 8:29–42
Pan Y, Suga K (2005) Numerical simulation of binary liquid droplet collision. Phys Fluids 17:1–14
Roisman L (2004) Dynamics of inertia dominated binary drops collisions. Phys Fluids 16:3438–3449

A Three-Dimensional SPH Approach for Modelling the Collision Process Between Liquid Drops: The Formation of Clusters of Unequal-Sized Drops

Alejandro Acevedo-Malavé

Abstract In this chapter, we present three-dimensional simulations of the coalescence collision and clustering of unequal-sized water drops in vacuum, using the method of Smoothed Particle Hydrodynamics (SPH). The thermodynamics of the problem is represented by a Mie-Grüneisen equation of state. Depending on the magnitude of the collision velocity three different outcomes are observed: permanent coalescence, permanent coalescence accompanied by fragmentation into satellite droplets, and drop clustering with no coalescence (flocculation). When the inertial forces prevail and the surface tension forces are too low permanent coalescence with or without fragmentation into satellite droplets is observed, but for low collision velocities of 0.5 mm/ms, or less, the simulations predict drop flocculation. In this latter case, the drops remain attached to one another, forming a drop clustering.

1 Introduction

The study of the coalescence collision of liquid drops has important natural and industrial applications as, for example, in emulsification, combustion of fuel sprays, spray coating, waste treatment, and raindrop formation among many others. The coalescence of liquid drops has been studied extensively both numerically and experimentally. However, most existing numerical simulations have considered binary drops in two-space dimensions or, in the best case, constrained by axi-symmetry (Foote 1974; Chen 1985; Li 1994; Nobari et al. 1995; Eow and Ghadiri 2003a,b; Mashayek et al. 2003; Meleán and Sigalotti 2005).

A first attempt to simulate head-on collisions of equal-sized drops for small Weber numbers (We $<$ 5) was reported by Foote (1974). The occurrence of bouncing and

A. Acevedo-Malavé (✉)
Centro Multidisciplinario de Ciencias, Instituto Venezolano de Investigaciones Científicas, IVIC, Mérida 5101, Estado Mérida, Venezuela
e-mail: alaceved@ivic.gob.ve

L. Di G. Sigalotti et al. (eds.), *Computational and Experimental Fluid Mechanics with Applications to Physics, Engineering and the Environment*, Environmental Science and Engineering,
DOI:10.1007/978-3-319-00191-3_19,

permanent coalescence from the head-on collision of binary drops was further investigated by Nobari et al. (1995). In more recent times, the coalescence produced by the collision of axi-symmetric binary drops was simulated by Mashayek et al. (2003), who investigated the effects of varying Reynolds numbers, impact velocity, drop size ratio, and internal circulation on binary coalescence for low Weber numbers ($\approx$1). These calculations were extended to off-centre collisions of binary drops by Meleán and Sigalotti (2005), using Smoothed Particle Hydrodynamics (SPH). On the other hand, Chen (1985) and Li (1994) studied the coalescence of two small bubbles or drops using a model for the dynamics of the film thinning behaviour, in which both London-van der Waals and electrostatic double-layer forces are taken into account. In particular, the model proposed by Chen (1985) describes the film profile evolution and predicts the film stability, timescale, and film thickness, depending on the radius of the drops and the physical properties of the fluids and surfaces, while Li (1994) proposed a general expression for the coalescence time in the absence of the electrostatic double-layer forces.

Experimental studies of the binary collision of alkane droplets were carried out by Ashgriz and Givi (1987, 1989). They found that for near head-on collisions with increasing We, the impact can result in permanent coalescence, bouncing, and permanent coalescence again followed by reflexive separation into two or more drops. As We is further increased, reflexive separation may result in a string of three or more drops, whereas for very high impact velocities the collision may result in shattering separation in which the colliding drops disintegrate into a cluster of much smaller droplets. Similar experimental studies of binary collision of equal-sized alcohol droplets were carried out by Brenn and Frohn (1989). Ashgriz and Poo (1990) conducted experimental studies of the off-centre and grazing binary collision of water droplets for a wide range of We and impact parameters. In general, they found permanent coalescence and bouncing at low and moderate We, reflexive separation for higher We and low impact parameters ($\leq$0.4), and stretching separation for both higher We and higher impact parameters. The collision of equal-sized water and normal-alkane drops for $1 \leq We \leq 100$ and radius of 0.1 mm have also been studied experimentally by Jiang et al. (1992). Although different fluids were used, these experiments were quite similar to those conducted by Ashgriz and Poo (1990). However, there was a discrepancy in the experimental observations on head-on collisions of water drops between the two studies. In the former case permanent coalescence was always observed, while in the latter experiments reflexive separation was observed.

In the petroleum refineries, coalescence of fine oil mist is sometimes attained using electric fields to break up the emulsions. With this in mind, Eow and Ghadiri (2003a) investigated the behaviour of a liquid-liquid interface and drop-interface coalescence under the influence of an electric field. The measurement of the electric current can be used to monitor and control the behaviour of a liquid-liquid interface, thus providing an optimum condition for instantaneous and single-staged drop-interface coalescence. In a companion chapter, Eow and Ghadiri (2003b) studied the effects of the direction of the applied electric field as well as the geometryof the electrodes.

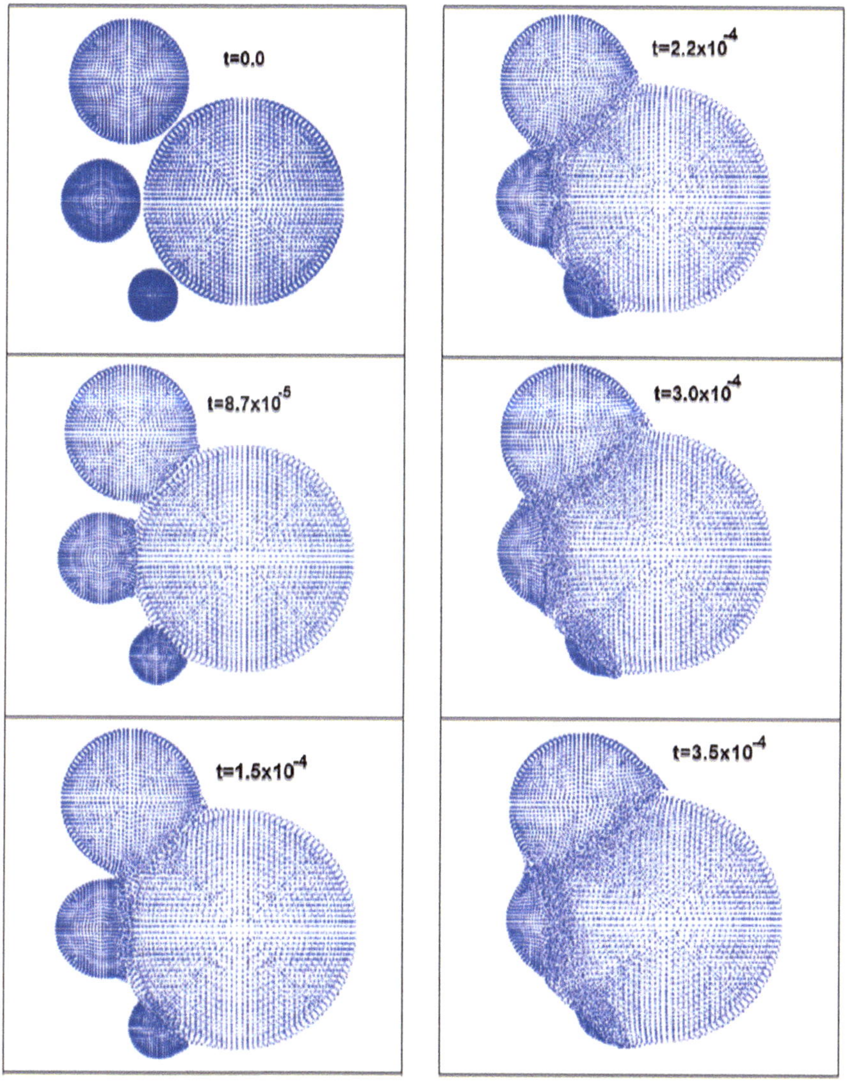

Fig. 1 Time sequence showing the evolution of the collision of four drops (permanent coalescence) with $V_{col} = 15.0$ mm/ms. Time is given in milliseconds. The origin of the coordinate system coincides with the *centre* of each *panel*

From this study it follows that the premature drop-drop coalescence in an electric field is influenced by the natural mechanical vibration and cavitation within the drops.

Recent numerical simulations by Acevedo-Malavé and García-Sucre (2011a,b,c, 2012) considered the coalescence process of liquid drops in three-space dimensions, using SPH methods. For binary drop collisions, they found that depending on the

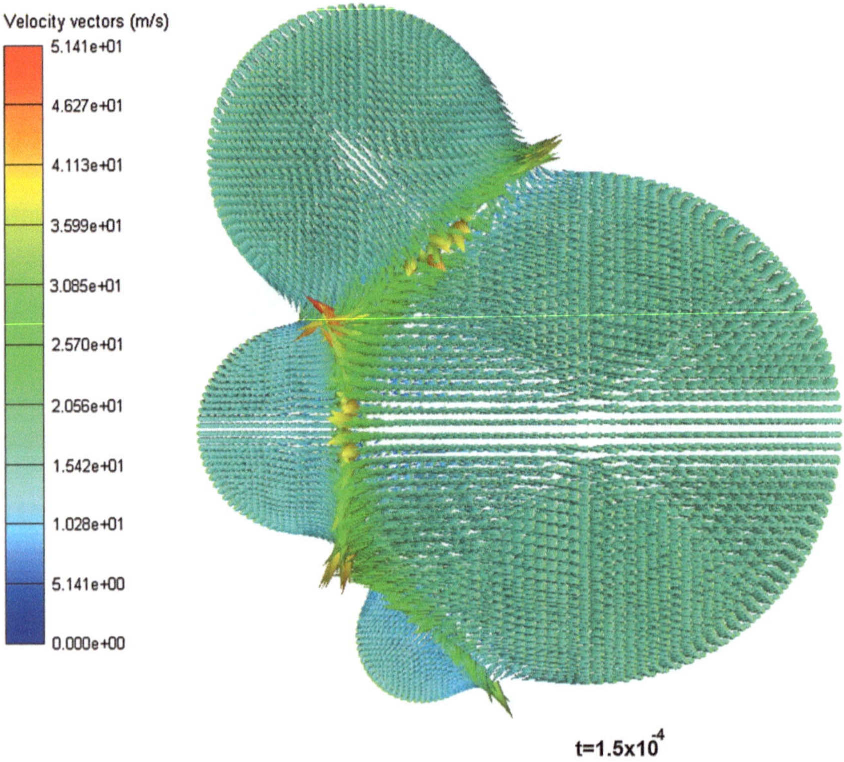

Fig. 2 Velocity vector field during the collision of four drops at $t = 1.5 \times 10^{-4}$ ms (permanent coalescence) with $V_{col} = 15.0$ mm/ms

impact velocity three possible outcomes may arise: drop flocculation, in which the drops remain attached without fragmenting, for low impact velocities; permanent coalescence into a bigger drop; and permanent coalescence with fragmentation into satellite droplets for higher impact velocities. In this chapter, we extend these previous simulations to multiple collisions of unequal-sized drops, where smaller drops are made to collide with a bigger one for impact velocities in the range $0.5 \leq V_{col} \leq 30$ mm/ms.

2 Basic Formulation and Numerical Methods

Here, we present only a brief description of the method, and refer the reader to Acevedo-Malavé and García-Sucre (2011a,b,c, 2012) for more details. SPH is a fully Lagrangian particle method used for simulations of discontinuous flows with large deformations (Monaghan 1985). It solves the laws of mass, momentum, and

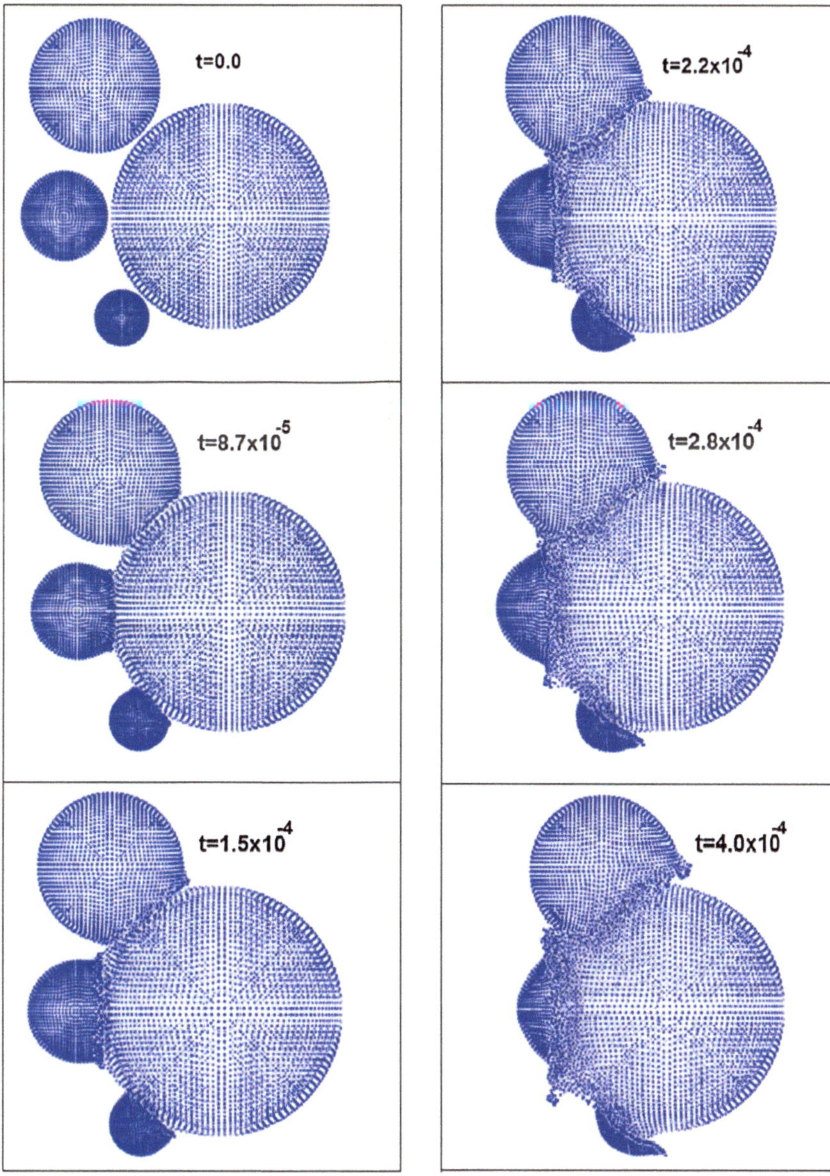

Fig. 3 Time sequence showing the evolution of the collision of four drops (fragmentation) with $V_{col} = 30.0$ mm/ms. Time is given in milliseconds. The origin of the coordinate system coincides with the *centre* of each *panel*

energy conservation. This equations are discretized through the use of an interpolating kernel function W that gives the estimate of the field variables at a set of particles

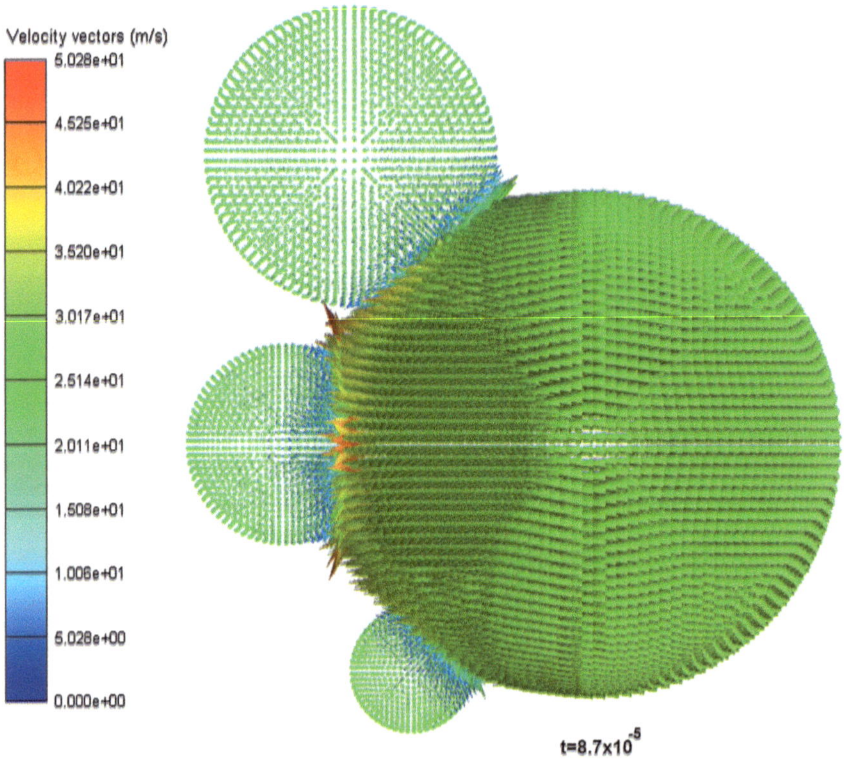

Fig. 4 Velocity vector field for the collision of four drops with $V_{col} = 30.0$ mm/ms (permanent coalescence with fragmentation) at $t = 8.7 \times 10^{-5}$ ms

suitably chosen to represent the fluid elements (Monaghan 1985). In practice, the mean value $A_s(\mathbf{r})$ of a function $A(\mathbf{r})$ is expressed by the convolution integral of the exact function with the kernel interpolant so that

$$A_s(\mathbf{r}) = \int A(\mathbf{r})\, W\left(\mathbf{r} - \mathbf{r}', h\right) d\mathbf{r}', \tag{1}$$

where the integration is performed over all space and h is the smoothing length which determines the spatial resolution. In the limit when h tends to zero, the smoothing function W becomes a Dirac delta function, and the smoothed representation $A_s(\mathbf{r})$ tends to $A(\mathbf{r})$.

In SPH, Eq. (1) is then approximated as a sum over all neighbouring particles. For instance, the mass density at the position of particle i, say $\mathbf{r}_i$ with $i = 1, \ldots, N$, where N is the total number of neighbours, is given by

$$\rho_i = \sum_j m_j W(\mathbf{r}_i - \mathbf{r}_j, h), \tag{2}$$

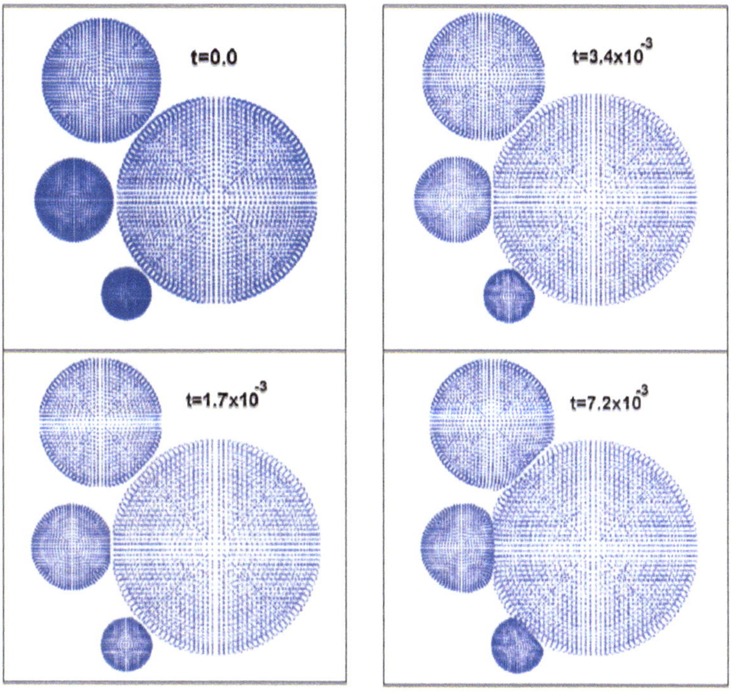

Fig. 5 Time sequence showing the evolution of the collision of four drops with $V_{col} = 0.5$ mm/ms (flocculation). Time is given in milliseconds. The origin of the coordinate system coincides with the *centre* of each *panel*

where the subscripts denote particle labels, m_j is the mass of neighbouring particle j, and $W(\mathbf{r}_i - \mathbf{r}_j, h)$ is a spherically symmetric kernel function here given by the cubic B-spline kernel of Monaghan (1985).

The position and velocity of particle i are given by the SPH representations

$$\frac{d\mathbf{r}_i}{dt} = \mathbf{v}_i, \tag{3}$$

$$\frac{d\mathbf{v}_i^{\alpha}}{dt} = \sum_{j=1}^{N} m_j \left(\frac{\sigma_i^{\alpha\beta}}{\rho_i^2} + \frac{\sigma_j^{\alpha\beta}}{\rho_j^2} \right) \cdot \nabla W_{ij}^h, \tag{4}$$

where σ is the total stress tensor. Here the Greek indices are employed to denote the components of the vector and tensor fields.

Water drops are simulated using a general Mie-Grüneisen equation of state with different analytic forms for the states of compression $(\rho/\rho_0 - 1) > 0$ and tension $(\rho/\rho_0 - 1) < 0$ (Acevedo-Malavé and García-Sucre 2011a).

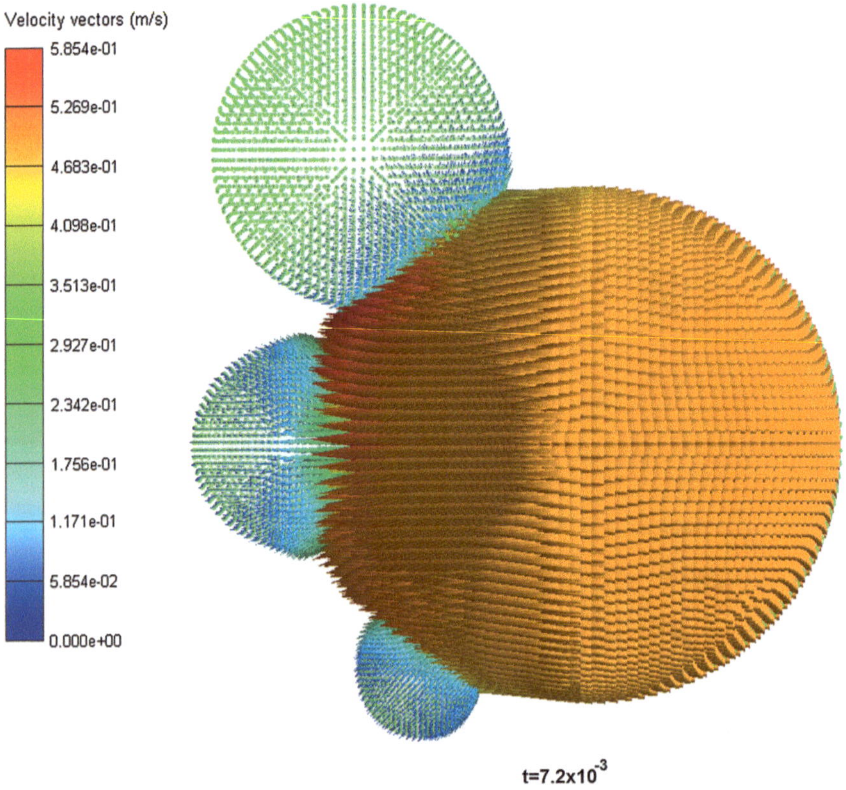

Fig. 6 Velocity vector field for the collision of four drops at $t = 7.2 \times 10^{-3}$ ms with $V_{col} = 0.5$ mm/ms (flocculation)

3 Coalescence, Fragmentation, and Flocculation of Drops

We simulate the impact of three unequal-sized water drops on a bigger drop. We consider spherical drops of diameters 5, 8, 12, and 20 μm represented by 4,456, 10,728, 15,504, and 34,040 SPH particles, respectively. The impacting drops have all collision velocities that are directed towards the centre of the coordinate system.

For a collision velocity of 15.0 mm/ms, Fig. 1 shows that at $t = 8.7 \times 10^{-5}$ ms a flat circular section appears after the drops enter in contact with one another, which then increases in diameter as coalescence proceeds. At $t = 1.5 \times 10^{-4}$ ms, the fluid of the smaller drops begins to penetrate into the bigger drop. By $t = 3.5 \times 10^{-4}$ ms the smallest drop has already coalesced with the bigger one, while the other two drops are already in the process of coalescing. Although we have terminated the calculation at this time, the evolution will eventually end with the formation of a big drop oscillating about a spherical shape (permanent coalescence). Figure 2 shows the velocity vector field inside the colliding drops as well as in the region of contact between them at $t = 1.5 \times 10^{-4}$ ms. Inside the drops, the fluid velocity tends to a value close to the impact velocity of 15.0 mm/ms, while in their areas of contact maximum velocities of about 30.0 mm/ms are observed.

For impact velocities as high as 30.00 mm/ms, the evolution is similar except that an outwardly directed flow develops in the contact region perpendicular to the direction of incidence of the drops, which may eventually condense into satellite droplets (see Fig. 3). While some of this liquid may return to the coalescing system, some other may condense into independent small droplets; the final process leading to permanent coalescence into a big drop accompanied by fragmentation into much smaller droplets. In Fig. 4 we show the velocity vector field inside the drops at 8.7×10^{-5} ms. Maximum velocities are 25 mm/ms in the bigger drop and 32 mm/ms in the smaller ones, while in the zones of contact between them the fluid reaches velocities as high as 40 mm/ms.

When the impact velocity is lowered to $V_{col} = 0.5$ mm/ms, the smaller drops do not coalesce but rather remain attached to the bigger one as shown in Fig. 5. In this case, the drops interact only through their deformed surfaces of contact, forming a clustering (drop flocculation). The stretching of the drop surfaces at contact is due to the fact that the surface tension forces prevail and the inertial forces are too low to induce permanent coalescence. Figure 6 shows the velocity vector field at $t = 7.2 \times 10^{-3}$ ms during the flocculation process. The maximum fluid velocity inside the drops is now around 0.47 mm/ms for the bigger one and 0.3 mm/ms for the smaller drops. In the zones of drop contact the maximum velocity is about 0.12 mm/ms.

4 Conclusions

In this chapter, we have presented numerical simulations of the coalescence collision of unequal-sized multiple drops in three-space dimensions, using the method of Smoothed Particle Hydrodynamics (SPH). For impact velocities in the range of $0.5 \leq V_{col} \leq 30$ mm/ms, the simulations predict three different outcomes: (a) the formation of a clustering of drops (flocculation), where the smaller drops do not undergo coalescence but remain attached to the bigger one, for low impact velocities; (b) permanent coalescence for moderate impact velocities; and (c) permanent coalescence accompanied by fragmentation into much smaller satellite droplets, as a consequence of an outflow of liquid along the plane of contact between the drops, for higher impact velocities. Future work in this direction will explore the ability of the present SPH scheme to follow the collision process for much higher impact velocities ($\leq$30 mm/ms) in order to study shattering separation for colliding drops of similar size and penetration of small drops into bigger ones.

References

Acevedo-Malavé A, García-Sucre M (2011a) 3D coalescence collision of liquid drops using smoothed particle hydrodynamics, vol 5. INTECH Publishers, Croatia, pp 85–106
Acevedo-Malavé A, García-Sucre M (2012) Head-on binary collisions of unequal size liquid drops with smoothed particle hydrodynamics, vol 9. Transworld Research Network, Kerala, pp 245–266
Acevedo-Malavé A, García-Sucre M (2011b) Coalescence collision of liquid drops I: off-center collisions of equal-size drops. AIP Advances 1:1–17
Acevedo-Malavé A, García-Sucre M (2011c) Coalescence collision of liquid drops II: off-center collisions of unequal-size drops. AIP Advances 1:1–12
Ashgriz N, Givi P (1987) Binary collision dynamics of fuel droplets. Int J Heat Fluid Flow 8:205–208
Ashgriz N, Givi P (1989) Coalescence efficiencies of fuel droplets in binary collisions. Int Commun Heat Mass Transfer 16:11–17
Ashgriz N, Poo JY (1990) Coalescence and separation of binary collisions of liquid drops. J Fluid Mech 221:183–204
Brenn G, Frohn A (1989) Collision and merging of two equal droplets of propanol. Exp Fluids 7:441–446
Chen JD (1985) A model of coalescence between two equal-sized spherical drops or bubbles. J Colloid Interface Sci 107:209–220
Eow JS, Ghadiri M (2003a) The behaviour of a liquid-liquid interface and drop-interface coalescence under the influence of an electric field. Coll Surf A: Physicochem Eng Aspects 215:101–122
Eow JS, Ghadiri M (2003b) Drop-drop coalescence in an electric field: the effects of applied electric field and electrode geometry. Coll Surf A: Physicochem Eng Aspects 219:253–269
Foote GB (1974) The water drop rebound problem: dynamics of collision. J Atmos Sci 32:390–401
Jiang YJ, Umemura A, Law CK (1992) An experimental investigation on the collision behaviour of hydrocarbon droplets. J Fluid Mech 234:171–190
Li D (1994) Coalescence between two small bubbles or drops. J Colloid Interface Sci 163:108–119
Mashayek F, Ashgriz N, Minkowycz WJ, Shotorban B (2003) Coalescence collision of liquid drops. Int J Heat Mass Trans 46:77–89
Meleán Y, Sigalotti L Di G (2005) Coalescence of colliding van der Waals liquid drops. Int J Heat Mass Transf 48:4041–4061
Monaghan JJ (1985) Extrapolating B splines for interpolation. J Comput Phys 60:253–262
Nobari MR, Jan YJ, Tryggvason G (1995) Head-on collision of drops: a numerical investigation. Phys Fluids 8:29–42

Numerical Simulations of Freely Oscillating Drops

Jorge Troconis, Armando Blanco, Dominique Legendre, Leonardo Trujillo and Leonardo Di G. Sigalotti

Abstract The non-linear oscillations of a viscous drop is a fundamental problem in diverse areas of science and technology. In this paper, we analyze the large-amplitude oscillations of an initially elongated liquid drop in two-dimensions by solving the free boundary problem comprised of the Navier-Stokes equations, using two different numerical codes. The drop models all start from the same deformation in vacuum with zero gravity and varied Reynolds numbers (Re). We find that non-isothermal drops undergo stronger damping than isothermal ones due to the additional dissipative effects of heat conduction. Regardless of the drop parameters and physical mechanisms of dissipation, the transition from periodic to aperiodic decay is seen to occur for $\mathrm{Re} \leq 1.5$ in good agreement with linear theory and previous numerical simulations.

J. Troconis (✉) · L. Di G. Sigalotti · L. Trujillo
Centro de Física, Instituto Venezolano de Investigaciones Científicas, IVIC, Apartado Postal 20632, Caracas 1020-A, Venezuela
e-mail: jorge.troconi@gmail.com

A. Blanco
Departamento de Mecánica,Universidad Simón Bolívar, USB, Caracas 1080, Venezuela
e-mail: ajblanco@usb.ve

D. Legendre
Institut de Mécanique des Fluides de Toulouse, IMFT,Université de Toulouse, CNRS-INPT-UPS, 2 Allée du Professeur Camille Soula, 31400 Toulouse, France
e-mail: dominique.legendre@imft.fr

L. Trujillo
The Abdus Salam International Centre for Theoretical Physics, ICTP, Strada Costiera 11, 34014 Trieste, Italy
e-mail: leonardo.trujillo@gmail.com

L. Di G. Sigalotti et al. (eds.), *Computational and Experimental Fluid Mechanics with Applications to Physics, Engineering and the Environment*, Environmental Science and Engineering, DOI:10.1007/978-3-319-00191-3_20,

1 Introduction

The free oscillations of liquid drops have been studied extensively over more than a century, both for the sake of basic scientific understanding as well as for their dumbfounding applications in the chemical, pharmaceutical, and food industry. The problem has applications in polymer processing, dispersion technologies, gene chip arraying, inkjet printing, catalyst production, containerless processing technology in space, and metereology among many others. In the absence of gravity, the infinitesimal-amplitude oscillations of a drop were first shown to be correlated to its surface tension, density, and size (Rayleigh 1879). Here by oscillations we refer to periodic changes of the drop surface shape from spherical to ellipsoidal and back. A few other analyses for viscous drops have been reported in the literature (Reid 1960; Miller and Scriven 1968; Prosperetti 1980). It was shown that the initial motion of a viscous drop is just that executed by a damped harmonic oscillator of natural frequency $\omega_n' = (\omega_n^2 - b_n^2)^{1/2}$, for which the amplitude decays exponentially with time, where ω_n is the Rayleigh frequency (Rayleigh 1879) and b_n is a damping parameter depending on the drop density, size, and dynamic viscosity. Moreover, a transition from periodic to aperiodic decay of the oscillations was found to occur between Re $\approx$ 1.3 and 1.768 (Prosperetti 1980), where Re is the Reynolds number.

For inviscid drops undergoing slightly non-linear oscillations, the oscillation frequency was shown to decrease with increasing initial amplitude (Tsamopoulos and Brown 1983). On the other hand, numerical simulations of large-amplitude oscillations of slightly viscous drops have revealed that even a small viscosity may have a relatively large effect on resonant-mode coupling (Lundgren and Mansour 1988). Simulations addressed to study the effects of finite viscosity on large-amplitude, oscillating drops were also considered by a number of authors (Basaran 1992; Becker et al. 1994; Mashayek and Ashgriz 1998; Meradji et al. 2001; Moran et al. 2003; López and Sigalotti 2006). These calculations have revealed that frequency modulation and mode coupling are dominant, even for small initial deformations (Becker et al. 1994), whereas internal circulation may have significant effects on the frequency and damping rate during the first few periods of oscillation (Mashayek and Ashgriz 1998). In the absence of internal circulation, the damping rate is essentially governed by the combined action of viscous and surface tension forces (López and Sigalotti 2006).

Experimental observations of acoustically levitated drops have confirmed qualitatively the behaviour predicted by the linear and non-linear theory (Trinh and Wang 1982). For instance, mode coupling and asymmetries in the oscillation amplitude of high-order modes have been observed in drops with initial $n = 2$ deformations larger than about 10 % of their spherical radius (Becker et al. 1991). Experiments of low-viscosity drops oscillating in the microgravity environment of a Space Shuttle flight have shown that the frequency shift of the oscillations agrees well with the predictions of inviscid non-linear theory (Wang et al. 1996). Unprecedented microgravity observations of the maximal shape oscillations of a surfactant-bearing water drop during a mission of Space Shuttle Columbia have also been documented by Apfel et al. (1997).

In this paper, we describe two-dimensional numerical simulations of initially elongated liquid drops undergoing free oscillations in vacuum with zero gravity, using two different numerical approaches. We consider two separate sequences of calculations: one where the drops are kept isothermal with associated Reynolds numbers in the range $0.5 \leq \mathrm{Re} \leq 50$ and the other where non-isothermal conditions are adopted for $0.5 \leq \mathrm{Re} \leq 1{,}000$. We consider arbitrary viscosity and limit our analysis to non-rotating drops.

In typical experiments of shape recovery of deformed drops, the drop is first distorted in a shear flow field or by acoustic levitation. After the shear flow is abruptly stopped or the levitating force is reduced to the strength necessary to maintain the drop suspended, the transient behaviour of the extended drop proceeds in one of two ways: the drop may relax back to its original spherical shape, or, if the extension was beyond a critical aspect ratio, it may break up into a number of smaller droplets. In the present study, we will only be concerned with situations where the initial drop deformations are below the critical elongation ratio, i.e., the drop will always relax back to spheres.

2 Isothermally Oscillating Drops

Calculations of a fluctuating liquid drop under isothermal conditions were carried out using the JADIM code for $0.5 \leq \mathrm{Re} \leq 50$, where $\mathrm{Re} = (\rho\sigma R)^{1/2}/\eta$ is the Reynolds number, ρ is the drop density, σ is the surface tension, R is the drop radius, and η is the dynamic (shear) viscosity. We restrict ourselves to two-space dimensions so that all variables are functions of the (x,y)-coordinates and time t. In this way, the drop is represented by an infinitely thin disk, with its oscillations about the spherical shape corresponding to deformations of the disk perimeter about its unperturbed circular shape.

The JADIM code is based on a finite-volume discretization method for solving the Navier-Stokes equations, where all spatial derivatives are approximated by second-order accurate central differences, coupled to the Volume of Fluid (VOF) method for tracking and locating free surfaces and fluid-fluid interfaces (Magnaudet et al. 1995; Legendre and Magnaudet 1998; Bonometti and Magnaudet 2007). Surface tension forces are handled by adding to the Navier-Stokes equations a body force per unit volume, $\mathbf{F}_s = -2\kappa\sigma\mathbf{n}$, where $\kappa = \nabla \cdot \mathbf{n}/2$ is the curvature of the interface and $\mathbf{n}$ is the unit normal to it. The unit vector $\mathbf{n}$ is evaluated according to $\mathbf{n} = \nabla c/[c]$, where c is the colour function (or volume fraction) identifying each fluid in the system and $[c]$ is the jump in c across the interface. In the VOF modulus of JADIM, the colour function is evolved by solving the transport equation

$$\frac{\partial c}{\partial t} + \mathbf{v} \cdot \nabla c = 0, \tag{1}$$

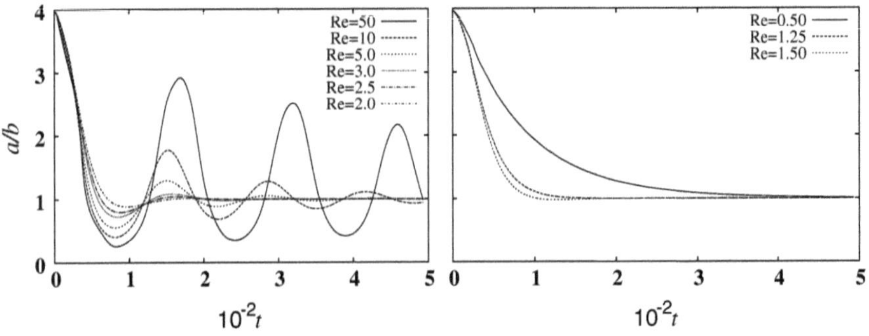

Fig. 1 Effect of Reynolds number on the large-amplitude oscillations when a drop is released from an elliptic elongation with aspect ratio $a/b = 4$ under isothermal conditions as calculated with the JADIM code. Underdamped oscillations are shown for $2.0 \leq \text{Re} \leq 50$ (*left panel*) and overdamped aperiodic returns to the circular rest state for $\text{Re} \leq 1.5$ (*right panel*)

where $c = 0$ if there is no traced fluid inside the cell volume and $c = 1$ otherwise, while $0 < c < 1$ when the interphasal surface cuts the cell volume. The time integration of the equations is performed using an implicit Runge-Kutta/Crank-Nicolson algorithm so that the overall code is also second-order accurate in time. A detailed description of the JADIM code and its VOF modulus can be found in the above references, while validation of the method on the motion and deformation of fluid-fluid interfaces can be found in Legendre et al. (2003) and Merle et al. (2005).

The initial model was constructed by mapping the elliptic drop on a square grid of 64×64 elements. The drop has density $\rho \approx 1.764$, uniform temperature $T \approx 0.2$, and internal pressure $p \approx 0.156$ in reduced units. The area of the ellipse was chosen to correspond to that of a circle of radius $R \approx 12.27$. The outer vacuum was approximated by assuming an ambient fluid (continuous phase) of density and viscosity three orders of magnitude lower than the drop values. On the line borders of the square grid no-slip boundary conditions are applied. All drops start with the same parameters except for their coefficient of shear viscosity, η, which was varied to provide a set of elongated drops with $0.5 \leq \text{Re} \leq 50$. Because of assumed reflection symmetry about the semi-minor and semi-major axes of the ellipse, only a quarter of the computational domain is effectively included in the calculations.

The variation of the drop aspect ratio with time is displayed in Fig. 1 for different Reynolds numbers. When Re is decreased from 50 to 2, the amplitude of the oscillations decreases as the strength of the viscous forces increases over the inertial ones. At low Reynolds ($2 \leq \text{Re} \leq 5$), the drop recovers its circular shape after about three to four periods, while several more are needed for the $\text{Re} \geq 10$ drops to relax back to circles. A change in the regime of the oscillations from underdamped (periodic), when $2 \leq \text{Re} \leq 50$ (left panel), to critically damped, when $\text{Re} = 1.5$ and 1.25 (right panel), corresponding to a fast aperiodic return to the circular shape, and then to overdamped when $\text{Re} = 0.5$, corresponding to a much slower aperiodic decay mode, is clearly observed. This result is in good agreement with linear theory which predicts an aperiodic decay mode for Re between ≈ 1.3 and 1.768 (Prosperetti 1980). For globular drops, Basaran (1992) and Meradji et al. (2001) found the same behaviour for $1.3 < \text{Re} < 1.4$ and $1.2 < \text{Re} < 1.4$, respectively, when the drop is released from a second-harmonic shape with initial aspect ratio $a/b \approx 1.015$.

3 Non-Isothermally Oscillating Drops

The non-linear oscillations of elongated drops under non-isothermal conditions are investigated using a smoothed particle hydrodynamics (SPH) code. As before, attention is focused on large-amplitude oscillations of drops that are released from a static elliptic shape with aspect ratio $a/b = 4$ in two-space dimensions. The SPH code solves the equations of mass, momentum, and energy conservation in Lagrangian form, including the effects of viscosity and heat conduction, coupled to the vdW equations of state

$$p = \frac{\rho \bar{k}_B T}{1 - \bar{\beta}\rho} - \bar{\alpha}\rho^2, \tag{2}$$

$$U = \frac{\xi}{2}\bar{k}_B T - \bar{\alpha}\rho, \tag{3}$$

for the pressure and thermal energy, respectively. In these equations, ξ is the number of degrees of freedom of the molecules (= 2 in two dimensions) and $\bar{k}_B = k_B/m$, where k_B is the Boltzmann's constant and m is the particle mass. Furthermore, $\bar{\alpha} = \alpha/m^2$ and $\bar{\beta} = \beta/m$, where α is the cohesive action and β is a constant parameter due to the finite size of the particles.

The effects of surface tension are simulated here with the aid of Eq. (2) by separating the cohesive term, $-\bar{\alpha}\rho^2$, from the remainder forces in the SPH representation of the momentum equation. The same applies to the energy term, $-\bar{\alpha}\rho$, in Eq. (3). The former term contributes with an attractive central force between the SPH particles, while the latter one contributes with an effective heating due to the work done by the cohesive pressure forces on the liquid within the free surface. A predictor-corrector leapfrog scheme is used to advance the position, velocity, and thermal energy of particles in time, from which updates of the density, temperature, and pressure are computed. Numerical stability is guaranteed by limiting the time step according to the CFL condition. A detailed account of the SPH code and its applications to model free-surface phenomena can be found in López and Sigalotti (2006), Sigalotti et al. (2006) and Sigalotti and López (2008).

In contrast to the isothermal models, the elliptic elongation is now obtained by deforming a stable circular drop by means of an area-preserving coordinate transformation (Twiss and Moores 1992). A circular drop is constructed numerically by starting the calculation with a square-cell array of 900 SPH particles of equal mass, separated along the x and y axes by a dimensionless distance $\Delta s = 0.78$. We adopt

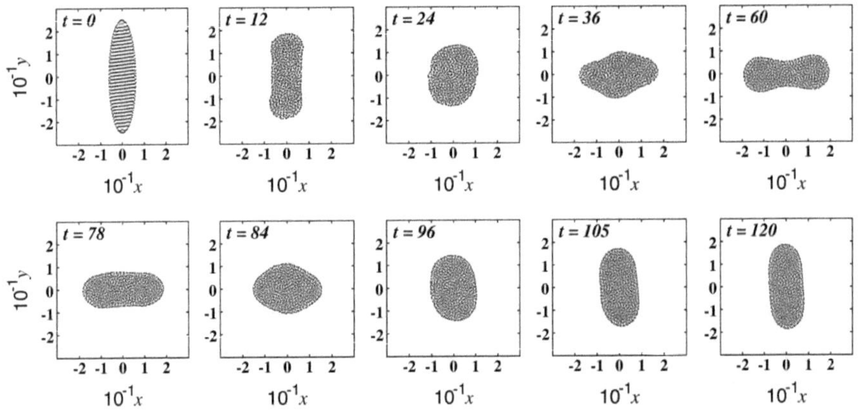

Fig. 2 Transient shapes during the first oscillation period of a drop released from an elliptic elongation with aspect ratio $a/b = 4$ at Re = 500 as calculated with the SPH code. The time is given in reduced units

$m = 1$, $\bar{\alpha} = 2$, $\bar{\beta} = 0.5$, and $\bar{k}_B = 1$ in reduced units. The initial density, ρ_0, and temperature, T_0, are chosen such that $\rho_0 < 1/\bar{\beta}$ and $\bar{k}_B T_0 > 2\bar{\alpha}\rho_0(1 - \bar{\beta}\rho_0)^2$ for thermodynamic stability. With a subcritical temperature $T_0 = 0.2$ and choosing $\eta = 1$, $\zeta = 1$, and $\kappa = 5$ in reduced units, where ζ is the bulk viscosity and κ is the coefficient of heat conduction, a stable circular drop of central density $\rho(0) \approx 1.769$, pressure $p(0) \approx 0.162$, temperature $T(0) \approx 0.423$, radius $R \approx 12.5$, and surface tension $\sigma = p(0)R \approx 2.02$, with no external atmosphere, is formed after $t = 600$. A set of equilibrium circular drops was constructed with the same parameters as before, except that the shear viscosity was varied in the range $0.0067 \leq \eta \leq 13.36$, corresponding to Reynolds numbers in the interval $0.5 < \text{Re} < 1{,}000$. Fluid motion is generated by deforming the reference circular drops into an elliptic shape via the density-conserving coordinate transformations (Twiss and Moores 1992):

$$x' = \frac{x}{1+\varepsilon}, \qquad y' = (1+\varepsilon)y, \tag{4}$$

where ε is the elongation given by $\varepsilon = (a/b)^{1/2} - 1$. An ellipse with aspect ratio $a/b = 4$ is obtained by setting $\varepsilon = 1$ in Eq. (4).

The time resolved evolution of a drop for Re = 500 is displayed in Fig. 2 during its first period of oscillation. The drop first contracts along its major axis as part of its surface energy is transformed into internal liquid movement, passing through a transient approximate circular shape ($t = 24$) before reaching a maximum elongation along the x-axis ($t = 60$). At this point, the rim pressure exceeds the stagnation pressure inside the drop, causing it to contract back under surface tension and reach a prolate shape after completion of the first oscillation period ($t = 120$). The variation of the drop aspect ratio with time is shown in Fig. 3 for all runs. At comparable Re the drops oscillate with lower amplitudes and undergo stronger dissipation than the

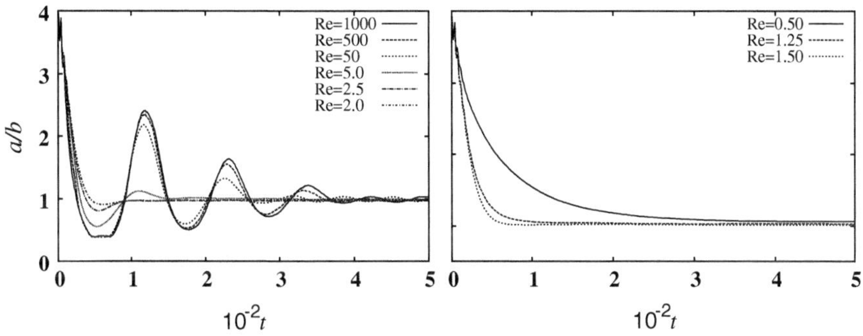

Fig. 3 Effect of Reynolds number on the large-amplitude oscillations when a drop is released from an elliptic elongation with aspect ratio $a/b = 4$ under non-isothermal conditions as calculated with the SPH code. Underdamped oscillations are shown for $2.0 \leq \mathrm{Re} \leq 1{,}000$ (*left panel*) and overdamped aperiodic returns to the circular rest state are obtained for $\mathrm{Re} \leq 1.5$ (*right panel*)

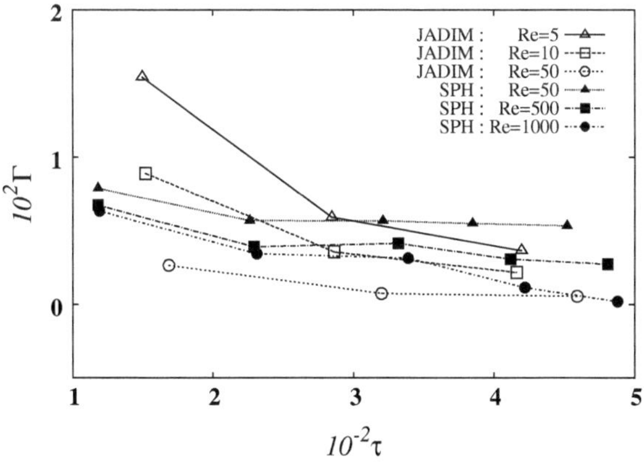

Fig. 4 Variation of the decay factor over several periods of oscillation for all runs: JADIM simulations (*open symbols*) and SPH calculations (*filled symbols*)

isothermal models of Fig. 1. The damping of the oscillations is mostly due to viscous dissipation and, to some extent, to the finite heat conductivity. The transition from periodic to aperiodic decay occurs at the same Re predicted by the isothermal models, suggesting that it is independent of the initial drop parameters and mechanisms of dissipation.

Finally, Fig. 4 shows how the decay factor (or damping rate), defined as

$$\Gamma_n = \frac{1}{\tau_n} \ln \left[\frac{(a/b - 1)_{\tau_{n-1}}}{(a/b - 1)_{\tau_n}} \right] \qquad n = 1, 2, ..., \tag{5}$$

varies as a function of the oscillation period for both the isothermal and non-isothermal models, where n is the period number. For all Reynolds numbers, the damping rate decreases with increasing period number, decaying most rapidly during the first period and changing only slightly with increasing time. Evidently, the lower is the Reynolds number, the higher is the damping rate.

4 Conclusions

We have presented two-dimensional calculations of the free oscillations of a viscous elongated drop surrounded by a vacuum in microgravity. The full Navier-Stokes equations with appropriate interfacial treatment were solved, using a finite-volume code under isothermal conditions and an SPH-based code under non-isothermal conditions, including the effects of heat conduction.

The main characteristics of drop relaxation back to its stable circular shape, with transition from periodic to aperiodic decay of the oscillations as viscosity is increased, are found in good agreement with linear theory and previous simulations. We find that heat conduction is an important additional mechanism for enhancing dissipation at moderate to high Reynolds numbers.

References

Apfel RE, Tian Y, Jankovsky J, Shi T, Chen X, Holt RG, Trinh E, Croonquist A, Thornton KC, Sacco A Jr, Coleman C, Leslie FW (1997) Free oscillations and surfactant studies of superdeformed drops in microgravity. Phys Rev Lett 78:1912–1915

Basaran OA (1992) Nonlinear oscillations of viscous liquid drops. J Fluid Mech 241:169–198

Becker E, Hiller WJ, Kowalewski TA (1991) Experimental and theoretical investigation of large-amplitude oscillations of liquid droplets. J Fluid Mech 231:189–210

Becker E, Hiller WJ, Kowalewski TA (1994) Nonlinear dynamics of viscous droplets. J Fluid Mech 258:191–216

Bonometti T, Magnaudet J (2007) An interface-capturing method for incompressible two-phase flow: validation and application to bubble dynamics. Int J Multiph Flow 33:109–133

Legendre D, Magnaudet J (1998) The lift force on a spherical bubble in a viscous linear shear flow. J Fluid Mech 368:81–126

Legendre D, Magnaudet J, Mougin G (2003) Hydrodynamic interactions between two spherical bubbles rising side by side in a viscous liquid. J Fluid Mech 497:133–166

López H, Sigalotti L Di G, (2006) Oscillations of viscous drops with smoothed particle hydrodynamics. Phys Rev E 73:051201

Lundgren TS, Mansour NN (1988) Oscillations of drops in zero gravity with weak viscous effects. J Fluid Mech 194:479–510

Magnaudet J, Rivero M, Fabre J (1995) Accelerated flows around a rigid sphere or a spherical bubble. Part I: steady straining flow. J Fluid Mech 284:97–135

Mashayek F, Ashgriz N (1998) Nonlinear oscillations of drops with internal circulation. Phys Fluids 10:1071–1082

Meradji S, Lyubimova TP, Lyubimov DV, Roux B (2001) Numerical simulation of a liquid drop freely oscillating. Cryst Res Technol 36:729–744

Merle A, Legendre D, Magnaudet J (2005) Forces on a high-Reynolds-number spherical bubble in a turbulent flow. J Fluid Mech 532:53–62

Miller AC, Scriven LE (1968) The oscillation of a fluid droplet immersed in another fluid. J Fluid Mech 32:417–435
Moran K, Yeung A, Masliyah J (2003) Shape relaxation of an elongated viscous drop. J Colloid Interface Sci 267:483–493
Prosperetti A (1980) Free oscillations of drops and bubbles: the initial-value problem. J Fluid Mech 100:333–347
Rayleigh JWS (1879) On the capillary phenomena of jets. Proc R Soc Lond 29:71–97
Reid WH (1960) The oscillations of a viscous liquid drop. Q Appl Math 18:86–89
Sigalotti L Di G, Daza J, Donoso A (2006) Modelling free surface flows with smoothed particle hydrodynamics. Condens Matter Phys 9:359–366
Sigalotti L Di G, López H (2008) Adaptive kernel estimation and SPH tensile instability. Comput Math Appl 55:23–50
Trinh E, Wang TG (1982) Large-amplitude free and driven drop-shape oscillations: experimental observations. J Fluid Mech 122:315–338
Tsamopoulos JA, Brown RA (1983) Nonlinear oscillations of inviscid drops and bubbles. Int J Fluid Mech 127:519–537
Twiss RJ, Moores EM (1992) Structural geology. Freeman, New York
Wang TG, Anilkumar AV, Lee CP (1996) Oscillations of liquid drops: results from USML-1 experiments in space. J Fluid Mech 308:1–14

Brownian Dynamics Simulation by Reticular Mapping Matrix Method

Eric Plaza

Abstract In this chapter, we present a method for the two-dimensional simulation of Brownian particles in a fluid with restrictions. The method combines characteristics of the cellular automata and Monte Carlo approaches, and is based on simple numerical rules that use two matrices for controlling the movement of the particles. One matrix serves to identify all particles on which statistical rules are adopted for their motion. This information is then mapped onto another matrix representing the positions of particles. The motion of the particles is governed by a statistical assignation mechanism, which allows to define either a random or non-random movement direction. The same probability of movement in each direction is assumed at each time step and for each particle to simulate the physical behaviour of Brownian movement in a two-dimensional network. For model validation, the predicted root-mean-square displacement of all particles along with their translational velocities are compared to theoretical values of the diffusion coefficient. The dependence of the computational time on the number of particles and concentration is calculated for the models.

1 Introduction

Brownian particle dynamics describes the presumable random motion of macrometric to nanometric suspensions in a fluid, and has direct applications in many important industrial processes as, for example, in the manufacturing of paints, pastes, and ceramics. The dynamics of suspensions in a fluid is also important in bio-molecular transport, chemical processes, erosion, and so on.

The behaviour of incompressible fluids is governed by the well-known Navier-Stokes equations (Landau and Lifshitz 1968). However, for most practical problems

E. Plaza (✉)
Instituto Zuliano de Investigaciones Tecnológicas, INZIT, La Cañada de Urdaneta, Km. 15, Vía Palmarejo Viejo, Estado Zulia, Venezuela
e-mail: eplaza@inzit.gob.ve

L. Di G. Sigalotti et al. (eds.), *Computational and Experimental Fluid Mechanics with Applications to Physics, Engineering and the Environment*, Environmental Science and Engineering,
DOI:10.1007/978-3-319-00191-3_21,

the analytical solution of these equations is mathematically unviable, and so different numerical techniques have been proposed for solving them, most of which rely on their spatial and temporal discretization (e.g., finite-difference, finite-element, and particle methods among others). Other discrete methods, such as those consisting of a regular grid of cells, simulate the motion of the particles in a fluid, according to rules that can be either random or predetermined. A well-known example of these methods was first described by (Wolfram 1983), which he called *elementary cellular automata*. Earlier attempts to simulate fluid flows based on lattices were proposed by (Hardy et al. 1973, 1976) and (Hardy and Pomeau 1977), who introduced the first Lattice-Boltzmann model (the HPP model). In this model, the lattice is square and motion of the particles is only allowed through any of the four sides whose cells share a common edge. Later on, this model was extended by Frisch et al. (1986) to include an hexagonal grid in the model, known as the FHP model after its inventors. The simple update rules of FHP models are chosen to conserve the particle number and momentum in the handling of particle collisions. Since the collision rules are not deterministic, the FHP models can reproduce the macroscopic Navier-Stokes equations with some simplifications. In contrast, other models that have been formulated for studying particle motion are completely mesh-free. Examples of them are the Smoothed Particle Hydrodynamics (SPH) method (Lucy 1977; Gingold and Monaghan 1977; Monaghan 1988), where the fluid is represented by a finite set of particles and field values are calculated by solving the hydrodynamics equations using a kernel function (in this case the particles can move arbitrarily in any direction restricted only by the physics), and the Molecular Dynamics method and its several variants, which solve the equations of motion for each particle for a large set, validating it in the thermodynamic limit (Haile 1992).

Most effort in computational dynamics has been addressed towards the development of numerical schemes capable of simulating real and complex physical systems at a relatively low computational cost. In some cases such refinements may either require the use of simplified algorithms to save computational resources or improve the efficiency of existing codes. In particular, examples of such efforts are given by Phillips et al. (2011) and Hellander and Lötstedt (2011), who have improved and reformulated existing codes to optimize the particulate simulations.

The Brownian motion is a physical phenomenon involving the random movement of small colloidal particles suspended in a liquid or gas medium as a consequence of the molecular thermal agitation over the surface of the particles. Understanding the microscopic properties of these displacements provides information on the macroscopic properties of the fluid where the particles are suspended like, for instance, the fluid diffusion coefficient (Einstein 1996). Its dynamics is usually described by two kinds of methods: Brownian dynamics (BD) (Ermak and McCammon 1978) and Stokes dynamics (SD) (Bossis and Brady 1987). The main idea behind these two methods is to divide the problem into two sub-problems: the particle motion and the fluid flow, where the former is described by the Langevin equation and the latter by the usual Navier-Stokes equations. However, one main disadvantage of the Langevin equation is that it is not easy to deal with particles of irregular shapes. As an alternative approach, it has been suggested to use the theory of fluctuating hydrodynamics (Landau and Lifshitz 1968), where the thermal fluctuations in the fluid, which in turn result in the Brownian motion of the particles, are included in the model. Such fluctuations have been recently modelled by adding a random stress tensor to the Navier-Stokes equations (Ladd 1994; Nie and Lin 2009).

In this chapter, we present a novel approach for simulating the Brownian motion of particles in a fluid. The model is based on a combination of two approaches: a cellular automaton in the sense that the method uses a regular uniform lattice (or "array") with a discrete variable at each site ("cell"). The state of the cellular automata is completely specified by the values of the variables at each site, which are simultaneously updated using information from their values at neighbouring cells at the preceding time step, according to a defined set of "local rules". In contrast to the cellular automata definition (Wolfram 1983), another lattice is used to control the rules for motion of the particles. In addition, the approach uses a Monte Carlo technique to randomly assign the movement direction of the particles.

In brief, the method uses a matrix called Mi, which represents the operations over all particles in the simulation, and a second one called Me representing the positions of these particles in a square grid. The motion of these particles is simulated by adding (subtracting) a random integer (related to the direction of movement) to (from) cells in the Mi matrix, and then by mapping these changes to positions in the Me matrix. This method is called *Reticular Mapping Matrix* (RMM) due to its conditioning to mapping operations between the matrix that generates the movement of the particles and the one that represents the space where the particles are moved. Using this approach, we study the Brownian motion of particles in a two-dimensional, adiabatic fluid confined in a closed recipient. The method is validated by comparing the model results with theoretical values of the mean-square displacement of the particles, the mean velocity of particles, and the self-diffusion coefficient.

2 Simulation Technique

As mentioned above, Brownian motion is the random movement observed in microscopic particles immersed in a fluid, resulting from their bombardment by the fast-moving atoms or molecules in the fluid. At the atomic scale, this bombardment is not uniform, producing pressure differences over the surface of the particles and causing their motion in a random direction. A mathematical description and a physical explanation of the phenomenon were provided by Einstein (1996), who related the microscopic properties of the medium to a macroscopic quantity, the diffusion coefficient D. For the self-diffusion coefficient of particles in a fluid, he derived the expression:

$$D = RT/6\pi a N\eta, \tag{1}$$

where R is the ideal gas constant, N is the number of particles, T is the temperature (in Kelvins), η is the fluid viscosity, and a is the radius of the Brownian particle. For Brownian motion, Einstein found that the mean displacement of the particles, λ, in

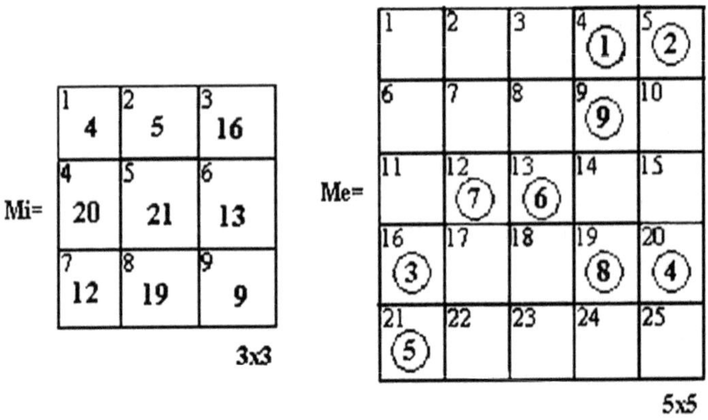

Fig. 1 Example: particle 1 of Mi (position 1×1) is mapped to cell 4 of Me (position 1×4). Particle 3 of Mi is mapped to position 16 of Me, and so on for all particles

one direction is given by

$$\lambda = \sqrt{2Dt}, \tag{2}$$

where t is the time.

The theory of Brownian motion was developed to describe the dynamic behaviour of particles whose mass and size are much larger than those of the solvent molecules. The equations governing the motion of these particles were written down by Langevin (1908) and a recent discussion on the validity of the theoretical results on Brownian motion is given in Chavanis (2010). In general, the results obtained shows that Brownian motion can be constructed as a scaling limit of a symmetric "random walk", in which the root-mean-square displacement of particles is a linear function of time after many steps (Hansen and McDonald 2006).

In this work, we present a method for simulating the self-diffusive behaviour of particles in a solution through a technique based on the mapping of results from a matrix of operations to a matrix of space. In the simulation, each particle is located randomly in a two-dimensional lattice. At each time step, a random movement direction is assigned to all particles. In order to do so the model uses two matrices, Me and Mi, as displayed schematically in Fig. 1. The numbers inside each cell of matrix Mi represent the positions of the particles in the matrix Me, while the index on the top left corner of each cell serves to identify each particle. In addition, in the Me matrix, the number in the cell centre identifies a particle and the index on the top left corner its position. (The dimensions of Mi and Me are $N1 \times N2$ and $N3 \times N4$, respectively, where the product $(N1)(N2)$ gives the total number of particles and $(N3)(N4)$ the total number of cells, see Fig. 1).

During each time step all particles within the network Me will be moved randomly in one of a number of possible directions. In two dimensions, we have a set of five directions of motion (i.e., rightward, leftward, upward, downward, and zero movement). The motion of each particle is governed by the matrix operations in Mi,

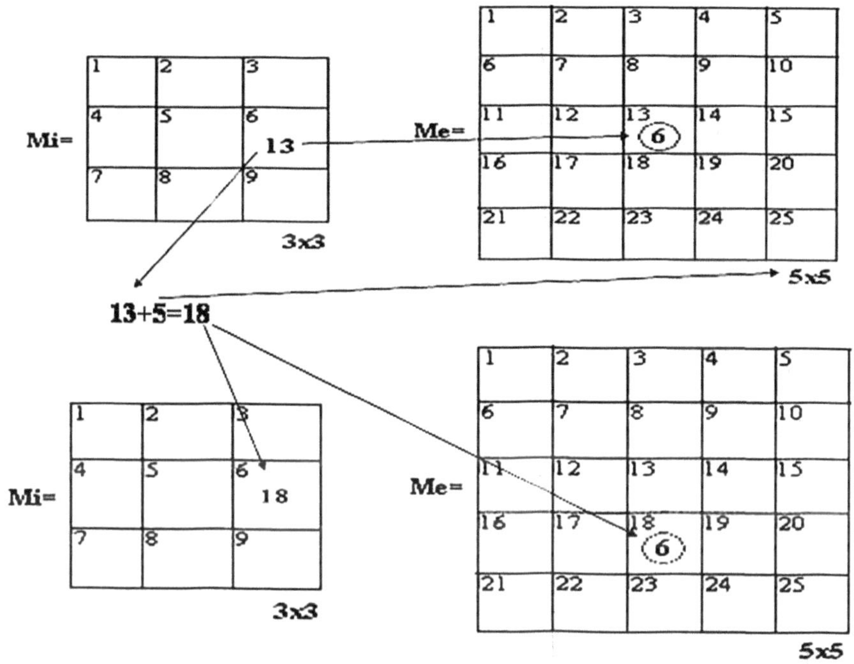

Fig. 2 Displacement of particle 6 from position 13 to position 18. Assuming the operation $13 + 5$ in cell $Mi(6)$, we perform a downward movement (the number five corresponds to the file number of matrix Me), that is the particle will be moved to position 18

where integers that are randomly added to each cell of Mi determine the direction of motion of the particle in Me, i.e., the sum will represent the direction of movement. If the sum is chosen to be +1 the particle will move towards the right. On the contrary, -1 means motion to the left, while $+N3$ and $-N3$ indicate downwardly and upwardly directed motion, respectively, and "0" stays for zero movement (see Fig. 2). This method is called *Reticular Mapping Matrix*, or simply RMM.

Brownian motion is therefore simulated by assigning to the particles a randomly chosen direction of motion at each time step. A mechanism in the program prevents the existence of particles with the same number within the matrix Mi, thus preventing particles to be in the same position and traffic problems. Boundary conditions are defined by imposing a set of rules at the grid borders. For the present simulations, we assume that the fluid is confined in a box so that particles cannot escape from the system. In addition, the method allows to easily change the weight of direction probabilities as well as the statistical behaviour of the particles, as shown in Fig. 3.

In the code, random numbers are generated by a Matlab function (rand), which returns uniformly distributed random numbers between 0 and 1. The choice of movement direction is controlled by the code, where a segment between 0 and 1 is divided into five intervals, with each segment corresponding to a specific direction. Once the random numbers are generated, each number is evaluated and depending on which

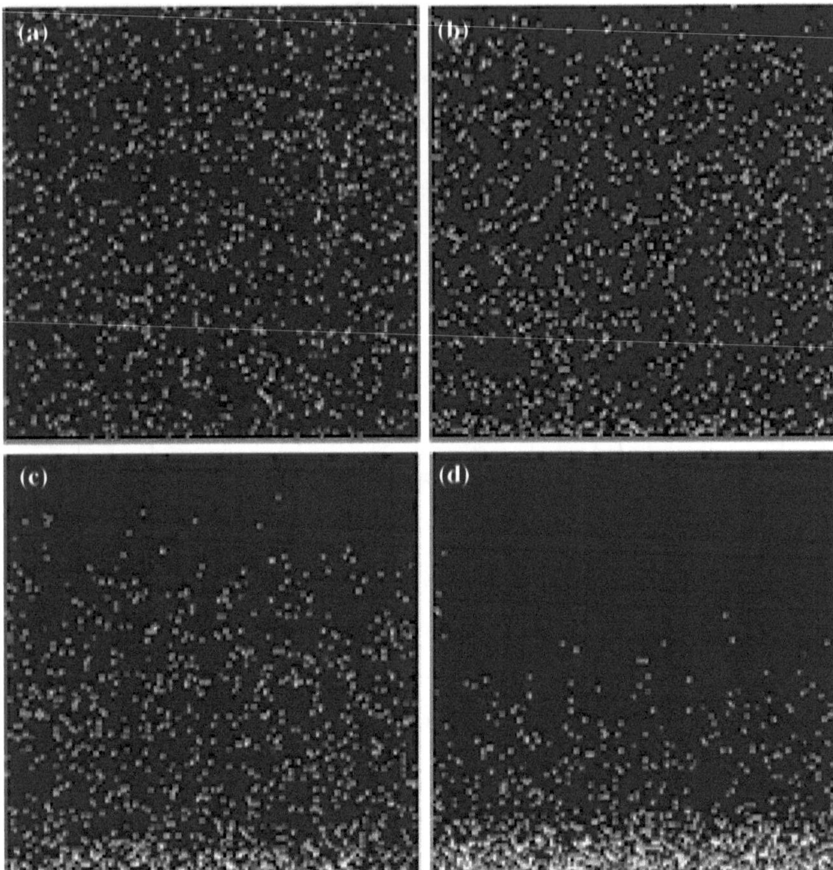

Fig. 3 Simulation performed for 1,000 time steps with $Me = 150 \times 150 = 22,500$ cells and $Mi = 50 \times 50 = 2,500$ particles. Using the following probability weights: 30 % downward, 20 % upward, 20 % rightward, 20 % leftward, and 10 % zero movement, we observe that after 1,000 time steps, the particles tend to agglomerate towards the bottom of the simulation area. **a** First time step; **b** time step 100; **c** time step 500; and **d** time step 1,000

interval it is, it will be moved in one randomly chosen direction. In a real suspension, the particles are not isolated but are coupled via hydrodynamic forces. In this model, however, it is assumed that there is complete separation between the timescales for the dynamics of the fluid and the motion of the solid particles, as in the classical theory of suspensions (Ermak and McCammon 1978).

3 Results and Discussion

In order to validate the results, we compute the mean-square displacement of all particles, λ, associated with the self-diffusion coefficient D given by the relation (Ermak and McCammon 1978; Hansen and McDonald 2006):

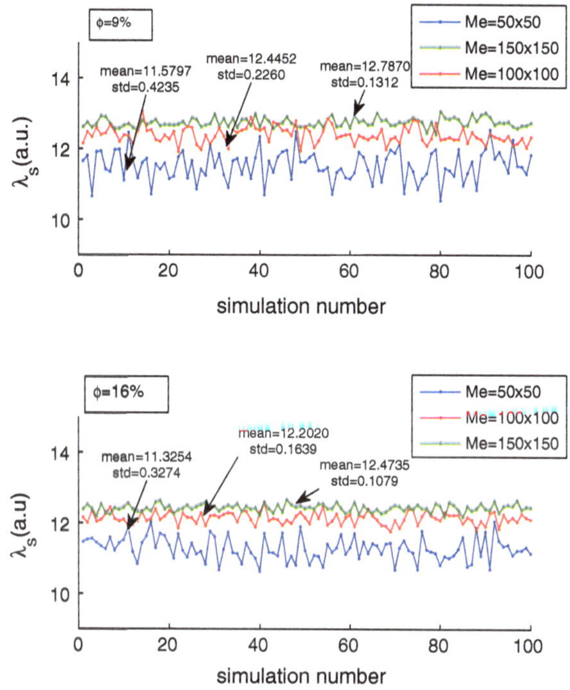

Fig. 4 Mean-square displacement of all particles, λ_s, for 100 simulations with two different particle concentrations (ϕ): 9 % and 16 %. λ_s follows a linear behaviour as predicted by theory

$$D_s = \frac{\langle |r_f(t) - r_i(0)|^2 \rangle}{4t}, \tag{3}$$

where the brackets denote the average displacement of all particles. This expression becomes a linear function of time after many random steps. The initial position of particles is labelled by the subscript i and the final one by the subscript f. The number of particles is $N1 \times N2$ (where $N1$ is the number of columns and $N2$ the number of files of Mi). After knowledge of the initial and final positions for each particle in the sample, we may then calculate D_s using espression (3).

Figure 4 shows the results for 100 simulations after 300 time steps each. The percentage of weight directions for particle movement was assumed to be 20 % for all five possible directions, while the volume fraction (or concentration) ϕ was set to 9 % and 16 % for each case. We see that the mean-square displacement, λ_s, behaves linearly for a representative number of simulations as predicted by theory. Note that all quantities are dimensionless.

Figure 5 shows the theoretical (λ_t) and the experimental (λ_e) mean-square displacements as functions of the time steps, for different concentrations and sizes of the network. Here λ_t was calculated using Eq. (2) with the diffusion coefficient given by Eq. (1), while λ_e was computed using the relation:

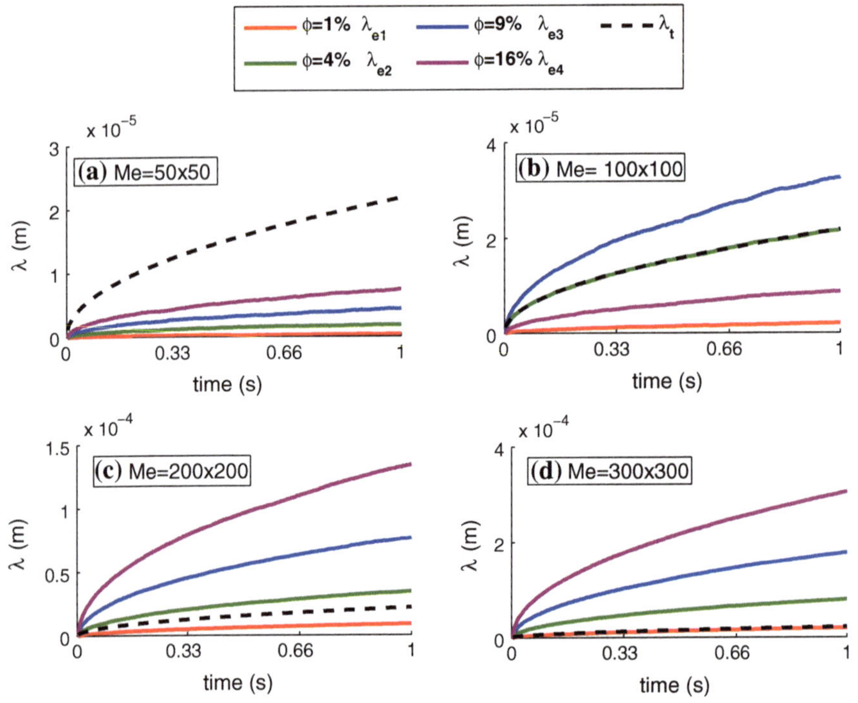

Fig. 5 λ_e and λ_t as a function of time for four different concentrations: $\phi = 1\,\%, 4\,\%, 9\,\%$, and $16\,\%$ and $Me = 200 \times 200$

$$\lambda_e = \lambda_s(a)(t/T), \tag{4}$$

where t is the total simulation time (in this case 1 s) and T is the total number of time steps (300 steps). The experimental mean-square displacements are shown for four different concentrations [i.e., λ_{e1} (ϕ: 1 %), λ_{e2} (ϕ: 5 %), λ_{e3} (ϕ: 9 %), and λ_{e4} (ϕ: 16 %)] and four different network sizes: $Me = 50 \times 50 = 2,500$ cells, $Me = 100 \times 100 = 10,000$ cells, $Me = 200 \times 200 = 40,000$ cells, and $Me = 300 \times 300 = 90,000$ cells. The theoretical value λ_t is the same for all plots because it does not depend on the particle and network size. There are clear differences between the experimental, λ_e, and the theoretical, λ_t, curves. The difference arises because there is no a theoretical relationship between the size of particles, the network size, and the concentration (Edelstein and Agmon 1993). Only for $Me = 100 \times 100$ at a concentration of 4 % and for $Me = 300 \times 300$ with a 1 % concentration, the experimental mean-square displacements are seen to match the theoretical values.

It is therefore necessary to correlate the size of the network to the sizes of the particles. To do so we introduce a simple method by assuming a square lattice, where a reference particle size is defined as the total length of the network d divided by the number of rows $N3$, that is

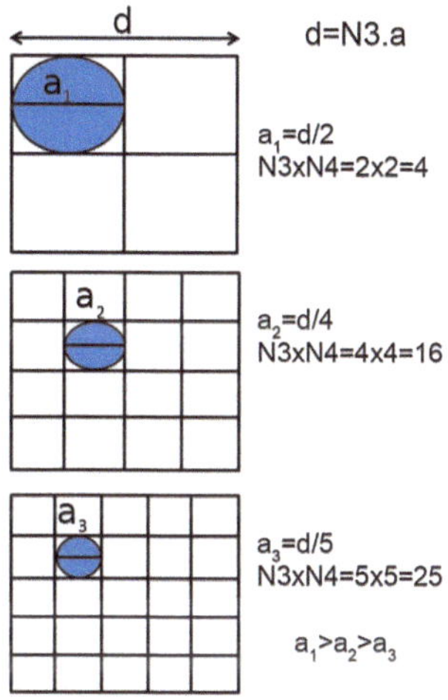

Fig. 6 Scheme adopted to define a reference particle size. Setting a reference value for the network size, say d, it is possible to define other particle sizes based on this scale by just increasing the number of cells of other equal-sized networks

$$a = \frac{d}{N3}. \tag{5}$$

Figure 6 shows graphically this relation for three equal-sized networks with increasing number of cells. We define a reference particle size and relates it to the total network size. Here we assume the particle size of Einstein's work, that is, 5×10^{-7} m and use a network with 100×100 cells as the reference case. If we multiply this particle size by the number of cells $N3 = 100$ in a file, we obtain the length of the network. Using this reference value and Eq. (5), we can obtain a relation for re-sizing all other particles.

Using the above scheme we performed a set of simulations for: $a_1 = 5 \times 10^{-7}$ m (reference case), $a_2 = 1 \times 10^{-6}$ m, $a_3 = 2.5 \times 10^{-7}$ m, and $a_4 = 2 \times 10^{-7}$ m, with network resolutions of a_1: 100×100 (reference case), a_2: 50×50, a_3: 200×200, and a_4: 250×250. The results of the simulations are shown in Fig. 7, where the theoretical (dashed lines) and experimental (solid lines) mean-square displacements are compared. As expected, an almost exact match is attained for particle size a_1, while for the other sizes the experimental and theoretical curves diverge by small amounts. These small errors may be caused by the shape of the particles and the limited number of directions allowed. We recall that in the simulations we have allowed for only four possible directions for particle motion (excluding movement along the cell diagonals) and the shape factor has not been taken into account.

We have also calculated the particle translational velocity as a function of the time steps (500) for three different cases. First, using particle sizes of $a_1 = 5 \times 10^{-7}$ m and $a_2 = 1 \times 10^{-6}$ m, a volume fraction of 10 %, and a weight probability of movement of 20 % in each direction (Brownian motion); second, the same values of a_1 and a_2, a volume fraction of 10 %, and varying the probability weights in the motion directions: 35 % downward, 15 % upward, 15 % rightward, 15 % leftward, and 20 % non-movement; and third, the same size of particles as before and weight

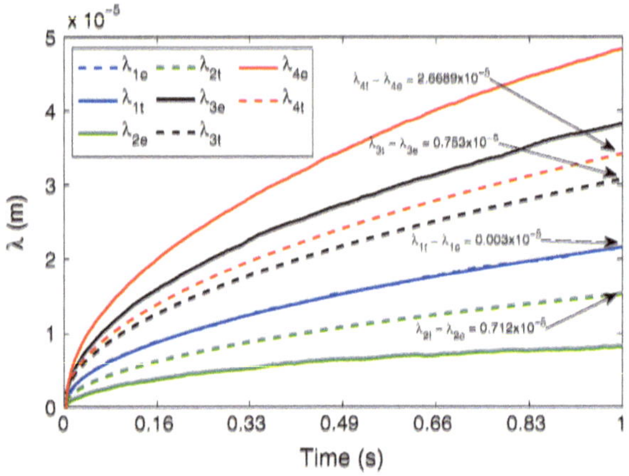

Fig. 7 Theoretical (λ_t, *dashed line*) and experimental (λ_e, *solid line*) mean-square displacements for all particles versus the time steps. The results show a close correspondence between the theoretical and experimental values after the re-size correlation

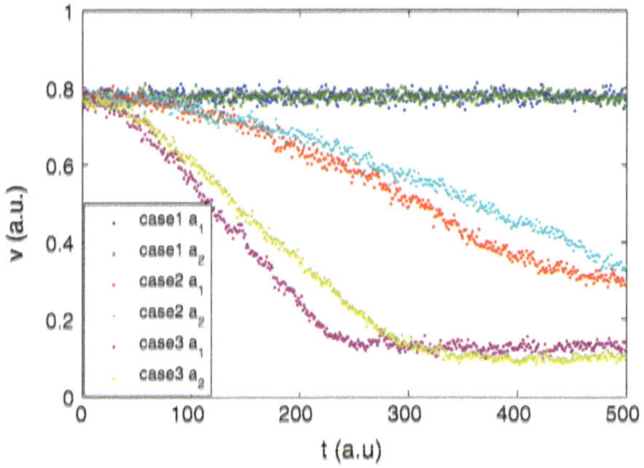

Fig. 8 Translational velocity for three different cases, with probability weights of movement: **a** Case 1: 20 % for all directions allowed (Brownian movement); **b** Case 2: 35 % down, 15 % up, 10 % right, 10 % left, 20 % non-movement; **c** Case 3: 50 % down, 10 % up, 10 % right, 10 % left, and 20 % non-movement

probabilistic directions of: 50 % downward, 10 % upward, 10 % rightward, 10 % leftward, and 20 % non-movement.

The evolution of a_1 and a_2 is shown in Fig. 8. In the first case, the system is in thermal equilibrium since the particle velocities are approximately constant in time (Haile 1992). Conversely, in the second case, the particles tend to occupy lower positions as time passes, agglomerating towards the bottom of the simulation area. This produces a decrease of the particle velocities with time. Finally, the third case is

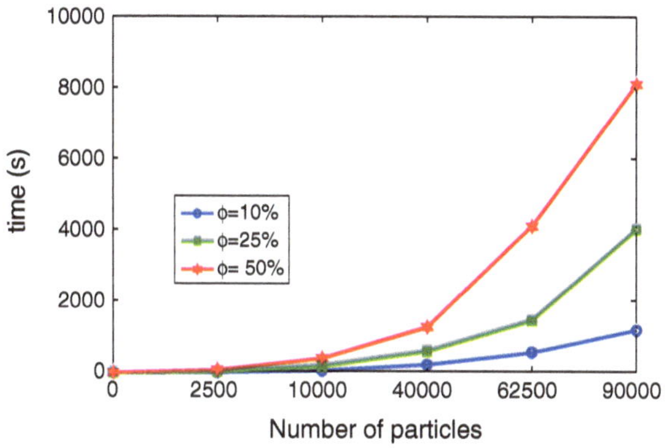

Fig. 9 Duration of the simulations for different values of the concentration and network size

similar to the second one, except for a faster equilibrium transition because a bigger probability of movement in the downward direction was assumed for this case.

Finally, the duration of the simulations was also calculated for three distinct concentration volumes ($\phi = 10\,\%$, $25\,\%$, and $50\,\%$) and $Me = 300 \times 300 = 90,000$, $250 \times 250 = 62,500$, $200 \times 200 = 40,000$, $100 \times 100 = 10,000$, and $50 \times 50 = 2,500$ cells. All runs were completed after 100 time steps, using a movement probability of 20 % for each direction. The calculations were carried out using Matlab on a personal computer (Intel Core 2 Duo processor and 3GB of RAM). The results are shown in Fig. 9, where the computational time is plotted against the number of particles. Independently of the number of particles, the heaviest calculations always corresponded to a concentration of 50 %, the trend being an increase of the computational time with increasing concentration and number of particles.

4 Conclusions

We have shown that the dynamics of Brownian particles can be very well modelled by the Reticular Mapping Matrix Method. Validation of the model was achieved by reproducing the typical mean-square displacements of Brownian particles and hence the values of the self-diffusion fluid coefficient. However, the model requires a correlation between the particles' size and the network size to achieve good agreement between the theoretical and the experimental results. To solve this issue a correlation spatial variable scheme was implemented. The design of the model is such that it can be easily extended to account for more complex physical situations, including tuning of different probability weights over different phases or groups of particles, the presence of obstacles, and the extension to three-space dimensions.

Acknowledgments I want to thank my colleagues José Arévalo and Rafael Martín for useful discussions and support.

References

Bossis G, Brady JF (1987) Self-diffusion of Brownian particles in concentrated suspensions under shear. J Chem Phys 87:5437–5448

Chavanis P-H (2010) Hydrodynamics of Brownian particles. Physica A 389:375–396

Edelstein AL, Agmon N (1993) Brownian dynamics simulations of reversible reactions in one dimension. J Chem Phys 99:5396–6404

Einstein A (1996) Investigations on the theory of the Brownian movement. Dover Publications, New York

Ermak DL, McCammon JA (1978) Brownian dynamics with hydrodynamic interactions. J Chem Phys 69:1352–1360

Frisch U, Hasslacher B, Pomeau Y (1986) Lattice-gas automata for the Navier-Stokes equation. Phys Rev Lett 56:1505–1508

Gingold RA, Monaghan JJ (1977) Smoothed particle hydrodynamics—theory and application to non-spherical stars. Mon Not R Astron Soc 181:375–389

Haile JM (1992) Molecular dynamics simulation: elementary methods. John Wiley and Sons, New York

Hansen JP, McDonald IR (2006) Theory of simple liquids. Academic Press, Amsterdam

Hardy J, Pomeau Y, de Pazzis O (1973) Time evolution of a two-dimensional classical lattice system. Phys Rev Lett 31:276–279

Hardy J, de Pazzis O, Pomeau Y (1976) Molecular dynamics of a classical lattice gas: transport properties and time correlation functions. Phys Rev A 13:1949–1961

Hardy J, Pomeau Y (1977) Microscopic model for viscous flow in two dimensions. Phys Rev A 13:720–726

Hellander S, Lötstedt P (2011) Flexible single molecule simulation of reaction-diffusion processes. J Comput Phys 230:3948–3965

Ladd AJC (1994) Numerical simulations of particulate suspensions via a discretized Boltzmann equation. Part 1 Theoretical foundation. J Fluid Mech 271:285–309

Landau LD, Lifshitz EM (1968) Fluid mechanics. Pergamon Press, Oxford

Langevin P (1908) Sur la théorie du mouvement brownien. Comptes Rendus de l'Académie des Sciences 146:530–533

Lucy LB (1977) A numerical approach to the testing of the fission hypothesis. Astron J 82: 1013–1024

Monaghan JJ (1988) An introduction to SPH. Comput Phys Commun 48:89–96

Nie D, Lin J (2009) A fluctuating lattice-Boltzmann model for direct numerical simulation of particle Brownian motion. Particuology 7:501–506

Phillips CL, Anderson JA, Glotzer SC (2011) Pseudo-random number generation for Brownian dynamics and dissipative particle dynamics simulations on GPU devices. J Comput Phys 230: 7191–7201

Wolfram S (1983) Statistical mechanics of cellular automata. Rev Mod Phys 55:601–644

Faraday Wave Patterns on a Triangular Cell Network

Franklin Peña-Polo, Iván Sánchez and Leonardo Di G. Sigalotti

Abstract We present experimental observations of the Faraday instability when an air/water interface is split over a network of small triangular cells for exciting frequencies in the range $10 \leq f \leq 30$ Hz. Just above the threshold for instability, waves appear on the water surfaces within all individual cells. After a transient state, adjacent cells progressively synchronize and self-organize to produce a regular pattern covering the whole grid. Collective cell behaviour is seen to lead to four different patterns depending on the forcing frequency range. Beyond $\approx$28 Hz, adjacent cells no longer interact as the vibration wavelength becomes smaller than half the altitudes of the triangular cells and so the waves remain constrained within individual cells in the form of localized harmonic oscillons.

1 Introduction

When a close receptacle containing liquid is submitted to vertical vibrations, a pattern of non-linear standing waves is often observed at the surface of the liquid. These waves, known as Faraday waves (Faraday 1831), are parametrically excited when the vertical vibrations exceed a critical frequency f_c, or critical acceleration Γ_c. The acceleration Γ is defined according to the relation $\Gamma = A\omega^2$, where A is the excitation amplitude and $\omega(=2\pi f)$ is the circular frequency. Faraday (1831) realized that these waves are sub-harmonic because they oscillate at half of the harmonic

F. Peña-Polo (✉) · I. Sánchez · L. Di G. Sigalotti
Centro de Física, Instituto Venezolano de Investigaciones Científicas, IVIC, Apartado Postal 20632, Caracas 1020-A, Venezuela
e-mail: franklin.pena@gmail.com

I. Sánchez
e-mail: ivanjo@gmail.com

L. Di G. Sigalotti
e-mail: leonardo.sigalotti@gmail.com

L. Di G. Sigalotti et al. (eds.), *Computational and Experimental Fluid Mechanics with Applications to Physics, Engineering and the Environment*, Environmental Science and Engineering, DOI:10.1007/978-3-319-00191-3_22,

excitation frequency. Modern experiments of single- and two-frequency forcing have revealed that not only spatially regular patterns of parallel lines, squares, circles, and hexagons may form but also many more complex symmetries such as quasi-patterns, superlattice patterns, and oscillons (Douady 1990; Edwards and Fauve 1994; Binks and van der Water 1997; Kudrolli et al. 1998; Arbell and Fineberg 2000a,b; Porter and Silber 2002; Westra et al. 2003).

Understanding the types of patterns that form is challenging. The threshold for instability and the observed patterns depend on the viscosity and surface tension of the fluid, the acceleration Γ, and the shape and size of the vessel. The full mathematical description of the problem involves the Navier-Stokes equations in a domain with a free surface, and the excitation makes the problem non-autonomous. In mathematics, a non-autonomous system is a system of ordinary differential equations which explicitly depends on the independent variable. In this case, non-autonomicity results from the external forcing that influences the fluid parameters when the oscillating behaviour initiates. On the other hand, the mechanisms of pattern selection have been investigated using the tools of symmetry and bifurcation theory (Silber et al. 2000; Rucklidge et al. 2003; Skeldon and Guidoboni 2007). The linear theory of this instability has been developed by Benjamin and Ursell (1954), who showed that the problem can be reduced to a set of Mathieu oscillators. However, the analysis relies on the potential flow approximation which is restricted to inviscid fluids only. If the instability is generated in a viscous liquid some mechanical energy is dissipated. These effects are usually treated by adding a heuristic damping in the Mathieu equation (Landau and Lifshitz 1987), which is proportional to the kinematic viscosity ν. The inclusion of such a term has been successively used in a number of linear analyses Müller (1993); Kumar and Tuckerman (1994); Kumar (1996) and Perlin and Schultz (2000). However, this approximation ignores viscous boundary layers along the vessel walls and beneath the surface, where additional dissipation occurs.

The most advanced theoretical investigation of the stability problem is fully numerical, which renders a physical understanding difficult Kumar and Tuckerman (1994). An analytic expression for the onset of sub-harmonic Faraday waves was obtained by Müller et al. (1997), which is applicable to a wide frequency range covering both shallow gravity and deep capillary waves. While this analysis is applicable in the limit of weak dissipation, an analytic treatment in the opposite limit was undertaken by Cerda and Tirapegui (1997). The linear aspects of the Faraday instability since Benjamin and Ursell (1954) were revisited by Müller (1998). It was not until very recently that the first numerical simulations of the dynamics of Faraday waves started to appear in the literature Périnet et al. (2009, 2012), involving the full solution of the Navier-Stokes equations in three-space dimensions coupled to a front-tracking method for resolving the free surface. In particular, these simulations have reproduced the square and hexagonal patterns seen in Kityk et al. (2005, 2009) with the same physical parameters. The hexagonal pattern was seen to be succeeded by recurrent alternation between quasi-hexagonal and beaded striped patterns.

In spite of recent progress, most work on Faraday waves assume that either the liquid bath has infinite extent or that the liquid surface is perfectly flat at the edge of the lateral walls where no-slip boundary conditions hold, which is unrealistic for experiments where the meniscus dynamics is important (Douady 1990). In general, as the system is shaken, the effective gravitational acceleration varies, making the meniscus length to become alternately large and small. In order to preserve the fluid mass, surface waves are emitted from the sidewalls of the vessel at the driving frequency f. Viscous dissipation is the primary cause for damping of these capillary waves.

In containers of small size there exists a strong coupling between the capillary waves generated by the meniscus and the Faraday waves, where the former extend all over the liquid surface. Recent experimental observations with cylindrical vessels of small diameters indicate that an increase of the Γ_c threshold is required for exciting Faraday waves in such small recipients Nguyem Thu Lam and Caps (2011). As previously for the viscosity, the addition of a phenomenological damping term, proportional to the thickness of the boundary layer, to the linear theory for modelling the viscous dissipation due to meniscus waves has successfully reproduced the experimental measurements of Γ_c for instability (Nguyem Thu Lam and Caps 2011).

Alternatively to experiments with single small containers, the formation of regular patterns has also been observed over a square network of centimeter-sized cells (Delon et al. 2010). After a transient state, just above the Faraday threshold, adjacent cells synchronize to form regular square lattices over the entire network, whose orientation with respect to the grid depends on the exciting frequency range. In this chapter, we extend these experimental observations to the case of an isometric grid consisting of equilateral triangular cells and study the effects of this geometry on the collective cell behaviour.

2 Experimental Set-Up

The experimental set-up consists of a transparent Plexiglas vessel with a base size of $15 \times 15\,\mathrm{cm}^2$ and 15 cm high, containing on its bottom plate a vinyl grid consisting of 60 equilateral triangular cells of sides 25 mm and depth 15 mm each. The cells are filled with coloured distilled water up to a height of 7 mm, and placed on an electromagnetic shaker which produces a clean vertical acceleration waveform. Small holes (of diameter 0.5 mm) have been drilled at the bottom of each cell to ensure equality of the fluid level in all cells. Fluid motion into these holes is prevented by the viscosity of the fluid and the low frequencies of oscillation, typically of a few Hz.

The vessel acceleration, which is the relevant bifurcation parameter, is measured by a piezoelectric accelerometer fixed onto the shaker table, receiving an oscillating voltage. The signal from the accelerometer is acquired using a multifunctional data acquisition board and processed by a host computer, where a software is run to give the oscillation amplitude A in millimeters and the frequency f in Hz from the maximum acceleration Γ normalized to the gravitational acceleration g_0 ($=9.81\,\mathrm{m\ s^{-1}}$). For pattern visualization the Plexiglas vessel is illuminated from above with white light using a pack of four halogen bulbs of 50 Watts each. The bulbs were placed at a height of 80 cm from the Plexiglas vessel and turned on only during video recording for not more than 20 s in order to avoid alterations of the density, viscosity, and surface tension of water due to heating. At such height, the halogen bulbs will transmit an average heat power to the water surface of about 10 W, which may produce a temperature increase of 1 °C only after 100 s of exposition. A CCD camera PL-B742 is used to observe the pattern from the top, and a second CCD camera (PL-B771) is positioned in front of one side of the vessel, tilted by an angle varied between 0 and 45° with respect to a plane perpendicular to the wall of the recipient, to record lateral and perspective views of the system. Pattern photographs from the top were also taken with a NIKON D60 digital SLR camera for presentation purposes. We varied the excitation frequency in the range between 10 and 30 Hz. For frequencies below $\approx$10 Hz, it is the maximum peak elevation which prevents the shaker from reaching the threshold acceleration Γ_c. Given the small size of the cells, the interaction between the meniscus and the Faraday waves are expected to have a stabilizing effect on the air/liquid interfaces (Nguyem Thu Lam and Caps 2011), thereby rising the instability threshold above the value required for non-confined liquids.

3 Observed Patterns

In a network of interconnected cells of small size, the relative effect of the numerous capillary menisci at the cell walls is an important factor. Due to the external forcing, the characteristic height of the menisci evolves according to $h = [\sigma/\rho g(t)]^{1/2}$ (Douady 1990), where σ is the surface tension of the liquid, ρ is its density, and $g(t)$ is the temporal modulation of gravity. This modification of the meniscus height leads to the generation of capillary waves, which dissipate by viscous shear and interact with the Faraday waves, excited sub-harmonically. This produces a stabilizing effect on the free surface so that more energy is required to excite Faraday waves than in large-recipient or non-confined fluids.

Just above $\approx$10 Hz, the wavelength of the forcing oscillations becomes smaller than the size of the cells, and after a transient state, adjacent cells progressively synchronize to form a regular pattern over the whole grid. Figure 1 shows top view photographs of the grid when bumps of fluid higher than the depth of the cells form at their intersections at $f = 10$ Hz and $\Gamma \approx 1.12 g_0$. If the driving force is maintained, the same pattern is recurrently repeated with periodic peak alternation occurring at the network scale along the horizontal direction every half a period as shown by the two images of Fig. 1. In this case, all adjacent cells collaborate synchronously to form a well-defined pattern by sharing nearly all of their liquid content into the emerging bumps. Some of them eventually pinch off at their ends and some liquid may be exchanged between adjacent cells. A similar collective behaviour was observed for higher exciting frequencies in the interval $10 \le f \le 14$ Hz. In particular, Fig. 2 displays a perspective view of this mode for $f = 12$ Hz. Synchronization is due to interacting cell waves converging at grid nodes as shown on the top left of Fig. 3.

Fig. 1 *Top* view images of the resulting pattern at $f = 10\,\text{Hz}$. Liquid bumps appear at grid nodes at the dual vertical and triple horizontal network scale. Waves inside adjacent cells interact for a global synchronization with alternation of the peak positions every half a period as seen by comparing the two images

Fig. 2 *Perspective* view showing the bumps forming at grid intersections when a forcing frequency of 12 Hz is applied. This mode is the same of that in Fig. 1, except that at this frequency the nodal bump distribution looks a bit more irregular

Note that six cells arranged in a regular hexagon contribute to each bump in the network.

We have observed three more modes of synchronized collective behaviour depending on the frequency range as shown schematically in Fig. 3. For $15 \leq f \leq 17\,\text{Hz}$, the liquid bumps form at the edges of adjacent cells (top right), with no peaks appearing on grid nodes. This pattern arises because waves inside four adjacent cells interact at their common edges. In this case, the four contributing cells form a larger triangle. For $18 \leq f \leq 20\,\text{Hz}$, the patterning consists of bumps appearing again at cell edges (bottom left). However, this time only two adjacent cells are allowed to contribute and their union forms a rhombus. A top view photograph of this mode is

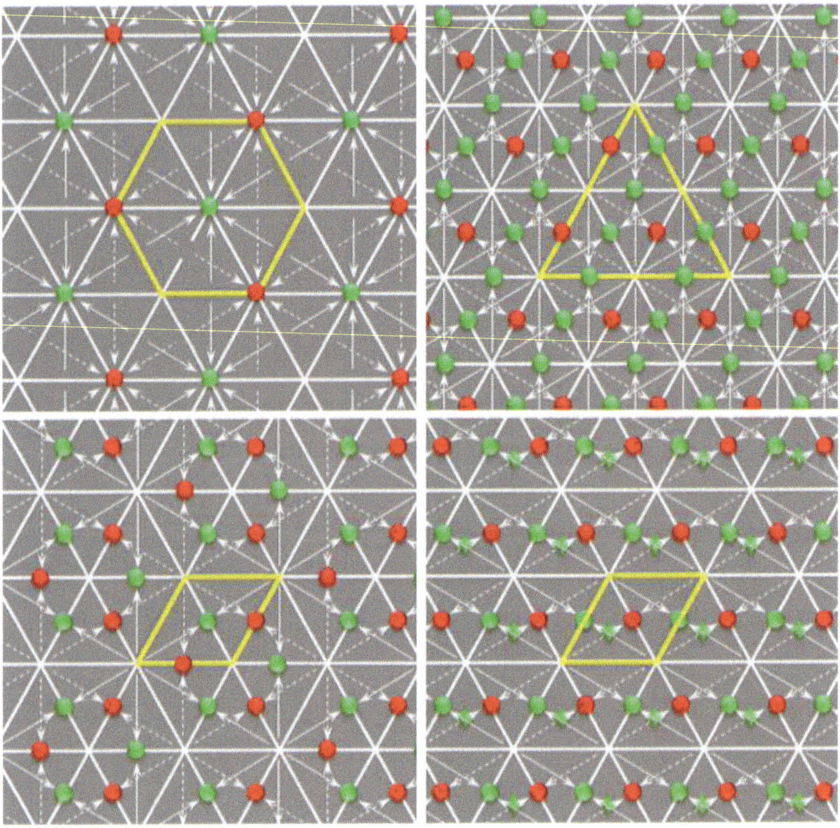

Fig. 3 Schematic of the four patterns observed according to the exciting frequency range: (i) $10 \leq f \leq 14$ Hz (*top left*), (ii) $15 \leq f \leq 17$ Hz (*top right*), (iii) $18 \leq f \leq 20$ Hz (*bottom left*), and (iv) $21 \leq f \leq 28$ Hz (*bottom right*). Alternation of the patterns is shown up by *green* and *red dots*, which mark the position of bumps at the beginning of and at half a period, respectively. The *arrows* indicate wave interaction between adjacent cells for one pattern (*continuous*) and its alternating counterpart (*dashed*). The *yellow* figures enclose the adjacent cells that work collectively

displayed in Fig. 4 at $f = 19$ Hz and $\Gamma \approx 2.38g_0$. The two images are separated in time by half a period. Occasionally, some bumps may appear at grid nodes. For $21 \leq f \leq 28$ Hz, peaks form synchronously at cell centres and edges owing to wave interaction between two adjacent cells (bottom right of Fig. 3). At such frequencies only part of the liquid inside a cell is shared with its neighbour, while the other part remains trapped within the cell to form a localized harmonic bump.

Beyond ≈28 Hz, the wavelength of the driving oscillations becomes smaller than half the triangular cell altitudes and so collective behaviour is no longer seen. The waves appear within each individual cell without interaction. The resulting pattern consists of localized harmonic oscillons forming at approximately the centre of each

Fig. 4 *Top* view images of the observed pattern at $f = 19\,\text{Hz}$. Liquid peaks appear at the edges between adjacent cells. Occasionally some bumps may form at grid nodes. The alternation of the bump positions is not easily discerned from the two images

Fig. 5 Example of non-collective behaviour at $f = 30\,\text{Hz}$. At this frequency the waves remain constrained within each individual cell without interaction. The pattern consists of localized harmonic oscillons at the centre of cells

cell as shown in Fig. 5 at $f = 30\,\text{Hz}$ ($\Gamma \approx 2.45 g_0$). Note that the size of the peaks is irregular and that no peaks are produced within some cells.

4 Conclusion

We have presented experimental observations of the Faraday instability on a network of triangular cells. For exciting frequencies between 10 and 28 Hz, we have observed four different mode patterns. Depending on the frequency range, collective cell behaviour may results in symmetric patterns of liquid bumps, forming either at grid nodes or at cell edges. Above ≈28 Hz, no collective behaviour is observed and temporally harmonic oscillons form at the centres of individual cells.

In previous experiments on a network of square cells, the liquid bumps formed square lattices at frequencies between 10 and 16 Hz due to diagonal wave interaction between adjacent cells (Delon et al. 2010). Evidently, changing the grid geometry from square to triangular not only adds an extra degree of freedom for wave interaction (due to the three altitudes in a triangular cell against the two diagonals in a square cell), but also doubles the number of patterns that forms along with the frequency range for which collective cell behaviour is observed.

References

Arbell H, Fineberg J (2000a) Two-mode rhomboidal states in driven surface waves. Phys Rev Lett 84:654–657

Arbell H, Fineberg J (2000b) Temporally harmonic oscillons in Newtonian fluids. Phys Rev Lett 85:756–759

Benjamin TB, Ursell F (1954) The stability of the plane free surface of a liquid in vertical periodic motion. Proc R Soc Lond A 225:505–515

Binks D, van der Water W (1997) Nonlinear pattern formation of Faraday waves. Phys Rev Lett 78:4043–4046

Cerda E, Tirapegui E (1997) Faraday's instability for viscous fluids. Phys Rev Lett 78:859–862

Delon G, Terwagne D, Adami N, Bronfort A, Vandewalle N, Dorbolo S, Caps H (2010) Faraday instability on a network. Chaos 20:041103

Douady S (1990) Experimental study of the Faraday instability. J Fluid Mech 221:383–409

Edwards WS, Fauve S (1994) Patterns and quasi-patterns in the Faraday experiment. J Fluid Mech 278:123–148

Faraday M (1831) On a peculiar class of acoustical figures; and on certain forms assumed by a group of particles upon vibrating elastic surfaces. Philos Trans R Soc London 121:299–318

Kityk AV, Embs J, Mekhonoshin VV, Wagner C (2005) Spatiotemporal characterization of interfacial Faraday waves by means of a light absorption technique. Phys Rev E 72:036209

Kityk AV, Embs J, Mekhonoshin VV, Wagner C (2009) Erratum: Spatiotemporal characterization of interfacial Faraday waves by means of a light absorption technique. Phys Rev E 79:029902

Kudrolli A, Pier B, Gollub JP (1998) Superlattice patterns in surface waves. Physica D 123:99–111

Kumar K, Tuckerman LS (1994) Parametric instability of the interface between two fluids. J Fluid Mech 279:49–68

Kumar K (1996) Linear theory of Faraday instability in viscous liquids. Proc R Soc Lond A 452:1113–1126

Landau L, Lifshitz EM (1987) Fluid Mechanics. Pergamon Press, New York

Müller HW (1998) Linear aspects of the Faraday instability. In: Parisi J, Müller SC, Zimmermann W (eds) A perspective look at nonlinear media from physics to biology and social sciences. Lecture Notes in Physics, vol 503. Springer, Berlin, pp 45–60

Müller HW (1993) Periodic triangular patterns in the Faraday experiment. Phys Rev Lett 71:3287–3290

Müller HW, Wittmer H, Wagner C, Albers J, Knorr K (1997) Analytic stability theory for Faraday waves and the observation of the harmonic surface response. Phys Rev Lett 78:2357–2360

Nguyem Thu Lam KD, Caps H (2011) Effect of capillary meniscus on the Faraday instability threshold. Eur Phys J E 34:112–116

Périnet N, Juric D, Tuckerman LS (2009) Numerical simulation of Faraday waves. J Fluid Mech 635:1–26

Périnet N, Juric D, Tuckerman LS (2012) Alternating hexagonal and striped patterns in Faraday surface waves. Phys Rev Lett 109:164501
Perlin M, Schultz WW (2000) Capillary effects on surface waves. Annu Rev Fluid Mech 32:241–274
Porter J, Silber M (2002) Broken symmetries and pattern formation in two-frequency forced Faraday waves. Phys Rev Lett 89:084501
Rucklidge AM, Silber M, Fineberg J (2003) Secondary instabilities of hexagons: a bifurcation analysis of experimentally observed Faraday wave patterns. In: Buescu J, Castro SBSD, Dias APS, Labouriau IS (eds) Bifurcations, symmetry and patterns. Birkhauser, Basel, pp 101–114
Silber M, Topaz CM, Skeldon AC (2000) Two-frequency forced Faraday waves: weakly damped modes and pattern selection. Physica D 143:205–225
Skeldon AC, Guidoboni G (2007) Pattern selection for Faraday waves in an incompressible viscous fluid. SIAM J Appl Math 67:1064–1100
Westra M-T, Binks DJ, van der Water W (2003) Patterns of Faraday waves. J Fluid Mech 496:1–32

Part III
Multiphase and Multicomponent Flow

Gas-Liquid-Solid Volumetric Phase Distribution Estimation in a Cold Slurry Bubble Column System for Hydro-Conversion Processes

Miguel V. Paiva-Rojas, Virginia Contreras-Andrade and Solange C. Araujo

Abstract Hydro-cracking slurry bubble column design, scale-up, and operation are strongly influenced by a fluid-dynamic parameter known as volumetric phase distribution. This parameter depends on the operating conditions (gas flow, liquid flow, pressure, and temperature) as well as on the gas, liquid, and solid physical properties. Experiments were carried out at ambient temperature and atmosphere pressure (cold conditions) in a 120 mm inner diameter plexiglas column (without any gas sparger) with air and CO_2, mineral oil, and coke with average particle sizes of 630 microns. The column was operated to up-flow continuous recirculation with superficial gas velocities ranging from 3 to 10 cm/s and a constant liquid-solid flow about 29 l/h. Experimental measurements were done by two methods: direct phase trapping and pressure drop. Measurement results indicate that the volumetric gas phase is highly affected by the superficial gas velocity. However, the superficial gas velocity effect on solid concentration is negligible. The experimental results were also compared with experimental data from other authors, obtaining a good agreement. A gas volumetric phase correlation was proposed.

M. V. Paiva-Rojas (✉) · S. C. Araujo
Refining and Industrialization Department, Instituto Tecnológico Venezolano del Petróleo (PDVSA-Intevep), Urbanización El Tambor, Los Teques 1201, Estado Miranda, Venezuela
e-mail: paivams@pdvsa.com

S. C. Araujo
e-mail: araujosc@pdvsa.com

V. Contreras-Andrade
Escuela de Ingeniería Química, Universidad de Los Andes (ULA),
Mérida 5101, Estado Mérida, Venezuela
e-mail: virginia.coa@gmail.com

L. Di G. Sigalotti et al. (eds.), *Computational and Experimental Fluid Mechanics with Applications to Physics, Engineering and the Environment*, Environmental Science and Engineering,
DOI:10.1007/978-3-319-00191-3_23,

1 Introduction

Slurry bubble column reactors (SBCR) are used nowadays as part of hidro-cracking units, having a notably importance for heavy crude oil processing. Several advantages of their use include: high heat and mass transfer rates, high conversion levels by strong gas-liquid-solid interactions, and improvement of catalyst recovery processes (Chengtian et al. 2008). Also, these reactors reduce maintenance costs due to the absence of moving parts or without any internal and improved control of the phase residence time inside the equipment (Shah et al. 1982).

The design and scale-up of these equipments is strongly influenced by the fluid-dynamic parameters. Some of the parameters with a major impact are the gas-liquid-solid phase volumetric distribution, the pressure drop, the flow pattern, and the mixing regime. Several of these variables, validated in an operation range and under certain physical conditions, have been cleared through empirical and semi-empirical correlations, which predominate as the bases of slurry bubble column reactor design. However, there is not enough to get from a universal methodology to describe the whole behaviour of a slurry bubble column (Hikita et al. 1981; Akita and Yoshida 1973; Shah et al. 1982). Additionally, process variables, like gas and liquid superficial velocities and material physical properties, influence the fluid-dynamic performance, which is a particular characteristic in each kind of process.

This work deals with the study of the gas-liquid-solid phase distribution based on two physical properties (specific particle range size and gas density) and one operation condition. For hydro-conversion processes, it is relevant to know the gas-liquid-solid volumetric phase distribution since it affects the conversion rate, which can be considered as the most important variable in these units.

2 Methodology and Experimental Setup

2.1 Experimental System and Materials

A schematic diagram of the slurry bubble column (SBC) used in this work is presented in Fig. 1. The core of the system is a plexiglas column of 120 mm internal diameter (ID) with an entrance at the bottom and an exit at the top, without any internal parts. The experimental equipment has a slurry piston pump for the feedstock, a slurry centrifugal pump for recycle, a mechanical agitated tank for the feedstock, and two discharge tanks. The measurement variable system has a mass flow-meter transmitter (FT) for gas Endress + Hauser Promass 83A 4–20 mA output range 0–450 kg/h error $\pm 0.50\,\%$, a pressure transmitter (PT) ABB 2600T model 261T dual wire 4–20 mA output range 13.7–397 mbar accuracy $\pm 0.15\,\%$, a pressure differential transmitter (DPT) Rosemount model 115T Smart 4–20 mA output range 0–20 bar error $\pm 0.25\,\%$, and videographer recorder ABB model Screen Master 200 with four input channel.

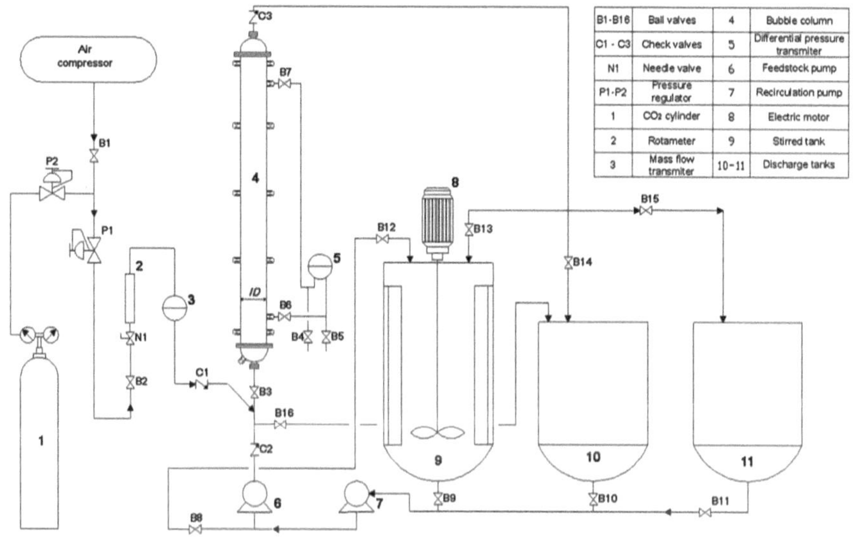

Fig. 1 Schematic diagram of slurry bubble column system

Table 1 Physical properties of fluid phases

Material	Properties		
	Viscosity (cSt)	Density (kg/m^3)	Superficial tension (dyn/cm)
Mineral oil (VASSA)	2.49	815	27.4
Air 25° C; 1 atm	0.018	1.185	
CO_2 25° C; 1 atm	0.015	1.808	

Commercial mineral oil VASSA LP-90 was used as the liquid phase because it has low toxicity and due to the fact that its physical properties (density and viscosity) at ambient temperature are similar to those of heavy oil, residual oil, or any other heavy hydrocarbon in operation conditions of hidro-cracking processes. Air and carbon dioxide were used at atmospheric pressure in the injection into the column. The air was supplied from the compression system from pilot plant services, which has about 6.8 bar, and carbon dioxide was supplied from gas cylinders, which have about 55.1 bar, connected to a pressure regulator to make pressure drop to 6.8 bar. The solid phase was ground petroleum coke. The particles sizes were measured by laser scattering of HORIBA device. The properties of fluids and solids are shown in Tables 1 and 2.

Table 2 Physical properties of solid phases

Material	Density (kg/m^3)	Particle size range (μm)	Mean size (μm)	Median size (μm)
Coke	1370	200–1100	629	617

2.2 Experimental Methodology

The unit was operating in recycling mode. In order to start the unit, a concentration of 1.5 wt % of solid with respect to the feedstock was mixed in the stirred tank. Then valves 2, 3, and 4 were opened (see Fig. 1), and after that pumps were started to begin the loading of the column.

The volumetric phase fraction was determined by phase trapping. This method consists in the rapid closing of the inlet and outlet valves of the column. This allows for an instantaneous and direct measure of the gas phase fraction inside the column. In operation, once the column reaches a steady value of pressure drop, the volume of gas phase is determined between the inlet and outlet valves of the column. This method has been used in vertical and horizontal pipes for oil production and transportation (Hagedorn and Brown 1965; Beggs and Brill 1973; Griffith et al. 1975) and also in the study of cold models in bubble column and slurry bubble column reactors (Tai-Ming et al. 1987; Wenge et al. 1995; Jena et al. 2008).

The experimental measurements were done when the stabilization time was reached, defined as the time when the differential pressure, between the fluidized zone, achieved an almost constant value. The estimated stabilization time was 12–16 h, approximately. Once the system has stabilized, the sampling procedure consists in closing the inlet valve just before the lower cone, and simultaneously opening the by-pass valve number 2. After that, it is necessary to keep the column for at least 12 h, until the phases have separated inside it. Finally, the volumes are measured directly through observation.

In addition, during the tests with air/oil/coke particles the gas density effect on the gas volumetric fraction was tested. After each test, the gas phase was switched to CO_2 during 30 min. This time was considered enough to get gas stabilization in this case.

3 Results and Analysis

The gas-liquid-solid phase distribution was determined in a slurry bubble column by means of cold modelling, following the experimental procedure described above. The temperature was kept at $25 \pm 2\,^{\circ}$C at atmospheric pressure. Figure 2 shows that the gas volume fraction increases with the gas flow. This agrees with the results reported by Akita and Yoshida (1973); Shah et al. (1982); Pino et al. (1990a,b, 1992); Hoefsloot and Krishna (1993); Wenge et al. (1995) and Jena et al. (2008).

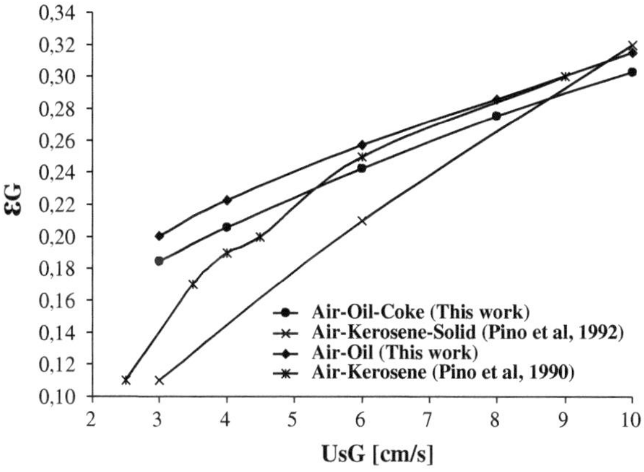

Fig. 2 Average gas volume fraction as a function of superficial gas velocity. The experimental results are compared with other experiments in the literature

The accumulated gas volume holds an average value between 0.17 and 0.30 v/v for all superficial gas velocities and solid phases tested.

Figure 2 also shows comparisons between two and three-phase studies reported by Pino et al. (1990a, 1992) and these experiments. The former authors used a bubble column of 100 mm internal diameter without gas sparger. The gas superficial velocity was varied in the range between 2 and 10 cm/s. In continuous mode operation, the average slurry superficial velocity was 0.5 cm/s. The solid concentration was 126 kg/m^3 with a particle range size of 3–180 microns and density of 4,530 kg/m^3. In the air-oil system mode, the gas volumetric phase is higher than in the air-kerosene system. This behaviour is due to the effect of the superficial liquid velocity, which is higher in the air-kerosene test. However, this effect is rather small in the two systems. On the other hand, the gas volumetric phase in the air-kerosene-solid system is lower than in the air-oil-coke testing. This performance is due to the fact that the solid concentration in the air-kerosene-solid system, is greater than in the air-oil-coke test, which increases the occurrence of bubble coalescence, thus inducing the formation of large bubbles with a high rise velocity and, consequently, a fast disengagement of the column, which then allows the gas volumetric phase to decrease (Kantarci et al. 2005; Behkish et al. 2007). Moreover, a lower average particle size facilitates particle movement throughout the bubble column, thus increasing the area of the solid-liquid contact. This also causes a decrease of the gas volumetric phase.

In Fig. 3, the effect of the gas density on the gas volumetric fraction is shown as a function of the gas velocity for the air/oil/coke and the CO_2/oil/coke systems. The trend shows that the gas volume increases with the gas velocity a little more (between 1 and 2 %) for the CO_2 system than for the air system, suggesting a small effect of the gas density between these two gases at the operation conditions tested. However, in systems with a pressure higher than atmospheric, it was found that the gas volumetric fraction increased and the number of fine bubbles formed with residence

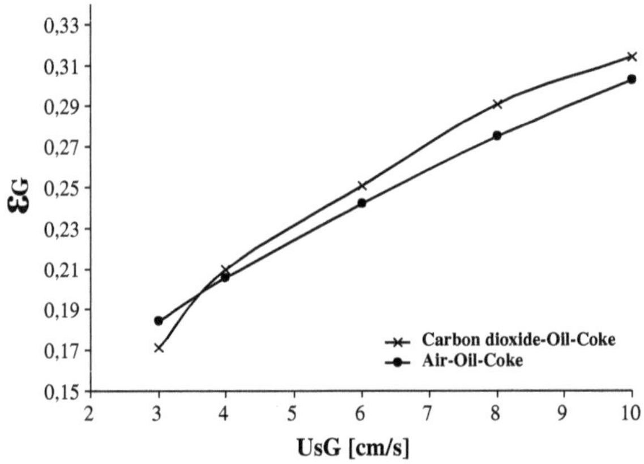

Fig. 3 Effect of the gas density on the gas volume fraction as a function of the superficial gas velocity

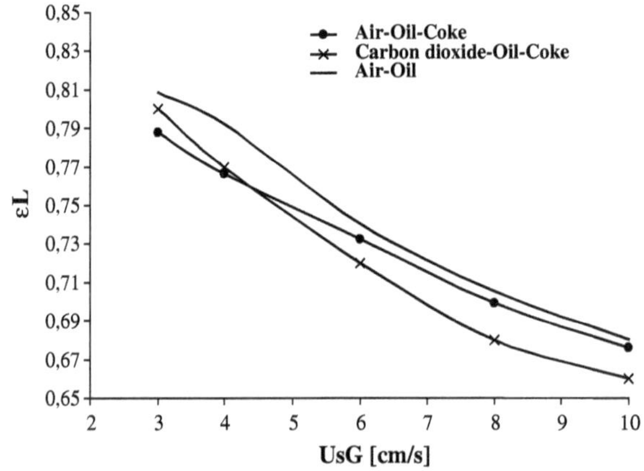

Fig. 4 Average liquid volume fraction as a function of the superficial gas velocity

time in the system was significantly higher. For columns with a pressure higher than atmospheric, the gas density can be affected in a higher proportion (Krishna et al. 1991). A large gas-phase density increases the kinetic energy and momentum of the bubbles, and this leads to an increase of the collision energy which then promotes bubble rupture (Inga and Morsi 1999). These parameters could increase appreciably the gas volumetric fraction.

According to the data plotted in Fig. 4, the average volume fraction of liquid decreases from 0.79 to 0.68 v/v as the gas velocity increases from 3 to 10 cm/s. There is a difference in the average of the liquid volume between 1 and 2 %, for the CO_2 system when compared with the air system, for gas superficial velocities of 6, 8, and 10 cm/s. This effect is, however, smaller at lower gas velocities of 3 and 4 cm/s.

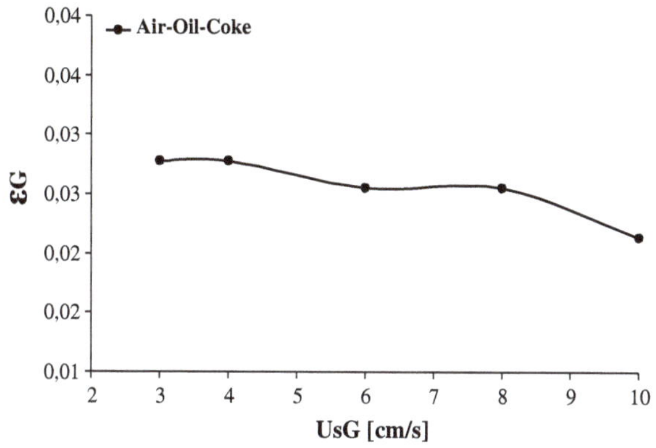

Fig. 5 Average solid volume fraction as a function of the superficial gas velocity

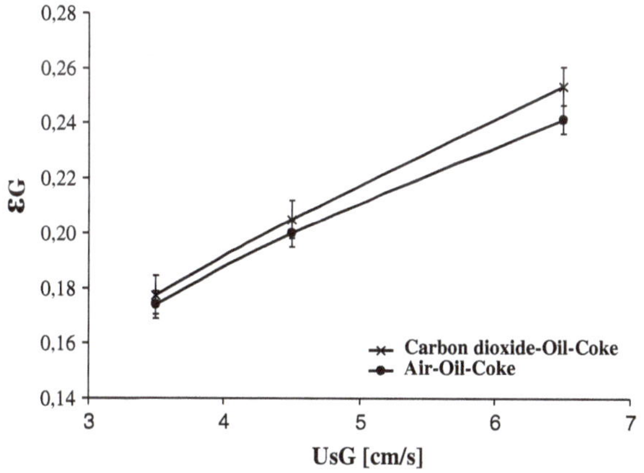

Fig. 6 Average gas volume fraction as a function of the superficial gas velocity with error bars

Looking at Fig. 5, we may see that the solid volume also decreases as the superficial gas velocity increases in a range between 0.03 and 0.02 v/v for carbon particles at the gas superficial velocity tested. The low accumulation of coke inside the column, for all gas superficial velocities tested here, indicates a high solid drag exerted by the liquid. Moreover, due to the low solid concentration inside the column, the influence on the average volume fractions of liquid is small. However, the solids might cause an antifoaming effect, decreasing the gas volume and increasing the liquid volume in the column (Guitian and Joseph 1997).

Figure 6 shows the error bars with three experimental points for the air/oil/coke and the CO_2/oil/coke tests. There can be seen a standard deviation of 0.5 % for the air/oil/coke test and of 0.7 % for the CO_2/oil/coke test. From the analysis of the experimental data obtained the following correlation was developed for the three-phase systems:

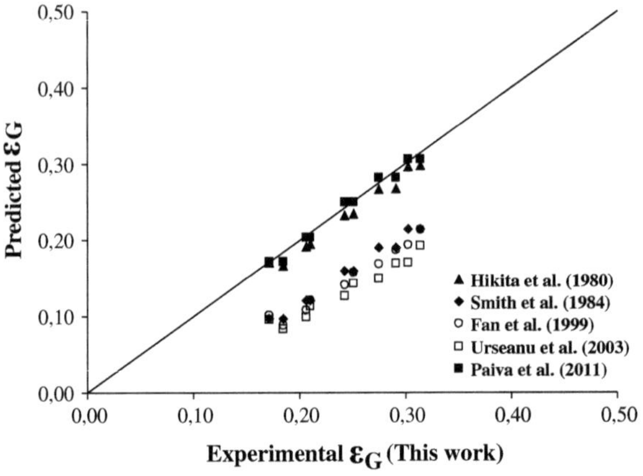

Fig. 7 Parity between the experimental (this work) and predicted average gas volume fractions

$$\varepsilon_G = 1.115 \times 10^{-1} \ln (UsG) + \left(\frac{UsG + \rho_s}{dp} \right)^{5.993 \times 10^{-3}} - 9.549 \times 10^{-1}.$$

This correlation takes into account particle properties, which appear to improve the correlation between the experimental and predicted gas volume fractions. It has an average error of 2 %, a standard deviation of 1 %, and a maximum error of 4 %, which is lower than the values obtained from other correlations in the literature as, for example, those reported by Hikita et al. (1980); Smith et al. (1984); Fan et al. (1999) and Urseanu et al. (2003) for different systems. Hikita et al. (1980) correlation had the lower average error (5 %), assuming for the gas volume fraction calculation that the slurry properties were given by the liquid properties, because their correlation applies only to two-phase systems (i.e., gas-liquid). As expected, from Fig. 7 we may see that the parity for the correlation developed here is higher than for the other correlations.

Finally, Fig. 8 shows a parity graph between the correlation obtained in this work and the experimental data of Pino et al. (1992). This latter correlation has an average error of 34.5 % in this case. This occurs because there are appreciable differences in the solid particle characteristics, the concentration of particles, and the superficial liquid velocity between both sets of data.

4 Conclusions

The gas, liquid, and solid phase distribution was estimated in a bubble column under cold conditions. It was found that the gas density does not affect significantly the volumetric fractions of gas and liquid. The low solid concentration obtained during the experiments denotes a high drag exerted by the liquid and a small effect of the

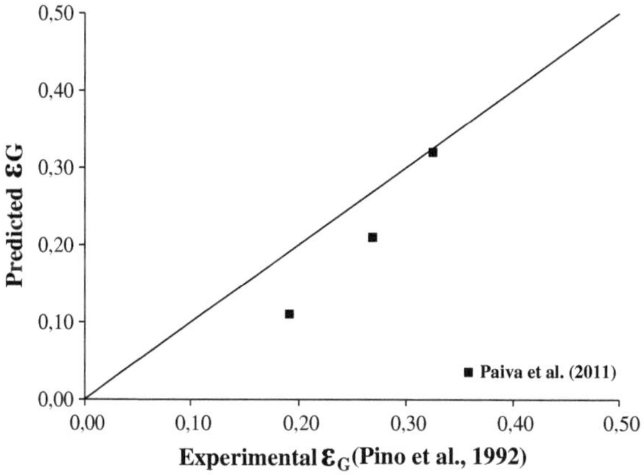

Fig. 8 Parity between experimental average gas volume fractions of Pino et al. (1992) and the predicted ones

superficial gas velocity on solid concentration. On the other hand, the comparison between two-phase and three-phase systems along with experimental data from the literature indicates that in systems with a tendency to foaming, the solids, even in small amounts, reduce the gas volume fraction. A new correlation for predicting the gas volume fraction in three-phase systems was developed, which has shown good agreement with experimental data that takes into account particle properties.

References

Akita K, Yoshida F (1973) Gas holdup and volumetric mass transfer coefficient in bubble columns. Ind Eng Chem Process Des Dev 12:76–80

Beggs HD, Brill J (1973) A study of two-phase flow in inclined pipes. J Petrol Technol 25:607–617

Behkish A, Lemoine R, Sehabiague L, Oukaci R, Morsi BI (2007) Gas holdup and bubble size behavior in a large-slurry bubble column reactor operating with an organic liquid under elevated pressure and temperatures. Chem Eng J 128:69–84

Chengtian W, Kelsey S, Muthanna H (2008) Bubble dynamics investigation in a slurry bubble column. American Ins Chem Eng 54:1203–1212

Chiu T-M, Hoh Y-C, Wang W-K (1987) Dispersed-phase hydrodynamic characteristics and mass transfer in three-phase (liquid-liquid) fluidized beds. Ind Eng Chem Res 26:712–718

Dargar P, Macchi A (2006) Effect of surface-active agents on the phase holdups of three-phase fluidized beds. Chem Eng Process 45:764–772

Fan L-S, Yang GQM, Lee DJ, Tsuchiya K, Luo X (1999) Some aspects of high-pressure phenomena of bubble in liquids and liquid-solid suspensions. Chem Eng Sci 54:4681–4709

Griffith P, Woou LC, Cheong HP, Pearson JF (1975) Two phase pressure drop in inclined and vertical oil wells. M.I.T. Cambrige, USA (SPE 5667-MS), p 1–22

Guitian J, Joseph D (1997) How bubbly mixture foam and foam control using a fluidized bed. Int J Multiph Flow 21:1–16

Hagedorn AR, Brown KE (1965) Experimental study of pressure gradients occurring during continuous two-phase flow in small-diameter vertical conduits. J Petroleum Tech 17:475–484
Hikita H, Asai K, Tanigawa K, Segawa K, Kitao M (1980) Gas hold-up in bubble columns. Chem Eng J 20:59–67
Hikita H, Asai K, Tanigawa K, Segawa K, Kitao M (1981) The volumetric liquid-phase mass transfer coeficient in bubble colums. Chem Eng J 22:61–69
Hoefsloot H, Krishna R (1993) Influence of gas density on the stability of homogeneous flow in bubble columns. Ind Eng Chem Res 32:747–750
Inga J, Morsi B (1999) Effect of operating variables on the gas holdup in a large-scale slurry bubble column reactor operating with an organic liquid mixture. Ind Eng Chem Res 38:928–937
Jena H, Sahho B, Roy G, Meikap B (2008) Prediction of gas holdup in a three-phase fluidized bed from bed pressure drop measurement. Chem Eng Res Des 86:1301–1308
Kantarci N, Borak F, Ulgen KO (2005) Bubble column reactors. Process Biochem 40:2263–2283
Krishna R, Wilkinson PM, van Dierendock LL (1991) A model for gas holdup in bubble columns incorporating the influence of gas density on flow regine transitions. Chem Eng Sci 46:2491–2496
Mitra-Majumadar D, Farouk D, Shah T, Macken N, Oh Y (1998) Two- and three-phase flows in bubble columns: tumerical predictions and measurements. Ind Eng Chem Res 37:2284–2292
Murray P, Fan L (1989) Axial solid distribution in slurry bubble columns. Ind Eng Chem Res 28:1697–1703
Pino LZ, Yépez MM, Sáez AE, De Drago G (1990) An experimental study of gas holdup in two-phase bubble columns with foaming liquids. Chem Eng Comm 89:155–175
Pino LZ, Yépez MM, Sáez AE (1990) Hydrodynamics of a semi-batch slurry bubble column a foaming liquid. AIChE J 36:1758–1962
Pino LZ, Solari RB, Siquier S, Estévez LA, Yépez MM, Saez AE (1992) Effect of operating conditions on gas holdup in slurry bubble columns with a foaming liquid. Chem Eng Comm 177:367–382
Shah YT, Kelkar BG, Godbole SP (1982) Design parameters estimations for bubble column reactors. AIChE J 28(3):353–379
Smith DN, Fuchs W, Lynn RJ, Smith DH, Hess M (1984) Bubble behavior in a slurry column reactor model. Chemical and catalytic reactor modeling. ACS Symp Ser 237:125–147
Tai-Ming C, Ying C, Wai Ko W (1987) Dispersed-phase hydrodynamic characteritics and mass transfer in three-phase (liquid-liquid) fluidized beds. Ind Eng Chem Res 26:712–718
Urseanu M, Guit RPM, Stankiewicz A, van Kranenburg G, Lommen JHGM (2003) Influence of operating pressure on the gas hold-up in bubble columns for high viscous media. Chem Eng Sci 58:697–704
Wenge F, Yusuf C, Murray M (1995) A new method for the measurement of solids holdup in gas-liquid-solid three-phase systems. Ind Eng Chem Res 34:928–935

Feasibility of Slug Flow Simulation Using the Commercial Code CFX

Mauricio A. Labarca, Juan J. González and Carlos Araujo

Abstract We evaluate the feasibility of simulating multiphase slug flow regimes in a horizontal pipe using Computational Fluid Dynamics (CFD) with a transient analysis and a Shear Stress Transport (SST) turbulence model available in the commercial code Ansys CFX, which is used as an improvement of the $k-\omega$ or $k-\varepsilon$ models. An Eulerian method is employed for solving the hydrodynamics of each fluid phase. To generate the flow regime, a sinusoidal geometric distribution of the phases is established in the computational domain, and a sinusoidal inlet time-dependent condition is used as a disturbance. Seventeen cases were simulated at different flow regimes. The results show that the slug pattern varies when the gas superficial velocity changes. The use of velocities corresponding to patterns such as the annular regime generated a phase distribution different from the slug flow even when using the same inlet function, tending to the expected morphology indicated by the Mandhane diagram in several cases. The effects of varying the amplitude of the sinusoidal-wave inlet function on the model were also analyzed. We find that a minimum of amplitude is required at the inlet to generate the slug flow pattern. An application of the model to the approximate calculation of safety factors for a pipe section subject to slug flow is given. In general, we find that it is feasible to simulate slug flow patterns with the proposed methodology using a commercial code such as Ansys CFX.

M. A. Labarca (✉) · J. J. González · C. Araujo
Universidad del Zulia, Facultad de Ingeniería, Avenida Goajira, Núcleo Técnico, Maracaibo, Venezuela
e-mail: mauriciolabarca4tw@gmail.com;

J. J. González
e-mail: gonzalezlopezjj@gmail.com

C. Araujo
e-mail: cdaraujo@fing.luz.edu.ve

L. Di G. Sigalotti et al. (eds.), *Computational and Experimental Fluid Mechanics with Applications to Physics, Engineering and the Environment*, Environmental Science and Engineering, DOI:10.1007/978-3-319-00191-3_24,

Nomenclature

r	Volume fraction (dimensionless)
ρ	Density (Kg/m ^{3})
U	Velocity (ft/s)
F	Drag force per unit volume (N m^{-3})
σ	Stress (kpsi)
$\mathbf{g}$	Gravity (m/s^2)
μ	Viscosity (m^2/s)
y	Liquid level (m)
C	Drag Coefficient (dimensionless)
Subscripts	
f	Fatigue Safety Factor
sl	Superficial
sg	Superficial gas velocity
α	Phase
y	Yield Safety Factor
l	Liquid
g	Gas liquid velocity

1 Introduction

There are many obstacles regarding the simulation of multiphase flows. For instance, it is well-known that energy, mass, and momentum transfer rates are sensitive to the geometric distribution of the phases, known as the flow regimes. One of these regimes is the slug flow pattern, which is of paramount importance in numerous industrial processes such as the production of oil and gas, the geothermal production of steam, the boiling and condensation processes, the handling and transport of cryogenic fluids, and the emergency cooling of nuclear reactors.

The primary characteristic of slug flow is its inherent intermittence of the fluid phases involved. For example, for a gas-liquid flow, an Eulerian observer looking along the axis of a pipe will see the passage of a sequence of slugs of liquid containing dispersed bubbles alternating with sections of separated flow within long bubbles. The flow is unsteady, even when the flow rates of gas and liquid are kept constant at the pipe inlet. The existence of slug flow can create severe problems for the designer or operator. For instance, the high momentum of the liquid slugs can produce considerable forces as they change direction through elbows and tees. In addition, severe damage can also take place along large piping structures as the low frequencies of slug flow can be in resonance with them. However, a number of practical benefits can also result from operating in the slug flow pattern. For example, due to the high liquid velocities, it is possible to move larger amounts of liquids in smaller lines than would otherwise be possible in two-phase flow. Moreover, these high velocities cause very high convective heat and mass transfers, resulting in very efficient transport operations.

Owing to its importance in the chemical and petroleum engineering industries, several theoretical and experimental studies have been conducted to find out and control slug flow parameters (e.g., Taitel and Dukler 1976; Nydal et al. 1992; Emerson and Leonardo 2005; Gu and Gue 2008). However, on the computational side only a few attempts have been made to simulate slug flow, most of which are related to the study of rising Taylor bubbles (Baritto and Segura 2008). Other existing numerical simulations are based on the so-called slug capturing technique in which the slug flow regime is predicted as a mechanistic and automatic outcome of the growth of hydrodynamic instabilities (Issa and Woodburn 1998; Issa and Kempf 2003). These simulation models rely on the two-fluid model (Ishii 1975), which has also been implemented in several industrial codes such as PLAC (Black et al. 1990) and OLGA (Bendiksen et al. 1991). However, it has never been demonstrated conclusively that the model can capture in a natural way the development of slug flow from the growth of instabilities in stratified flow. On the other hand, Frank (2005) and Vallée et al. (2007) have used an Eulerian model to simulate the slug flow regime in horizontal channels, noting a dependence of the results on the inlet boundary conditions. In this work, we rely on the two-phase fluid model implemented by Frank (2005) to simulate slug flow in horizontal pipes with the commercial code Ansys CFX.

2 Governing Equations

The basis of a two-fluid model is the formulation of two sets of conservation equations for the balance of mass, momentum, and energy for each of the phases. As was pointed out by Frank (2005), slug flow may include gas entrainment in the liquid phase (bubbles) and so a realistic treatment of this type of flow requires the numerical solution of these sets of partial differential equations.

The present study is based on the transport equations for an isothermal flow. Hence the equations that are solved for the conservation of mass and momentum for the gas and liquid phases, written in Eulerian form, are:

$$\frac{\partial}{\partial t}(r_\alpha \rho_\alpha) + \nabla \cdot (r_\alpha \rho_\alpha \boldsymbol{U}_\alpha) = 0, \quad (1)$$

$$\frac{\partial}{\partial t}(r_\alpha \rho_\alpha \boldsymbol{U}_\alpha) + \nabla \cdot (r_\alpha \rho_\alpha \boldsymbol{U}_\alpha \boldsymbol{U}_\alpha) = -r_\alpha \nabla p_\alpha + r_\alpha \rho_\alpha \boldsymbol{g} + \nabla \cdot (r_\alpha \mu_{\alpha\tau} (\nabla \boldsymbol{U}_\alpha + (\nabla \boldsymbol{U}_\alpha)^\tau)) + \boldsymbol{F}_\alpha, \quad (2)$$

where the subscript α refers to the gas (g) and liquid (l) phases such that $r_l + r_g = 1$. It is implicit in Eqs. (1) and (2) that there is no mass transfer between the liquid and gas phases. The term F_α in Eq. (2) stands for the frictional forces per unit volume between each phase and the walls of the pipe. The turbulent viscosity is calculated with the Shear Stress Transport model, which represents an improvement over the traditional $k - \omega$ approach since it has a better prediction of flow separation under adverse pressure gradients, and it has been used in other simulations of multiphase flow (Al Issa et al. 2007). The only interfacial force present is the drag force, which is calculated using a drag coefficient $C_D = 0.44$.

3 Boundary Conditions and Models

The boundary geometry considered here is a horizontal pipe of total length $L = 8$ m and circular cross-section of diameter $D = 54$ mm, split by a symmetry plane. The resulting mesh has 153,298 tetrahedral and prismatic elements, with four inflation layers in the circular wall. This mesh was the result of a sensitivity analysis. To do so we have used a single-phase transient model with a water inlet velocity of 1 m/s and 140 time steps. Successive runs were performed using a finer mesh each time, until the pressure drop variation was below 1 %.

A total number of 17 transient cases were chosen with different superficial velocities for the air and water, each of them corresponding to different flow regimes in the Mandhane flow diagram. The superficial velocities are used to calculate an inlet mixture velocity U_m for each case. The level of liquid at the inlet and the liquid volume fraction were initialized using the following function:

$$y_L = y_0 + A_L sin\left(2\pi \frac{U_m t}{p_L}\right). \tag{3}$$

where $y_0 = 0$, the amplitude $A_L = 0.25\ D$, and $p_L = 0.25\,L$. The wavelength and amplitude were as determined by Lex (2003). A time-independent form of this function that depends only on the geometry was used for initialization. The walls of the pipe were set to be hydraulically smooth with no-slip conditions for both phases. Outlet boundary conditions at the exit of the pipe were specified. In all cases, the velocity fields were initialized with inlet values in the whole domain area. For all time steps with 70 coefficient loops, a maximum error of 0.001 RMS is guaranteed for the numerical solutions of Eqs. (1) and (2). Such an error is sufficiently small to achieve a quantitative accurate understanding of the flow field and variables. For all runs, the time step size was set to 0.005 s, implying about 1,600 steps per run.

4 Results for Slug and Annular Flow

The results for slug flow are shown in Figs. 1 and 2, where the volume fraction and pressure gradients are depicted. The free surface is represented as an iso-surface where $r = 0.5$. The volume fraction gradients show the expected shape for the chosen velocity in the Mandhane diagram. Slug flow appears after 1.5 s and lasts for all the evolution up to 7 s. When the superficial velocity U_{sg} is decreased, the gradients

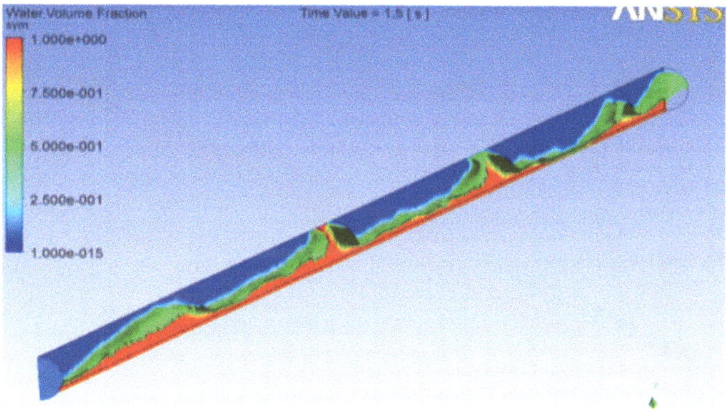

Fig. 1 Volume fraction gradients of slug flow at 1.5 s, with U_{sl} = 5 ft/s and U_{sg} = 12 ft/s

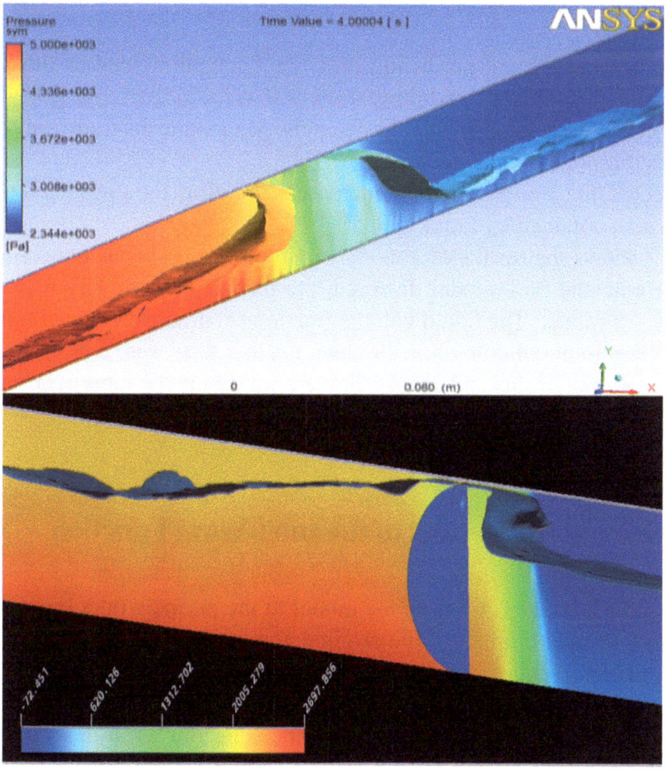

Fig. 2 Pressure gradients of slug flow at 4 s, *top* U_{sl} = 5 ft/s and U_{sg}= 12 ft/s, *bottom* U_{sl} = U_{sg}= 3.28 ft/s

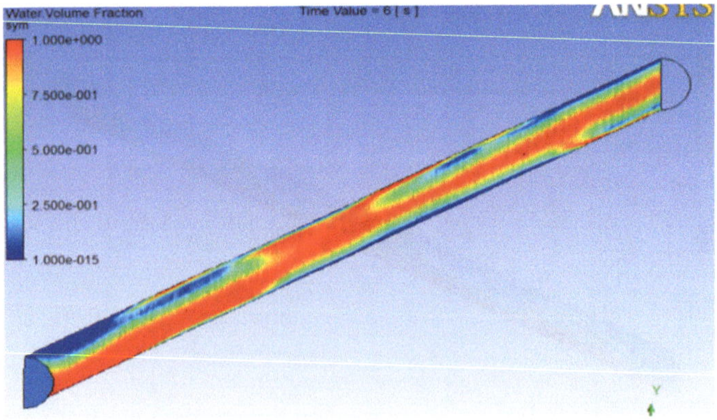

Fig. 3 Water fraction gradients for annular flow at 6 s, U_{sl} = 1 ft/s, U_{sg} = 300 ft/s

for slug flow are affected, producing pressure fluctuations due to the limitations imposed by the boundary conditions used. All velocity pairs (U_{sl}, U_{sg}) for slug behaviour come from experimental limits in the Mandhane diagram; the pairs out of these limits did not exhibit slug flow.

As expected, the pressure gradient increases just before the slug (Fig. 2). A similar result was also obtained by Frank (2005).

Figure 3 shows the results of a model simulation for annular flow. Since in this case a run with the SST model diverged, the turbulence viscosity was calculated using the $k - \varepsilon$ model. The liquid volume fraction (r) gradient evolved fully into the annular pattern in the middle of the domain. Further tests were employed to predict the boundary between slug and annular flow by increasing the superficial gas velocity U_{sg}.

5 Sensitivity of the Results to the Inlet Wave Function

In order to check the sensitivity of the model to the periodic inlet boundary condition, a case was run with half of the original amplitude chosen by Lex (2003). The superficial velocities were the same as in a previous case run (i.e., U_m = 2 ft/s), and only the amplitude was varied. The results show that there is a minimum amplitude necessary for the inlet function to generate the slug flow as in the previous cases. Figure 4 displays how, after 1.5 s, the reduced amplitude can prevent the formation of a complete slug as compared to the case where the normal amplitude is employed.

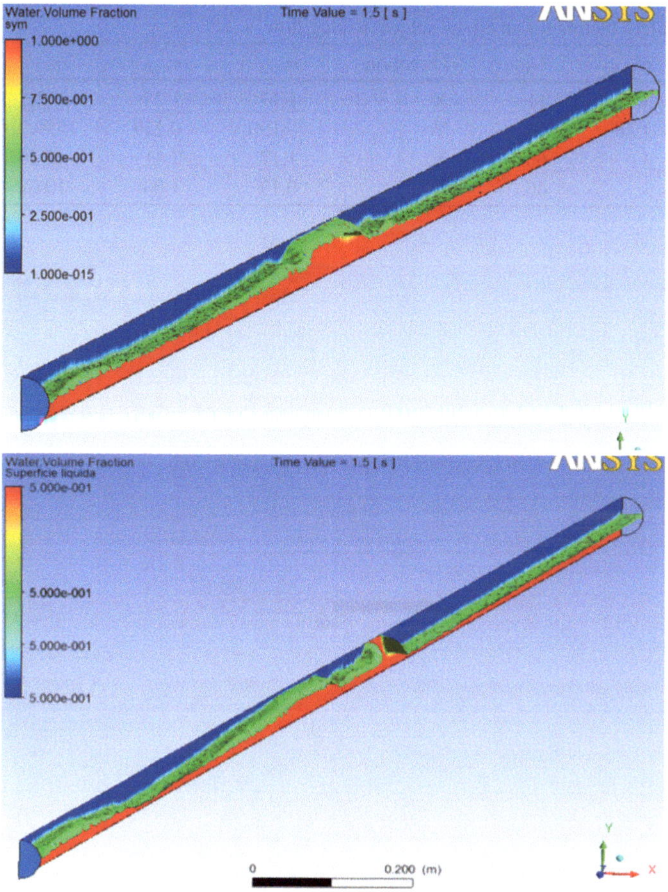

Fig. 4 *Top* slug flow, reduced amplitude. *Bottom* Slug flow, normal amplitude. Both at 1.5 s and with $U_m = 2$ ft/s

6 An Application Example

One useful application of the present simulations of slug flow is just to analyze the impact of the implicit pressure increases that are generated. Using a commercial code, it is possible to generate plots with a wall force as a function of time and this information can be used to calculate safety factors. To illustrate this, a simulation of an elbow subject to slug flow was run for two different cases. Assuming the wall stress to be a pure shear stress, in the framework of Gerber's fatigue theories, AISI 4340 properties, and a thickness of 2 mm, it is possible to calculate static and dynamic safety factors, as shown in Table 1. Figure 5 shows the geometry employed for the pipe plus elbow model.

Table 1 Safety factors for an elbow subject to slug flow

Case	U_{sl}	U_{sg}	Direction	σ_{min}	σ_{max}	n_y	n_f
C-A	3.28	11	X	1.17	1.27	111.5	112.9
C-A	3.28	11	Y	0.190	0.239	559.2	508.3
C-B	3	60	X	1.17	2.41	48.46	102.5
C-B	3	60	Y	0.19	1.00	104.8	55.70

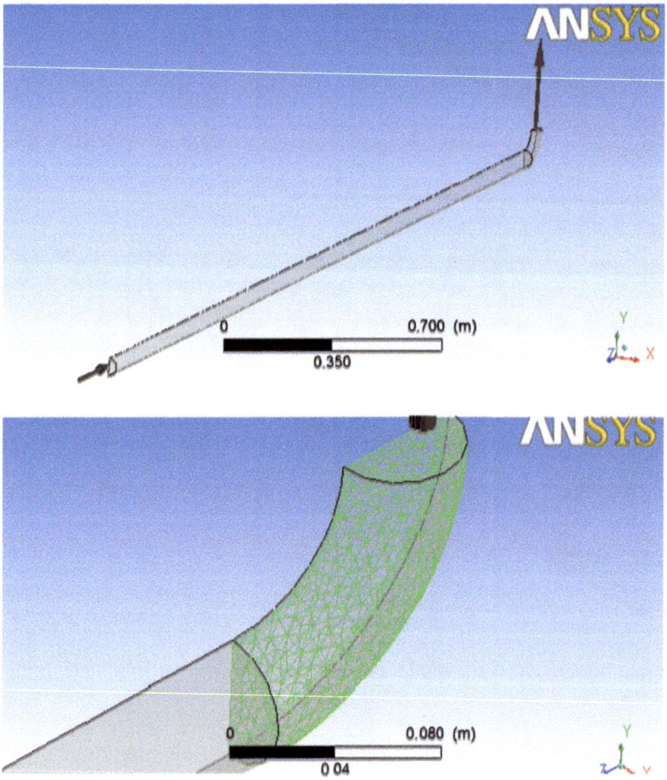

Fig. 5 *Top* Pipe and elbow geometry. *Bottom* Elbow geometry and mesh

The columns in Table 1, starting from the second, list the superficial liquid and gas velocities in units of ft/s, the direction in a Cartesian coordinate frame in which the safety factors were calculated (note that the flow is towards the z-axis), the minimum and maximum stresses exerted on the pipe walls, the yield safety factor, and the fatigue safety factor, respectively. Although the assumptions used here are ill-advised for a complete structural analysis, these data clearly show how a slight increase in the superficial liquid velocity can affect the stresses and the safety factors regardless of the gas superficial velocities.

7 Conclusion

In this chapter, we have shown preliminary results of slug flow simulations along a pipe, using a two-fluid model approach. The results of the simulations show that it is possible to obtain realistically slug flow patterns by using the commercial code Ansys CFX, with a periodic inlet function and the SST model for the calculation of the turbulent viscosity. For a sequence of runs with increasing superficial velocity, we have also possibly identified the boundary between slug and annular flow. Finally, the model was applied to determining structural safety factors in pipeline systems.

References

Al Issa S, Beyer M, Prasser H-M, Frank T (2007) Reconstruction of the 3D velocity field of the two-phase flow around a half moon obstacle using wire-mesh sensor data. In: 6th international conference on multiphase flow (ICMF 2007), Leipzig, Germany (Paper No 661), pp 14

Baritto M, Segura J (2008) Estudio numérico del ascenso de burbujas de Taylor en mini-conductos verticales de secci6n no-circular: Parte II. Revista de la Facultad de Ingeniería de la Universidad Central de Venezuela 23:37–44

Bendiksen KH, Malnes D, Moe R, Nuland S (1991) The dynamic two-fluid model OLGA: theory and application. SPE Prod Eng 6:171–180

Black PS, Daniels LC, Hoyle NC, Jepson WP (1990) Studying transient multiphase flow using the pipeline analysis code (PLAC). J Energy Resour Technol 112:25–29

Emerson DR, Leonardo GJ (2005) A non-intrusive probe for bubble profile and velocity measurement in horizontal slug flows. Flow Meas Instrum 16:229–239

Frank T (2005) Advances in computational fluid dynamics (CFD) of 3-dimensional gas-liquid multiphase flows. In: NAFEMS seminar: simulation of complex flows (CFD), Niedernhausen/Wiesbaden, Germany, pp 18

Gu H, Gue L (2008) Experimental investigation of slug development on horizontal two-phase flow. Chin J Chem Eng 16:171–177

Ishii M (1975) Thermo-Fluid dynamic theory of two-phase flow. Eyrolles, Paris

Issa RI, Woodburn P (1998) Numerical prediction of instabilities and slug formation in horizontal two-phase flows. In: 3rd international conference on multiphase flow (ICMF 1998), Lyon, France

Issa RI, Kempf MHW (2003) Simulation of slug flow in horizontal and nearly horizontal pipes with the two-fluid model. Int J Multiph Flow 29:69–95

Lex Th (2003) Beschreibungeines Testfallszurhorizontalen Gas-Flüssigkeitsströmung. Technische Universität München, Lehrstuhlfür Thermodynamik, Internal Report 3 pp

Nydal OJ, Pintus S, Andreussi P (1992) Statistical characterization of slug flow in horizontal pipes. Int J Multiph Flow 18:439–453

Taitel Y, Dukler AE (1976) A model for predicting flow regime transitions in horizontal and near horizontal gas-liquid flow. AIChE J 22:47–55

Vallée C, Höhne T, Prasser H-M, Sühnel T (2007) Experimental investigation and CFD simulation of slug flow in horizontal channels. Technical Report (Forschungszentrum Dresden-Rossendorf), pp 579–596

Heavy Oil Transportation as a Solid-Liquid Dispersion

Adriana Brito, H. Salazar, Ramón Cabello, Jorge Trujillo, L. Mendoza and L. Alvarez

Abstract Traditionally, heavy oil pipelines are designed to handle liquids with effective viscosity below 0.5 Pa s at the pump outlet, in order to minimize the frictional pressure gradient and obtain a pipeline size and economically optimum pumping requirements. Asphaltenes and resins are the components of crude oil which have the highest molecular weights and are, also, the more polar ones. It has been determined that the characteristics of the asphaltenes play an important role in the high viscosity of heavy oils of the Orinoco Oil Belt. This chapter presents an experimental investigation of the behaviour of a potential transport method for heavy oils based on precipitation and conditioning of asphaltenes, followed by an ulterior reincorporation into the de-asphalted oil to obtain a solid-liquid dispersion (slurry) with a lower effective viscosity than the one of the original crude oil. The study comprises two steps: an analysis under static conditions to identify the rheological behaviour of the

A. Brito (✉) · R. Cabello · L. Mendoza
Instituto Tecnológico Venezolano del Petróleo (PDVSA-Intevep), Urbanización El Tambor, Los Teques 1201, Estado Miranda, Venezuela
e-mail: britoah@pdvsa.com

R. Cabello
e-mail: cabellor@pdvsa.com

L. Mendoza
e-mail: mendozale@pdvsa.com

H. Salazar · L. Alvarez
Escuela de Ingeniería y Ciencias Aplicadas, Universidad de Oriente, UDO, Núcleo de Anzoátegui, Avenida Argimiro Gabaldón, Municipio Bolívar, Barcelona, Estado Anzoátegui, Venezuela
e-mail: salazar.heider@gmail.com

L. Alvarez
e-mail: lalvarezm57@gmail.com

J. Trujillo
Multiphase Solutions Inc. (MSi) Kenny Australia, Level 3, Wood Group House, 432 Murray Street, Perth, WA 6000, Australia
e-mail: jorgenicolas71@gmail.com

L. Di G. Sigalotti et al. (eds.), *Computational and Experimental Fluid Mechanics with Applications to Physics, Engineering and the Environment*, Environmental Science and Engineering, DOI:10.1007/978-3-319-00191-3_25,

slurry for different solid contents, from 0 to 12 % (weight basis), and a fluid dynamic study to characterize the effectiveness of the solid–liquid dispersion method in a laminar flow regimen in a 1 inch horizontal pipeline, for mixture velocities between 0.2 and 2.3 m/s, corresponding to Reynolds Number values $<1{,}400$. A maximum effective viscosity of 0.15 Pa s @ 20 °C was measured 24 h after conducting the dynamic test, which implies a significant reduction compared to a typical viscosity range of 100–1,000 Pa s @20 °C for an original crude oil of similar API density and SARA composition. As expected, dispersion viscosity increases with time as asphaltenes are progressively reabsorbed into the de-asphalted oil as a colloidal suspension. The investigated transport method can be implemented together with a low pressure–low temperature de-asphalting process to improve transport properties of the heavy oils of Orinoco Oil Belt.

1 Introduction

Heavy oil production and transportation are particularly difficult processes due to the high viscosities of the heavy oil and the complexity of the multiphase flow involved. Most heavy oil transport methods are focused on reducing the liquid—wall shear stress. To achieve this, some methods as core annular flow and oil in water emulsions induce the water phase to be in contact with the wall, as water viscosity is several orders of magnitude lower than heavy oil viscosity. Other methods as, for example, partial or total upgrading and slurry transportation focus on reducing the viscosity of heavy oil itself, by removing or modifying the nature of the crude compounds, which are the main cause of the high viscosity.

Venezuela has 296,500 million barrels of heavy and extra heavy oil certified reserves, located in the Orinoco Oil Belt. It is expected that several of the new joint ventures established to exploit the heavy and extra heavy oil from the Orinoco Oil Belt include as a transportation method an integration of dilution and upgrading of the oil in order to convert extra heavy crude oil of approximately 8° API into synthetic crude of 42° API. This upgraded oil will be mixed with an original virgin heavy oil to obtain a final crude oil of 22° API to be transported to the Punta Araya terminal for export (Petróleos de Venezuela 2011; Brito and Trujillo 2011). The main strategy to be developed for the Orinoco Oil Belt is a combination of dilution and upgrading of the extra heavy oil. However, in order to make feasible all the developments, it is necessary to explore other alternatives for heavy oil transportation. This chapter presents a study of a potential transportation method based on transforming the heavy oil into a solid–liquid suspension with lower effective viscosity than the original crude oil. A simple alternative to the heavy oil transportation method was developed by Argillier et al. (2006). They presented a solid-liquid dispersion (slurry), where the asphaltenes are precipitated by n-alkanes as n-heptanes and reincorporated to the oil to obtain a slurry or suspension of non-colloidal particles with relatively low viscosity compared to the original extra heavy oil (Martínez-Palou et al. 2011).

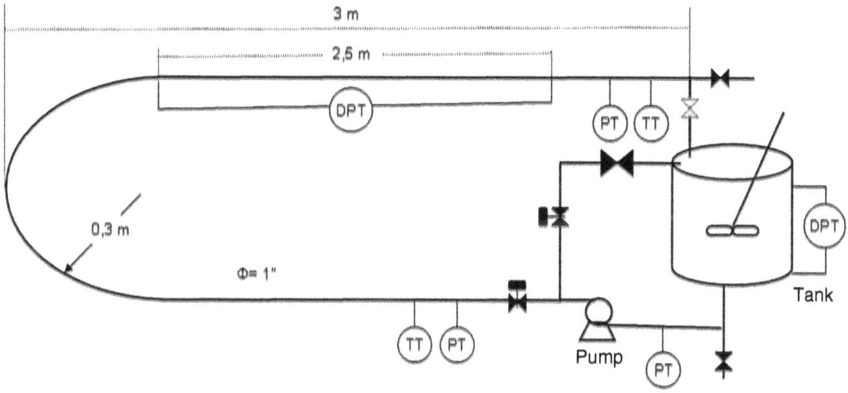

Fig. 1 Experimental test loop diagram

One of the main characteristics of heavy oil is its large content of asphaltenes that contribute to its high viscosity. For instance, Sanière et al. (2004) found that the slurry exhibits approximately 75 % less viscosity than the original crude oil. However, they observed the slurry during 146 days and observed that with time the suspension returns progressively to a colloidal stage. This transformation is quite slow and gives rise to a correlated increase of viscosity (at 20 °C) that can be accelerated at higher temperatures (40 °C). They also report that there is a critical concentration for asphaltenes (around 12 %): under this critical concentration the asphaltenes behave like colloidal particles dispersed in oil as a semi-dilute domain, while above this value the particles of asphaltenes are overlapped and the viscosity increases dramatically.

2 Experimental Test

Experiments were carried out in an Ø0.025 m test loop facility at PDVSA INTEVEP. Figure 1 show the experimental facility diagram, the total length of the loop is 6 m of carbon steel pipeline and is composed of two sections: the first one is a 3.5 m length section for flow development; the second one is the test section, which is 2.5 m long and equipped with transmitters for measurements of pressure, temperature, and differential pressure. Differential pressure transducers are used to measure the pressure drop between pressure taps. Downstream to the test section, the dispersion returns to the tank in which the slurry is continuously agitated. The slurry is pumped using a gear pump with a variable speed control to regulate the flow rate, the maximum dispersion flow rate is approximately $3.6\,m^3/hr$ and it is measured by deviating the flow to a small tank with a level transmitter.

The slurry is prepared with de-asphalted oil (also known as maltenes) as the liquid phase and the asphaltic residue as the solid phase, with both phases being obtained from a proprietary de-asphalting process at relatively low pressure and temperature.

The solid phase, composed mainly of asphaltenes and resins, is prepared and sieved with a special procedure in which the particle size is kept up to 500 μm, then the solids are incorporated in the liquid under a mechanical stirring for 20 min. Dispersion properties are analyzed for each solid concentration and mixture velocities between 0.2–2.3 m/s at three stages, i.e., before starting the dynamic test (initial stage), just after running the dynamic test (middle stage), and 24 h after finalizing it. The dispersion morphology is analyzed using optical microscopy. The viscosity of the dispersion is assessed using a viscometre Anton Paar model MCR 301 according to the ASTM D 7,483 standard test method. Finally, the dispersion density is quantified using a densitometre Anton Paar DMA 4,500 according to the ASTM D 4,052 standard test method.

3 Experimental Results

Figures 2 and 3 show that the precipitated asphaltenes become progressively colloidal with time due to the viscosity and opacity of the samples, increasing both with time and solid concentration in agreement with the findings of Sanière et al. (2004). This evolution is probably due to the fact that the thermodynamic equilibrium state promotes the colloidal state rather than the precipitated one.

Similarly to the results obtained by Sanière et al. (2004), the slurry sample remains Newtonian and does not exhibit yield strength, which is quite unusual for slurry products. During the dynamic tests, it was observed that the viscosity and density of the fluid increase with the solid concentration, as expected, and decrease as the flow rate is increased, as is shown in Figs. 4 and 5. This last phenomenon is believed to be due to the significant viscous dissipation of energy inside the pump and pipes, which promotes an increase of the temperature.

As mentioned above, the temperature rises in the system because of the increasing viscous dissipation accompanying the higher flow rates. For a given concentration, the higher viscosities correspond to the lower flow rates and accordingly, the lower viscosities correspond to the higher flow rates (Fig. 5). As expected, the pressure drop in the pipeline increases with mixture velocity and solid concentration in the slurry (see Fig. 6).

The dispersion viscosity was measured at 20 and 26 °C. When comparing the results obtained during the initial, middle, and final stages, small differences in the viscosity are found for the solid concentrations, which are below 12 %. However, for a 12 % solid concentration, the viscosity differences with time are clearly noticeable. This could be explained by a re-dissolution of asphaltenes into the maltenes, thus returning to their colloidal state. The 12 % threshold seems to coincide with the findings of Sanière et al. (2004) for the critical asphaltene concentration in the Orinoco Belt heavy oils.

Concentration (% w/w)	Initial	Final
1		
3	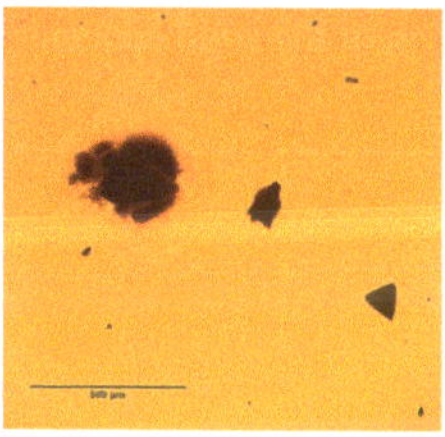	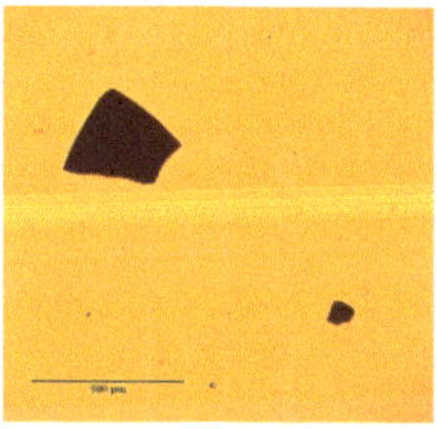
6	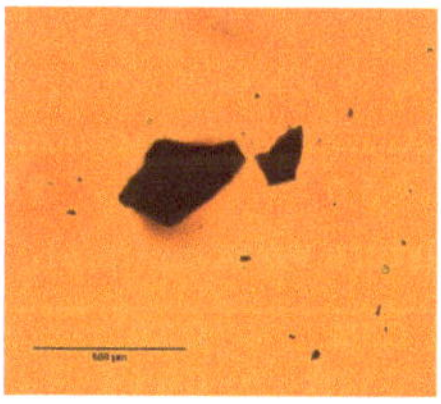	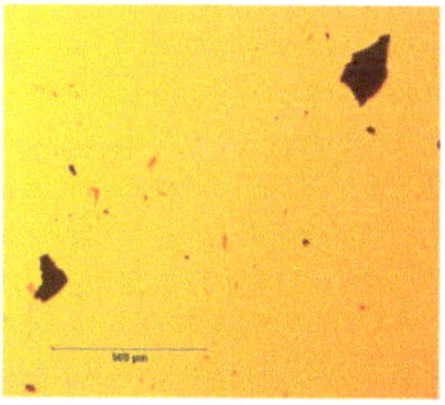
8	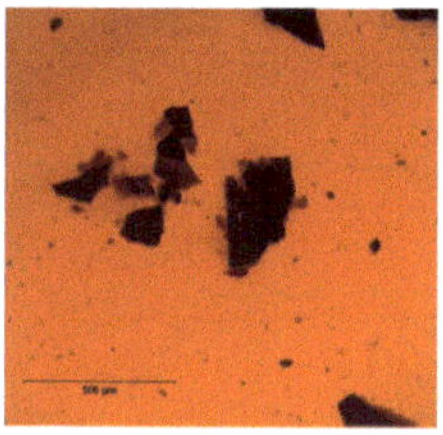	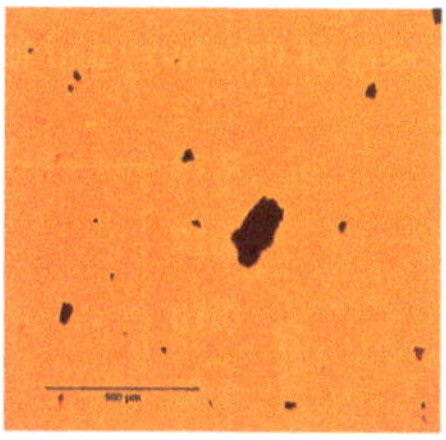
12	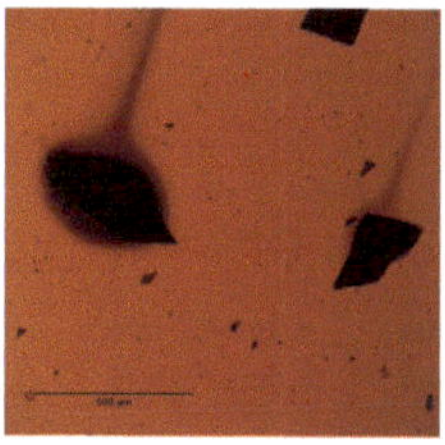	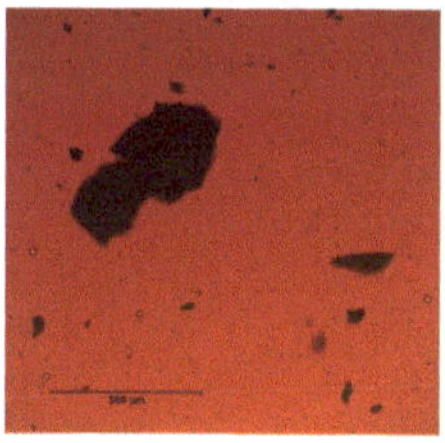

Fig. 2 Evolution with time of the morphology of the slurry sample

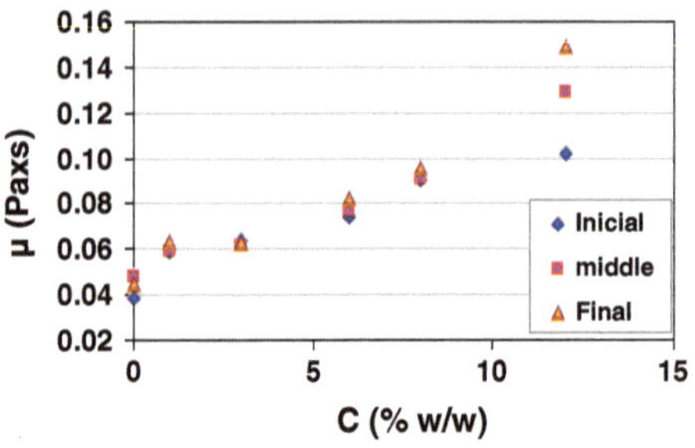

Fig. 3 Viscosity of the slurry as a function of the solid concentration at 20 °C

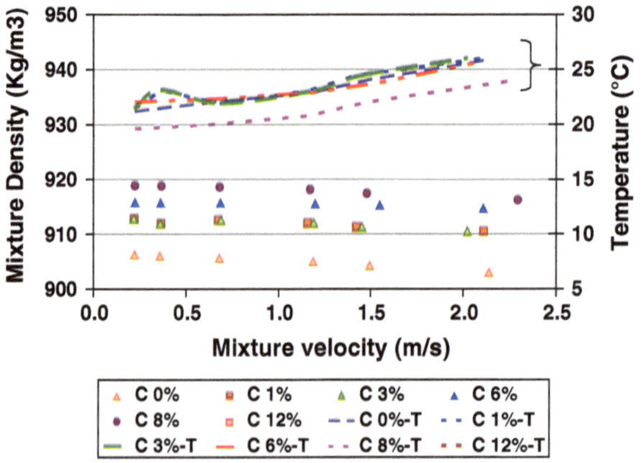

Fig. 4 Density of the slurry and temperature of the system as a function of the mixture velocity

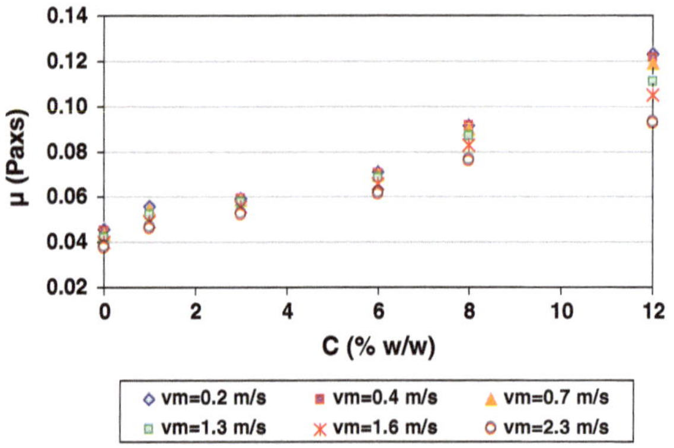

Fig. 5 Slurry viscosity as a function of the solid concentration in the slurry for different mixture velocities

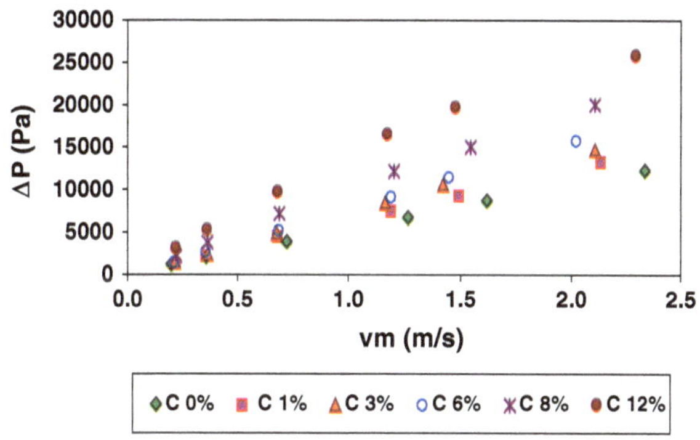

Fig. 6 Pressure drop in the horizontal pipeline as a function of the mixture velocity and solid concentration in the slurry

4 Conclusions

- The morphology and viscosity of the slurry are time-dependent. The solid asphaltenes obtained from a proprietary de-asphalting process are reincorporated as a solid into the maltenes and tend to return to a colloidal state, increasing the viscosity of the slurry with time.
- The viscosity of the slurry is at least three orders of magnitude lower than the typical viscosity of the heavy oil of the Orinoco Oil Belt.
- This transportation method could be implemented downstream of a de-asphalting process, in order to obtain the components needed to prepare the dispersion.
- Additional residue processing is necessary depending on the de-asphalting method used for obtaining the residue as a solid phase, which must be easy to handle and sieve in order to be reincorporated into the de-asphalted oil to prepare the dispersion.
- This transportation method could be used not only for the transport of heavy oils. It could be applied to transport other solid residues of hydrocarbon processes as is the case of coke, either in water or in synthetic oil.

Acknowledgments The authors thank the support of PDVSA Intevep.

References

Argillier J, Hénaut I, Gateau P (2006) Method of transporting heavy crude oils in dispersions. US Patent Application 20,060,118,467
Brito A, Trujillo J (2011) Considering multiphase flow issues for selection of heavy oil transportation methods in Venezuela. WHOC11-107-2011

Martínez-Palou R, Mosqueira M, Zapata-Rendón B, Mar-Juárez E, Bernal-Huicochea C, Clavel-López J, Aburto J (2011) Transportation of heavy and extra-heavy crude oil by pipeline: a review. J Pet Sci Eng 75:274–282

Petróleos de Venezuela (2011) http://www.pdvsa.com/index.php?tpl=interface.sp/design/readsearch.tpl.html&newsid_obj_id=8975&newsid_temas=0. Accessed 15 Feb 2011

Sanière A, Hénaut I, Argillier J (2004) Pipeline transportation of heavy oils, a strategic, economic and technological challenge. Oil Gas Sci Technol Revue d'IFP 59(5):455–466

Comprehensive Evaluation of Gas-Liquid Cyclonic Separation Technologies

Yessica Arellano, Adriana Brito, Jorge Trujillo and Ramón Cabello

Abstract PDVSA-Intevep has developed a portfolio of technologies for gas–liquid phase separation based on centrifugal forces effects on fluids of different densities. Research has been focused on both separation technologies cylindrical–conical cyclonic (CYCINT®) and multiple cylindrical cyclones (CIMCI®), contemplating numerical modeling, construction, and extensive experimental tests conducted for a wide range of inflow rates and multiphase mixture properties (Brito et al. 2001, 2003, 2009; González et al. 2002; Martínez 2002; Carrasco 2008; Matson and Brito 2008; Cáliz et al. 2009; Valdez et al. 2009; Martínez 2010). Cyclonic separators are centrifugal technologies whose geometry construction promotes rotational flow within them. Centrifugal forces generated inside the separators conduct the fluid to follow a spiral trajectory with the heavier phase forced to flow nearby the separator walls, whilst the lighter phase is directed to the centre of the equipment ascending to the top of the device. This paper presents a comprehensive quantitative evaluation methodology based on a thorough parametric matrix developed to screen the most promising technologies based on experimental essays results. As a consequence, an optimal allocation of resources will allow further development of the top ranked technologies to conduct further field tests. The processing of experimental data from laboratory tests conducted on cyclonic technologies comprises parameters of great

Y. Arellano (✉) · A. Brito · R. Cabello
Instituto Tecnológico Venezolano del Petróleo (PDVSA-Intevep),
Urbanización El Tambor, Los Teques 1201, Estado Miranda, Venezuela
e-mail: arellanoy@pdvsa.com

A. Brito
e-mail: britoah@pdvsa.com

R. Cabello
e-mail: cabellor@pdvsa.com

J. Trujillo
Multiphase Solutions Inc. (MSi) Kenny Australia, Level 3, Wood Group House,
432 Murray Street, Perth, WA 6000, Australia
e-mail: jorgenicolas71@gmail.com

L. Di G. Sigalotti et al. (eds.), *Computational and Experimental Fluid Mechanics with Applications to Physics, Engineering and the Environment*, Environmental Science and Engineering,
DOI:10.1007/978-3-319-00191-3_26,

interest for the purpose of this evaluation. Gas carry under, liquid carry over, pressure loss, and generated G forces, in hand with liquid level control strategies, operational envelope width, operability, and compact design are some of the parameters used for the evaluation of technologies considered in this study. The evaluation of parameters was conducted through group categorization followed by variables grading on a 0–8 scale by means of a binary comparison methodology. The evaluation of technologies was conducted based on the results obtained during experimental tests and further analysis. As a result, an unbiased technology ranking was obtained, in which the multi-cylindrical technology (CIMCI®) provides an overall best performance with emphasis in a superior gas separation efficiency and easier constructability, whilst the cylindrical-conic cyclonic technology (CYCINT®), on the other hand, presented the upmost liquid separation efficiency and wider operational envelope. Further efforts will focus on continuous development of these two technologies to provide more compact, efficient, and economical gas–liquid separation solutions.

1 Methodology

Technology evaluation was conducted through a multiple binary decision method that reduces the subjectivity involved in decision making processes by providing binary parametric evaluation through matrix construction.The methodology assigns weight factors to the different parameters by means of a one-to-one comparison providing a matrix from which the proposed alternatives can be selected based on the highest scores obtained.

Experimental data processing from laboratory tests conducted on cyclonic technologies comprise the parameters shown in Fig. 1.

Further parametric evaluation was conducted through a two stages process: first a group categorization, and then variable grading, following the multiple binary decision methodology.

2 Cyclonic Technologies

CYCINT® cyclonic separation device is a vertical cylinder attached to a conical section. Its working principle is based on centrifugal forces that induce vortex formation, generated when the fluid enters the inlet nozzle, inducing a significant angular momentum that will not allow the heavier phase to turn as rapidly as the lighter phase, and then separating the liquid from the gas. Figures 2 and 3 present versions of the CYCINT® and CYCINT ER®, respectively.

The main difference between both devices is the inlet and gas regions geometry. CYCINT® inlet geometry emulates a 90° bend that allows the tangential entrance of the fluid to the conic section. The CYCINT ER®, on the other hand, incorporates a straight inlet in addition to a vortex finder at the top of the cylinder.

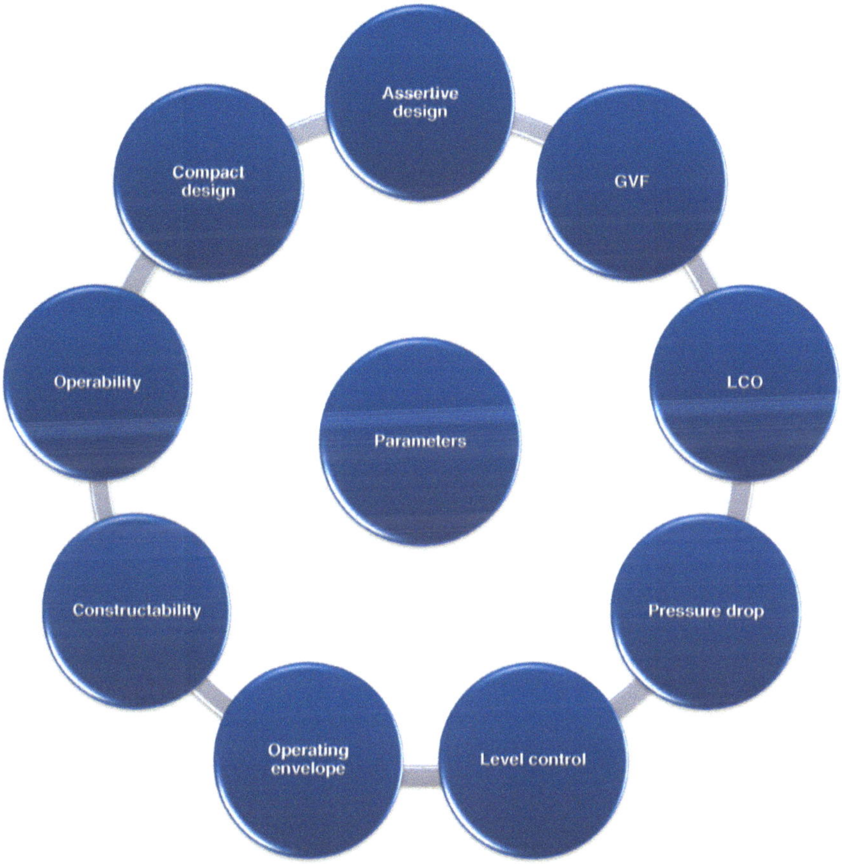

Fig. 1 Selected parameters for technology evaluation

CIMCI® separator is a device conformed by multiple cylinders; the inlet configuration is variable and depends on the type of flow: straight-through or reverse flow. Centrifugal forces generated within the separator force the fluid to flow following a spiral trajectory diminishing phase re-entrainment. Figures 4 and 5 show versions of the CIMCI® and CIMCI UP® whose main geometric difference is the inlet position located either in the middle section or the bottom of the cylinders for the CIMCI® and CIMCI UP®, respectively.

The paths of the fluid phases are different for both multiple cylindrical separators: within the CIMCI® the liquid attaches to the walls and descends to the bottom, meanwhile the gas near the walls follows a descending trajectory but as it migrates from the walls to the core zone, it reverses its flow direction following and ascending trajectory in the centre (reverse flow type), whilst in the CIMCI UP® the heavier phase is forced to ascend through the cylinder and leave the device through the annular space between the cylinder body and the vortex finder, meanwhile the gas also follows an ascending trajectory but leaves the device through the vortex finder (straight through type).

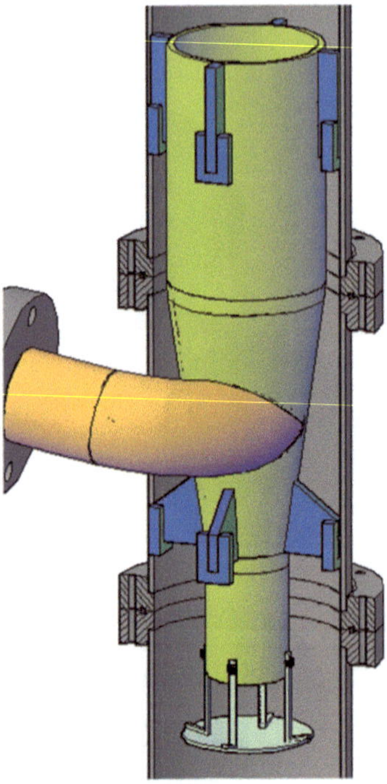

Fig. 2 CYCINT®

3 Technology Evaluation

3.1 Experimental Results

Table 1 summarizes experimental test results conducted on the four cyclonic prototypes with two different two-phase mixtures (water-air and oil-air) at different inlet flow rates. The experimental measurements have focused on determining actual device operational envelope, gas carry under, liquid carry over, total pressure drop, tangential velocities, centrifugal forces, and predominating flow patterns. The experimental test results are shown in Table 1. Elements shaded in gray and blue represent the best and worst figures respectively for the given parameter.

3.2 General Matrix Weighting

Experimental data processing comprises not only direct measurements from laboratory tests but also parameters of great interest for the purpose of this evaluation. Following this, gas carry under (GCU), expressed as gas void fraction (GVF) in the liquid stream, liquid carry over (LCO) in the gas stream, pressure drop in the

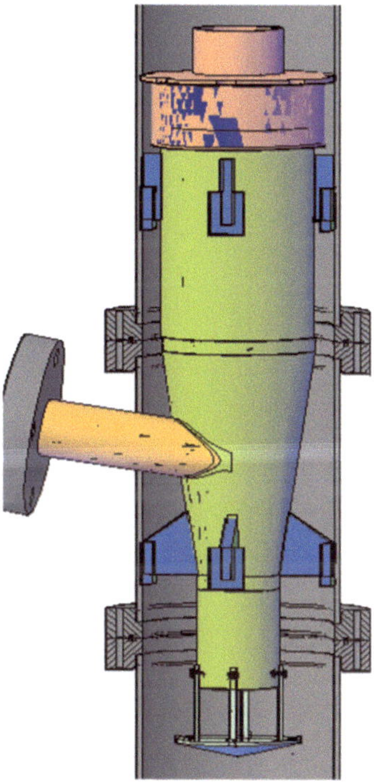

Fig. 3 CYCINT ER®

separator, and generated centrifugal force, expressed as G forces, were considered for the technologies evaluation in hand with liquid level control, operational envelope width, operability, and compact design.

The evaluation of parameters was conducted through group categorization followed by variables grading on a 0–8 scale by means of a binary comparison methodology (we refer to Appendix A for details). On the other hand, technology evaluation was conducted based on the results obtained during experimental tests and further analysis, resulting in the technology weighting matrix shown in Table 2.

Based on the unbiased technology ranking obtained, a percentage distribution graph was constructed to better illustrate the technology ranking (see Fig. 6). Evaluation results show that the multi–cylindrical technology (CIMCI®) provides an overall best performance with emphasis in a superior gas separation efficiency and easier constructability, whilst the cylindrical–conical technology (CYCINT®), on the other hand, presented the upmost liquid separation efficiency and wider operational envelope.

Fig. 4 CIMCI®

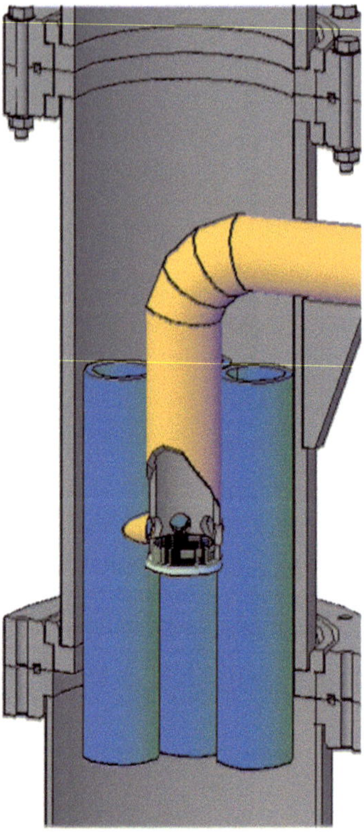

4 Analysis of the Parameters and Geometry Relations

A separation device classification was conducted based on the most representative geometric characteristics by region (inlet, gas, and liquid regions). From the analysis conducted here it was possible to identify the parameters whose magnitude was greatly influenced by the main geometric characteristics.

According to the results of the different experimental tests conducted with each cyclonic technology, it was observed a clear correlation between the devices geometry and their performance. Regarding the inlet area, there is a close relationship between the geometry of the devices and their performance when inlet operational conditions change; promotion of stratified flow in the inlet has been proven to contribute to the vortex formation and subsequent enhanced separation efficiency. Laboratory experiences also indicate that the use of a straight inlet for the CYCINT ER® combined with a vortex finder device results in a 40 % increase in gas handling capacity and lower gas carry under figures.

Similarly, the geometry of the gas region is closely related to the amount of liquid entrainment into the gas stream. After installing a vortex finder at the top of the cylindrical section of the CYCINT®, its latter version (CYCINT ER®) presented a

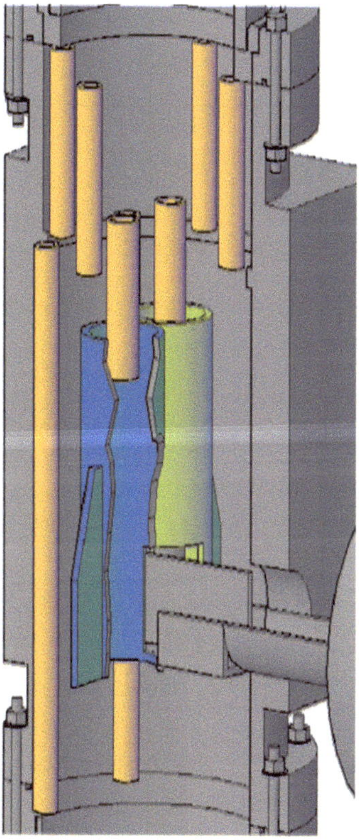

Fig. 5 CIMCI UP®

decrease in liquid carry over and better overall performance. Vortex finder installation, however, constricts the gas flowing area, promoting higher pressure drops.

5 Conclusions

The main conclusions can be summarized as follows:

- Multi-cylindrical technology (CIMCI®) provides an overall best performance.
- Multi-cylindrical technologies (CIMCI® and CIMCI UP®) provide superior gas separation efficiency than cylindrical–conical technologies.
- Multi-cylindrical technologies present significant advantages regarding adaptability, constructability, and compact design.
- Cylindrical—conical cyclonic technologies (CYCINT® and CYCINT ER®) present the best liquid separation efficiency and wider operational envelope.
- There is a clear correlation between the geometry of the devices and their performance.

Table 1 Experimental test results

MODEL	CYCINT®	CYCINT ER®		CIMCI®	CIMCI®	CIMCI UP®
Feeding	Inclined	Inclined		Horizontal	Horizontal	Horizontal
Inlet	Tangential	Straight		Tangential middle zone	Tangential middle zone	Tangential lower zone
Fluids	Water-Air	Water-Air Qg = 70	Oil-Air Qg = 26	Water-Air	Oil-Air	Water-Air
Operational envelope	Qg = 70	Ql = 706	Ql = 1500	Qg = 50	Qg < 50	Q = 50
	Ql = 1000	Qg = 50 Ql = 1200	Qg = 50 Ql = 91	Ql = 800	Ql = 800	Ql < 700
Gas void fraction	<10 % @ Ql < 1000	10 % @ Ql < 1000	16 % @ Qg = 70 Ql = 1500	<5 %	10 % Qg = 50 Ql = 800	<5 %
Liquid carry over	Qg = 50.9 Ql = 1000	Qg > 50	Qg > 50 Ql = 500	Qg = 50 Ql = 1000	Qg = 50,9	Qg = 50 Ql = 500 >5@ Ql = 700
Pressure drop	2 psi	4 psi	2 psi	<5 psi	<5 psi	<5@ Ql = 500
Tangential velocity /	2–24 ft/s	2–24 ft/s	2.6–84 ft/s	8–37 ft/s	2.6–7.5 ft/s	7–130 G
G Force	0.9–10.5 G	0.9–10.5 G	1.1–49.9 G	5–22 G	1.1–3.2 G	
Flow pattern	Stratified	Stratified	Wavy stratified	Wavy stratified/slug	Wavy stratified/slug	

Qg: MSCFD
Ql: BPD

Table 2 Technology weighting results

Parameter		Weight	CYCINT®	CYCINT ER®	CIMCI®	CIMCI UP®
A	GVF	16.7	0.0	2.8	8.3	5.6
B	LCO	22.2	4.4	13.3	4.4	0.0
C	Pressure drop	11.1	5.6	3.7	1.9	0.0
D	Level control	19.4	0.0	3.2	9.7	6.5
E	Operational envelope	13.9	4.6	6.9	2.3	0.0
F	Constructability	5.6	0.0	0.9	2.8	1.9
G	Operability	0.0	0.0	0.0	0.0	0.0
H	Compact design	2.8	0.5	0.5	0.9	0.9
	Score	100	16	31	35	18

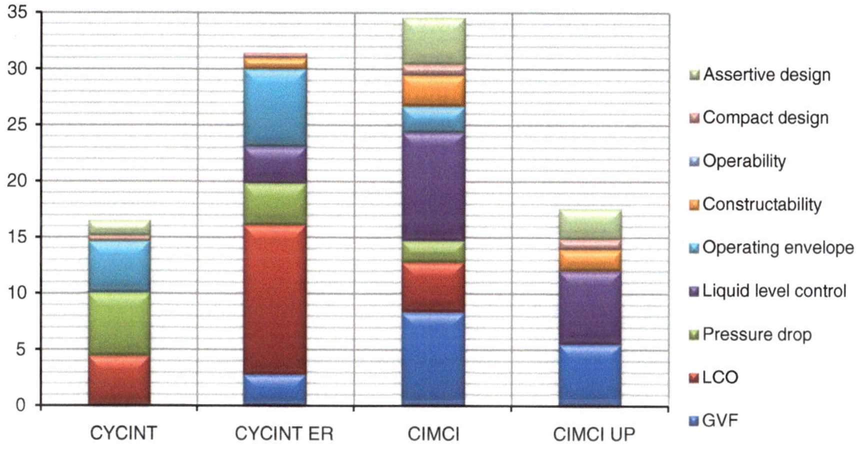

Fig. 6 Technology evaluation results

Table A.1 Parameters' comparison matrix (MBDM)

	A	B	C	D
A	–	0	1	1
B	1	–	1	0
C	0	0	–	1
D	0	1	0	–

Appendix A: Multiple Binary Decision Method

The binary comparison methodology employed for the technology evaluation is the Multiple Binary Decision Method (MBDM). The MBD method is used to assign weighting factors to different parameters comprised in an evaluation matrix and selecting, amongst different alternatives, the one that best qualifies according to the scores obtained. The procedure is detailed below and explained through a generic example:

(1) Selection of the more relevant parameters to be considered. These parameters should be precisely defined in order to quantitatively assess the alternatives under evaluation.
(2) Each selected parameter is assigned a weight resulting from a one-to-one comparison. This comparison determines which one of the evaluated parameters is the most important, by assigning it the value of '1' and the least important resulting with a '0' weight; following this procedure each parameter is compared to the remaining parameters. An illustration of the matrix obtained is shown in Table A.1.

Table A.2 Parameters' comparison matrix with weights (MBDM)

	A	B	C	D		SW	Weight (%)
A	–	0	1	1		2	33.3
B	1	–	1	0		2	33.3
C	0	0	–	1		1	16.6
D	0	1	0	–		1	16.6
					ST =	6	

Table A.3 Alternatives' comparison matrix for parameter A (MBDM)

	I	II	III		SW	Weight (%)
I	–	0	1		1	33.3
II	1	–	1		2	66.7
III	0	0	–		0	0.0
				ST =	3	

Table A.4 Alternatives' comparison matrix for parameter B (MBDM)

	I	II	III		SW	Weight (%)
I	–	0	1		1	33.3
II	1	–	1		2	66.7
III	0	0	–		0	0.0
				ST =	3	

(3) Once the one-to-one comparison is completed and the indicative 'ones' and 'zeros' are obtained, the parameter weighting factors are computed by applying the following equation:

$$weight = \frac{SW}{ST} \times 100, \tag{A.1}$$

where SW represents the weight of each parameter and ST is the total sum of the parameters' scores.

Table A.2 is complemented to illustrate the weighting distribution.

(4) Once the parameter weighting factors are obtained, the alternatives are evaluated. For the purpose of this illustration, three alternatives are proposed (I, II, and III). To obtain the most favourable alternative, all alternatives are compared to one another in reference to an alternate defined parameter. This way, alternatives I and II are compared to each other for parameter A, the alternative with the best performance gets a '1'; later alternatives II and III are compared and so on. Applying the weighting equation, the procedure is repeated, obtaining the alternatives' scores by parameter. Tables A.3, A.4, A.5 and A.6 illustrate the procedure.

Table A.5 Alternatives' comparison matrix for parameter C (MBDM)

	I	II	III		SW	Weight (%)
I	–	1	1		2	66.7
II	0	–	0		0	0.0
III	0	1	–		1	33.3
				ST =	3	

Table A.6 Alternatives' comparison matrix for parameter D (MBDM)

	I	II	III		SW	Weight (%)
I	–	1	1		2	66.7
II	0	–	0		0	0.0
III	0	1	–		1	33.3
				ST =	3	

Table A.7 Technology comparison general matrix (MBDM)

	Weight (%)	ALT. I (%)	ALT. II (%)	ALT. III (%)
Parameter A	33.3	(33.3 × 33.3) = 11.1	(66.7 × 33.3) = 22.2	(0 × 33.3) = 0
Parameter B	33.3	(33.3 × 33.3) = 11.1	(66.7 × 33.3) = 22.2	(0 × 33.3) = 0
Parameter C	16.6	(66.7 × 16.6) = 11.1	(0 × 16.6) = 0	(33.3 × 16.6) = 5.5
Parameter D	16.6	(66.7 × 16.6) = 11.1	(0 × 16.6) = 0	(33.3 × 16.6) = 5.5
Score	100.0	44.4	44.4	11.0

Following the previously described steps the alternative's partial score by parameter is obtained.

(5) Scores obtained in Tables A.3, A.4, A.5 and A.6 are then weighted by the specific weight computed for each parameter within the Parameters' Comparison Matrix (Table A.2). To exemplify this, take alternative II's weight for parameter A (66.7 %), parameter A weights 33.3 % according to Table A.2, thereafter alternative II score within the general matrix is computed as follows:

$$\frac{66.7 \times 33.3}{100} = 22.2\,\text{points}. \tag{A.2}$$

Scores obtained from Eq. (A.2) are later tabulated and added together to obtain the general score for every alternative. The alternative with the highest score will be the preferred one. Table A.7 illustrates the general matrix of technology selection.

References

Brito A, Colmenares J, Perdomo Y, Ortega P, Corrales F, Zapata N, Chacón L, Flores R, Lara J, Castillo F (2003) Evaluación del comportamiento del separador gas-líquido ciclónico CYCINT en función de las propiedades del fluido. INT-10139, 2003. Intevep, Los Teques.

Brito A, Vielma J, Pereyra E, Trallero J, Garcia S, Cabello R, Comenares J, Terán V (2001) Ventana de operación del separador gas-líquido cilíndrico ciclónico GLCC. INT-9058, 2001. Intevep, Los Teques.

Brito A, Trujillo J, López J, Blanco W (2009) Bases de diseño de un separador ciclónico de múltiples cilindros de flujo ascendente. INT-12904, 2009. Intevep, Los Teques.

Cáliz L, Valdez J, Cabello R, Brito A, López J, Trujillo J (2009) Evaluación de los separadores ciclónicos CYCINT y ciclones múltiples con un sistema agua/aire en el laboratorio de flujo multifásico Norte 6. INT-12687, 2009. Intevep, Los Teques.

Carrasco Y (2008) Identificación y selección de tecnologías de separación mecánica en mezclas de crudo-agua para crudos livianos y medianos. Universidad Central de Venezuela, Caracas

González J, Rojas L, Casique L, Ramírez J (2002) Informe de clausura del Proyecto 6007: Evolución del desarrollo y diseño del separador CYCINT para crudos de gravedad °API>18-agua-sólidos (>100cP). INT-7088, 2002. Intevep, Los Teques.

Martínez M (2002) Diseño conceptual de separadores. Ingenieros consultores y Asociados, Venezuela

Matson J, Brito A (2008) Evaluación del desempeño de dos prototipos de separadores ciclónicos gas-líquido. INT-11884, 2008. Intevep, Los Teques.

Martínez M (2010) Evaluación a escala banco de un prototipo de separador CIMCI de flujo ascendente con aire y agua. Universidad Central de Venezuela, Caracas

Valdez J, López J, Naveda R (2009) Evaluación de los separadores ciclónicos CYCINT y ciclones múltiples con un sistema aceite/aire en el laboratorio de flujo multifásico Norte 6. INT-12688, 2009. Intevep, Los Teques.

Geometric Design Optimization of a Prototype Axial Gas-Liquid Cyclonic Separator

Luis D. Peréz Guerra, Jorge Trujillo and William Blanco

Abstract Oil and gas industry faces new challenges these days: new off-shore fields are located in harsher environments, at longer distances from shore, in deeper waters, demanding more compact and efficient process facilities, to optimize investment costs and then, to guarantee the economic feasibility of these new projects. On the other hand, brown fields with decaying production experience significant changing process conditions which usually impose constrains in existing facilities. The bottlenecking of these facilities requires process improvements to increase their capacity and efficiency, minimizing at the same time any production deferment which could translates into unwanted higher operational costs. Usually, in both cases there are severe space limitations to deploy solutions, demanding these solutions to become more and more compact. PDVSA-Intevep has identified the need for a compact, high efficiency, and high capacity separation technology to address potential gas scrubbing problems in both green and brown fields, and started the development of an axial gas liquid cyclone as an answer to these needs. The separator consists of a flow conditioning section, a swirl generator section, and a segregating section with a discharge for gas and liquid phases in the outlet. An extensive planning, design, construction, and further experimental validation process of a prototype was conducted in the multiphase flow loop facilities of PDVSA-Intevep to demonstrate the axial cyclone concept. As a result of the experimental validation, several aspects of geometrical design were identified to be susceptible to improvements in order to achieve

L. D. Pérez Guerra (✉) · W. Blanco
Instituto Venezolano Tecnológico del Petróleo, (PDVSA-Intevep), Production,
Urbanización El Tambor, Los Teques 1201, Estado Miranda, Venezuela
e-mail: perezlu@pdvsa.com

W. Blanco
e-mail: blancowa@pdvsa.com

J. Trujillo
Multiphase Solutions Inc. (MSI) Kenny, Wood Group House, Level 3, 432 Murray Street,
Perth, WA 6000, Australia
e-mail: jorgenicolas71@gmail.com

L. Di G. Sigalotti et al. (eds.), *Computational and Experimental Fluid Mechanics with Applications to Physics, Engineering and the Environment*, Environmental Science and Engineering,
DOI:10.1007/978-3-319-00191-3_27,

target separation efficiency. The geometric variables identified and addressed in order to improve the performance of this equipment are: incorporating a pre-separation chamber to remove segregated flow incoming to the device, a static mixer to homogenize the gas liquid mixture incoming to the swirl generator, improvement of swirl generator configuration for constructability purposes, improvement of the liquid annular outlet, gas recycle, and outlet gas flow conditioner configurations. The new design is the result of a comprehensive process of revisiting and evaluating the state-of-the-art of axial separation technologies, incorporating lessons learned during the concept demonstration tests and mechanistic modelling of the prototype. Design was conducted considering the operating envelope of the multiphase flow loop of PDVSA-Intevep, to carry out an experimental performance assessment of the incorporated improvements.

1 Introduction

A very important issue in the oil and gas industry is to assure the production of hydrocarbons from reservoir to surface facilities through the entire life of the field. This ambitious goal is partially fulfilled by identifying any potential or existing bottlenecks, their root causes and then developing solutions to those issues, usually by improving process efficiency and installed capacity.

The Venezuelan hydrocarbon development plan for 2012–2018, associated with the natural gas business, was conceived considering the impact of the global economic crisis on the expectations of the economic growth as well as the projection of the supply and demand of hydrocarbons. The largest Venezuelan gas reserves are located in off-shore reservoirs in the Eastern and Western Caribbean Sea in the northern region of the country, and will be developed through the Deltana Platform, the Rafael Urdaneta, and the Mariscal Sucre projects, which are planned to drive the increase on national gas production to 11,839 MMPCSD by 2018, (Fig. 1) satisfying the domestic market and exporting to strategic markets, thereby driving the development of the country and ensuring absolute sovereignty over the gas resources (PDVSA 2011).

Intevep, PDVSA subsidiary responsible for generating technology solutions with a focus on the exploration, production, and refining of hydrocarbons in the country, has set as its project portfolio the development of compact and efficient gas-liquid separation technologies. Axial Cyclonic Scrubbers, offering a solution due to its high separation efficiency and small footprint required for installation, are ideal for developments on offshore platforms, where the available space is usually very limited. The successful design of an enhanced prototype will advance to the second phase of the research and development plan, which includes capturing lessons learned from previous experimental stage, incorporating them into the design, construction, and experimental evaluation of the new improved prototype.

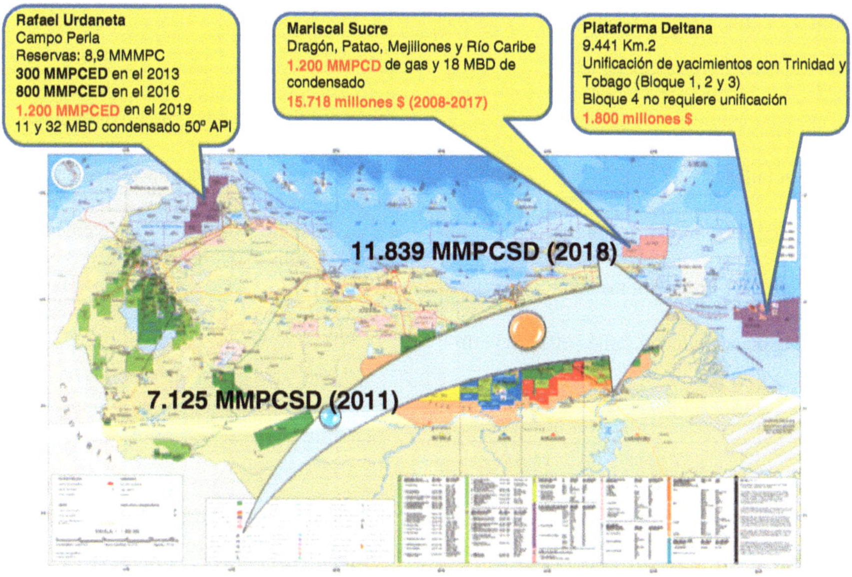

Fig. 1 Gas reservoirs in the Eastern and Western Venezuelan Caribbean Sea

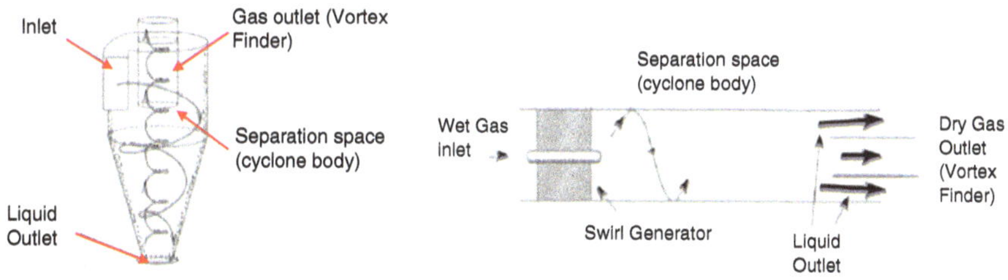

Fig. 2 Type of cyclone separator. *Left* Reverse flow cyclone. *Right* Axial flow cyclone

2 Background

According to Rawlins (2003), the concept of compact separation is commonly applied in the separation unit processes that do not strongly depend on the sedimentation caused by gravitational forces present in these devices. Due to the improvement obtained in the performance of the process by saving space and weight of this equipment, the compact separation promises to revolutionize the design of the facilities.

The use of cyclone separators in the oil and gas industry is a relatively recent phenomenon. However, they have been used successfully for some time in other industrial applications, offering advantages over conventional gravity separators. Its use in the oil industry is now of great interest to many companies and research centres around the world in order to face the great challenge of the extraction and processing of oil and gas.

Table 1 Comparison of reverse flow cyclone and axial flow cyclone

Advantages	Disadvantages
• Less susceptible to re-entrainment, for fewer changes in direction	• Necessity of careful design of the outlet section, since the gas and liquid separation is carried out very close
• Low turbulence	• Need to perform a secondary cleaning of the purge gas stream (recycling)
• Improved performance for high flows	• Erosion in the swirl-generating element (stator) when solids are present in the gas
• Low pressure drop with the same collection efficiency	
• More compact with the same collection efficiency	
• Ideal for grouping in multiple cyclone arrangements (no danger of poor distribution)	

There is abundance of published references in the literature regarding the design and operation of reverse flow cyclones, yet the amount of work associated with axial flow cyclones is significantly lower. Hoffman and Stein (2002) present the most comprehensive information compilation regarding the principles of design and operation for gas cyclones and swirl tubes and the comparison between reverse flow cyclone and axial flow cyclone.

Several papers describe axial flow applications, thus increasing their popularity (Swanborn 1988; Verlaan 1991; Dickson 1998; Austrheim 2006; Trujillo and Ulloa 2007).

Intevep began the development of an Axial Cyclonic Scrubbers in 2007 thanks to the work of Ulloa and Trujillo (2007), who identified trends and technological advances on the different types of axial cyclonic scrubbers, identifying main design parameters and operating conditions for inline axial cyclonic scrubbers developed and installed worldwide (see Fig. 3).

On the other hand, Ruiz et al. (2009) refined the conceptual design proposed by Demey and Trujillo (2008) and built the first prototype of an axial cyclonic scrubber. Additionally, they evaluated and identified the operational envelope through an extensive experimental campaign in the Norte 6 multiphase flow loop of PDVSA-Intevep, and identified the key geometric parameters to assess: departure angle of the blades, stator-vortex finder distance, and the annular opening formed between the vortex finder and the inner diameter of the separator (see Fig. 4).

Delgado (2010) conducted an extensive experimental campaign to assess the best geometric configuration in order to maximize the separation efficiency of the axial cyclonic scrubber prototype proposed by Ruiz et al. (2009) and concluded that the most appropriate geometrical configuration for the prototype—having the highest pressure drop—corresponded to a departure angle of the blades of 75°, with a performance separation between 83 and 93 %, for gas flow rates of 120–190 MCFSD, respectively. The increase in the departure angle of the blades from 45 to 75°, results in a marked improvement in the separation efficiency (see Fig. 5) at the expense of

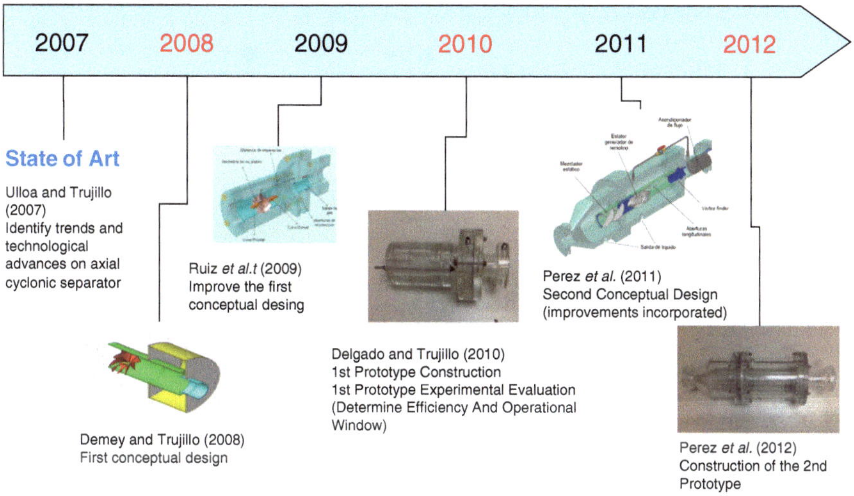

Fig. 3 Development stages technology in PDVSA-Intevep

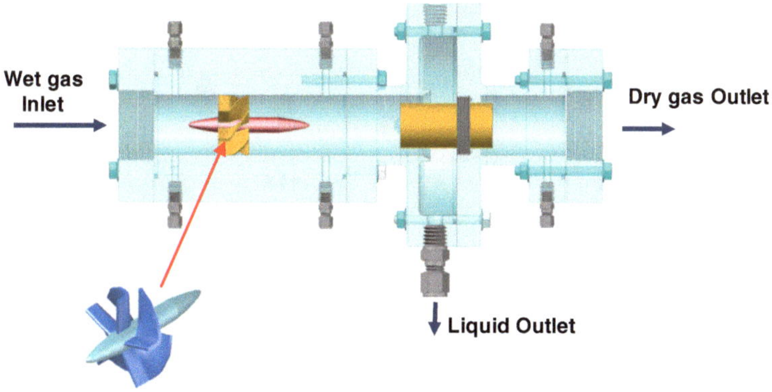

Fig. 4 First prototype designed by Intevep

a higher pressure drop. However, the pressure drop through the device is relatively low (Fig. 6).

Figures 5 and 6 show how Delgado (2010) determined by experimental evaluations, the best geometrical configuration based in the separation efficiency of the first axial cyclone scrubber prototype scale proposed by Ruiz et al. (2009). They concluded that the most appropriate geometry for the prototype, despite having the higher pressure drop, corresponded to a departure angle of 75° blades, creating a performance separation between 83 and 93 %, for a range of gas flow between 120 and 190 MPCSD, respectively. Increasing the departure angle of the vanes from 45 to 75°, results in a notable improvement in the separation efficiency.

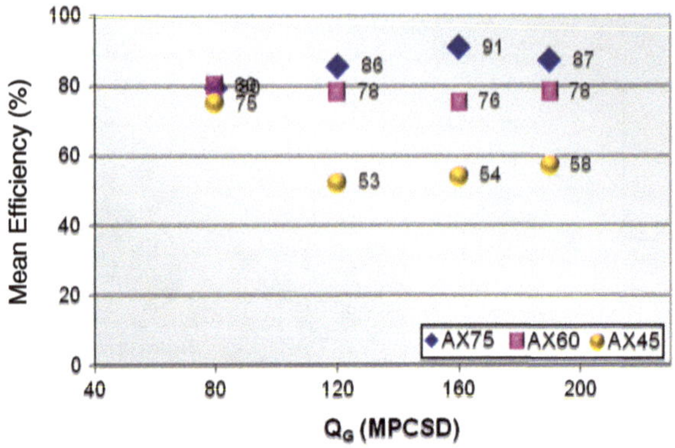

Fig. 5 Separation efficiency for the first prototype

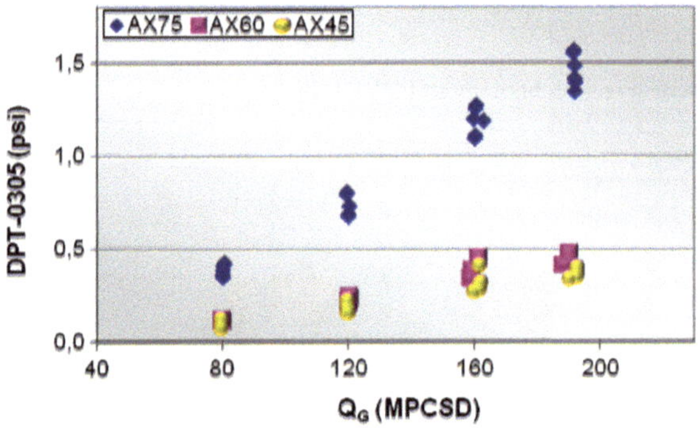

Fig. 6 Pressure drop for the first prototype

3 Description

3.1 Desing Variables

The most important geometric variables considered in the optimization of a prototype axial gas-liquid cyclonic separator are based on previous work by renowned researchers. In particular, Swanborn (1988), Verlaan (1991), and Austrheim et al. (2007) in their experimental and numerical evaluations achieved a high collection efficiencies with the proposed geometries (greater than 95 %). In addition, detailed geometric information on the design of the swirl generator stator can also be found in the literature.

The parameters considered for an improved design of the axial cyclonic scrubber prototype are illustrated in Fig. 7.

The change in any of the above mentioned variables can significantly affect the overall device performance. The impact will be discussed in more detail below.

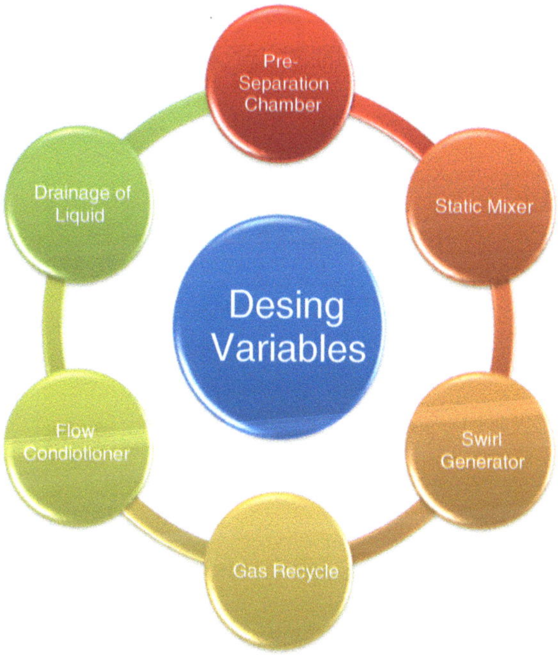

Fig. 7 Key design parameters

3.2 Equipment Description

The equipment is conformed of a pre-separation chamber to disengage the larger liquid droplets by gravitational separation, as well to separate any stratified liquid entering the device. A static mixer is then responsible for ensuring a homogeneous entry of the mixture to the swirl generating element (stator). The stator, whose design is based on a series of stationary helical guide vanes, creates a rotational flow, producing the necessary centrifugal force to facilitate the phase segregation caused by phase density differences, i.e., the heavier phase (dispersed phase liquid droplets) will form a thin annular film to be sent to an accumulation chamber through four longitudinal slots located in the cyclone walls, and an axial opening located at the outlet, which is a concentric cylinder of smaller diameter called vortex finder; meanwhile the lighter phase (continuous gas phase) will form a swirling core of smaller diameter inside the cyclone and will leave it through the vortex finder to a flow conditioner element, aimed at converting the angular momentum into linear momentum, thereby recovering pressure energy and improving mobility of the gas at the outlet. Furthermore, the liquid separated in the pre-separation chamber will be sent to the liquid outlet line. Another feature of this equipment is the use of a pipe for recycling the gas flow, which improves the efficiency (see Fig. 8).

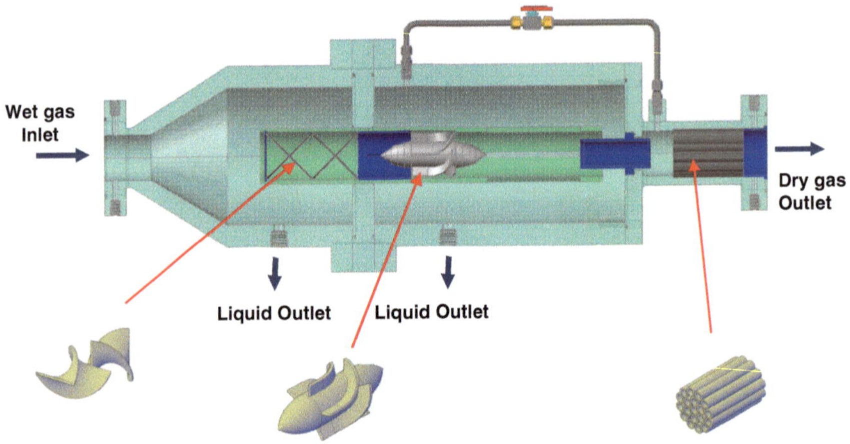

Fig. 8 Sketch of the new geometrical configuration of axial cyclonic scrubber prototype

4 Characterization of the Axial Cyclone Separator

4.1 Static Mixer

Thakur et al. (2003), in their review of static mixers for the process industry, indicate that one of the general parameters for choosing a static mixer is the pressure drop, which many manufacturers and researchers reported as a ratio between the pressure drop through the pipe with static mixer to the pressure drop through the empty (no static mixer) pipe:

$$Z = \frac{f_{\text{mixer}}}{f_{\text{empty}}} = \frac{\Delta P_{\text{mixer}}}{\Delta P_{\text{empty}}},$$

where Z is the pressure drop radius, ΔP_{mixer} is the pressure drop with the mixer (in units of psi), and ΔP_{empty} is the pressure drop without mixer (in psi).

Figure 9 shows how the pressure drop ratio Z increases as the Reynolds number increases. The Kenics static mixer has a lower Z factor, and based on the experience and results obtained by Regner et al. (2006), the Kenics design was used as a model to sizing the static mixer required in the prototype (see Fig. 10).

4.2 Swirl Generator (Stator)

Hoffman and Stein (2002) suggest the most important aspects to consider in the design of the stators or veins swirl generators. They assert that it is the entry and exit angles and the thickness of the blades, which will define the area available to the

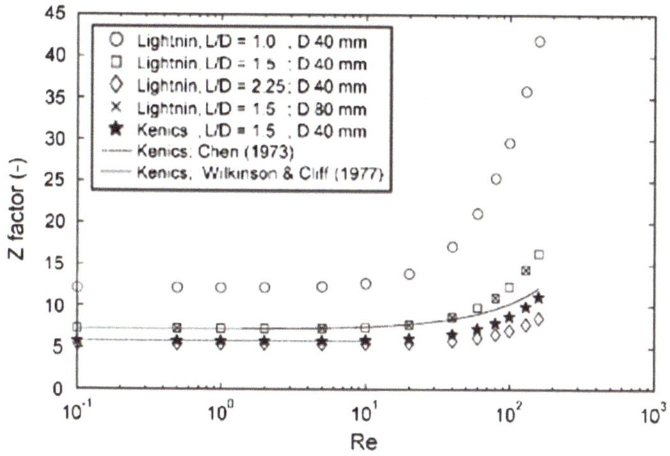

Fig. 9 Influence of the Reynolds number and aspect ratio on the Z factor for the Lightnin and the Kenics KM mixer

Fig. 10 Static mixer

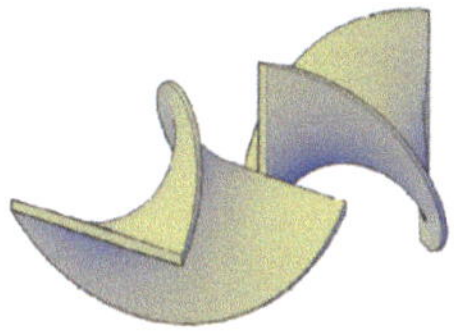

flow (throat area) and therefore the swirl velocity required to reach the required G force to achieve phase separation. In the conventional tangential inlet cyclones, the velocity of the vortex near the wall is determined by the velocity and the construction coefficient of the inlet duct, while in the axial cyclones, this component is determined by the exit velocity through the available area of the stator (throat area) in conjunction with the exit angle of the blades. The equations that are used to calculate the throat area and tangential velocity, given the stator geometry, are:

$$D_{\text{mid}} = \sqrt{D_{ov} D_{iv}},$$

$$A_{th} = N_v \left(\frac{\pi D_{\text{mid}}}{N_v} - \frac{t}{\sin(\beta)} \right) \left(\frac{D_{ov} - D_{iv}}{2} \right) \sin(\beta),$$

$$V_{th} = 10^6 \cdot \frac{Q_m}{A_{th}},$$

$$V_o = V_{th} \cos(\beta),$$

where D_{mid}, D_{ov}, and D_{iv} are the mean, outer, and internal diameter of the throat (in mm), respectively, N_v is the number of blades, β is the departure angle, t is the

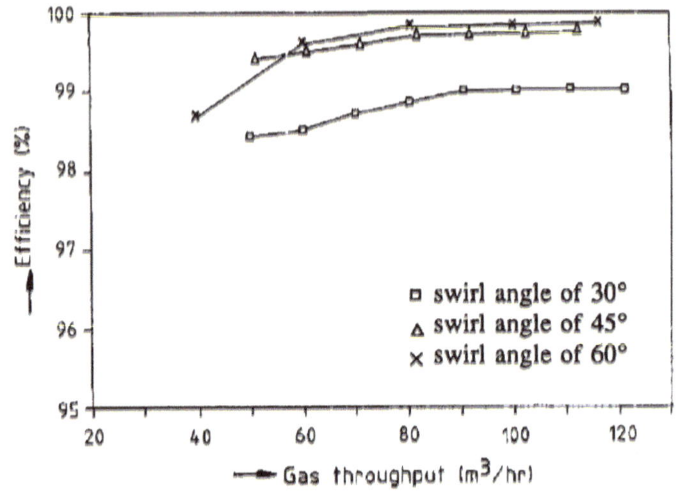

Fig. 11 Influence of the stator blade angle on the separation efficiency axial cyclone (experiment with an air/water body of diameter 3 cm)

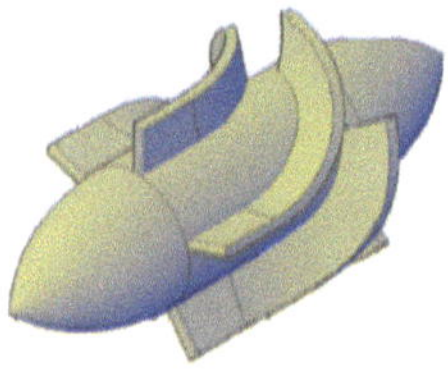

Fig. 12 Swirl generator (stator)

thickness of the blade (in mm), V_{th} is the axial velocity of the throat (in m/s), Q_m is the flow rate of the mixture (in m^3/s), A_{th} is the flow area of the throat (in mm^3), and V_0 is the tangential velocity (in m/s).

The geometry of the blades affects the separation efficiency and the pressure drop of the axial cyclone separator (see Fig. 11). An increase in the angle between the blades and the axial direction produces increased tangential velocity and therefore the centrifugal force increases, improving phase separation. However, there is a limit in increasing this angle, since as the output angle increases, also increases the pressure drop of the device because the cross section available for the flow decreases, so that large angles formed between the blades and the axial, produce high pressure drops (see Fig. 12).

The efficiency of the cyclonic separation depends considerably on the G force generated by the centrifugal acceleration imposed on the fluid due to the high centrifugal accelerations, which cause the breaking of the foam bubbles in the mixture, allowing for the liquid droplets to coalesce in the wall of the cyclone and be extracted from the gaseous phase of the fluid, thereby generating separation (Chin et al. 2002). The G forces generated can be determined by the following relation:

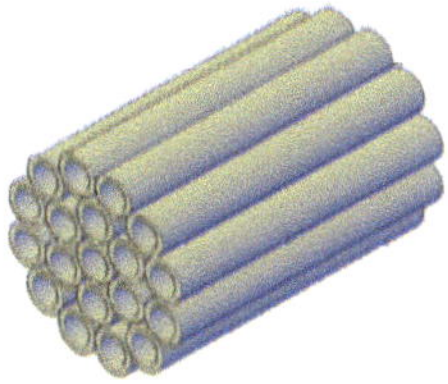

Fig. 13 Flow conditioner

$$G = \frac{2 \cdot (v_t)^2}{g \cdot D},$$

where D is the diameter of the cyclone, v_t is the tangential velocity, and g is the gravity. Many researchers report recommended G-force ranges, and the most commonly used are:

- Between 300–500 G (Swanborn et al. 1995) reverse flow
- Between 5000–6000 (Swanborn et al. 1995) axial flow
- Between 56–100 G (Gomez et al. 1999)
- Between 50–1000 G (Frankiewicz et al. 2001)
- Between 50–100 G (Barbuceanu and Scott 2001)
- Between 100–150 G (Chin et al. 2002)
- Between 10–5000 G (Rawlins 2003)

4.3 Flow Conditioner

The flow conditioner used in the prototype design is shown in Fig. 13 and was based primarily on the recommendations outlined in the GPSA Engineering Data Book (GPSA 1998). Dimensions of cylindrical fins that conform the flow conditioner are determined by the following equations:

$$Sin\left(15^\circ\right) = \frac{r}{R - r},$$

$$De_{\mathrm{alt}} = De_{af}\left(\frac{Sin\left(\frac{\pi}{12}\right)}{1 + Sin\left(\frac{\pi}{12}\right)}\right),$$

$$Di_{\mathrm{alt}} = De_{\mathrm{alt}} - 2t_{\mathrm{alt}},$$

where R is the outer radius of the flow conditioner (in mm), r is the outer radius of the cylindrical fin (in mm), De_{alt} is the external diameter of each cylindrical fin (in mm), De_{af} is the external diameter of the flow conditioner (in mm), Di_{alt} is the inner diameter of the cross section of each cylindrical fin (in mm), and t_{alt} is the thickness of each cylindrical fin (in mm).

5 Conclusions

The more relevant conclusions of this work can be summarized in the following points:

- Compact separation technologies do not depend heavily on the sedimentation caused by the gravitational force. Instead, the main separation principle is the inertia, based on the change of flow direction, to induce centrifugal forces on the fluids.
- Separation compact technologies promise to revolutionize the off-shore facilities due to significant space and weight savings, while improving process performance.
- The efficiency of a cyclone separator is associated with two important variables, which are usually categorized primarily by the separation efficiency related to the amount of liquid collected in terms of the amount of liquid fed to the equipment and secondly by the drop pressure through the device.
- The Kenics KM design with $L/D = 1.028$ was used as a model for the static mixer geometry required in the prototype because it has shown better results compared to the model Lightnin Series 45.
- The geometries designed for the three stators helical swirl generators were based mainly on the design proposed by Verlaan (1991) and validated by Austrheim et al. (2007).
- The cyclone geometry and the inlet flow rates define the G forces generated and the flow pattern at the inlet.
- The flow conditioner used in the design was based primarily on the recommendations in the GPSA standard for fabrication of these devices.
- The proposed design considers the incorporation of a gas recirculation line from the top of the liquid collection chamber to the section between the output of the vortex finder and the flow conditioner.
- The lack of mathematical models for defining the geometric dimensions of axial cyclonic scrubbers constrains designers to the use of extrapolated design criteria from other cyclonic devices and of semi-empirical equations and geometric relationships resulting from experimental and computational evaluations.
- With the development of this technology PDVSA will help ensure a reliable production of hydrocarbons from reservoir to surface facilities, increasing the separation efficiency, the installed capacity of surface facilities, and reducing the required footprint and therefore, the capital and operational expenditures.

6 Further Work

In order to continue with the development of this axial cyclonic scrubber technology, significant resources have been allocated to two ongoing main activities:

- Evaluation of the current design in the multiphase flow loop facilities of PDVSA-Intevep to experimentally verify and validate the selected geometrical configuration of the prototype.
- Use of Computational Fluid Dynamics simulations to reduce the costs and time spent on the experimental research for the development of the final design before conducting further tests with "real" fluids in the PDVSA-Intevep industrial scale experimental facilities.

References

Austrheim T (2006) Experimental characterization of high-pressure natural gas scrubbers. Ph.D. Thesis, Univertsity of Bergen, Norway.

Barbuceanu N, Scott S (2001) Novel inlet design expands range of operability for compact separator. In: SPE annual technical conference and exhibition, New Orleans, Louisiana, 30 September-3 October 2001, paper SPE 71555.

Chin RW, Stanbridge DI, Schook R (2002) Increasing separation capacity with new proven technologies. In: SPE annual technical conference and exhibition, San Antonio, Texas, 29 September-2 October 2002, San Antonio, paper SPE 77495.

Delgado G (2010) Test bench scale evaluation of an axial cyclonic scrubber prototype development by PDVSA Intevep. Universidad Central de Venezuela, Venezuela, Thesis

Demey H, Trujillo J (2008) Conceptual design of a scrubber axial flow cyclone gas-liquid. Universidad de Carabobo, Venezuela, Thesis

Dickson PJ (1998) Gas/Liquid separation within a novel axial flow cyclone separator. Ph.D. Thesis, School of Mechanical Engineering, Cranfield University, UK.

Frankiewicz T, Browne M, Lee CM (2001) Reducing separation train sizes and increasing capacity by application of emerging technologies. In: Offshore technology conference, Houston, Texas, 30 April-3 May 2001, paper OTC 13215.

Gómez L (1998) A state of the art simulator and field application design of gas-liquid cylindrical cyclone separators. The University of Tulsa, USA, Master Thesis

GPSA (1998) Engineering Data Book, FPS Version, Vol. I.

Hoffman A, Stein L (2002) Gas cyclones and swirl tubes: principles: design and operation. Springer-Velarg, Berlin

PDVSA (2011) Annual management report. Ministerio del Poder Popular para la Energía y Petróleo, Venezuela

Rawlins CH (2003) The case for compact separation. Technology Today Series, paper SPE 80994:77–80

Ruiz R, Trujillo J, López J (2009) Design of a test bench prototype of an axial cyclonic separation technology for high gas-liquid Ratio. Technical Report INT-12922, Intevep, Los Teques, Venezuela.

Swanborn RA (1988) A new approach to the design of gas-liquid separators for the oil industry. PhD Thesis, Delf University of Technology, The Netherlands.

Thakur RK, Vial C, Nigam KDP, Nauman EB, Djelveh G (2003) Static mixers in the process industries–a review. Chem Eng Res Des 81:787–826

Ulloa J, Trujillo J (2007) Frontiers and trends identification of axial gas-liquid cyclonic separation. Technical Report SIT-0362-2007, Intevep, Los Teques, Venezuela.
Verlaan CCJ (1991) Performance of novel mist Eliminators. Ph.D. Thesis, Delf University of Technology, The Netherlands.

Effect of Hydrotreating Reaction Conditions on Viscosity, API Gravity and Specific Gravity of Maya Crude Oil

Yanet Villasana, Sergio Ramírez, Jorge Ancheyta and Joaquín L. Brito

Abstract The hydrotreatment of Maya crude oil was carried out in a Parr batch reactor, using alumina-supported catalysts based on NiMo and CoMo sulfides, carbides, and nitrides, which were sulfided ex situ with a mixture of H_2/CS_2, prior to reaction. Hydrotreating reactions were carried out under the following conditions: temperature: 320 °C, pressure: 70–80 kg/cm^2, time: 4 h, stirring: 500 rpm, and catalyst mass: 2 g. The products of reaction were analyzed by simulated distillation, and the physical properties of the hydrotreated crude were obtained, such as the specific weight and viscosity, at different temperatures, and these values were used to determine specific gravity (SG) and API. In this contribution, we illustrate changes in the physical properties of Maya crude oil before and after hydrotreatment reaction with variations on residue conversion when different hydrotreatment catalysts were used.

1 Introduction

In order to satisfy the world demand of petroleum derivatives, and also meet the stringent environmental regulations and product quality standards, many studies have been directed to improve the quality of heavy crude oils motivated by the decreasing

Y. Villasana (✉) · J. L. Brito
Laboratorio de Fisicoquímica de Superficies, Centro de Química, Instituto Venezolano de Investigaciones Científicas, IVIC, Apartado Postal 20632, Caracas 1020-A, Venezuela
e-mail: ytvillas@gmail.com

J. L. Brito
e-mail: joabrito@ivic.gob.ve

S. Ramírez · J. Ancheyta
Laboratorio de Síntesis e Hidrotratamiento, Instituto Mexicano del Petróleo, IMP, Eje Central Lázaro Cárdenas Norte 152, Colonia San Bartolo Atepehuacan, 07730 México D. F., Mexico
e-mail: sramire@imp.mx

J. Ancheyta
e-mail: jancheyt@imp.mx

L. Di G. Sigalotti et al. (eds.), *Computational and Experimental Fluid Mechanics with Applications to Physics, Engineering and the Environment*, Environmental Science and Engineering, DOI:10.1007/978-3-319-00191-3_28,

availability of conventional crudes as a consequence of the increase of global population. Heavy crude oils are more difficult to recover from the sub-surface reservoir than light oils. In addition, this type of feedstocks have a much higher viscosity (and lower API gravity) than conventional petroleum, and primary recovery of these petroleum types usually requires thermal stimulation of the reservoir, which strongly affects trade price. The generic term heavy oil is often applied to a crude oil that has less than 20° API (Speight 1999), and this type of feed can be described as a colloidal solution consisting of three fractions: oils, resins, and asphaltenes, in which micelles are present. Micelles are aggregates of resins and asphaltenes held together by weak physical interactions, the latter being in the core and the former in the periphery as a dispersing agent (Ancheyta et al. 2005). This behaviour is caused by the high complexity, high molecular weight, and highly aromatic nature of asphaltenes molecules, characterized by a large aromatic sheet, which could pile up on each other to form a large unit cell in the absence of resins (during hydroconversion). This fact could lead to instability problems by coagulation and precipitation during transportation and processing (Dickie and Yen 1967; Tynan and Yen 1969; Rana et al. 2007).

Moreover, it is worth mentioning that API gravity is related to its specific gravity (the ratio of the density of the crude oil to the density of water under specific conditions). In addition, viscosity appears to be the most important physical property of heavy feeds, due to transportation and handling implications (Ancheyta et al. 2005). A common way to improve the quality of heavy feedstocks is through hydroprocessing (Shuetze and Hoffman 1984; Dickenson et al. 1997; Ancheyta et al. 2005) using hydrogen and a proper active catalyst. The term hydroprocessing is broad, and includes hydrocracking, hydrotreating (HT), and hydrorefining. All these reasons have motivated the study of the effect of HT reaction with different catalysts on the physical properties of Maya crude oil such as specific gravity, viscosity, and API gravity. Although the Maya crude oil that has been used as feed in this investigation has 23°API, which is considered to be a medium crude oil, the results obtained here represent an interesting approach to the correlation between the variations of these physical properties and the residue conversion and gas products yield, and therefore give an idea of the behaviour of heavier feeds during hydroprocessing.

2 Experimental Procedure

For the HT reaction of Maya crude oil, we have used alumina-supported catalysts based on NiMo and CoMo sulfides, carbides, and nitrides (metal loading 15wt. %, 0.33 of atomic ratio = $(X/(X+\text{Mo}))$, with $X=$ Co or Ni), which were synthesized for FeW (Villasana et al. 2013a) and NiMo (Villasana et al. 2013b) carbides and nitrides. It is worth mentioning that prior to hydrotreatment reaction, each catalyst, including commercial catalysts *Cat-3*, were presulfided ex situ with a mixture of H_2/CS_2 in order to activate the catalytic phases on the surface (300°C/3h). Simultaneously, 200 g of Maya crude oil was set on a Parr batch reactor vessel (model 4842 of

1.0 L of capability). After pre-treatment, the activated catalyst was transferred into a catalyst basket and then introduced into the vessel to avoid catalyst mass loss during reaction. Hydrotreating reactions were carried out under the following conditions: temperature: 320°C, pressure: 70–80 kg/cm^2, time: 4 h, stirring: 500 rpm, and catalyst mass: 2 g. The products of the reaction were recovered and separated from the catalyst after reaction, and then analyzed by simulated distillation in an Agilent HP 7890A chromatograph, Model G3340A following the ASTM D7169-11 method (ASTM 2011). The physical properties of the hydrotreated crude obtained, such as the specific weight and viscosity (kinematic and dynamic), at different temperatures (15, 20, and 37.8 °C), were set using a Parr viscometer model SVM3000. Specific weight values at 20 °C were used to determine the SG related to the specific weight of water at 4 °C (0.999840 g/cm^3). Moreover, specific weight values at 15 °C were used to determine the SG values related to the specific weight of water at this temperature (0.999012 g/cm^3) in order to estimate the API gravity at 15 °C (60 °F) as follows:

$$API\ Gravity\ at\ 60^{\circ} F = \left(\frac{141.5}{SG}\right) - 131.5. \quad (1)$$

3 Results and Discussion

Table 1 shows the yield of gas and liquid products, where we may see the cracking effect of NiMo and CoMo catalysts on the hydrocarbon molecules present in Maya crude oil, indicated by the high yield of gas products, which is even superior to that obtained for commercial catalysts when sulfide catalysts were used, while NiMo and CoMo carbides and nitrides revealed almost the same behaviour as Cat-3. Remarkably, the catalysts with high levels of residue conversion showed the highest values of yield in gas products (see Table 1). These results are related to the cracking reactions during the HT process, as mentioned earlier, or could be due to the dealkylation of aromatic or polyaromatic rings, associated to asphaltene or resin molecules, as was reported previously for similar feedstocks (Acevedo et al. 2004; Villasana et al. 2011).

Noteworthy, the differences between the results obtained by different catalysts are not the target of this contribution, but it could be said that each catalyst follows different reaction pathways which lead to different product yields, which strongly affect the physical properties of the obtained hydrotreated crude. The details of the NiMo catalysts were reported by Villasana et al. (2013b).

Figure 1 illustrates the SG values, revealing a decrease in all cases with respect to the Maya crude oil, except for the reaction using NiMoC as catalyst in which case this value reached a maximum. The lowest values in this property were exhibited by the products obtained during the reaction in the presence of sulfide catalysts of NiMo and CoMo (NiMoS and CoMoS). In addition, the values obtained for the API gravity are shown in Fig. 2, which revealed that this property increased when we used NiMo and CoMo sulfides as catalysts, as confirmed by the SG results at 20 °C/4 °C. In contrast, those reactions in which both the carbide and nitride catalysts and the commercial

Table 1 Yield of gas and liquid products and residue conversion after HT reaction with different catalysts compared to commercial catalyst "Cat-3"

Catalyst	Yield in gas products (wt. %)	Yield in liquid products (wt. %)	Residue conversion 538^{+}(%)
Cat-3	1.50	87.49	11
CoMoS	3.50	76.11	41
CoMoC	1.00	89.70	8
CoMoN	2.00	91.41	10
NiMoS	2.49	82.25	36
NiMoC	2.00	90.91	8
NiMoN	1.49	91.29	12

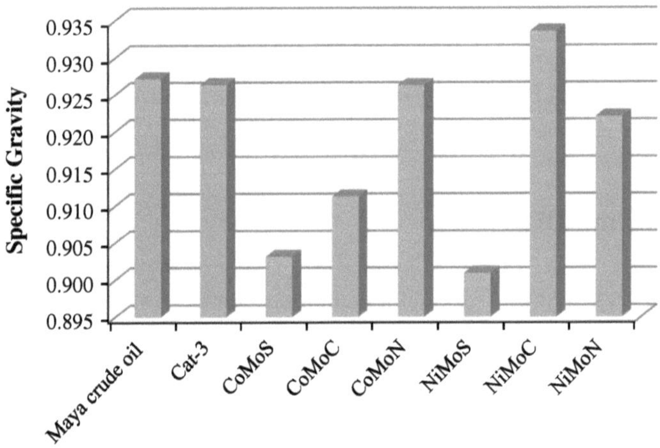

Fig. 1 Specific Gravity (@20/4 °C) of Maya crude oil before and after the HT reactions with different catalysts

catalyst were used, a decrease in this property was observed in this order: NiMoN > CoMoN > Cat-3 > CoMoC > NiMoC. The specific and the API gravity could be correlated to the high conversion achieved with the sulfide catalysts in which the yield of gas products is higher.

Figures 3 and 4 show the kinematic and dynamic viscosities at different temperatures for the Maya crude oil before and after the HT reactions. As expected, they decrease with increasing temperature with respect to the values obtained for the Maya crude oil. It is important to notice that both values have relevance due to transportation implications because the dynamic viscosity is the resistance of the oil to flow, for example, through a pipeline, while the kinematic viscosity is related to the density of the fluid. According to the commonly accepted rule, the viscosity should be lower than 120 cSt at 20 °C, whereas in the southern hemisphere a greater viscosity can be tolerated. This is the reason why refiners must ensure an adequate

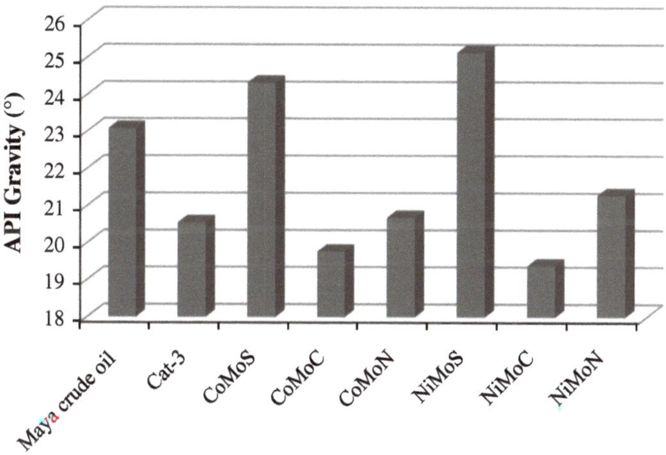

Fig. 2 API Gravity (@60 °F) of Maya crude oil before and after the HT reactions with different catalysts

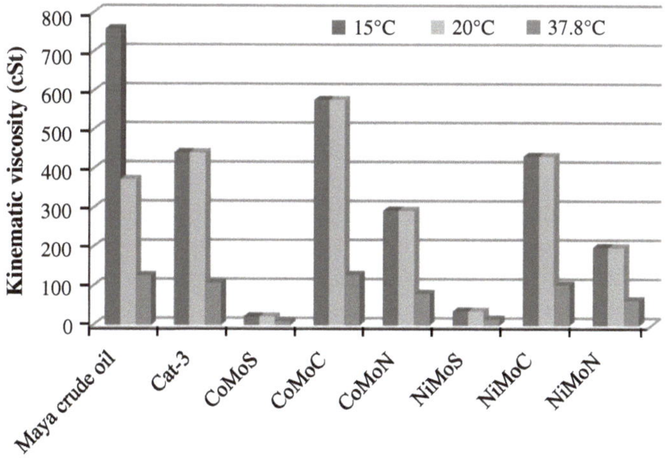

Fig. 3 Kinematic viscosity of Maya crude oil before and after the HT reactions with different catalysts at several temperatures

pumpability and reduce transportation difficulties, by maintaining a sufficiently low viscosity during processing and refining of crude oil. Notably, at room temperature, heavy crude oils are in a semi-solid form, which leads to problems on their handling and processing. Thus, by increasing the temperature, the viscosity decreases because of a gradual change in the colloidal structure of semi-solids caused by an increased mobility of micelles in the oil fraction. Consequently, in order to minimize costs in refineries, the reheating of heavy feeds is accomplished by integrating distillation units to the catalytic reactors, thus ensuring that distillation residues with sufficiently

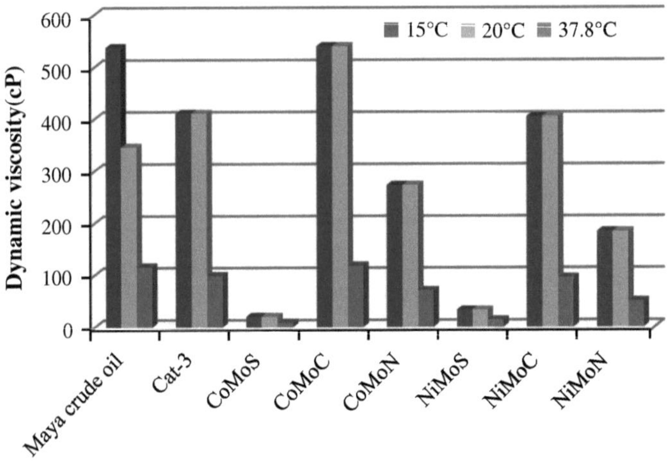

Fig. 4 Dynamic viscosity of Maya crude oil before and after the HT reactions with different catalysts at several temperatures

low viscosity are fed directly to a reactor for hydroprocessing. Usually, this process is carried out on site with the aim to achieve sufficient pumpability for pipelining (Jacquin et al. 1983; Peries et al. 1988).

On the other hand, Fig. 3 shows that the catalysts, corresponding to NiMo and CoMo sulfides, reach values of the kinematic viscosity below 120cSt at 20°C, followed by NiMoN with approximately 200cSt and CoMoN around 300cSt. Similarly, Fig. 4 shows that low values of the dynamic viscosity were achieved during hydrotreating with NiMo and CoMo sulfides, followed by their corresponding nitrides.

Additionally, the catalysts that showed high viscosities also had low values of API gravity. However, it is known that there is no correlation between viscosity and specific gravity. Thus, feeds varying widely in viscosity may have similar specific gravity, which is not surprising because different factors are involved in determining the values of viscosity and specific gravity, as reported by Ancheyta et al. (2005). This assertion could also be extended to API gravity since this value is determined from the specific gravity. In this regard, we notice that those catalysts with high values of conversion, which are NiMo and CoMo sulfides, also shows the lowest values of viscosity (dynamic and kinematic) and high API and specific gravity values.

4 Conclusions

After the hydrotreatment reaction of Maya crude oil, we have observed changes in the physical properties with variations of the residue conversion when different hydrotreatment catalysts are used. The simulated distillation revealed that catalysts with high levels of residue conversion showed the highest values of yield in gas products, which are related to cracking of molecules or dealkylation of aromatic or polyaromatic rings, associated to asphaltene or resin molecules. The specific gravity seems to decrease in all cases with respect to the Maya crude oil, except for the reaction using NiMoC as catalyst. The API gravity increased when NiMo and CoMo sulfide catalysts were used. In contrast, the rest of the catalysts, including the commercial one, exhibited a decrease in this property, with the behaviour of commercial catalysts being comparable to that of NiMo and CoMo nitrides. These results were correlated to the high conversion achieved during hydrotreating reaction with different catalysts. As expected, the kinematic and dynamic viscosities decrease with increasing temperature with respect to the value obtained for the Maya crude oil. Finally, it is notable that those catalysts that showed low viscosities also had high values of the API gravity.

Acknowledgments One of us (Y. V.) acknowledges financial support from FONACIT and IVIC. We also thank E. Rodriguez, A. Quitian, G. Felix, R. Rivera, J. Antonio Montes, F. Hernández, and F. Alonso for experimental, technical, and analysis support.

References

Acevedo S, Escobar O, Echevarría L, Gutiérrez LB, Méndez B (2004) Structural analysis of soluble and insoluble fractions of asphaltenes isolated using the PNP method. relation between asphaltene structure and solubility. Energy Fuels 18:305–311

Ancheyta J, Rana MS, Furimsky E (2005) Hydroprocessing of heavy petroleum feeds: tutorial. Catal Today 109:3–15

Standard ASTM, D7169–11, 2005, (2011) Standard Test Method for Boiling Point Distribution of Samples with Residues Such as Crude Oils and Atmospheric and Vacuum Residues by High Temperature Gas Chromatography. ASTM International, West Conshohocken, PA 2011: doi:10.1520/D7169-11, http://www.astm.org

Dickenson RL, Biasca FE, Schulman BL, Johnson HE (1997) Refiners options for converting and utilizing heavy fuel oil. Hydrocarbon Process 76:57–62

Dickie JP, Yen TF (1967) Macrostructures of the asphaltic fractions by various instrumental methods. Anal Chem 39:1847–1852

Jacquin Y, Toulhoat H, Quignard A, Le Page JF (1983) Développements récents dans le domaine de l'hydrotraitement des fractions lourdes. Oil Gas Sci Technol - Rev. IFP 38:371–385

Peries JP, Renard P, Des Courires T, Rossarie J (1988) ASVAHL new routes for processing heavy oils. In: Proceedings of the 4th UNITAR/UNDP International Conference on Heavy Crude and Tar Sands, Edmonton, Canada, Aug. 18, 1988; paper 95, pp 1–21

Rana MS, Sámano V, Ancheyta J, Díaz JA (2007) A review of recent advances on process technologies for upgrading of heavy oils and residua. Fuel 86:1216–1231

Shuetze B, Hoffman H (1984) How to upgrade heavy feeds. Hydrocarbon Process 2:75–82

Speight JG (1999) The chemistry and technology of petroleum. Marcel Dekker, New York

Tynan EC, Yen TF (1969) Association of vanadium chelates in petroleum asphaltenes as studied by ESR. Fuel 43:191–208

Villasana Y, Luis-Luis MA, Labrador H, Brito JL (2011) Optimización del tiempo de reacción de hidrotratamiento del crudo Furrial usando como catalizador $NiMo/\gamma - Al_2O_3$ In: Barbosa AL, Villa AL, Ramírez A (eds) VII Simposio Colombiano de Catálisis (ISBN: 978-958-8736-16-7) Editorial Universidad de Cartagena. Cartagena de Indias. Paper N° CA-P0-5

Villasana Y, Ruscio-Vanalesti F, Pfaff C, Méndez FJ, Luis-Luis MA, Brito JL (2013a) Atomic ratio effect on catalytic performance of FeW-based carbides and nitrides on thiophene hydrodesulfurization. Fuel 110:259–267

Villasana Y, Escalante Y, Rodríguez-Nuñez JE, Méndez FJ, Ramírez S, Luis-Luis MA, Ancheyta J, Brito JL (2013b) Maya crude oil hydrotreating reaction in a batch reactor using NiMo carbide and nitride as catalysts. Catalysis Today, submitted

Mechanistic Model for Eccentric Annular Gas-Liquid Flow in Horizontal Pipelines

Adriana Brito, Nelson MacQuhae, Francisco García, Nelson Fernández and José Colmenares

Abstract A mechanistic model for the prediction of pressure drop in horizontal pipelines is presented for annular flow. A new empirical correlation for the liquid/wall interfacial friction is proposed, where the effects of the annular flow eccentricity, due to the difference between the fluid density and viscosity, are accounted for. The model is compared to three different correlation models and five mechanistic models in current use. Its accuracy has been validated against experimental data for annular gas-liquid flow in horizontal pipelines, taken from different sources. A number of 240 experiments were carried out with superficial liquid velocities between 0.003 and 5.96 m/s, superficial gas velocities between 9 and 69.6 m/s, liquid viscosities between 1 and 1200 cP, and pipeline diameters between 0.0261 and 0.0953 m. We find that the mechanistic model proposed here reduces the absolute error of the pressure drop prediction by approximately 20 % compared to other mechanistic models.

1 Introduction

The drop of pressure in gas-liquid segregated flow patterns is perhaps the most difficult parameter to predict, while annular flow is one of the most common two-phase flow patterns that arise in practice. The most widely used mechanistic models

A. Brito (✉) · J. Colmenares
Instituto Tecnológico Venezolano del Petróleo (PDVSA-Intevep), Urbanización El Tambor, Los Teques 1201, Estado, Miranda, Venezuela
e-mail: britoah@pdvsa.com

J. Colmenares
e-mail: jcolmenares@cantv.net

N. MacQuhae · F. García · N. Fernández
Escuela de Ingeniería Mecánica, Instituto de Mecánica de Fluidos, Universidad Central de Venezuela, Caracas 1051, Venezuela
e-mail: nelsonm@microsoft.com

N. Fernández
e-mail: nelson.fernandez@ucv.ve

L. Di G. Sigalotti et al. (eds.), *Computational and Experimental Fluid Mechanics with Applications to Physics, Engineering and the Environment*, Environmental Science and Engineering, DOI:10.1007/978-3-319-00191-3_29,

for predicting pressure drops in pipelines are those reported by Xiao et al. (1990), Ouyang et al. (1998), Gómez et al. (1999), and Holden (2002). However, the accuracy of these predictive models against the experimental data is over 45 %.

In annular flow patterns, the relevant parameters for the prediction of the pressure drop are the film distribution, the droplet entrainment in which small drops of one phase remain trapped in the other phase, and the fluid-wall friction factor. Therefore, the aim of this study is to develop a mechanistic model that takes into account all these parameters and reduces the uncertainty in the pressure gradient prediction.

The improved accuracy of the model has been validated against gas-liquid annular flow data from Beggs (1972), Mukherjee (1979), and Andritsos (1986), as well as from experimental data provided by different companies related to the database of the Stanford University and from Petróleos de Venezuela S. A. (PDVSA) for the case of air and heavy oil.

2 Annular Flow Models in Pipelines

In horizontal pipelines, where the film distribution is around the pipe wall and the gas is characterized by its continuity along the core of the pipe, the annular flow tends to be eccentric. The level of eccentricity depends on the density and viscosity of the fluid, as well as on the flow rates of liquid and gas. The liquid film is thinner in the upper than in the lower part of the pipe and the liquid phase moves in a wavy manner close to the gas-liquid interface so that droplets are entrained in the gas core.

In models of gas-liquid eccentric annular flow, a Newtonian two-fluid approach is usually employed, where the liquid film is the liquid phase and the gas-droplet mixture is considered to be the gas phase. We assume that the flow is stationary, incompressible, isothermal, and one-directional. Moreover, we consider the simple case in which there is no mass transfer between the phases and assume that the pressure gradients in the gas and liquid film are the same.

If we start from the continuity equation:

$$\frac{\partial \rho_f}{\partial t} + \nabla \cdot \left(\rho_f \vec{v}\right) = 0, \tag{1}$$

where ρ_f is the mass density and v is the fluid velocity vector, the continuity equation for steady-state flow written in a generalized orthogonal coordinate system reduces to

$$\frac{1}{h_1 h_2 h_3}\left[\frac{\partial}{\partial z}\left(h_1 h_2 v_z\right)\right] = 0, \tag{2}$$

where h_1, h_2, and h_3 are the components of the metric tensor of the orthogonal coordinate system defined as

$$
\begin{aligned}
h_1^2 &= \left(\tfrac{\partial x}{\partial \rho}\right)^2 + \left(\tfrac{\partial y}{\partial \rho}\right)^2 + \left(\tfrac{\partial z}{\partial \rho}\right)^2, \\
h_2^2 &= \left(\tfrac{\partial x}{\partial \phi}\right)^2 + \left(\tfrac{\partial y}{\partial \phi}\right)^2 + \left(\tfrac{\partial z}{\partial \phi}\right)^2, \\
h_3^2 &= \left(\tfrac{\partial x}{\partial z}\right)^2 + \left(\tfrac{\partial y}{\partial z}\right)^2 + \left(\tfrac{\partial z}{\partial z}\right)^2,
\end{aligned}
\tag{3}
$$

where (x, y, z) are the three Cartesian-coordinate axes, and $\rho = \rho(x, y), \phi = \phi(x, y)$ are some orthogonal coordinates to be specified later below. In Eq. (2), v_z is the velocity component in the axial (z-axis) direction. In addition, for one-directional flow along the z-axis, the momentum equation becomes

$$
0 = -\left(\frac{\partial P}{\partial z}\right) + \rho_f g + \frac{1}{h_1 h_2 h_3}\left[\frac{\partial}{\partial \rho}\left(h_2 h_3 \tau_{\rho z}\right)\right], \tag{4}
$$

where P is the pressure, g is the acceleration of gravity, and $\tau_{\rho z}$ is the shear stress in the axial direction. We write the momentum equation in dimensionless form by introducing the following dimensionless parameters:

$$
\tilde{\rho} = \frac{2}{D}\bar{\rho} \;\; ; \;\; \tilde{P} = \frac{D}{2v_0\mu}P \;\; ; \;\; \tilde{V} = \frac{v_i}{v_0} \;\; ; \;\; \tilde{\tau}_{\rho z} = \frac{D}{2v_0\mu}\tau_{\rho z} \;\; ; \;\; \tilde{z} = \frac{2}{D}z \tag{5}
$$

where $\bar{\rho}$ is the inner radius derived from the hydraulic diameter of the gas phase, v_i is the velocity at the gas-liquid interface, v_o is the characteristic velocity of the system, μ is the viscosity, and D is the pipe diameter. In the momentum equation for the liquid phase, the parameters v_o and μ are the superficial velocity and the viscosity of the liquid, while in the momentum equation for the gas phase they correspond to the superficial velocity and viscosity of the gas.

For the liquid phase, we use an orthogonal coordinate system, i.e., $(\tilde{\rho}(x, y), \phi(x, y))$, based on the coordinate system proposed by González (1998). This coordinate system arises from a bilinear transformation in non-dimensional form, where the fluid domain is the space confined between the inner diameter (formed by the gas phase), which is less than one, and the outer diameter, which is equal to one.

The eccentric annular flow is mainly affected by the floatation effect of the gas phase and the viscosity of the liquid. In order to represent mathematically this effect it is necessary to use the modified bilinear transformation (see Fig. 1):

$$
w = \frac{z^* - ai}{az^* - i}, \tag{6}
$$

where a is the transformation pole. Following the procedure given by González (1998), we rotate the bilinear coordinate system and transform to rectangular Cartesian coordinates (*x*,*y*) such that $(x(\tilde{\rho}, \phi), y(\tilde{\rho}, \phi))$, where

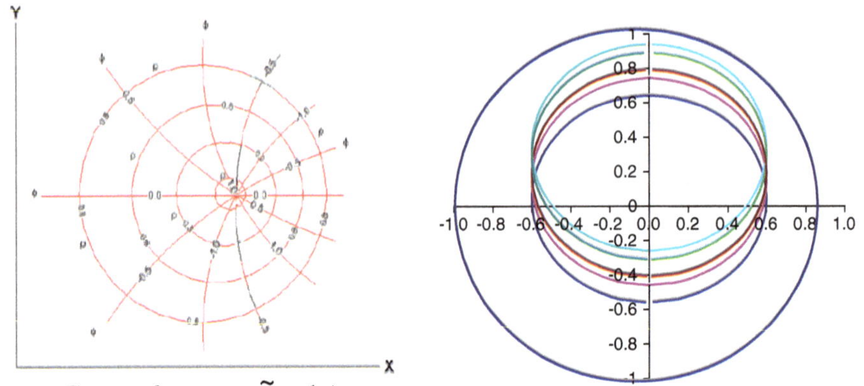

Curves of constant $\tilde{\rho}$ and ϕ

Fig. 1 Orthogonal coordinate system

$$
\begin{aligned}
x &= \frac{\tilde{\rho}\sin\phi\left(1-a^2\right)}{a^2\tilde{\rho}^2-2a\tilde{\rho}\cos\phi+1}, \\
y &= \frac{a\left(\tilde{\rho}^2+1\right)-\tilde{\rho}\cos\phi\left(1+a^2\right)}{a^2\tilde{\rho}^2-2a\tilde{\rho}\cos\phi+1}.
\end{aligned} \tag{7}
$$

Here$\tilde{\rho}$ is the dimensionless hydraulic inner radius of the gas phase and ϕ is the angle formed by the radial lines from the centre of the circumference ($0 < \varphi < 2\pi$; see Fig. 1).

For the proposed coordinate system the scale factors $h_{1,} h_{2,}$ and h_3 for the liquid phase are given by

$$
h_1 = \frac{\left(1-a^2\right)}{\left(a^2\tilde{\rho}^2-2a\tilde{\rho}\cos\phi+1\right)}; \quad h_2 = \frac{\tilde{\rho}\left(1-a^2\right)}{\left(a^2\tilde{\rho}^2-2a\tilde{\rho}\cos\phi+1\right)}; \quad h_3 = 1. \tag{8}
$$

Substituting the scale factors (8) into the dimensionless form of Eq. (4), we obtain for the momentum equation of the liquid phase

$$
\begin{aligned}
0 = -\left(\frac{\partial\tilde{P}}{\partial\tilde{z}}\right)_L - \frac{Re_{SL}}{4\mathrm{Fr}_{SL}}\sin\alpha + \frac{\left(a^2\tilde{\rho}^2-2a\tilde{\rho}\cos(\phi)+1\right)^2}{\left(1-a^2\right)^2\tilde{\rho}} \\
\left[\frac{\partial}{\partial\tilde{\rho}}\left(\frac{\left(1-a^2\right)\tilde{\rho}}{\left(a^2\tilde{\rho}^2-2a\tilde{\rho}\cos(\phi)+1\right)}\tilde{\tau}_{\rho z}\right)\right],
\end{aligned} \tag{9}
$$

where Re_{SL} is the Reynolds number of the liquid phase, Fr_{SL} is the liquid Froude number, and $\tilde{\tau}_{\rho z}$ is the dimensionless shear stress in the axial direction, ie.,

$$
Re_{SL} = \frac{\rho_L v_{SL} D}{\mu_L}; \quad \mathrm{Fr}_{SL} = \frac{v_{SL}^2}{gD}; \quad \tilde{\tau}_{\rho z} = \frac{D}{2v_{SL}\mu_L}\tau_{\rho z}. \tag{10}
$$

Here v_{SL} is the velocity of the liquid phase, while ρ_L and μ_L are the density and viscosity of the liquid. In order to determine the pressure gradient in the pipe, we integrate Eq. (9) to obtain:

$$\frac{\left(R_i^2-1\right)\left(1-a^4R_i^2\right)}{(aR_i-1)^2\,(aR_i+1)^2}\left(\frac{\partial\tilde{P}}{\partial\tilde{z}}+\frac{Re_{SL}}{4\mathrm{Fr}_{SL}}\sin\alpha\right)_L=\left[2\tau_{zw}-\frac{2R_i\left(1-R_i^2\right)}{\left(1-a^2R_i^2\right)}\tilde{\tau}_{iL}\right], \tag{11}$$

where R_i, $\tilde{\tau}_{zw}$, and $\tilde{\tau}_{iL}$, are, respectively, the dimensionless hydraulic radius of the gas phase in the annular core, the shear stress at the pipe wall for the liquid, and the interfacial shear stress given by

$$R_i=\frac{(e-R_1)-a}{a\,(e-R_1)-1};\quad \tilde{\tau}_{zw}=\frac{D}{2v_{SL}\mu_L}\tau_{zw};\quad \tilde{\tau}_{iL}=\frac{D}{2v_{SL}\mu_L}\tau_i. \tag{12}$$

For the gas phase, we write the momentum equation in cylindrical coordinates and assume that the gas phase is flowing in a perfect cylinder, where its diameter is given by the hydraulic diameter of the gas core in the annular flow pattern. After substitution of the scale factors $h_1=1$, $h_2=\rho$, and $h_3=1$ into Eq. (4), it is possible to obtain the momentum equation for the gas phase in dimensionless form

$$\left(\frac{\partial\tilde{P}}{\partial\tilde{z}}+\frac{Re_{SG}}{4\mathrm{Fr}_{SG}}\sin\alpha\right)_G=-\frac{2}{R_i}\tilde{\tau}_{iG}, \tag{13}$$

where Re_{SG}, Fr_{SG}, and $\tilde{\tau}_{iG}$ are, respectively, the Reynolds and Froude numbers of the gas phase and the dimensionless interfacial shear stress, defined as

$$Re_{SG}=\frac{\rho_C v_{SG} D}{\mu_G};\quad \mathrm{Fr}_{SG}=\frac{v_{SG}^2}{gD};\quad \tilde{\tau}_{iG}=\frac{D}{2v_{SG}\mu_G}\tau_i, \tag{14}$$

where now v_{SG} is the velocity of the gas phase and ρ_G and μ_G refer to the density and viscosity of the gas. The gas core density in the annular flow, ρ_c, is a no-slip density because the core is considered a homogeneous mixture of gas and entrained liquid droplets flowing at the same velocity (Ansari et al. 1994), that is

$$\rho_c=\rho_L H_{Lc}+\rho_G(1-H_{Lc}). \tag{15}$$

The no-slip holdup in the gas core, H_{Lc}, is given by

$$H_{Lc}=\frac{v_{SL}\mathrm{Fe}}{v_{SG}+v_{SL}\mathrm{Fe}}, \tag{16}$$

where Fe is the fraction of the total liquid entrained in the gas core and given by (Oliemans et al. 1986)

$$\frac{\mathrm{Fe}}{1-\mathrm{Fe}} = 10^{\beta_0} \rho_L^{\beta_1} \rho_G^{\beta_2} \mu_L^{\beta_3} \mu_G^{\beta_4} \sigma^{\beta_5} D^{\beta_6} V_{LS}^{\beta_7} V_{GS}^{\beta_8} g^{\beta_9}. \tag{17}$$

In this relation, the exponents β are numbers corresponding to fitted parameters.

On eliminating the pressure gradient from Eqs. (11) and (13), it is possible to obtain the combined momentum equation

$$\tau_i \frac{S_L}{A}\left(\frac{1}{R_i} + \frac{R_i\left(1-a^2\right)}{\left(1-a^2R_i^2\right)}\lambda\right) - \tau_{wL}\frac{S_L}{A}\lambda + g\sin\alpha\left(\rho_G - \rho_L\right) = 0, \tag{18}$$

where A is the cross-sectional area of the pipe, S_L is its perimeter, and λ is a geometric factor that takes into account the effects of the annular flow eccentricity

$$\lambda = \frac{\left(aR_i - 1\right)^2\left(aR_i + 1\right)^2}{\left(1-R_i^2\right)\left(1-a^4R_i^2\right)}. \tag{19}$$

Moreover, the shear stress at the pipe wall for the liquid, τ_{wL}, is defined by

$$\tau_{wL} = \frac{1}{2} f_{wL}\rho_L v_L^2, \tag{20}$$

where f_{wL} is the wall-liquid friction factor, which obeys the experimentally obtained correlation for annular flow

$$f_{wL} = 0.0063 + 53.4662\mathrm{Re}_M^{-1}. \tag{21}$$

In the above relation, Re_M is the Reynolds number proposed by García et al. (2003):

$$Re_M = \frac{\rho_L v_M D}{\mu_L}. \tag{22}$$

Furthermore, the interfacial shear stress, τ_i, is given by

$$\tau_i = \frac{1}{2} f_i \rho_c \left(v_G - v_L\right)^2, \tag{23}$$

where the gas-liquid interfacial friction factor, f_i, is given by Whalley and Hewitt (1978) correlation. They determine the interfacial friction factor by considering an interface roughness ($k = C \pm \Delta h_f$), using Colebrook (1939) equation

$$\frac{1}{\sqrt{f_i}} = -4\log\left[\frac{k/D}{3.7} + \frac{1.255}{Re_{SG}\sqrt{f_i}}\right], \tag{24}$$

where Δh_f is the apparent roughness or wave height and the factor C is the density ratio $C \cong 0.3\left(\rho_L/\rho_G\right)^{0.33}$.

Table 1 PDVSA-Intevep experimental data

Source	Points	Fluids	μ_L (cP)	v_{SL} (m/s)	v_{SG} (m/s)	D (m)	ε/D
Exp-A (2001)	5	Air-kerosene	1	0.11-1.6	28.2-45.7	0.0508	0
Exp-B (2000)	14	Air-oil	500	0.01-0.3	10.1-38.2	0.0508	0
Exp-C (2001)	7	Air-oil	1200	0.85-0.9	7.2-24.4	0.0508	0

Table 2 Stanford University data

Source	Points	Fluids	μ_L (cP)	v_{SL} (m/s)	v_{SG} (m/s)	D (m)	ε/D
Govier and Omer (1962)	5	Air-water	1	0.01-0.1	6.3-16.6	0.0261	0
Ansari et al. (1994)	3	Air-oil	80	0.02-0.3	6.1-13.2	0.0266	1.7E-3
Companies[a]	36	Air-oil	15	0.04-0.5	18.7-69.6	0.0502	3.0E-5
	17	Air-HL	1-25	0.02-2.2	8.0-24.1	0.0381	1.2E-3
	12		3-22	0.03-0.6	23.1-59.5	0.0909	1.7E-5
	4		19	0.11-0.6	40.5-63.4	0.0232	6.5E-5
	8		19	0.10-1.0	34.9-57.1	0.0237	6.5E-5
	73	Air-water	1	0.01-0.5	16.9-61.3	0.0455	0

[a] Data sets are identified as: SU24, SU25, SU28, SU29, SU184, SU187, SU199

Table 3 Tulsa University data

Source	Points	μ_L (cP)	v_{SL} (m/s)	v_{SG} (m/s)	D (m)	ε/D
Andritsos (1986)	36	1-70	0.001-0.56	12.15-30.09	0.0252	0
	3	80	0.004-0.02	14.04-24.65	0.0953	0
Beggs (1972)	5	1	0.02-0.56	15.96-24.97	0.0254	0
	3	1	0.02-0.11	14.85-15.12	0.0381	0
Mukherjee (1979)	9	1	0.03-0.56	11.40-24.06	0.0381	0

3 Experimental Data

The experimental data for the gas-liquid annular flow is made of 240 experimental points of a database containing information from the Stanford Multiphase Flow Database (SMFD), the Tulsa University Fluid Flow Project (TUFFP), and PDVSA-Intevep experiments. The range of operation conditions and fluid properties of each database are summarized in Tables 1, 2, and 3.

In each table, the last colum lists the absolute roughness, ε, in terms of the pipe diameter. The statistical parameters employed in this study are listed in Table 4. They are given by: the average percentage error (E_1) and the average error (E_5), which are related to the agreement between predicted and measured data; the average absolute percent error (E_2) and average absolute error (E_6), which are two of the most important statistical parameters because the negative and positive values do not cancel out; the standard percent deviation (E_3) and standard deviation (E_7), which are related to the scatter of the errors with respect to the average error of the experimental data; and the mean root square percent error (E_4) and the root

Table 4 Statistical parameters

Statistical parameter	Definition	Unit
E_1	$\frac{1}{N}\sum_{I=1}^{N}\left(\frac{\Delta P_{cal}-\Delta P_m}{\Delta P_m}\right) * 100$	%
E_2	$\frac{1}{N}\sum_{I=1}^{N}\left\|\left(\frac{\Delta P_{cal}-\Delta P_m}{\Delta P_m}\right)\right\| * 100$	%
E_3	$\sqrt{\frac{1}{N-1}\sum_{I=1}^{N}\left(\left(\frac{\Delta P_{cal}-\Delta P_m}{\Delta P_m}\right)-E_1\right)^2} * 100$	%
E_4	$\sqrt{\frac{1}{N-1}\sum_{I=1}^{N}\left(\left(\frac{\Delta P_{cal}-\Delta P_m}{\Delta P_m}\right)\right)^2} * 100$	%
E_5	$\frac{1}{N}\sum_{I=1}^{N}(\Delta P_{cal}-\Delta P_m)$	Pa/m
E_6	$\frac{1}{N}\sum_{I=1}^{N}\|(\Delta P_{cal}-\Delta P_m)\|$	Pa/m
E_7	$\sqrt{\frac{1}{N-1}\sum_{I=1}^{N}((\Delta P_{cal}-\Delta P_m)-E_5)^2}$	Pa/m
E_8	$\sqrt{\frac{1}{N-1}\sum_{I=1}^{N}((\Delta P_{cal}-\Delta P_m))^2}$	Pa/m

mean square error (E_8), which indicate how close the model prediction is to the experimental data.

4 Results and Discussion

In order to develop a model that takes into account the eccentricity of the annular flow in pipelines, we have carried out a series of experiments in PDVSA-Intevep with an air-oil flow of 400 cP in horizontal pipes. We found that for a constant liquid rate, when the gas rate increases, the eccentricity of the annular flow decreases and eventually tends to zero, as shown in Fig. 2, where the eccentricity is plotted as a function of the gas superficial velocity.

We observe that the eccentricity has no relevant effects on the pressure gradient because when we force the model to predict the pressure gradient with an eccentricity equal to zero, the difference obtained is about ± 1 %. Figure 3 compares the predicted values of the pressure gradient for annular flow with the experimental data listed in Tables 1, 2 and 3 for a gas-liquid system. The average absolute error is 30.5% with an estimated standard deviation of 28.9 %. As we may see, the average absolute error of 60 %, corresponding to the experimental database (147 points), has a deviation less than 30 % over a wide range of fluid properties and pipe diameters.

We also compare the performance of the our model with that of other mechanistic models (MM) as proposed by Xiao et al. (1990), Ouyang et al. (1998), Gómez et al.

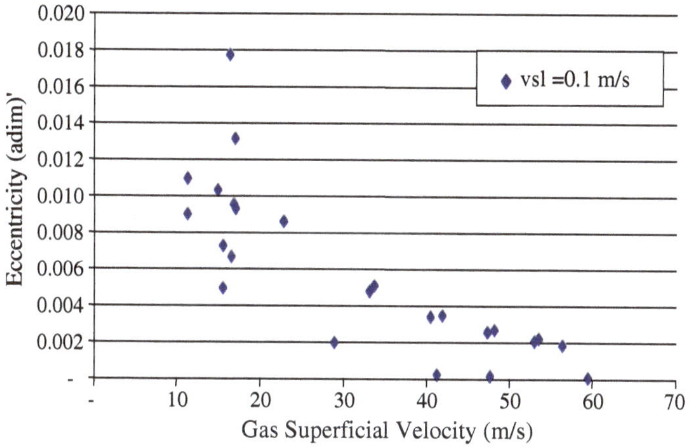

Fig. 2 Eccentricity behaviour of the annular flow

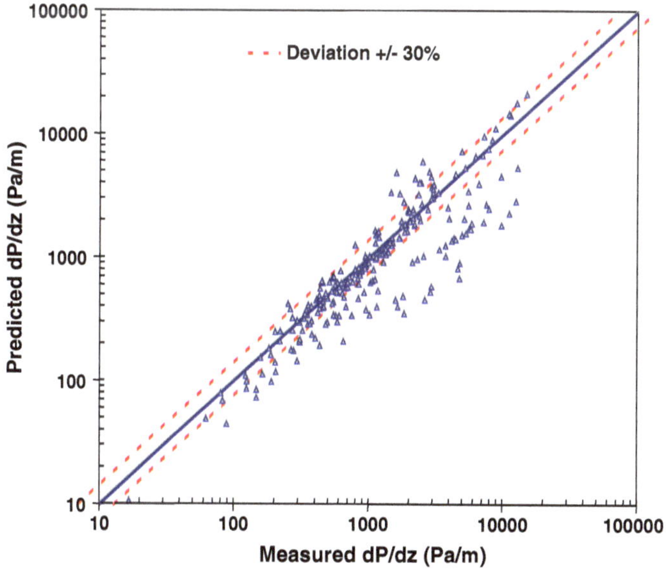

Fig. 3 Comparison of the predicted pressure gradients with existing experimental data

(1999), and Holden (2002) and the correlation models (CM) of Dukler et al. (1964), Beggs and Brill (1973), and García et al. (2003) for the same experimental data of Tables 1, 2 and 3.

The accuracy of the different mechanistic and correlation models is listed in Table 5 and is expressed in terms of the statistical parameters displayed in Table 4. All these models were compared against the experimental data of Tables 1, 2 and 3. We may see that for most existing models in current use the absolute error E_2 is more

Table 5 Comparison of the accuracy of pressure gradient prediction against the experimental data for different mechanistic and correlation models

	E_1 (%)	E_2 (%)	E_3 (%)	E_4 (%)	E_5 (Pa/m)	E_6 (Pa/m)	E_7 (Pa/m)	E_8 (Pa/m)
Garcia CM	−11,8	29,4	31,7	33,8	−127,3	690,6	1444,4	1450,0
Brito MM	−10,3	30,5	40,8	42,1	−354,2	862,0	1780,9	1815,9
Dukler CM	−27,6	31,8	26,6	38,4	−803,3	878,2	1527,8	1726,8
Holden MM	28,4	47,9	118,0	121,4	667,9	1048	5883,6	5921,5
Ouyang MM	−52,8	53,9	25,8	58,7	−1007	1120	1788,9	2053,8
Xiao MM	62,0	92,6	183,7	193,9	33,7	1137	1960,1	1960,4
Beggs and Brill CM	128,9	141,1	554,9	569,8	9578	10108	73313	73939
Gomez MM	164,8	178,9	2023	2030	416,7	1076	4458	4478

than 50 % in the prediction of the pressure gradients for annular flow in pipelines. In contrast, the overall performance of our mechanistic model (Brito MM) yields an absolute error of 30.5 % and so it has a superior accuracy compared to all quoted mechanistic models. We note, however, that the correlation models of Dukler et al. (1964) and García et al. (2003) also produced deviations of about 30 %, similar to our model.

5 Conclusions

A new mechanistic model for the prediction of gas-liquid annular flow in horizontal pipelines has been presented. The accuracy of the model has been assessed by comparing its performance with that of seven different models for a set of 240 experimental points. The main conclusions can be summarized as follows:

- The eccentricity has no important effects on the prediction of the pressure gradients in annular flow patterns, with a difference less than 1 % when concentric and eccentric patterns are considered.
- The limiting case of the model proposed is when the eccentricity is equal to zero (concentric flow).

The present model reduces the absolute error in the prediction of pressure gradients in pipelines by 17 % compared to Holden (2002) mechanistic model, which is considered to be the mechanistic model with better performance. The overall performance of the model is around a 30 % absolute error, similar to the performance obtained with correlation models by Dukler et al. (1964) and García et al. (2003). These results clearly indicate that more studies are indeed required to improve the accuracy of prediction of the physical parameters relevant to annular flow in pipelines.

Acknowledgments This work was performed under the auspices of PDVSA Intevep with the support of the Escuela de Ingeniería Mecánica of the Universidad Central de Venezuela.

References

Andritsos N (1986) Effect of pipe diameter and liquid viscosity on horizontal stratified flow. University of Illinois, Champaign-Urbana, USA, PhD dissertation

Ansari A, Sylvester N, Sarica C, Shoham O, Brill J (1994) A comprehensive mechanistic model for upward two-phase flow in wellbores. SPE Prod Facil 9:142–152

Alves G (1954) Concurrent liquid-gas flow in a pipe-line contactor. Chem Eng Prog 50:449–456

Beggs H (1972) An experimental study of two-phase flow in inclined pipes. University of Tulsa, USA, PhD dissertation

Beggs HD, Brill JP (1973) A study of two-phase flow in inclined pipes. J Petrol Technol 25:607–617

Colebrook CF (1939) Turbulent flows in pipes with particular reference to the transition region between smooth and rough pipe laws. J Inst Civil Eng 11:133–156

Dukler AE, Wicks M, Cleveland RG (1964) Frictional pressure drop in two-phase flow: B. An approach through similarity analysis. AIChE J 10:44–51

García F, García R, Padrino J, Mata C, Trallero J, Joseph D (2003) Power law and composite power law friction factor correlations for laminar and turbulent gas-liquid flow in horizontal pipelines. Int J Multiph Flow 29:1605–1624

Gómez L, Shoham O, Schmidt Z, Chokshi R, Brown A, Northung T (1999) A unified model for steady state two-phase flow in wellbores and pipelines. SPE 56520. In: The 1999 SPE technical conference and exhibition, Houston, Texas, Oct 3–6.

González J (1998) Non-dimensional coordinate system suitable to study flow between two concentric or eccentric cylinders. MSc thesis, The Graduate School, The University of Pittsburgh, USA.

Govier G, Omer M (1962) The horizontal pipeline flow of air-water mixture. Canadian Journal of Chemical Engineering. 40:93

Holden Z (2002) Unified model for gas-liquid pipe flow. Fluid Flow Projects. Ninety ninth research report. Department of Petroleum Engineering, University of Tulsa, USA. Jan 2002.

Mukherjee H (1979) An experimental study of two-phase flow. University of Tulsa, USA, PhD dissertation

Oliemans RV, Pots BFM, Trompé N (1986) Modeling of annular dispersed two-phase flow in vertical pipes. Int J Multiph Flow 12:711–732

Ouyang LB, Arbabi S, Petalas N, Aziz K (1998) Analysis of horizontal well experiments–experiments conducted by Marathon oil for Stanford University's SUPRI-HW program. (SUPRI-B-SUPRI-HW), Stanford University, CA.

Wallis G (1969) One-dimensional two-phase flow. McGraw-Hill, New York

Whalley PB, Hewitt GF (1978) The correlation of liquid entrainment rate in annular two-phase flow. UKAEA report, AERE-R9187, Harwell.

Xiao JJ, Shoham O, Brill JP (1990) A comprehensive mechanistic model for two-phase flow in pipelines. In: The 65th SPE annual technical conference and exhibition, New Orleans, LA. Paper SPE 20631. 167–180, Sept 23–26.

Scaling Properties in the Adsorption of Ionic Polymeric Surfactants on Generic Nanoparticles of Metallic Oxides by Mesoscopic Simulation

Estela Mayoral and Eduardo Nahmad-Achar

Abstract We study the scaling of adsorption isotherms of polyacrylic dispersants on generic surfaces of metallic oxides $XnOm$ as a function of the number of monomeric units, using Electrostatic Dissipative Particle Dynamics simulations. The simulations show how the scaling properties in these systems emerge and how the isotherms rescale to a universal curve, reproducing reported experimental results. The critical exponent for these systems is also obtained, in perfect agreement with the scaling theory of de Gennes. Some important applications are mentioned.

1 Introduction

Polyelectrolyte solutions have properties quite different from those observed in solutions of uncharged polymers, and their behaviour is less well known (de Gennes 1976; Odijk 1979; Dobrynin et al. 1995). In particular, the scaling of some quantities could present a different behaviour and so atypical scaling exponents could be found. In most cases, the statistical properties of these interesting systems cannot be obtained analytically because of the long-range Coulombic repulsion produced by the presence of small mobile counterions in the bulk, which interact both with the charge in the polymer and with one another. The use of simulation methodologies have shown, however, to be a promising tool in the study of very complex systems (Fermeglia and Pricl 2007).

E. Mayoral (✉)
Departamento de Física, Instituto Nacional de Investigaciones Nucleares, ININ, Km. 36.5, Carretera México-Toluca, 52750 La Marquesa , Estado de México, Mexico
e-mail: estela.mayoral@inin.gob.mx

E. Nahmad-Achar
Instituto de Ciencias Nucleares, Universidad Nacional Autónoma de México, UNAM, Apartado Postal 70-543, 04510 México D. F., Mexico
e-mail: nahmad@nucleares.unam.mx

L. Di G. Sigalotti et al. (eds.), *Computational and Experimental Fluid Mechanics with Applications to Physics, Engineering and the Environment*, Environmental Science and Engineering, DOI:10.1007/978-3-319-00191-3_30,

In our case of study, the presence of big charged molecules (such as polymers) and small ones (like counterions and solvents) involving different length and time scales, makes an electrostatic mesoscopic approach a good alternative. One of these mesoscopic approaches is *Dissipative Particle Dynamics* (DPD), which is a Langevin dynamics approximation where the fluid is represented by virtual interacting particles through three forces: conservative, random, and dissipative. The conservative force includes repulsive and electrostatic interactions, and determines the equilibrium state of the system, whereas the dissipative and random forces act as a thermostat and allow transport properties, preserving the thermodynamic equilibrium. The electrostatic interactions in DPD simulations were first incorporated by Groot (2003), who solved the electrostatic field locally on a lattice. An alternative way to solve the electrostatic problem in DPD was developed later on by González-Melchor et al. (2006), where the calculation of the electrostatic interactions employs the standard Ewald sum method and, in order to prevent the artificial ionic pair formation, charge distributions are included on the DPD particles.

In this work, we study using electrostatic mesoscopic dissipative particle dynamics simulations, the adsorption of dispersants onto pigments and show the resulting density profiles, the adsorption isotherms, and their scaling properties.

2 Mesoscopic Approach

One of the main problems in many industrial and academic areas is that the systems of interest are often constituted by many particles of different length scales, interacting in different time scales. In order to simplify the study of these systems, in the early 1990 s Hoogerbrugge and Koelman (1992) introduced a mesoscopic simulation technique. This is known as Dissipative Particle Dynamics (DPD) and is a coarse-graining approach, which consists of representing complex molecules as soft spherical beads interacting through a simple pair-wise potential, and thermally equilibrated through hydrodynamics (Groot and Warren 1997). In this formalism, the beads obey Newton's equations of motion

$$\frac{d\boldsymbol{r}_i}{dt} = \boldsymbol{v}_i, \qquad \frac{d\boldsymbol{v}_i}{dt} = \boldsymbol{f}_i, \tag{1}$$

where $\boldsymbol{r}_i$ and $\boldsymbol{v}_i$ are the position and the velocity of the ith particle, respectively, and the force $\boldsymbol{f}_i$ is given by three components:

$$\boldsymbol{f}_i = \sum_j \left(\boldsymbol{f}^C{}_{ij} + \boldsymbol{f}^D{}_{ij} + \boldsymbol{f}^R{}_{ij}\right), \tag{2}$$

corresponding to the conservative, dissipative, and random contributions, respectively. The sum runs over all neighbouring particles within a certain distance R_c. The conservative force $\boldsymbol{f}^C$ derives from a soft interaction potential and there is no hard-core divergence as in the case of a Lennard-Jones potential, thus providing a more efficient scheme of integration; it has the form

$$f^C{}_{ij} = a_{ij}\,\omega^C(\boldsymbol{r}_{ij})\,\frac{\boldsymbol{r}_{ij}}{|\boldsymbol{r}_{ij}|}\,. \tag{3}$$

When we need to introduce a more complex molecule, such as a polymer, we use beads joined by springs, so we also have an extra spring force given by $\boldsymbol{f}^S{}_{ij} = k\,\boldsymbol{r}_{ij}$ if i is connected to j. The dissipative $\boldsymbol{f}^D$ and random $\boldsymbol{f}^R$ standard DPD forces are given by

$$\boldsymbol{f}_{ij}^D = -\gamma\,\omega^D(\boldsymbol{r}_{ij})\,\frac{(\boldsymbol{r}_{ij}\cdot\boldsymbol{v}_{ij})\,\boldsymbol{r}_{ij}}{|\boldsymbol{r}_{ij}|^2}, \tag{4}$$

and

$$\boldsymbol{f}^R{}_{ij} = -\sigma\,\omega^R(\boldsymbol{r}_{ij})\,\frac{\theta_{ij}}{\delta_t^{1/2}}\,\frac{\boldsymbol{r}_{ij}}{|\boldsymbol{r}_{ij}|}\,. \tag{5}$$

Here, δ_t is the time step, $\boldsymbol{v}_{ij} = \boldsymbol{v}_i - \boldsymbol{v}_j$ is the relative particle velocity, θ_{ij} is a random Gaussian number with zero mean and unit variance, γ and σ are the dissipation and noise strengths, respectively, while $\omega^C(\boldsymbol{r}_{ij})$, $\omega^D(\boldsymbol{r}_{ij})$, and $\omega^R(\boldsymbol{r}_{ij})$ are dimensionless weight functions. Not all these quantities are independent: some of them are related through the fluctuation-dissipation theorem (Español and Warren 1995) by $\gamma = \sigma^2/2\kappa_B T$ and $\omega^D(\boldsymbol{r}_{ij}) = [\omega^R(\boldsymbol{r}_{ij})]^{1/2}$, with κ_B being the Boltzmann constant and T the temperature.

The methodology used in our mesoscopic simulations and, specifically, the electrostatic DPD methodology, is briefly described in the following subsection.

2.1 Mesoscopic Simulation: Electrostatic Dissipative Particle Dynamics

We consider in our study an ionic polymeric dispersant, for example polyacrylic acid (*PAA*) or a salt derived from it, in water and in the presence of substrate particles which we assume to be metallic oxides, such as TiO_2, Al_2O_3, CeO_2, etc. We map the polymer chain into beads which we call *DPD beads* as shown by the label *A*—in Fig. 1. Each DPD bead has a volume $v_{DPD} = 90$ Å^3 and radius $r_{DPD} = 2.78$ Å, which correspond to the volume of three water molecules. We can represent a *PAA* chain by N_{DPD} beads of carboxylate monomeric units joined by springs with some spring constant k. In this case $N_{DPD} = v_{mon}N/v_{DPD}$, where v_{mon} is the volume of a carboxylate monomeric unit and N is the number of monomeric units in the chain.

As was mentioned in the Introduction, here we replace the point charge at the centre of the DPD particle by a charge distribution throughout the particle. This is in order to avoid the formation of artificial clusters from oppositely charged ions. Groot (2003) solved the problem by calculating the electrostatic field on a grid.

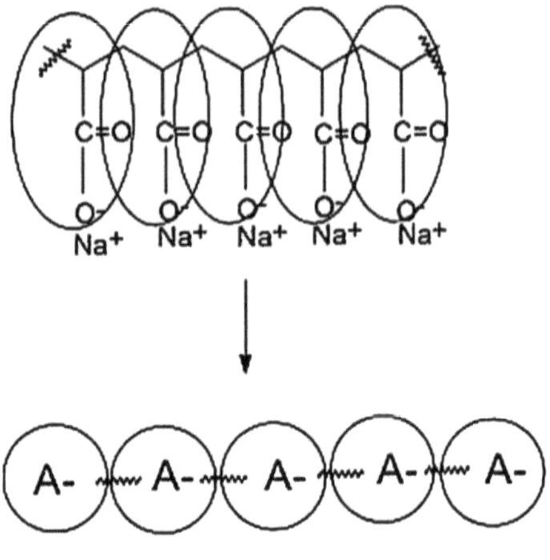

Fig. 1 Mesoscopic identification for a polyelectrolyte such as the sodium salt of *PAA*

The algorithm is known as the particle-particle-particle mesh (PPPM) algorithm. In González-Melchor et al. (2006), we solved this problem by combining the standard method with charge distributions on particles, adapting the standard Ewald method to DPD particles. In the present work, we use the latter method because the Ewald sum technique is the most employed route to calculate electrostatic interactions in microscopic molecular simulations. We take, as in González-Melchor et al. (2006)

$$\omega^C(r) = \omega^R(r) = \omega^D(r)^{1/2} = \omega(r), \tag{6}$$

with

$$\omega(r) = \begin{cases} 1 - r/R_c & : \quad r \leq R_c, \\ 0 & : \quad r > R_c, \end{cases} \tag{7}$$

where R_c is the cut-off distance, here assumed to be 6.46 Å (the simulation characteristic length). We also take $\sigma = 3\,\mathrm{kg\,m\,s^{-3/2}}$. We represent the PAA^{N-} with N DPD beads, each one having $v_{mon} = 90$ Å^3 and bonded by a spring with $k = 100$. The Na^+ ions were simulated by one DPD bead each with charge 1^+, and three water molecules per neutral DPD particle. These values reproduce the isothermic compressibility of water in standard conditions. All other quantities, including k, are dimensionless quantities given in reduced units. This is accomplished as follows: since we keep $T = const$ in our simulations, we may take $\kappa_B T = 1$ as the unit of energy; the distance r is measured in units of R_c, i.e., $r^* = r/R_c$; the force has therefore no units as $[E]/[r^*] = [1]$, and neither does k; the density is given as the number of molecules per unit volume and has no units, and the *mass* does not ever enter into the model. Since the force is dimensionless so is the time. The integration time step is taken to be $\Delta t^* = 0.02$ and the total average density is $\rho^* = 3$ (i.e., three water molecules per DPD particle; one may also see that it has the unit of mass as being defined by the mass of a DPD particle with three water molecules).

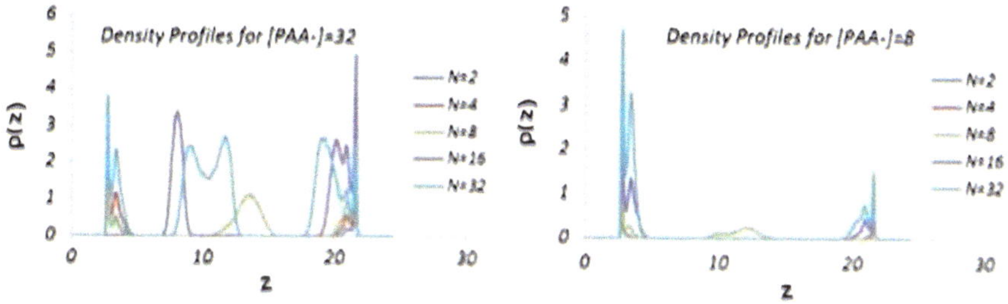

Fig. 2 Density profiles for $[PAA] = 32$ (*left*) and 8 (*right*)

3 Results and Discussion

3.1 Results for Adsorption Isotherms

The DPD electrostatic simulations were performed using our mesoscopic model as described in the last section, in order to obtain the adsorption isotherms for $[PAA^{N-}][Na^{+}]_N$ on generic surfaces of metallic oxides $XnOm$ at a basic pH. The length of the *PAA*-DPD molecule was varied as $N = 2, 4, 8, 16$, and 32 DPD particle units. The repulsive constants a_{ij} in the DPD model were set to $a_{W-PAA^-} = 100$, $a_{W-Na^+} = 100$, $a_{W-H_2O} = 100$, $a_{H_2O-PAA^-} = 82$, $a_{H_2O-Na^+} = 25$, and $a_{PAA^- -Na^+} = 25$. These values can be obtained from solubility parameters, and a more refined calculation can be made by using activity coefficients (cf. Mayoral and Nahmad-Achar 2012).

The resulting density profiles $\rho(z)$, describing the spatial organization of the molecules as a function of one of the spatial coordinates, are shown in Fig. 2 for two different concentrations of the polyelectrolyte, namely $[PAA] = 32$ and $[PAA] = 8$. They show that larger molecules tend to adsorb at the edges of the box (which represent the metallic substrate), and remain less in the aqueous medium (in between the box walls), where smaller molecules can be found.

To obtain the adsorption isotherms we calculate the amount of polyelectrolyte Γ carried by the particle, by integrating the density profile according to

$$\Gamma = \int_0^{L_z} [\rho(z) - \rho_{bulk}]\, dz, \tag{8}$$

where L_z is the width of the first adsorbed layer and ρ_{bulk} the bulk density. Figure 3 shows the number Γ_{mol} of *PAA*-molecules adsorbed on a TiO_2 surface *vs.* the number Γ^b_{mol} of non-adsorbed molecules, by considering a single adsorbed layer. As expected, the saturation on the surface is reached earlier for larger molecules.

However, we may easily renormalize these curves by plotting the number of independently adsorbed DPD beads Γ_{DPD} *vs.* non-adsorbed DPD beads Γ^b_{DPD}, using $N\,\Gamma_{mol} = \Gamma_{DPD}$. The behaviour will then be that of a universal isotherm conformed

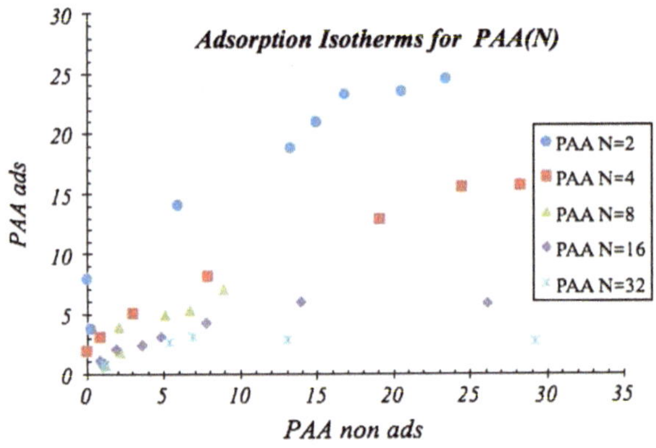

Fig. 3 Adsorption isotherms for *PAA* on a TiO_2 surface for different N

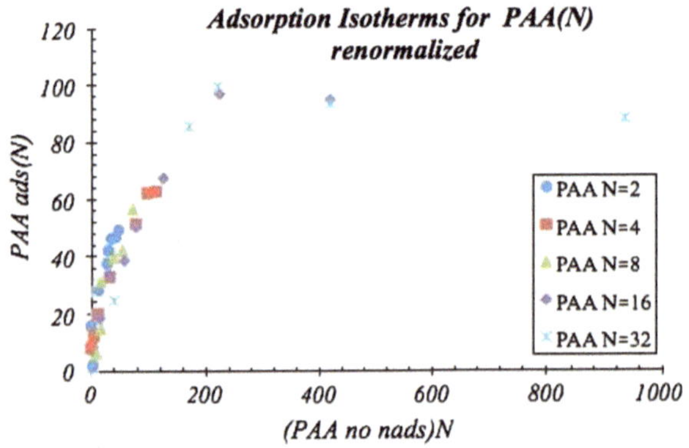

Fig. 4 Universal adsorption isotherm for *PAA* on a TiO_2 surface, renormalized

by the contribution of all sizes, as shown in Fig. 4. Assuming that only one layer is adsorbed on the surface (the self similar region) and that all adsorption positions are equivalent, we can extract the maximum concentration at equilibrium and the adsorption-desorption constant for each isotherm, which is given by the Langmuir isotherm. The dynamic equilibrium is given by $A + N \Leftrightarrow AN$ with velocity constants K_a for the adsorption and K_d for the desorption. The expression for this kind of adsorption model, in the case of *neutral* species, is given by the Langmuir isotherm expressed by

$$\frac{1}{\Gamma} = \frac{1}{\Gamma_M} + \frac{1}{\Gamma_M\, K\, C}, \tag{9}$$

where $K = K_a/K_d$ and C is the concentration in the bulk (Γ^b). Γ is the adsorbed quantity and Γ_M is the maximum adsorbed quantity. The linear fit for this isotherm is shown in Fig. 5, and is given by $1/\Gamma_M = 0.0094$ and $1/(\Gamma_M\, K) = 0.5432$, from which $\Gamma_M = 106.38\, PAA_{DPD}$ and $K = 0.0173$.

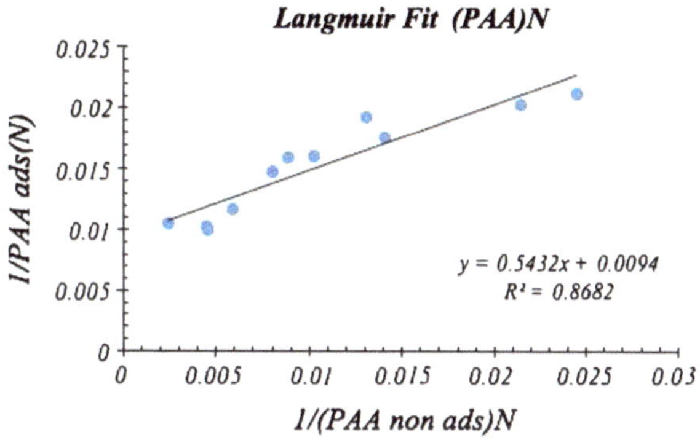

Fig. 5 Langmuir fit for the adsorption isotherms of *PAA* on TiO_2

Table 1 Scaling for Γ_{max} as a function of N

ln N	ln Γ_{max}
0.6931	3.3077
1.3863	2.8134
2.0794	2.4581
2.7726	1.5626
3.4657	1.1907

3.2 Scaling for Γ_{max}

In the light of these results, it is interesting to study the behaviour of Γ_{max} with N. We can do this, once more, via DPD electrostatic simulations. Γ_{max} is obtained by fitting each isotherm in Fig. 3 with the Langmuir model, which we have shown to be adequate (*vide supra*). Table 1 shows the results for the fit in each case. When we plot Γ_{max} *versus* N we obtain the behaviour shown in Fig. 6 and the scaling function is $\Gamma_{max} \propto N^{-0.79} \simeq N^{-4/5}$. This result is in perfect agreement with de Gennes et al. (1976).

The scaling theory in the weak adsorption regime indicates that in the flat plateau, i.e., at maximum saturation

$$\gamma_p \sim N^{1/5}, \tag{10}$$

where γ_p is the number of monomers adsorbed in the flat plateau, $\gamma_p = \Gamma_{max} N$. Equation (10) then implies $\Gamma_{max} \sim N^{-4/5} = N^{-0.8}$, which agrees very well with our result. It is interesting that the renormalized behaviour adjusts itself to the scaling model in the weak adsorption regime, *even though we are dealing here with charged molecules*.

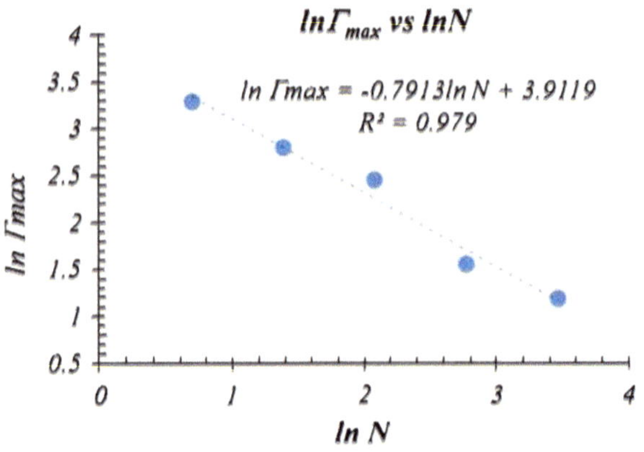

Fig. 6 Scaling of Γ_{max} with N

4 A Simple Application to Colloid Stability

Since *PAA* and its salt derivatives tend to be very hydrophilic, the adsorbed segments will see the substrate as a flat surface when in a good solvent. If R is the effective radius of the substrate particle and $R_g = a_f N^{\nu}$ is the radius of gyration of the polymer chain, with ${a_f}^3$ proportional to the Flory volume, the flat surface regime is given, according to the *loops and tails* model, by $R > a_f\, N^{3/5}$ (Aubouy et al. 1993; Aubouy and Raphaël 1998). Lately, and in order to improve the performance properties of coatings, CeO_2 and Al_2O_3 nanoparticles have been used in their formulations (Mayoral et al. 2012). For these we have $R_{CeO_2} \simeq 10\,\text{nm}$ and $R_{Al_2O_3} \simeq 20\,\text{nm}$ which would give, for *PAA* and its salt derivatives, $N < 400$ ($Mw < 40{,}000\,\text{gr/mol}$) for CeO_2 and $N < 1{,}250$ ($Mw < 125{,}000\,\text{gr/mol}$) for Al_2O_3. This accommodates even the higher molecular weight dispersants, so that a flat substrate approximation is appropriate in our mesoscopic approach.

Let us consider a small particle of diameter $d_1 = 2r_1$ and area a_1, and a larger particle with diameter $d_2 = 2r_2 > d_1$ and area $a_2 > a_1$, to be stabilized in an aqueous medium. We know from the previous section that the number of monomers adsorbed in the flat plateau is $\gamma_p = \Gamma_{max} N$, with these quantities scaling as $\gamma_p \sim N^{1/5}$ and $\Gamma_{max} \sim N^{-4/5}$.

Γ_{max} is the number of chains of size N per unit area needed in order to cover satisfactorily some amount, say 1 mol, of material. If we want to cover a surface of area a_1, then $c = a_1 \Gamma_{max,1}$ chains are needed. Now suppose that the weight of one monomeric unit is 1 unit of mass, then $a_1 \Gamma_{max,1}[chains] = a_1 \Gamma_{max,1} N_1 = a_1 \gamma_{p,1}$ and for the same amount of material but with area a_2, we will require $a_2 \Gamma_{max,2}[chains] = a_2 \gamma_{p,2}$. If κ is the amount of mass needed to cover the surface of particles of diameter $2r_1$ divided by the mass of dispersant necessary to cover the surface of particles of diameter $2r_2$, then

$$\kappa = a_1\gamma_{p,1}/a_2\gamma_{p,2} = (a_1 N_1^{1/5})/(a_2 N_2^{1/5}), \tag{11}$$

or

$$\kappa = (r_2/r_1)\,(N_1/N_2)^{1/5}. \tag{12}$$

We can make use of this expression to analyze two interesting cases:

Case 1. If we want to use the same amount of dispersant, taking dispersants with different lengths N_1 and N_2 and having the same chemistry, $\kappa = 1$ and $1 = (r_1/r_2)\,(N_1/N_2)^{1/5}$, that is $N_1 = (r_1/r_2)^5 N_2$. We would then need a dispersant with a very small degree of polymerization compared with N_2 for $r_1 << r_2$. In this case the smallest and the best dispersant will be $N_1 = 1$ (monomeric dispersant), in agreement with the results of Goicochea et al. (2009). If, on the other hand, $(r_1/r_2)^5 << 1$, a change in the chemistry of the dispersant would be a better option.

Case 2. In the limit of a flat approximation, we can consider $N = (R/a_f)^{5/3}$ and have $\kappa = (r_2/r_1)\,[(r_1/r_2)^{5/3}]^{1/5} = (r_2/r_1)^{2/3}$. For Al_2O_3 nanoparticles as compared to ordinary TiO_2 particles used in coatings, for example, we have $R_{TiO_2} \simeq 125$ nm and $R_{Al_2O_3} \simeq 20$ nm, and so $\kappa = (125/20)^{2/3} = 3.3993$. Comparing this result to an estimation based on purely geometric arguments (Mayoral et al. 2012), where 6.25 times the dispersant amount was needed for Al_2O_3-nanoparticles, we observe that by choosing a dispersant with an adequate length N we would need a much smaller quantity.

5 Conclusions

Langmuir isotherms were calculated for polyacrylate dispersants adsorbed on metallic oxides, while their scaling properties as a function of the number of monomeric dispersant units were obtained via DPD-simulations. The critical exponent for the renormalized isotherms was obtained, and this agrees perfectly well with the scaling theory of de Gennes et al. (1976), even though polyelectrolytes are being considered.

The results presented here suggest a methodology for estimating the amount of dispersant necessary in different scenarios and for a better choice of the appropriate dispersants. The particular case of the stabilization of metallic nanoparticles is interesting, as their inclusion in many formulations to improve performance properties is presently a major area of research. Problems arise because the dimensions of the nanoparticles and polymeric dispersants are similar, and because of the large total surface area to be covered. However, excessive amounts of any surfactant will cause the property degradation of the material, and new especially designed surfactants circumvent the need for large quantities. Here it was shown that our simulation results improve upon the experimental values obtained by Mayoral et al. (2012).

Acknowledgments This work was partially supported by DGAPA-UNAM (under project IN102811). Valuable support in computing resources was obtained from DGTIC-UNAM.

References

Aubouy M, di Meglio J, Raphaël E (1993) Irreversible adsorption of a polymer melt on a colloidal particle. Europhys Lett 24:87–92

Aubouy M, Raphaël E (1998) Scaling description of a colloidal particle clothed with polymers. Macromolecules 31:4357–4363

Dobrynin AV, Colby RH, Rubinstein M (1995) Scaling theory of polyelectrolyte solutions. Macromoleules 28:1859–1871

Español P, Warren PB (1995) Statistical mechanics of dissipative particle dynamics. Europhys Lett 30:191–196

Fermeglia M, Pricl S (2007) Multiscale modeling for polymer systems of industrial interest. Prog Org Coat 58:187–199

de Gennes PG, Pincus P, Velasco RM, Brochard F (1976) Remarks on polyelectrolyte conformation. Journal de Physique 37:1461–1473

de Gennes PG (1976) Scaling theory of polymer adsorption. Le Journal de Physique 37:1445–1452

Goicochea AG, Nahmad-Achar E, Pérez E (2009) Colloidal stability dependence on polymer adsorption through disjoining pressure isotherms. Langmuir 25:3529–3537

González-Melchor M, Mayoral E, Velázquez ME, Alejandre J (2006) Electrostatic interactions in dissipative particle dynamics using Ewald sums. J Chem Phys 125:224107

Groot RD, Warren PB (1997) Dissipative particle dynamics: Bridging the gap between atomistic and mesoscopic simulation. J Chem Phys 107:4423–4435

Groot RD (2003) Electrostatic interactions in dissipative particle dynamics–simulation of polyelectrolytes and anionic surfactants. J Chem Phys 118:11265

Hoogerbrugge PJ, Koelman JMVA (1992) Simulating microscopic hydrodynamic phenomena with dissipative particle dynamics. Europhys Lett 19:155–160

Mayoral E, Nahmad-Achar E (2012) Study of interfacial tension between an organic solvent and aqueous electrolyte solutions using electrostatic dissipative particle dynamics simulations. J Chem Phys 137:194701

Mayoral E, Nahmad-Achar E, Rodriguez J (2012) Diameter rules demand: nanoparticle dispersant adsorption and effects evaluated. Eur Coat J 12(2012):84–88

Odijk T (1979) Possible scaling relations for semidilute polyelectro-lytesolution. Macromolecules 12:688–693

Effect of Mixtures of Polysorbate 80 and Low Molecular Weight Alcohols on Density and °API Gravity of Treated Venezuelan Extra Heavy Oil

Efrén D. J. Andrades, Ledys Y. Sánchez, Hilda C. Grassi, Erick A. Pacheco, Silvia E. Andrades-Grassi and Gerardo E. Medina-Ramírez

Abstract Formulations of extra heavy oil with biocompatible polyethoxilated compounds have not received much attention. We investigate the behaviour of biocompatible mixtures in the treatment of Venezuelan extra heavy oil, using the non-ionic surfactant Polysorbate 80 (Tween 80) and low molecular weight linear n-alcohols with even and odd number of carbon atoms, in order to predict the best fit in the ethoxide—alcohol interaction. Venezuelan extra heavy oil was recovered from mixtures that contained water, NaCl, polysorbate 80, and n-alcohols ranging from 1 to 8 carbon atoms. The density, °API gravity (American Petroleum Institute gravity), and other properties were measured and compared for the even and odd numbered n-alcohols. We found a significant difference in density and °API gravity values in the treated and recovered extra heavy oil, for n-alcohols with even and odd number of carbons, in the presence of polysorbate 80. This finding suggests that the ether

E. D. J. Andrades (✉) · H. C. Grassi · E. A. Pacheco · S. E. Andrades-Grassi · G. E. Medina-Ramírez
UBIP Project, Sección de Biotecnología, Instituto de Investigaciones, Escuela de Farmacia, Facultad de Farmacia y Bioanálisis, Universidad de Los Andes, Mérida 5101, Estado Mérida, Venezuela
e-mail: andradesefren@cantv.net

H. C. Grassi
e-mail: cristinagrassi@gmail.com

E. A. Pacheco
e-mail: alejandropc@gmail.com

S. E. Andrades-Grassi
e-mail: silvigrassi@gmail.com

G. E. Medina-Ramírez
e-mail: medinag@ula.ve

L. Y. Sánchez
Laboratorio de Cinética y Catálisis, Departamento de Química, Facultad de Ciencias, Universidad de Los Andes, Mérida 5101, Estado Mérida, Venezuela
e-mail: ledys@yahoo.com

L. Di G. Sigalotti et al. (eds.), *Computational and Experimental Fluid Mechanics with Applications to Physics, Engineering and the Environment*, Environmental Science and Engineering, DOI:10.1007/978-3-319-00191-3_31,

within the ethoxide-repeating units of polysorbate 80 is the hydrogen bond acceptor of the n-alcohol donor. However, this interaction is favoured for the even number alcohols that interact in an "in-frame" manner with the ethoxide. We propose the formation of a micellar nanoparticle that promotes the improvement of Venezuelan extra heavy oil.

1 Introduction

The use of biocompatible molecules in the treatment of extra heavy oil, bitumens, asphalts, and other hydrocarbons, has been limited mainly due to the relatively high operative costs, the difficulty to recover separately the biocompatible molecules and the hydrocarbon after treatment, and the time required for the procedure, especially if indigenous microorganisms or designed microorganisms are expected to participate. Furthermore, knowledge of the potential roles of chemicals and biosurfactants in accelerating hydrocarbon accession to microorganisms is still very limited (Van Hamme et al. 2003; Tunio et al. 2011; Al-Sulaimani et al. 2011). Toxic polyethoxilated derivatives have been used for the improvement, transportation, and storage, and even for the design of new and lower cost fuels, such as Orimulsion® (Rivas 1999).

Due to these limitations, toxic polyethoxilated molecules, such as those of Orimulsion®, have been more widely used instead of their biocompatible counterparts. Thus, formulations of extra heavy oil with biocompatible polyethoxilated compounds have not received much attention. These molecules, which are composed of ethoxide-repeating units with ether bonds between them, have several oxygen atoms which may be potential H-bond acceptors. Ethers are less polar than alcohols and they participate in hydrogen bonds as H-bond acceptors, but cannot act as donors, unlike their parent alcohols (March 1992).

The presence of two lone pairs of electrons on the oxygen atoms makes hydrogen bonding with water (or with alcohols) possible. On the other hand, alcohols can be hydrogen bond donors (H donor) or H-bond acceptors (lone pair), forming hydrogen bonds with themselves and with other hydrogen bond donor/acceptors. Studies on H-bonding in alcohols show that the H-bond strength with primary and secondary alcohols as donors should serve as a tool for the characterization of these important interactions in chemical and biochemical systems (Gawlita et al. 2000). In view of these possible molecular interactions among alcohols and ethoxide moieties, and based on a formulation previously proposed by Grassi (2001), we investigate the behaviour of biocompatible mixtures in the treatment of Venezuelan extra heavy oil, using the non-ionic surfactant Polysorbate 80 (Tween 80) and low molecular weight linear n-alcohols with even and odd number of carbon atoms, in order to find the best fit in the ethoxide—alcohol interaction. Several properties of the treated extra heavy oil, such as density and °API gravity, were studied in order to understand this idea.

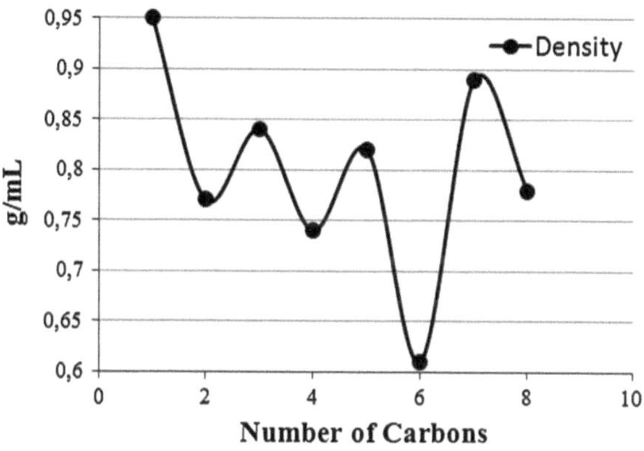

Fig. 1 Density (in units of g/mL) of the recovered heavy crude oil after treatment, as a function of the number of carbon atoms in the n-alcohol

2 Materials and Methods

10 mL of extra heavy oil from Morichal, Estado Monagas, Venezuela, with 9° API gravity and density 1.01 g/mL were mixed with 5 mL of distilled water containing 17,800 ppm NaCl and left at 61°C for 24 h. 100 μL of polysorbate 80 (Tween 80 pharmaceutical grade) and 10 μL of the test alcohol (analytical grade) were added (optimal formulation 10:5:0.1:0.01 oil:brine-water:non-ionic surfactant:alcohol). In some experiments, Tween 80 was substituted by I200 (from Orimulsion®), which was kindly supplied by the UBIP Project. This was mixed at 200 rpm at 61°C for 90 min and then left at room temperature without mixing, allowing for the separation of the respective aqueous phases from the recovered oil phases. Density (in units of mg/mL) was measured by weighing several times a fixed volume of the recovered oil phase. The other parameters were measured as a control and will be reported elsewhere. The significance of the values obtained with alcohols with even and odd number of carbon atoms was calculated by the Student's t-test.

3 Results and Discussion

Figure 1 and Table 2 show the density of the crude oil recovered after treatment with Polysorbate 80 (Tween 80) and the density of each of the linear n-alcohols containing from one to eight saturated carbons. Alcohols with an even number of Carbon atoms (i.e., 2, 4, 6, and 8) yield lower density values than odd alcohols (i.e., 1, 3, 5, and 7). Applying the Student's t-test on the results of density for both groups of data (crude oil recovered from mixtures of even and odd alcohols), the null hypothesis, stating that the means of the densities of the two groups are equal, is rejected. Therefore, the alternative hypothesis indicating that the means of the densities of even alcohols is proven to be statistically different and lower than that obtained with the odd alcohols, is accepted, as shown in Table 1. In the same test, for other parameters such as viscosity, conductivity, drop weight, oil retained, and water retained, the null hypothesis is accepted indicating that there is no significant difference in the results obtained with even and odd alcohols (see Table 1).

From this data, the °API gravity of the recovered oil was calculated and tabulated (see Table 2). As expected, linear n-alcohols with an even number of carbon atoms yield values that are higher than 50 °API gravity, while odd alcohols yield values that are lower than 50 °API gravity. A possible explanation for the oscillation is that the polysorbate 80 (Tween 80) may be allowing for the formation of hydrogen bonds with the ether being the acceptor and the alcohol being the donor. This interaction would be better for all the even alcohols and would be impaired for odd alcohols.

Aizawa (2009) reported the formation of a micellar nanoparticle in aqueous solutions with polysorbate 80 and 1,4-dioxane. It is possible to hypothesize a similar interaction between polysorbate, even alcohols, and some of the components of the extra heavy oil. Isopropanol (2-propanol) is an odd short alcohol with the hydroxyl group in an even and symmetric position (second carbon in any direction), and it yields a °API gravity value of the recovered oil, higher than 50, as expected from an even alcohol (see Table 2). On the other hand, although sec-butanol, iso-butanol, and tert-butanol have an even number of carbon atoms, the position of the hydroxyl group resembles an odd alcohol, and they yield °API gravity values of the recovered oil lower than 50, as expected from odd alcohols (see Table 2). The same results are obtained with 2-hexanol which yields a °API gravity value of the recovered oil that is close to the value of n-pentanol and far from the value of n-hexanol (see Table 2).

In order to evaluate whether the presence of the polyethoxides is the only condition necessary for the improvement of extra heavy oil, other known formulations were assayed. The use of I200 (containing nonyl phenol ethoxylated used in Orimulsion®) with ethanol yields a °API gravity value of the recovered oil much lower than 50 (see Table 2). This may be due to the presence of the aromatic phenol group. In order to test this, an aromatic alcohol was assayed with either Tween 80 or I200. The recovered oil yields °API gravity values of 22 for Tween 80 and 83 for I200 (Table 2).

These results indicate that the presence of the aromatic groups (one from benzyl alcohol and one from nonyl phenol ethoxylated) may provide an additional interaction that is not achieved with linear n-alcohols or with polysorbate 80, suggesting two different types of interaction: on one hand, polysorbate 80 with linear n-alcohols and, on the other hand, aromatic groups from nonyl phenol ethoxylated and benzyl alcohol. Therefore, for the treatment and improvement of extra heavy oil, we propose that the present results should be considered so as to take advantage of the interaction of the polyethoxilated molecule with the appropriate alcohol.

The proposed interaction for the treatment of extra heavy oil must have special requirements for the surfactant molecule, for the alcohol, and for the hydrocarbon and is different from other formulations previously used with extra heavy oil. We propose the interaction of surfactant-alcohol in the extra heavy oil through binding that occurs mainly by an “in-frame” hydrogen-bonding interaction between the oxyethylene and

Table 1 Student's t-test of the values obtained for different properties of the oil recovered after treatment with mixtures of Polysorbate 80 and low molecular weight linear even and odd n-alcohols

Variable	Density	Viscosity	Conductivity	Drop weight	Oil retained	Water retained
T	5.6032	0.7908	0.546	0.3216	−0.7792	0.8431
Df	3	3	2	3	3	3
P-value	0.0011	0.4868	0.6398	0.7688	0.4927	0.4611
Lower confidence bound	−0.0064	−13639.27	−16.6046	−0.8580	−2.8318	−2.9637
Upper confidence bound	0.3064	17910.94	18.5380	0.9580	2.1652	3.9637
Difference between two means	0.15	2135.838	0.9667	0.05	−0.3333	0.5
Interpretation	$P < 0.001$ The null hypothesis is rejected, the alternative hypothesis is accepted	$P > 0.001$ The null hypothesis is accepted	$P > 0.001$ The null hypothesis is accepted	$P > 0.001$ The null hypothesis is accepted	$P > 0.001$ The null hypothesis is accepted	$P > 0.001$ The null hypothesis is accepted

Table 2 Density and °API gravity of heavy crude oil after treatment and recovery from mixtures

n-ALCOHOL	N° of carbon atoms	Polyethoxilated molecule	Density g/mL	Calculated °API gravity
Methanol	1	Tween 80	0.95	17
Ethanol	2	Tween 80	0.77	52
n-propanol	3	Tween 80	0.84	37
n-butanol	4	Tween 80	0.74	60
n-pentanol	5	Tween 80	0.82	41
n-hexanol	6	Tween 80	0.61	100
n-heptanol	7	Tween 80	0.89	27
n-octanol	8	Tween 80	0.78	50
Iso-Propanol	3	Tween 80	0.78	50
Sec-butanol	4	Tween 80	0.89	27
Iso-butanol	4	Tween 80	0.94	19
Tert-butanol	4	Tween 80	0.91	24
2-hexanol	6	Tween 80	0.82	41
Benzyl alcohol	7	Tween 80	0.92	22
Ethanol	2	I200	0.96	16
Benzyl alcohol	7	I200	0.66	83

the hydroxyl group of the even alcohol. Interactions have been proposed for Tween 80 and the carboxylic groups of poly (acrylic acid) in water, which are usually weak except for the association of the carboxylic acid with ethoxylated surfactants. In these systems, hydrogen bonding between the carboxylic groups and the oxygen of the ethyleneoxide chain contributes to the aggregation of both molecules (Barreiro-Iglesias et al. 2003).

Aizawa (2009) has also reported the formation of core–shell cylindrical micelles, core–shell discus micelles, and core–shell elliptic discus micelles in aqueous mixtures of polysorbate 80 with 0–20 %, 30–40 %, and 50 % of 1,4-dioxane, respectively. Since 1,4-dioxane may be considered as two fused ethoxides in a ring, these results are in good agreement with our findings for even alcohols. Hydrogen bonding of the hydroxyl of ethanol, propanol, and butanol with a dipalmitoylphosphatidylcholine lipid bilayer has been studied using molecular dynamics simulations. According to this study, butanol behaves in a different manner from ethanol and propanol in response to an increase of the alcohol concentration: as the ethanol and propanol concentrations increase, the fluidity increases, whereas the opposite trend is seen for butanol, where an increase in concentration results in a drop of the fluidity. Furthermore, ethanol also decreases in density (Dickey and Faller 2007).

4 Conclusions

In this chapter, we have examined how n-alcohol length influences the behaviour of the interaction of polysorbate 80 (Tween 80) with extra heavy oil. Density and °API gravity were examined in the oil recovered after treatment and separation of the aqueous phase. We find that there is a significant difference in density and °API gravity values in the treated and recovered extra heavy oil, for alcohols with even and odd number of carbons, in the presence of polysorbate 80. The fluctuation in density between odd and even numbered alcohols suggests that the latter are allowing for the formation of hydrogen bonds that fit in an "in-frame" interaction with the ethoxides of polysorbate 80. We propose that a micellar nanoparticle similar to the one proposed by Aizawa (2009) may be forming by the interaction of the even n-alcohols, polysorbate 80, and extra heavy crude oil, improving the density and the °API gravity of the recovered oil.

Acknowledgments This work was supported in part by the UBIP Project.

References

Aizawa H (2009) Morphology of polysorbate 80 (Tween 80) Micelles in Aqueous 1,4-dioxane Solutions. J Appl Cryst 42:592–596

Al-Sulaimani H, Joshi S, Al-Wahaibi Y, Al-Bahry S, Elshafie A, Al-Bemani A (2011) Microbial Biotechnology for enhancing oil recovery:current developments and future prospects. Biotechnol Bioinf Bioeng 1(2):147–158

Barreiro-Iglesias R, Alvarez-Lorenzo C, Concheiro A (2003) Poly (acrylic acid) microgels (carbopol®934)/surfactant interactions in aqueous media part I: nonionic surfactants. Int J Pharm 258:165–177

Dickey AN, Faller R (2007) How alcohol chain-length and concentration modulate hydrogen bond formation in a lipid bilayer. Biophys J 92:2366–2376

Gawlita E, Lantz M, Paneth P, Bell AF, Tonge PJ, Anderson VE (2000) H-bonding in alcohols is reflected in the CαH bond strength: variation of C-D vibrational frequency and fractionation factor. J Am Chem Soc 122(47):11660–11669

Grassi HC (2001) Patent Pending Code 224288. Producto y Proceso para Mejoramiento, Recuperación y Biotransformación de Crudo y sus Derivados. Request # 2001–000538 www.patentesonline.com.ve

March J (1992) Advanced organic chemistry, 4th edn. J. Wiley and Sons, New York

Rivas H (1999) Patent EP0892035 A1. Intevep SA Multiple emulsion and method for preparing same

Tunio SQ, Tunio AH, Ghirano NA, El Adawy ZM (2011) Comparison of different enhanced oil recovery techniques for better oil productivity. Int J Appl Sci Tech 1(5):143–153

Van Hamme JD, Singh A, Ward OP (2003) Recent advances in petroleum microbiology. Microbiol Mol Biol R 67(4):4503–4549

Part IV
Granular and Porous-Media Flow

On the Construction of a Continuous Theory for Granular Flows

Juan C. Petit, Juan F. Marín and Leonardo Trujillo

Abstract We present a brief introduction of the coarse graining technique, which we shall use to construct a continuous matter field for discrete particulate systems. In particular, we address the problem of the micro to macro transition in the theoretical framework of granular hydrodynamics. The equations for momentum conservation, elastic stress tensor, and elastic energy are obtained introducing a harmonic interaction between the particles. These equations are compared with previous work, where the microscopic discrete nature of granular matter has not been considered. The microscopic description of a granular system leads to a matrix for the definition of the elastic moduli, which depends on position and coarse graining resolution. This provides some insight into the mathematical description of the origin of force chains in granular packings.

1 Introduction

A granular material is composed of a huge collection of particles, whose sizes are larger than about one micron and so their thermal energy is negligible compared to their mechanical potential energy. Due to their sizes, the particles can only interact by

J. C. Petit · J. F. Marín · L. Trujillo
Centro de Física, Instituto Venezolano de Investigaciones Científicas, IVIC, Apartado Postal 20632, Caracas 1020-A, Venezuela
e-mail: jcpetit71@gmail.com

J. F. Marín
Escuela de Física, Facultad de Ciencias, Universidad Central de Venezuela (UCV), Avenida Los Ilustres, Los Chaguaramos, Apartado Postal 20513, Caracas 1020-A, Venezuela
e-mail: juanfmarinm@gmail.com

L. Trujillo (✉)
The Abdus Salam International Centre for Theoretical Physics (ICTP), Strada Costiera 11, 34014 Trieste, Italy
e-mail: leonardo.trujillo@gmail.com

L. Di G. Sigalotti et al. (eds.), *Computational and Experimental Fluid Mechanics with Applications to Physics, Engineering and the Environment*, Environmental Science and Engineering, DOI:10.1007/978-3-319-00191-3_32,

contact forces and so they collide inelastically. As it has been observed experimentally (Eshuis et al. 2007), this kind of systems may exhibit different behaviours: solid-, liquid-, and gas-like (Jaeger et al. 1996). However, at the present time there is no a theory for granular materials capable to describe these solid and fluid states within a unified framework.

In a first insight a theory of granular matter must be based on a continuous description. In the literature, a lot of work has been devoted to the construction of continuous mathematical descriptions for discrete systems (Irving and Kirkwood 1950; Murdoch and Bedeaux 1994; Babic 1997; Rudd and Broughton 1998; Glasser and Goldhirsch 2001), which have been applied to molecular and granular systems. It is well-known that slowly varying spatial and temporal disturbances of a mechanical system can be described in terms of its local thermodynamical variables. A notorious example of such a description is given by ordinary fluid mechanics, in which the slowly varying non-equilibrium behaviour of a fluid is described in terms of five partial differential equations, one for each conservation law (i.e., the mass, the three components of momentum, and the energy).

The hydrodynamic approach has been successfully employed to account for many mechanical systems, such as liquid crystals and superfluids. Recently, a continuum hydrodynamic theory for granular solids was introduced by Jiang and Liu (2009) (referred to as *Granular Solid Hydrodynamics*). This theory includes expressions for the energy current and the entropy production. The underlying notion in granular solid hydrodynamics is: granular media are elastic when at rest, but turn transiently elastic when the grains are agitated. The microscopic forces are taken into account through the definition of some coefficients, which stand for the contact forces. However, the rigorous passage from the microscopic to the macroscopic (continuum) mechanical description of matter fields, as was introduced by Jiang and Liu (2009), is still missing. For example, dense granular flows exhibit a large number of force chains which lacks a mathematical representation. On the other hand, experimental evidence and numerical calculations show that the macroscopic quasi-static response of granular matter is consistent with an elastic description (Geng et al. 2001; Reydellet and Clément 2001; Serero et al. 2001; Goldenberg and Goldhirsch 2002). In some cases, the classical theory of elasticity has been applied to describe heterogeneous systems, including granular materials (Nedderman 1992). However, some criticisms on this approach have been raised by Goldhirsch and Goldenberg (2002), who suggested that at small scales the continuous elasticity theory is no longer applicable and that it is necessary to introduce a grain scale to describe the elasticity properties of a material based on microscopic entities. A theory of elasticity for static granular systems based on a coarse-grained microscopic description was reported in Zhang et al. (2010), which is consistent with the continuous theory of elasticity.

The aim of this work is to sketch the transition from the micro to the macro representation of a continuous matter field, and obtain the balance equations using the coarse-graining method. To compare with previous work, we introduce a harmonic interaction between the particles that allows for the definition of the elastic stress tensor and the elastic free energy for the particular case of linear elasticity. The chapter is presented as follows: in Sect. 2 we discuss on the non-existence of a transition from the discrete (micro) to the continuous (macro) description in the context of granular hydrodynamics. In Sect. 3 we introduce the microscopic description using the coarse-graining method, and in Sect. 4 we summarize the conclusions.

2 Missed Justification for the Micro to Macro Transition in Granular Hydrodynamics

In Jiang and Liu (2009), a microscopic description for the elastic free energy was introduced through the dependence of the elastic moduli (defined in terms of some exponents) on compressional and shear deformations. The value of the exponents depends on the type of contact between the grains. For instance, the free elastic energy is given by (Jiang and Liu 2003):

$$f_{el} = \frac{1}{2} K_b \varepsilon_{nn}^2 + K_a \varepsilon_{kl}^0 \varepsilon_{kl}^0, \tag{1}$$

where

$$K_a = \tilde{K}_a \Delta^a, \quad K_b = \tilde{K}_b \Delta^b. \tag{2}$$

Here, K_b and K_a are the elastic constants, ε_{nn} is the trace of the strain tensor, ε_{kl}^0 is its traceless part, $\Delta = -\varepsilon_{nn}$, and $\tilde{K}_b$ and $\tilde{K}_a$ are constants that satisfy the conditions $\tilde{K}_b > 0$ and $\tilde{K}_a > 0$ for $\Delta \geq 0$ and $\tilde{K}_b = \tilde{K}_a = 0$ for $\Delta < 0$. This latter condition holds when the particles loose contact with one another. For a Hertz contact model, the exponents are $a = b = 1/2$, which result in $K = \tilde{K}\sqrt{\Delta}$. From the elastic free energy one can calculate the elastic stress tensor:

$$\pi_{ij} = -K_b \varepsilon_{nn} \delta_{ij} - K_a \varepsilon_{ij}^0 + \delta^{-1} \left(\frac{1}{2} b K_b \varepsilon_{nn}^2 + a K_a \varepsilon_{kl}^0 \varepsilon_{kl}^0 \right) \delta_{ij}. \tag{3}$$

Replacing the expressions for K_b and K_a given by Eq. (2) into Eq. (1), the elastic free energy in terms of the exponents a and b takes the form

$$f_{el} = \frac{1}{2} \tilde{K}_b \Delta^b \varepsilon_{nn}^2 + \tilde{K}_a \Delta^a \varepsilon_{kl}^0 \varepsilon_{kl}^0. \tag{4}$$

We see that relations (3) and (4) are non-linear in the deformation and that linear elasticity is readily recovered as long as $a = b = 0$. The construction of the elastic stress tensor and the elastic free energy allows us to define a conserved energy density, $\omega(s, s_g, \rho, g_i, \varepsilon_{ij})$, where its differential variation is given by

$$d\omega = T ds + \bar{T}_g ds_g + (\mu_0 - v^2/2) d\rho - \pi_{ij} d\varepsilon_{ij} + v_i dg_i. \tag{5}$$

Here, s and T are the entropy and temperature of the system, s_g is the granular entropy and $\bar{T}_g$ is related to the granular temperature, ρ is the density, g_i is the momentum density, v_i is the velocity, and μ_0 is the chemical potential (see Jiang and Liu 2009).

The energy density, mass density, momentum density, entropy, and granular entropy conservation laws along with the dynamics for the stress tensor are given by the following differential equations:

$$\frac{\partial \omega}{\partial t} + \nabla_i Q_i = 0, \quad \frac{\partial \rho}{\partial t} + \nabla_i(\rho v_i) = 0, \quad \frac{\partial g_i}{\partial t} + \nabla_i \Pi_{ij} = 0, \tag{6}$$

$$\frac{\partial s}{\partial t} + \nabla_i f_i = \frac{R}{T}, \quad \frac{\partial s_g}{\partial t} + \nabla_i F_i = \frac{R_g}{\bar{T}_g}, \tag{7}$$

$$\frac{\partial u_{ij}}{\partial t} - v_{ij} - X_{ij} = -\Big[(u_{ik}\nabla_j v_k + \nabla_i y_j/2) + (u_{jk}\nabla_i v_k + \nabla_j y_i/2)\Big]. \tag{8}$$

The derivation of these equations lacks of any microscopic justification for the micro (discrete) to the macro (continuous) description of the system. Let us remark that the microscopic interaction between the particles has been introduced by Jiang and Liu (2009) through the dependence of the elastic constants on the exponents a and b in Eq. (2).

3 Coarse-Grained Microscopic Description

In this section, we show that it is possible to derive the equations of hydrodynamics for granular media through a statistical average over a system with many grains. In this way, continuous and smooth matter fields can be obtained. To do so we first define a kernel function, also called smoothing kernel or interpolant kernel, which depends on the finite spatial and/or temporal resolution. In order to introduce the coarse-graining procedure, lets us start with the following definition for the mass density

$$\rho(\mathbf{r}, t) = \int dt' \Gamma(t - t', \tau) \sum_{j=1}^{N} m_j \phi(\mathbf{r} - \mathbf{r}_j(t'), h), \tag{9}$$

where we have taken temporal and spatial averages by means of the kernel functions $\Gamma(t - t', \tau)$ and $\phi(\mathbf{r} - \mathbf{r}_j(t'), h)$. Here τ is the temporal resolution and h is the spatial resolution. If we take the temporal derivative of Eq. (9) we obtain the equation

$$\frac{\partial \rho}{\partial t} + \frac{\partial p_\alpha}{\partial r_\alpha} = 0, \tag{10}$$

which corresponds to the mass conservation law. Greek indices are used to denote the coordinate components for the vector and tensor fields. The spatial derivative of $\phi(\mathbf{r} - \mathbf{r}_j(t'), h)$ commutes with all variables that describe the degrees of freedom of the particles, while the temporal derivative of the kernels are $\partial\Gamma/\partial t = 0$ and $\partial\phi/\partial t = -v_{j\alpha}(t)\partial\phi/\partial r_\alpha$. Similarly, for the momentum density p_α we have the following spatially and temporally averaged expression

$$p_\alpha(\mathbf{r}, t) = \int dt'\Gamma(t - t', \tau) \sum_{j=1}^{N} m_j v_{j\alpha}(t')\phi(\mathbf{r} - \mathbf{r}_j(t'), h). \tag{11}$$

Now, differentiating Eq. (11) with respect to time and invoking the Leibniz's rule for differentiation under the integral sign, we obtain

$$\begin{aligned}\frac{\partial p_\alpha}{\partial t} = -\frac{\partial}{\partial r_\beta}\Bigg(\frac{1}{2}\int dt'\Gamma(t - t', \tau) \sum_{i,j=1,i\neq j}^{N} f_{ij\alpha} r_{ij\beta} \int_0^1 ds\phi(\mathbf{r} - \mathbf{r}_j(t') + s\mathbf{r}_{ij}(t'), h)\\ + \int dt'\Gamma(t - t', \tau) \sum_{j=1}^{N} m_j v_{j\alpha}(t') v_{j\beta}(t')\phi(\mathbf{r} - \mathbf{r}_j(t'))\Bigg),\end{aligned} \tag{12}$$

where $f_{ij\alpha}$ is the force between particles i and j. The fluctuation velocity of particle j is defined as $v'_{j\beta}(\mathbf{r}, t) := v_{j\beta}(t) - V_\beta(\mathbf{r}, t)$, where $V_\beta(\mathbf{r}, t)$ is the field velocity given by $V_\beta(\mathbf{r}, t) = p_\beta(\mathbf{r}, t)/\rho(\mathbf{r}, t)$. Replacing the fluctuation velocity in Eq. (12), grouping terms, and considering the condition $\sum_{j=1}^{N} m_j v'_{j\alpha}(t)\phi(\mathbf{r} - \mathbf{r}_j(t) = 0$, we obtain the momentum conservation law:

$$\frac{\partial p_\alpha}{\partial t} + \frac{\partial}{\partial r_\beta}\left(\rho V_\alpha V_\beta + \sigma_{\alpha\beta}\right) = 0. \tag{13}$$

Here, the stress tensor $\sigma_{\alpha\beta}$ is defined by

$$\begin{aligned}\sigma_{\alpha\beta} := -\frac{1}{2}\int dt'\Gamma(t - t', \tau) \sum_{i,j=1,i\neq j}^{N} f_{ij\alpha} r_{ij\beta} \int_0^1 ds\phi(\mathbf{r} - \mathbf{r}_j(t') + s\mathbf{r}_{ij}(t'), h)\\ - \int dt'\Gamma(t - t', \tau) \sum_{j=1}^{N} m_j v'_{j\alpha}(t') v'_{j\beta}(t')\phi(\mathbf{r} - \mathbf{r}_j(t')),\end{aligned} \tag{14}$$

where the first term on the right-hand side is the contact stress and the second is the kinematic stress.

As an example, let us now consider the case of a harmonic interaction between the particles, that is

$$\mathbf{f}_{ij} = -\kappa_{ij}(\hat{\mathbf{r}}^o_{ij} \cdot \mathbf{u}_{ij})\hat{\mathbf{r}}^o_{ij}, \tag{15}$$

where κ_{ij} is the elastic constant related to the interaction between particles i and j. The displacement from the equilibrium is given by $\mathbf{u}_{ij} = (|\mathbf{r}_{ij}| - l_{ij})\hat{\mathbf{r}}^o_{ij}$, where l_{ij} is the equilibrium separation distance between particle i and j, and $\hat{\mathbf{r}}^o_{ij}$ is a unit vector. The superscript o stands for the unstressed reference configuration. Replacing Eq. (15) into Eq. (14), the linear stress tensor takes the form

$$\sigma_{\alpha\beta} = \frac{1}{2}\int dt'\Gamma(t-t',\tau)\sum_{i,j=1,i\neq j}^{N} \kappa_{ij}\hat{r}^o_{ij\mu}u_{ij\mu}\hat{r}^o_{ij\alpha}r_{ij\beta}\int_0^1 ds\phi(\mathbf{r}-\mathbf{r}_j(t')+s\mathbf{r}_{ij}(t'),h)$$
$$-\int dt'\Gamma(t-t',\tau)\sum_{j=1}^{N} m_j v'_{j\alpha}(t')v'_{j\beta}(t')\phi(\mathbf{r}-\mathbf{r}_j(t')), \tag{16}$$

Since the system is linear, the same method used by Goldhirsch and Goldenberg (2002) can be applied to estimate the displacement difference between particles i and j. Such a method uses the principle of the Green's function, i.e., a source gives rise to a perturbation in space. Therefore, the displacement $u_{ij\mu}$ can be approximated as

$$u_{ij\mu} \approx L_{ij\mu J\nu}\frac{\partial u_\nu}{\partial r_\gamma}(r_{J\gamma}-r_\gamma), \tag{17}$$

where $L_{ij\mu J\nu}$ is related to the Green's function through $L_{ij\mu J\nu} := G_{i\mu J\nu} - G_{j\mu J\nu}$. Replacing Eq. (17) into Eq. (16) we finally obtain that

$$\sigma_{\alpha\beta} \approx \left[\frac{1}{2}\int dt'\Gamma(t-t',\tau)\sum_{i,j=1,i\neq j}^{N} \kappa_{ij}\hat{r}^o_{ij\mu}L_{ij\mu J\nu}(r^0_{J\gamma}-r_\gamma)\hat{r}^o_{ij\alpha}r_{ij\beta}\int_0^1 ds\phi_{ij}\right]\varepsilon_{\nu\gamma}$$
$$-\int dt'\Gamma(t-t',\tau)\sum_{j=1}^{N} m_j v'_{j\alpha}(t')v'_{j\beta}(t')\phi(\mathbf{r}-\mathbf{r}_j(t')), \tag{18}$$

where we have used $\phi_{ij} = \phi(\mathbf{r}-\mathbf{r}_j(t')+s\mathbf{r}_{ij}(t'),h)$ and defined the strain tensor $\varepsilon_{\nu\gamma}$ as

$$\varepsilon_{\nu\gamma} := \frac{1}{2}\left(\frac{\partial u_\nu}{\partial r_\gamma}+\frac{\partial u_\gamma}{\partial r_\nu}\right) = \frac{1}{2\rho^2}\sum_{i,j=1,i\neq j}^{N} m_i m_j\phi_j\left[\frac{\partial\phi_i}{\partial r_\gamma}u_{ij\nu}+\frac{\partial\phi_i}{\partial r_\nu}u_{ij\gamma}\right], \tag{19}$$

where $\phi_i = \phi(\mathbf{r}-\mathbf{r}_i(t'),h)$, which is calculated from the microscopic displacements of particles i and j. The first term on the right-hand side of Eq. (18) is just the elastic stress tensor $\pi_{\alpha\beta}$, that is

$$\pi_{\alpha\beta} := \left[\frac{1}{2}\int dt'\Gamma(t-t',\tau)\sum_{i,j=1,i\neq j}^{N} \kappa_{ij}\hat{r}^o_{ij\mu}L_{ij\mu J\nu}(r^0_{J\gamma}-r_\gamma)\hat{r}^o_{ij\alpha}r_{ij\beta}\int_0^1 ds\phi_{ij}\right]\varepsilon_{\nu\gamma}. \tag{20}$$

The terms between the brackets in the above expression are the elastic moduli, which as we may see depend on the position and the scale resolution of the coarse-graining function. We may further write the above expression as $\pi_{\alpha\beta} = K_{\alpha\beta\nu\gamma}\varepsilon_{\nu\gamma}$, where $K_{\alpha\beta\nu\gamma}$ is defined by

$$K_{\alpha\beta\nu\gamma} := \frac{1}{2}\int dt'\Gamma(t-t',\tau)\sum_{i,j=1,i\neq j}^{N} \kappa_{ij}\hat{r}^{o}_{ij\mu}L_{ij\mu J\nu}(r^{0}_{J\gamma}-r_{\gamma})\hat{r}^{o}_{ij\alpha}r_{ij\beta}\int_{0}^{1} ds\phi_{ij}, \tag{21}$$

which is a fourth-rank tensor with 81 elements. This tensor can be reduced to 21 elements by the symmetries of the stress and strain tensors. Moreover, the elastic stress tensor can be decomposed into two terms as

$$\pi_{\alpha\beta} = -\kappa_b\varepsilon_{nn}\delta_{\alpha\beta} - 2\kappa_a\varepsilon_{\alpha\beta}, \tag{22}$$

where κ_b and κ_a are, respectively, the compressional and shear elastic moduli defined by the relations

$$\kappa_b = K_{\alpha\alpha\alpha\alpha} = \frac{1}{2}\int dt'\Gamma(t-t',\tau)\sum_{i,j=1,i\neq j}^{N} \kappa_{ij}\hat{r}^{o}_{ij\mu}L_{ij\mu J\alpha}(r^{0}_{J\alpha}-r_{\alpha})\hat{r}^{o}_{ij\alpha}r_{ij\alpha}\int_{0}^{1} ds\phi_{ij}, \tag{23}$$

$$\kappa_a = K_{\alpha\beta\alpha\beta} = \frac{1}{2}\int dt'\Gamma(t-t',\tau)\sum_{i,j=1,i\neq j}^{N} \kappa_{ij}\hat{r}^{o}_{ij\mu}L_{ij\mu J\alpha}(r^{0}_{J\beta}-r_{\beta})\hat{r}^{o}_{ij\alpha}r_{ij\beta}\int_{0}^{1} ds\phi_{ij}, \tag{24}$$

with $\alpha \neq \beta$.

From Eq. (22) we can calculate the elastic energy through the Maxwell relation $\pi_{\alpha\beta} = -\partial E_{el}/\partial\varepsilon_{\alpha\beta}$:

$$E_{el} = \frac{1}{2}\kappa_b\varepsilon^2_{nn} + \kappa_a\varepsilon^0_{\alpha\beta}\varepsilon^0_{\alpha\beta}, \tag{25}$$

where ε_{nn} is the trace of the strain tensor and $\varepsilon^0_{\alpha\beta}$ is its traceless part. Equations (22) and (25) are similar in form to Eq. (3) and (4) for the case of linear elasticity in the continuum granular hydrodynamics theory (i,e., $a = b = 0$). The energy density containing the microscopic information of the system can be written as $\omega(s, s_g, \rho, g_\alpha, \varepsilon_{\alpha\beta}) = \omega_1(s, s_g, \rho, g_i) - \pi_{\alpha\beta}\varepsilon_{\alpha\beta}$. As we have shown here, $\pi_{\alpha\beta}$ and $\varepsilon_{\alpha\beta}$ can be constructed from the microscopic interactions between the particles. We further note that the conserved energy density includes the microscopic information whenever ω_1 depends on the local thermodynamic variables.

4 Conclusions

In this chapter, we have presented a brief introduction to the Granular Solid Hydrodynamics, as derived by Jiang and Liu (2009), and outlined the coarse-graining method to describe granular solids. We showed that the elastic energy and the elastic stress tensor are congruent in both formulations. The analogies were established using a harmonic interaction at the scale of the granular particles. The results presented in this article pave the way towards a formal derivation of continuous matter fields for particulate discrete systems, such as granular materials. Further work is required to include more realistic microscopic models for the interaction forces between grains. A first step in this direction could be to include the force model formulations employed in molecular dynamics simulations of granular materials, e.g., the Hertz's contact model. This will allow for a general analysis of the continuous matter field equations introduced by Jiang and Liu (2003, 2009) along with the emergent bulk properties implicit in the exponents a and b.

References

Eshuis P, van der Weele K, van der Meer D, Bos R, Lohse D (2007) Phase diagram of vertically shaken granular matter. Phys Fluids 19:123301

Geng J, Howell D, Longhi E, Behringer RP, Reydellet G, Vanel L, Clément E, Luding S (2001) Footprints in sand: The response of a granular material to local perturbations. Phys Rev Lett 87:035506

Glasser BJ, Goldhirsch I (2001) Scale dependence, correlations, and fluctuations of stresses in rapid granular flows. Phys Fluids 13:407–420

Goldenberg C, Goldhirsch I (2002) Force chains, microelasticity, and macroelasticity. Phys Rev Lett 89:084302

Goldhirsch I, Goldenberg C (2002) On the microscopic foundations of elasticity. Eur Phys J E 9:245–251

Babic M (1997) Average balance equations for granular materials. Int J Eng Sci 5:523–548

Irving JH, Kirkwood JG (1950) The statistical mechanics theory of transport processes. IV. The equations of hydrodynamics. J Chem Phys 18:817–829

Jaeger HM, Nagel SR, Behringer RP (1996) Granular solids, liquids, and gases. Rev Mod Phys 68:1259–1273

Jiang Y, Liu M (2003) Granular elasticity without the Coulomb condition. Phys Rev Lett 91:144301

Jiang Y, Liu M (2009) Granular solid hydrodynamics. Granular Matter 11:139–156

Murdoch AI, Bedeaux D (1994) Continuum equations of balance via weighted averages of microscopic quantities. In: Proceedings of the Royal Society A, vol. 445, pp 157–179.

Nedderman RM (1992) Statics and kinematics of granular materials. Cambridge University Press, Cambridge

Reydellet G, Clément E (2001) Green's function probe of a static granular piling. Phys Rev Lett 86:3308–3311

Rudd RE, Broughton JQ (1998) Coarse-grained molecular dynamics and the atomic limit of finite elements. Phys Rev B 58:R5893–R5896.

Serero D, Reydellet G, Claudin P, Clément E, Levine D (2001) Stress response function of a granular layer: Quantitative comparison between experiments and isotropic elasticity. Eur Phys J E 6:169–179

Zhang J, Behringer RP, Goldhirsch I (2010) Coarse-graining of a physical granular system. Prog Theor Phys Suppl 184:16–30

Integral Representation for Continuous Matter Fields in Granular Dynamics

Juan F. Marín, Juan C. Petit, Leonardo Di G. Sigalotti and Leonardo Trujillo

Abstract We introduce a mathematical formalism towards the construction of a continuum-field theory for particulate fluids and solids. We briefly outline a research program aimed at unifying the fundamentals of the coarse-graining theory and the numerical method of Smoothed Particle Hydrodynamics (SPH). We show that the coarse-graining functions must satisfy well defined mathematical properties that comply with those of the SPH kernel integral representation of continuous fields. Given the appropriate dynamics for the macroscopic response, the present formalism is able to describe both the solid and fluid-like behaviour of granular materials.

J. F. Marín · J. C. Petit · L. Di G. Sigalotti · L. Trujillo (✉)
Centro de Física, Instituto Venezolano de Investigaciones Científicas, IVIC, Apartado Postal 20632, Caracas 1020-A, Venezuela
e-mail: juanfmarinm@gmail.com

J. C. Petit
e-mail: jcpetit71@gmail.com

L. Di G. Sigalotti
e-mail: leonardo.sigalotti@gmail.com

L. Trujillo
e-mail: leonardo.trujillo@gmail.com

J. F. Marín
Escuela de Física, Facultad de Ciencias, Universidad Central de Venezuela, UCV, Avenida Los Ilustres, Los Chaguaramos, Apartado Postal 20513, Caracas 1020-A, Venezuela

L. Trujillo
The Abdus Salam International Centre for Theoretical Physics, ICTP, Strada Costiera 11, 34014 Trieste, Italy

L. Di G. Sigalotti et al. (eds.), *Computational and Experimental Fluid Mechanics with Applications to Physics, Engineering and the Environment*, Environmental Science and Engineering, DOI:10.1007/978-3-319-00191-3_33,

1 Introduction

Continuum mechanics in its simplest form has been the paragon of field theory to describe the response of solid bodies and fluids. The mathematical robustness of continuum theories rests on the field of partial differential equations (PDEs), which is one of the main and mature cores of mathematical physics. These mathematical models rationalize, explain, and predict very well the macroscopic behaviour of isotropic and homogeneous solid bodies. However, there is still no general continuum model for the description of discrete microscopic, mesoscopic, and nanoscale systems. For example, granular materials are composed of large collections of small grains and exhibit many interesting collective phenomena emerging from the many-body classical dynamics of their constituents (Jaeger et al. 1996). This kind of materials can exhibit solid-like as well as fluid-like behaviours. Nevertheless, a complete theoretical description of granular materials as well as particulate amorphous and heterogeneous solids is still missing (Jaeger et al. 1996; Kadanoff 1999; de Gennes 1999; Goldhirsch 2003). An analytical framework is required to bridge the gap between the experimental and theoretical studies in this area, and develop numerical algorithms for new and more efficient simulation techniques.

The study of continuum concepts and field values related to local (scale-dependent) space-time averages began in 1946 with the celebrated statistical mechanical theory of transport processes by Kirkwood (1946) and Irving and Kirkwood (1950). They assumed that any ensemble average of space-time averages could be equated with a space-time average of an ensemble average computed at given scales of length and time. This approach represents an alternative to the Gibbs measure (Murdoch and Bedeaux 1994; Babic 1997) (associated with the Boltzmann distribution and the notion of canonical ensemble in equilibrium statistical thermodynamics, connecting measurable macroscopic properties of materials to the properties of their constituent particles and the interactions between them).[1] Another interesting development was reported by Glasser and Goldhirsch (2001), who based on the ideas worked out in Kirkwood (1946); Irving and Kirkwood (1950) and Murdoch and Bedeaux (1994), constructed a field theory for granular fluids.

The purpose of this paper is to present a further step towards the construction of a macroscopic continuous-field mathematical description of granular materials that fulfils the condition of reproducing microscopic properties in a macroscopic representation of physical quantities. We address the problem of the estimation of coarse-graining functions and show that they must satisfy well-defined mathematical conditions for the requirement of compact support of the smoothing function.

---

[1] For granular systems this approach seems doomed from the outset: because energy is lost through internal friction, and gained by a non-thermal source such as tapping or shearing, the dynamical equations do not leave the canonical or any other known ensemble invariant.

2 Starting Definitions and Postulates of the Theory

2.1 Domains and Observables

Consider a physical system whose properties may vary in time and space. Denoting by $\mathbb{R}^3$ the three-dimensional coordinate space (and by $\mathbf{r}$ its elements), we shall then simply denote by $\mathbb{R}^4$ the four-dimensional space-time vector field whose elements $\mathbf{R}$ have three spatial coordinates and a time-coordinate t:

$$\mathbf{R} = (\mathbf{r}, t) \in \mathbb{R}^4. \tag{1}$$

This space is the stage on which our theory is developed, so herein we will refer to such a space as the *physical domain*:

Definition 1 ***(Physical domain)**. The* physical domain *Ω of any problem is the set of all points in $\mathbb{R}^4$.*

In a discrete solid, for instance, the physical domain consists of all points occupied by the grains of the solid as well as the empty spaces (pores) inside the material.

Following Murdoch and Bedeaux (1994), we introduce the next definition:

Definition 2 ***(Material system)**. A material system $\mathscr{M} \subseteq \Omega$ is an instant-by-instant identifiable set of fundamental discrete entities.*

In other words, a material system is a set of fundamental disjoint sets in space that belongs to the geometrical locus of the material to be described (atoms, ions, molecules, pores, etc). In our problem of granular materials, such discrete entities are modelled as interacting particles (grains), indexed by $\{i\}$. Since a granular material is a large system of grains with macroscopic sizes ($> 1\,\mu$m), we have for such materials that $\mathscr{M} \subset \Omega$. In the case of a continuous solid or fluid, there will be a unique identifiable discrete entity, for which we have $\mathscr{M} = \Omega$.

Definition 3 ***(Observable)**. An observable is any real square-integrable function defined over the entire physical domain or a subset of it:*[2]

$$\begin{aligned} &f : \mathscr{D} \subseteq \Omega \to \mathbb{R}, \\ &f \in L^2[\mathscr{D}]. \end{aligned} \tag{2}$$

Definition 4 ***(Microscopic field)**. Any function ψ is a microscopic field if it is an observable defined only in the material system, i. e.:*

[2] *Remark* 1 In general, an n-rank tensor function $G : \mathscr{D} \to \mathbb{R}^4 \otimes^{[n]} \mathbb{R}^4$ (where $\otimes^{[n]}$ denotes the n-rank tensor product) is not an observable. Nevertheless, each component of the tensor is an observable. Typical examples are the components of the stress tensor ($\sigma^{\alpha\beta}$, where Greek indices denote Cartesian coordinates), or the components of the velocity field (v^{α}).

$$\begin{aligned}\psi &: \mathscr{M} \rightarrow \mathbb{R}, \\ \psi &\in L^2[\mathscr{M}].\end{aligned} \quad (\mathscr{M} \subseteq \Omega) \tag{3}$$

Definition 5 *(Macroscopic field). Any function* $\tilde{\psi}$ *is a macroscopic field if it is an observable defined over the entire physical domain, i. e.:*

$$\begin{aligned}\tilde{\psi} &: \Omega \rightarrow \mathbb{R}, \\ \psi &\in L^2[\Omega].\end{aligned} \tag{4}$$

We may characterize a granular material with the set of fundamental microscopic fields: $\{m_i \,;\, \rho_i \,;\, r_{\alpha_i}(t)\}$, which denote the mass, the density, and the α-component of the centre-of-mass position of the ith grain at time t, respectively. Obviously, in such a fundamental set, there are as many components of the centre-of-mass positions as spatial dimensions of the physical domain (which, for the sake of generality, we have taken as three-dimensional).

2.2 Principles

Consider the problem of building a *macroscopic field* $\tilde{\psi} : \Omega \rightarrow \mathbb{R}$ from a microscopic one $\psi(\mathbf{r}_i, t)$ (with $i = 1, 2, \ldots, N$; where N is the total number of grains). This has to be done in such a way that the microscopic information of the material is accurately represented in the macroscopic field. This problem can be formally stated as:

$$\mathscr{L}\tilde{\psi}(\mathbf{r}, t) = \psi(\mathbf{r}, t), \tag{5}$$

where $\mathscr{L} : L^2[\Omega] \rightarrow L^2[\mathscr{M}]$ is some operator that when applied to a macroscopic field, returns the microscopic field. Thus, the sources of the macroscopic field $\tilde{\psi}(\mathbf{R})$ are the values of the microscopic field $\psi(\mathbf{R})$ evaluated in each grain.

Postulate 1 *(Linearity) The problem* (5) *satisfies the superposition principle.*

The above postulate allows us to write the output of Eq. (5) (the macroscopic field) as a linear functional of the input (the microscopic field), i. e.:

$$\tilde{\psi}(\mathbf{R}) = \int_{\mathscr{M}} \Phi(\mathbf{R}, \mathbf{R}')\psi(\mathbf{R}')d\mathbf{R}', \tag{6}$$

where Φ must belong to $L^2[\Omega]$.

Postulate 2 *(Space-time translation invariance) The problem* (5) *is space-time translation invariant.*

So, if the input (the microscopic field) is shifted along some displacement vector $\mathbf{D} = (\Delta\mathbf{r}, \Delta t)$ in $\mathscr{M}$, the output (macroscopic field) is also shifted in the same manner: if $\tilde{\psi}(\mathbf{R})$ corresponds to $\psi(\mathbf{R})$, then $\tilde{\psi}(\mathbf{R}+\mathbf{D})$ corresponds to $\psi(\mathbf{R}+\mathbf{D})$. This can only be possible if the function Φ in Eq. (6) depends on the difference $\mathbf{R}-\mathbf{R}'$ of its arguments: $\Phi(\mathbf{R},\mathbf{R}') = \Phi(\mathbf{R}-\mathbf{R}')$. Then, Eq. (6) turns into the convolution product:

$$\tilde{\psi}(\mathbf{R}) = \int_{\mathcal{M}} \Phi(\mathbf{R}-\mathbf{R}')\psi(\mathbf{R}')d\mathbf{R}' = (\Phi \times \psi)_{(\mathbf{R})}. \tag{7}$$

Setting $\Phi(\mathbf{R}) = \Phi(\mathbf{r},t) \equiv \phi(\mathbf{r})F(t)$, where $F(t)$ and $\phi(\mathbf{r})$ are the temporal and spatial parts of Φ, respectively, Eq. (7) becomes:

$$\tilde{\psi}(\mathbf{r},t) = \int_T dt' F(t-t') \int_{\mathcal{M}_s} d\mathbf{r}'\phi(\mathbf{r}-\mathbf{r}')\psi(\mathbf{r}',t), \tag{8}$$

where $T \subset \Omega$ is the time subspace of $\mathcal{M}$ and $\mathcal{M}_s \subseteq \mathbb{R}^3$ is the spatial subspace of $\mathcal{M}$:

$$\begin{aligned} \mathcal{M}_s &= \{\mathbf{r} \in \mathbb{R}^3 : (\mathbf{r},t) \in \mathcal{M}\}, \\ & \qquad\qquad\qquad\qquad (\mathcal{M}_s \otimes T = \mathcal{M}) \\ T &= \{t \in \mathbb{R} : (\mathbf{r},t) \in \mathcal{M}\}. \end{aligned} \tag{9}$$

We see that the solution of problem (5) is given by Eq. (8), where $\phi : \mathcal{V} \to \mathbb{R}$ and $F : \mathcal{T} \to \mathbb{R}$ are the *spatial and temporal parts of the Green's function* Φ associated to the operator $\mathcal{L}$, respectively, with $\mathcal{V}$ and $\mathcal{T}$ being the spaces of all possible displacements in the Euclidean spaces $\mathbb{R}^3$ and $\mathbb{R}$, respectively. We see that ϕ has dimensions of the inverse of volume and F of the inverse of time. For a granular material, $\mathcal{M}_s$ is a disconnected subset of $\mathbb{R}^3$, and so the integral over $\mathcal{M}_s$ in Eq. (8) is, in fact, a discrete sum. Since the infinitesimal volume elements of size $d\mathbf{r}'$ in Eq. (8) lies in $\mathcal{M}_s$, such a sum must be taken over the small volumes ΔV_i of each disconnected set (i.e. over the volume of each particle). Setting $\Delta V_i = m_i/\rho_i$, Eq. (8) can be written as:

$$\tilde{\psi}(\mathbf{r},t) = \int_T dt' F(t-t') \sum_{i=1}^{N} \frac{m_i}{\rho_i}\psi(\mathbf{r}_i,t)\phi(\mathbf{r}-\mathbf{r}_i). \tag{10}$$

For static granular systems (granular packings in which the grains remain static but the fields could still vary in time), we have that $F(t-t') = \delta(t-t')$ and Eq. (10) becomes the weighted average sum:

$$\tilde{\psi}(\mathbf{r},t) = \sum_{i=1}^{N} \frac{m_i}{\rho_i}\psi(\mathbf{r}_i,t)\phi(\mathbf{r}-\mathbf{r}_i). \tag{11}$$

Postulates 1 and 2 has led us to a formulation via weighted-average sums of the same kind as those introduced heuristically by Irving and Kirkwood (1950); Murdoch and Bedeaux (1994); Babic (1997) and Glasser and Goldhirsch (2001), referring to the Green's function Φ either as the *coarse-graining function*, the *probability density function*, or simply as the *weighting function*. The continuous representation of fields like the density, velocity, and stress tensor field components obtained using Eq. (11) are, indeed, identical to those obtained by Murdoch and Kowalski (1992); Babic (1997) and Glasser and Goldhirsch (2001). For instance, for the mass density we have from Eq. (10) that:

$$\tilde{\rho}(\mathbf{r}, t) = \int_T dt' F(t - t') \sum_{i=1}^{N} m_i \phi(\mathbf{r} - \mathbf{r}_i), \tag{12}$$

which is the well-known *coarse-grained mass density formula* (Babic 1997; Glasser and Goldhirsch 2001).

3 Properties of the Spatial Green's Function

The solution of problem (5) is then given by Eq. (8). The next step is to find the Green's function. Since we do not know in advance the exact form of the operator $\mathscr{L}$, we require an alternative method to the conventional one using delta-type sources for calculating the Green's function. This is exactly the same kind of mathematical problem that was solved by Lucy (1977) and Gingold and Monaghan (1977) when they developed the method of *Smoothed Particle Hydrodynamics* (SPH) for the discretization of PDEs. Here, we postulate that *one can use the method employed for constructing SPH smoothing functions to calculate the spatial Green's function.*

A kernel interpolant, as in Eq. (8), with $F(t - t') = \delta(t - t')$ is also used in SPH (called *smoothed* or *integral representation*) and the domain of the PDE is discretized using a particle generator, leading to a sum over particles of masses m_i and densities ρ_i identical to Eq. (11). This defines a Lagrangian mesh-free, semi-discrete version of the PDE (Monaghan 1992; Liu and Liu 2003, 2010; Li and Liu 2007). For the integral macroscopic representation $\tilde{\psi}$ in Eq. (8) to be consistent with the microscopic field ψ, the kernel ϕ must satisfy some well established mathematical conditions. In establishing such properties, one needs to consider some additional information about the macroscopic behaviour of the system (this is due to the lack of information about the exact form of the operator $\mathscr{L}$). For granular flows, we require the system to satisfy macroscopically the equations of granular hydrodynamics derived in Glasser and Goldhirsch (2001). Assuming that the macroscopic behaviour of the system is described by a second-order PDE (as is the case for granular hydrodynamics), Liu et al. (2003) demonstrated that the integral representation will have a consistency of order n if the kernel ϕ fulfils the following conditions:

$$\begin{cases} \phi(\mathbf{r}-\mathbf{r}')\big|_{\partial\mathbb{S}(\mathbf{r})} = 0, \\ \phi'(\mathbf{r}-\mathbf{r}')\big|_{\partial\mathbb{S}(\mathbf{r})} = 0, \\ M_i = \delta_{i0} \quad i = 0, 1, 2, \ldots, n \ , \end{cases} \tag{13}$$

where:

$$M_i \equiv \int_{\mathbb{S}(\mathbf{r})} (\mathbf{r}-\mathbf{r}')^i \, \phi\,(\mathbf{r}-\mathbf{r}')\, d\mathbf{r}', \tag{14}$$

is the ith momentum of the kernel, and $\mathbb{S}(\mathbf{r})$ is a *support domain of the point* $\mathbf{r}$, a subset of $\mathscr{M}_s$ outside of which the kernel's value is zero. The first and second of Eqs. (13) are referred to as the *compactness condition* for the kernel and its first derivative. Without such conditions, much more conditions would be needed over ϕ to guarantee consistency between the discrete and the continuous description, as was shown explicitly in Liu et al. (2003) and Liu and Liu (2006). The domain of the spatial Green's function is no longer the space of all possible displacements in Euclidean space (as was stated in Sect. 2.2), but rather the space of those displacements such that the argument of ϕ lies within the support domain $\mathbb{S}(\mathbf{r})$. When this support domain is spherically symmetric, its radius is called the *smoothing length* or *spatial resolution* in the SPH literature. We note that by considering this compactness property of the Green's function in the sum (11), only those particles that are within the support domain of the point $\mathbf{r}$ makes a contribution to the value of the macroscopic field. We also note that the kernel's momentum equation for $i = 0$ imposes a normalization condition over ϕ, while for $i = 1$ it establishes a parity condition. Note that all these properties were imposed heuristically over the spatial coarse-graining function by Glasser and Goldhirsch (2001), while here we have given a formal justification of them.

4 Conclusions

In this paper, we have given, by means of physical and mathematical arguments, a formal justification of the description of discrete complex systems using weighted average sums. This leads to a unification of the various formulations that have been separately treated so far, namely the *weighted average sums* introduced by Irving and Kirkwood (1950), the *coarse-graining* representation by Glasser and Goldhirsch (2001), and the *SPH integral representation* by Lucy (1977) and Gingold and Monaghan (1977). With this formulation we can guarantee that the macroscopic fields are consistent with the microscopic properties of the material, satisfying the equations of granular hydrodynamics in its macroscopic description up to a certain order n in its Taylor's series expansion around any point in the material system $\mathscr{M}$. The method developed by Liu et al. (2003) and Liu and Liu (2006) becomes one for the estimation of the spatial Green's function of the unknown operator $\mathscr{L}$. Using this method for the calculation of such kernels, we can assure that (8) interpolates the microscopic (discrete) field ψ. We propose that, given the appropriate macroscopic dynamics, this representation can be used to successfully model both the liquid and the solid phase of a granular material. This is tightly connected to the fact that the SPH method is able to model the dynamics of fluids and elastic solids the same way (Gray et al. 2001).

References

de Gennes PG (1999) Granular matter. A tentative view. Rev Mod Phys 71:S374–S382

Gingold RA, Monaghan JJ (1977) Smoothed particle hydrodynamics–Theory and application to non-spherical stars. Monthly Not R Astron Soc 181:375–389

Glasser BJ, Goldhirsch I (2001) Scale dependence, correlations, and fluctuations of stresses in rapid granular flows. Phys Fluids 13:407–420

Goldhirsch I (2003) Rapid granular flows. Annu Rev Fluid Mech 35:267–293

Gray JP, Monaghan JJ, Swift RP (2001) SPH elastic dynamics. Comput Methods Appl Mech Eng 190:6641–6662

Babic M (1997) Average balance equations for granular materials. Int J Eng Sci 5:523–548

Irving JH, Kirkwood JG (1950) The statistical mechanical theory of transport processes. IV. The equations of hydrodynamics. J Chem Phys 18:817–829

Jaeger HM, Nagel SR, Behringer RP (1996) Granular solids, liquids, and gases. Rev Mod Phys 68:1259–1273

Kadanoff LP (1999) Built upon sand: theoretical ideas inspired by granular flows. Rev Mod Phys 71:435–444

Kirkwood JG (1946) The statistical mechanical theory of transport processes I. General theory. J Chem Phys 14:180–201

Li S, Liu WK (2007) Meshfree particle methods. Springer-Verlag, Berlin

Liu GR, Liu MB (2003) Smoothed particle hydrodynamics: a meshfree particle method. World Scientific, Singapore

Liu MB, Liu GR, Lam KY (2003) Constructing smoothing functions in smoothed particle hydrodynamics with applications. J Comp Appl Math 155:263–284

Liu MB, Liu GR (2006) Restoring particle consistency in smoothed particle hydrodynamics. Appl Num Math 56:19–36

Liu MB, Liu GR (2010) Smoothed particle hydrodynamics (SPH): an overview and recent developments. Arch Comput Methods Eng 17:25–76

Lucy LB (1977) A numerical approach to the testing of the fission hypothesis. Astro J 82:1013–1024

Monaghan JJ (1992) Smoothed particle hydrodynamics. Ann Rev Astron Astrophys 30:543–574

Murdoch AI, Kowalski SJ (1992) On fluid-fluid coupling within porous media: a mixture theoretic approach based upon molecular considerations. Transp Porous Media 8:47–70

Murdoch AI, Bedeaux D (1994) Continuum equations of balance via weighted averages of microscopic quantities. Proc R Soc A 445:157–179

Numerical SPH Calculations of Fluid Flow Through Saturated and Non-saturated Porous Media

Estela Mayoral, Mario A. Rodríguez-Meza, Eduardo de la Cruz-Sánchez, Jaime Klapp, Francisco Solórzano-Araujo, César Ruiz-Ferrel and Leonardo Di G. Sigalotti

Abstract The fluid flow through saturated and non-saturated homogeneous porous media is studied numerically using a modified version of a Smoothed Particle Hydrodynamics (SPH) code. The modifications implemented in the original SPH code to model the incompressible flow at low Reynolds numbers through a porous medium are described. The performance of the model is demonstrated for three-dimensional flow through idealized porous media consisting of regular square and hexagonal arrays of solid spheres. For each of these configurations we consider a set of flow calculations through saturated and non-saturated porous matrices differing in the magnitude of the z-component of the hydraulic gradient. For the saturated case, the

E. Mayoral (✉) · M. A. Rodriguez-Meza · E. de la Cruz-Sánchez · J. Klapp
Departamento de Física, Instituto Nacional de Investigaciones Nucleares, ININ, Km 36.5, Carretera México-Toluca, 52750 La Marquesa, Estado de México, Mexico
e-mail: estela.mayoral@inin.gob.mx

M. A. Rodríguez-Meza
e-mail: marioalberto.rodriguez@inin.gob.mx

J. Klapp
Departamento de Matemáticas, Cinvestav del Instituto Politécnico Nacional (I.P.N.), 07360 México, D. F., Mexico
e-mail: jaime.klapp@hotmail.com

J. Klapp
e-mail: jaime.klapp@inin.gob.mx

F. Solórzano-Araujo · C. Ruiz-Ferrel
Universidad Tecnológica del Valle de Toluca, Carretera del Departamento del D. F. Km 7.5, Santa María Atarasquillo, Lerma, Estado de México, Mexico
e-mail: frank_trov17@hotmail.com

C. Ruiz-Ferrel
e-mail: cesarunder90@gmail.com

L. Di G. Sigalotti
Centro de Física, Instituto Venezolano de Investigaciones Científicas, IVIC, Apartado Postal 20632, Caracas 1020-A, Venezuela
e-mail: leonardo.sigalotti@gmail.com

L. Di G. Sigalotti et al. (eds.), *Computational and Experimental Fluid Mechanics with Applications to Physics, Engineering and the Environment*, Environmental Science and Engineering, DOI:10.1007/978-3-319-00191-3_34,

Darcy's law is recovered and the hydraulic conductivity is calculated for both geometries. The numerical results are consistent with previous two-dimensional simulations in that the square case has a lower hydraulic conductivity than the hexagonal case. Finally, for the non-saturated case the relaxation time is calculated for different body forces. In this case, the system never reaches steady-state conditions.

1 Introduction

The flow of complex fluids and the dispersion of solids in multiphase fluids through porous media play a fundamental role in the current environment: in oil industry applications, in the pollution of soils and aquifers by industrial products, in the designing of remediation techniques for sites with mixed contamination, etc. In particular, fluids contaminated with hazardous particles and their mobility through a porous medium (soil) have important environmental implications due to the risk of ground water contamination. Specifically, radioactive contamination of the environment, including soils and water, is a relevant problem that has occurred in many parts of the world as a by-product of nuclear activities such as defense-related operations, power production, research, medical and industrial applications, among others. In order to evaluate and elaborate adequate strategies for controlling and remediating sites with this kind of contamination, the transport of solid and dissolved particles in multiphase fluids and their flow through porous media need be understood. With this aim, more realistic and accurate numerical models must be developed.

In order to reproduce real situations, these numerical models must not only include information concerning the geological structure as, for example, the complex structure of the solid matrix, consisting of different stratus with different geometries, the presence of fractures, and mobile or unstable boundaries originated due to dynamical changes caused by geochemical and biochemical processes, but also physicochemical information (i.e., chemical reactions, adsorption–desorption processes, and changes in the thermodynamic characteristics of the medium).

In general, flow through saturated media has been modelled using Darcy's law given by $\mathbf{v} = -k\mathbf{i}$, where $\mathbf{v}$ is the velocity of the fluid, k is the hydraulic conductivity, and $\mathbf{i}$ is the hydraulic gradient, which is related to the body force $\mathbf{F}$ and the gravitational acceleration g by $\mathbf{i} = \mathbf{F}/g$ (Zhu et al. 1999). However, this macroscopic approach does not consider the characteristics of the medium at the pore scale. In typical numerical models, the structure of the porous medium is mimicked by a distribution of solid particles and to reproduce the fundamental physics the models must involve changes in the flow and transport processes. Although Darcy's law is statistically equivalent to the Navier-Stokes equations, it does not include this kind of information. For this reason, it is demanding to propose a more detailed approximation to model the flow through porous media.

As a first step to model the flow at the pore scale, we need to represent in detail the geometry of the solid matrix. Recent efforts to study pore-scale flow phenomena in porous media have been made using the method of Smoothed Particle Hydrodynamics (SPH) (Tartakovsky and Meakin 2006; Tartakovsky et al. 2007; Tartakovsky 2010), where at these scales the flow is assumed to be incompressible and at low Reynolds numbers. One advantage of SPH is that it is a mesh-free, Lagrangian method for solving the equations of fluid dynamics in which additional physics and irregular and/or mobile solid boundaries can easily be included. High spatial resolution seems to be an unavoidable requirement for any numerical method trying to solve the flow at the scale of the smallest pores. For example, the use of the new computational technologies available today, such as the GPU processors, allow for much higher spatial resolution at a much lower computational cost, making SPH to be a promising tool for simulating flow through porous media with sufficient detail. The method does not only allow for calculation of the fluid velocity and pressure distribution, but also for fluid particle path lines and discharge velocities (Gesteira et al. 2010), the former being related to a well-known and still unresolved problem called *hydrodynamic dispersion* in heterogeneous porous media (Allen 1985).

2 The SPH Method

As was outlined above, SPH is a gridless, Lagrangian method for solving the equations governing the motion of fluids. It essentially consists of two approximations: the integral and the particle approximation (Monaghan 1982, 1992, 2005; Benz 1990; Liu and Liu 2003).

In integral form any function $A : \mathbf{R}^3 \rightarrow \mathbf{R}$ can be written as $A(\mathbf{x}) = \int A(\mathbf{x}')\delta(\mathbf{x}-\mathbf{x}')d^3\mathbf{x}'$, where $\delta(\mathbf{x}-\mathbf{x}')$ is the Dirac delta function. In SPH, we approximate this exact integral to second-order by replacing the Dirac delta function by a kernel function W so that

$$A(\mathbf{x}) = \int_V A(\mathbf{x}')W(\mathbf{x}-\mathbf{x}')d^3\mathbf{x}' + O(h^2), \tag{1}$$

where W is an interpolating function of compact support, defined inside a sphere of radius h, obeying the normalization condition: $\int_{V(h)} W d^3\mathbf{x}' = 1$, where $V(h)$ is the spherical volume of influence of the kernel.

The particle approximation consists of replacing the continuous fluid by a set of particles, each having its own volume of influence ΔV. In this way, the above integral can be approximated by a sum over all neighbouring particles so that

$$A(\mathbf{x}) \approx \int_V A(\mathbf{x}')W(\mathbf{x}-\mathbf{x}')d^3\mathbf{x}' \approx \sum_{j=1}^{N} A(\mathbf{x}_j)W(\mathbf{x}-\mathbf{x}_j)\Delta V_j, \tag{2}$$

where the subscript j identifies all N particles inside the volume V centred at the field position $\mathbf{x}$.

For numerical work, the summation on the right-hand side of Eq. (2) can be rewritten as

$$A(\mathbf{x}) = \sum_j \frac{m_j}{\rho_j} A_j W(|\mathbf{x} - \mathbf{x}_j|, h), \tag{3}$$

where m_j and ρ_j are, respectively, the mass and density of particle j and $A_j = A(\mathbf{x}_j)$ is the field value of A evaluated at the position of particle j. Here $\Delta V_j \to m_j/\rho_j$, i.e., the inverse of the number density of particle j is assumed to be equal to the volume of fluid associated with it. In the following, we will write for simplicity $W_{ij} = W(|\mathbf{x}_i - \mathbf{x}_j|, h)$, where the subscript i now refers to the position of particle i, where the field must be evaluated.

The kernel function is a monotonically decreasing function of the distance between pairs of particles and behaves as a delta function as the smoothing length, h, tends to zero. We refer the interested reader to Benz (1990), Monaghan (1992) and Liu and Liu (2003) for a detailed discussion on the kernel functions. Most existing numerical codes consider four possible kernel functions (Gesteira et al. 2010): the Gaussian kernel (Monaghan 1992), the cubic spline kernel, and the higher-order quartic and quintic spline kernels (Gesteira et al. 2010). In this work, we use the cubic spline kernel given by

$$W(r,h) = \alpha_D \begin{cases} 1 - \frac{3}{2}q^2 + \frac{3}{4}q^3 & ; 0 \le q < 1 \\ \frac{1}{4}(2-q)^3 & ; 1 \le q < 2 \\ 0 & ; q \ge 2, \end{cases} \tag{4}$$

where $\alpha_D = 10/(7\pi h^2)$ in two-dimensions (2D) and $\alpha_D = 1/(\pi h^3)$ in three-dimensions (3D). We note that the tensile correction is automatically activated when using kernel functions with first derivatives that tend to zero with decreasing inter-particle spacing (Monaghan 2000).

Using Eq. (3), the density field at the position of particle i can be written in SPH notation as

$$\rho_i = \sum_j m_j W_{ij}, \tag{5}$$

where, as we mentioned above, the sum includes only the nearest neighbours to particle i. SPH expansions for the gradient and divergence of the fluid velocity vector, $\mathbf{v}(\mathbf{x})$, are also required for the discretization of the Navier-Stokes equations. In particular, we adopt the commonly used expressions

$$\nabla \mathbf{v}(\mathbf{x}) = \sum_j \frac{m_j}{\rho_j} \mathbf{v}_j \nabla W(|\mathbf{x} - \mathbf{x}_j|, h), \tag{6}$$

for the gradient and

$$\nabla \cdot \mathbf{v}(\mathbf{x}) = \sum_j \frac{m_j}{\rho_j} \mathbf{v}_j \cdot \nabla W(|\mathbf{x} - \mathbf{x}_j|, h). \tag{7}$$

for the divergence of $\mathbf{v}(\mathbf{x})$.

In the present formulation, we evolve the continuity equation rather than using the standard summation (5) for the density, which in discretized SPH form reads (Monaghan 1992):

$$\frac{d\rho_i}{dt} = \sum_j m_j \mathbf{v}_{ij} \cdot \nabla_i W_{ij}, \tag{8}$$

where d/dt is the total derivative and $\mathbf{v}_{ij} = \mathbf{v}_i - \mathbf{v}_j$, with $\mathbf{v}_i$ and $\mathbf{v}_j$ being the velocity vectors of particles i and j, respectively. This form is better suited when dealing with fluids in the presence of solid boundaries. Moreover, instead of using Darcy's law, we solve the momentum conservation equation

$$\frac{d\mathbf{v}}{dt} = -\frac{1}{\rho}\nabla P + \mathbf{g} + \Theta, \tag{9}$$

where P is the pressure, $\mathbf{g}$ is the Earth gravitational acceleration, i.e., $\mathbf{g} = (0, 0, -9.81)$ m/s^2, and Θ accounts for the diffusion terms. Three different options for diffusion can be used: (1) an artificial viscosity, (2) a laminar viscosity, or (3) the full viscosity, i.e., laminar viscosity plus sub-particle scale turbulence. Here we use the artificial viscosity proposed by Monaghan (1992) and refer the reader to Gesteira et al. (2010) for a more detailed account on the choice of the diffusion term. In SPH notation, Eq. (9) can be written as

$$\frac{d\mathbf{v}_i}{dt} = -\sum_j m_j \left(\frac{P_j}{\rho_j^2} + \frac{P_i}{\rho_i^2} + \Pi_{ij} \right) \nabla_i W_{ij} + \mathbf{g}, \tag{10}$$

where the fluid acceleration due to the pressure gradients has been approximated using the standard symmetrized SPH representation

$$\left(-\frac{1}{\rho}\nabla P \right)_i = -\sum_j m_j \left(\frac{P_j}{\rho_j^2} + \frac{P_i}{\rho_i^2} \right) \nabla_i W_{ij}, \tag{11}$$

which leads to exact conservation of linear and angular momentum. The artificial viscous term, Π_{ij}, in Eq. (10) is given by Monaghan (1992)

$$\Pi_{ij} = \begin{cases} -\frac{\alpha \bar{c}_{ij} \mu_{ij}}{\bar{\rho}_{ij}} & ;\ \mathbf{v}_{ij} \cdot \mathbf{x}_{ij} < 0, \\ 0 & ;\ \mathbf{v}_{ij} \cdot \mathbf{x}_{ij} > 0, \end{cases} \tag{12}$$

where $\mu_{ij} = h\mathbf{v}_{ij} \cdot \mathbf{x}_{ij}/(\mathbf{x}_{ij}^2 + \eta^2)$, $\mathbf{x}_{ij} = \mathbf{x}_i - \mathbf{x}_j$, $\bar{c}_{ij} = (c_i + c_j)/2$ is the average sound speed between particles i and j, $\bar{\rho}_{ij} = (\rho_i + \rho_j)/2$, and α is a free parameter typically of order unity. The parameter $\eta^2 \approx 0.01$ is added in the denominator of the definition of μ_{ij} to prevent numerical singularities. Implicit in the form of Eq. (12)

is that the viscosity is zero for particles moving away from each other and positive for particles approaching each other. This form of the artificial viscosity is Galilean invariant and so it allows SPH to perform well for slowly moving shocks. In addition, it vanishes under solid-body rotation, conserves total linear and angular momentum, and guarantees that the entropy change due to dissipation is positive definite.

The time rate of change of the specific internal energy, e, is determined by the thermal energy equation, which is derived from the first law of thermodynamics: $de = -Pd(1/\rho) + Tds$, where T is the temperature and ds is the change of specific entropy, which includes all non-adiabatic effects. The symmetrized standard representation (Monaghan 1994)

$$\frac{de_i}{dt} = \frac{1}{2}\sum_j m_j \left(\frac{P_j}{\rho_j^2} + \frac{P_i}{\rho_i^2} + \Pi_{ij} \right) \mathbf{v}_{ij} \cdot \nabla_i W_{ij}, \tag{13}$$

is employed to evolve the thermal energy of particle i. This form is consistent with Eq. (10). Moreover, the position of particle i is determined by means of the equation (Monaghan 1989):

$$\frac{d\mathbf{r}_i}{dt} = \mathbf{v}_i + \varepsilon \sum_j \frac{m_j}{\overline{\rho}_{ij}} \mathbf{v}_{ij} W_{ij}, \tag{14}$$

where $\varepsilon = 0.5$. This expression guarantees that each SPH particle moves with a velocity that is close to the average fluid velocity in its neighborhood. Finally, the above SPH equations are solved by specifying an equation of state for the pressure. Here we treat the fluid as weakly compressible and adjust the compressibility to slow down the speed of sound and maintain reasonable values of the time step, which is here determined using the Courant condition. The compressibility is limited by the fact that the sound speed should be about ten times faster than the maximum fluid velocity in order to keep variations of the density to within less than one percent. In order to do so we use the following equation of state (Monaghan and Kos 1999; Batchelor 2000):

$$P = B\left[\left(\frac{\rho}{\rho_0} \right)^{\gamma} - 1 \right], \tag{15}$$

where $\gamma = 7$ and $B = c_0^2 \rho_0/\gamma$, with $\rho_0 = 1{,}000\,\mathrm{kg/m^3}$ being the reference density and $c_0 = c(\rho_0)\sqrt{(\partial P/\partial \rho)}|_{\rho_0}$ the sound speed at the reference density. A more detailed account of the SPH method and its numerical implementation can be found in Liu and Liu (2003) and Gesteira et al. (2010).

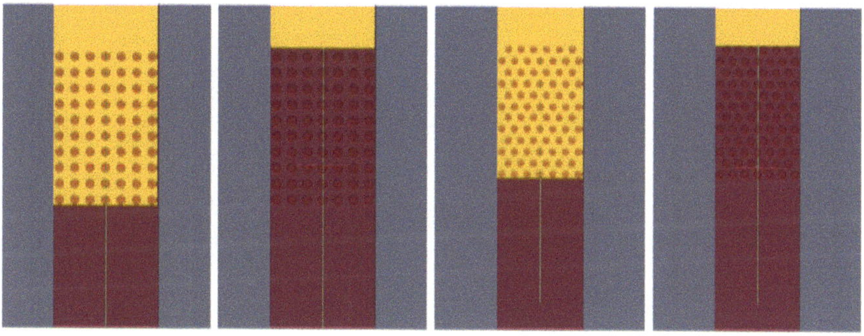

Fig. 1 Initial configuration for the saturated and non-saturated square (*first two panels*) and hexagonal arrays (*last two panels*) of particles. The fluid is shown in *yellow*

3 Simulation Models

We study the flow of a fluid through saturated and non-saturated homogeneous porous media. To do so we first consider a column of fluid with density ρ_0, which is introduced at the top of a regular 3D monolayer consisting of either a square or an hexagonal lattice of solid spherical particles as shown in Fig. 1. The solid spherical particles are assumed to remain static during the simulation. This configuration is then employed to represent the structure of our porous media.

For all simulations the porosity is given by $\theta = V_v/V_T = 0.8$, where $V_v = V_T - n_s 4\pi r_0^3/3$ is the void volume and $V_T = l_x l_y l_z$ is the total volume of the lattice representing the porous medium. This relationship fixes the radius r_0 of the spheres as a function of the l_x, l_y, and l_z lattice dimensions and the number of spheres n_s. For the square lattice we take $n_s = 70$, $l_x = 8.08$, $l_y = 1$, and $l_z = 11.6$, while for the hexagonal array we take $n_s = 78$, $l_x = 8.08$, $l_y = 1$, and $l_z = 12.3$. In order to facilitate comparisons with other simulations we employ dimensionless units by means of the following transformations: $\rho_0 = \{\rho_0\}\,\mathrm{kg/m^3} = 1$, $r_0 = \{r_0\}$ m $= 1$, and $c_0 = \{c_0\}$ m/s $= 1$, where ρ_0 is the fluid density, r_0 is the radius of the spheres, and c_0 is the speed of sound. Here $\{A\}$ represents the numerical value of quantity A. Thus, the physical units of mass, length, and time are recovered by making kg $= 1/\{r_0\}^3\{\rho_0\}$, m $= 1/\{r_0\}$, and s $= \{c_0\}/\{r_0\}$.

We analyze two different cases. In the first case, we consider a saturated medium in which the fluid particles are distributed uniformly between the solid particles in the lattice, while in the second case a non-saturated medium is defined where no fluid particles are placed between the spherical particles (see Fig. 1).

For the saturated models we use $n_f = 1{,}019{,}013$ fluid particles for the square lattice and $n_f = 919{,}217$ for the hexagonal lattice, with a total number of $n_p = 1{,}447{,}675$ and $n_p = 1{,}354{,}295$ particles, respectively, while in the non-saturated models $n_f = 663{,}520$ fluid particles are used for the square array and $n_f = 563{,}295$ for the hexagonal lattice, with a total number of $n_p = 1{,}092{,}182$ and $n_p = 998{,}758$, respectively. In each case, the system is confined by four lateral solid walls: two

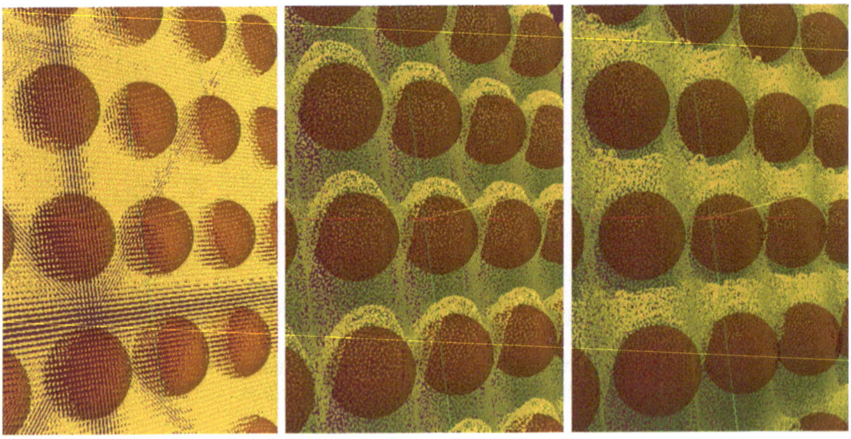

Fig. 2 Amplified views of the saturated square system at three different times: $t = 0$ (*left*) and during the evolution (*middle* and *right*). The fluid is shown in *yellow* and the spherical solid particles are shown in *red*

coinciding with the xz-plane and two with the yz-plane. At the bottom of the porous medium (coinciding with the xy-plane), the liquid is allowed to flow freely (outlet boundary conditions). Fluid motion is started by applying a body force $\mathbf{F} = g\mathbf{i}$ in the z-direction, where $\mathbf{i}$ is the hydraulic gradient and $g = 9.8\,\text{m/s}^2$ is the gravitational acceleration. This force is maintained constant during a simulation. A modified version of the Dual-SPHysics code (Gesteira et al. 2010) was employed and the simulations were carried out on GPU processors. Details of the evolution for the saturated model in a square lattice is shown in the amplified images of Fig. 2. Initially (left panel), the fluid particles are uniformly distributed in a regular grid and as the system evolves they become disordered (middle and right panels).

4 Results

4.1 Saturated Case

A set of calculation models were carried out for a saturated medium for both the square and hexagonal lattices by varying the magnitude of the body force $\mathbf{F}$ in the z-direction. This was accomplished by varying the magnitude of the hydraulic gradient $\mathbf{i}$ in the z-direction, with $i_z = 0.01, 0.006, 0.005, 0.0025$, and 0.00125 for the square lattice and $i_z = 0.04, 0.02, 0.01, 0.005$, and 0.0025 for the hexagonal array. For each run we calculate the discharge velocity v_z in an arbitrary point close to the centre of the porous medium. The discharge velocity field at three different times

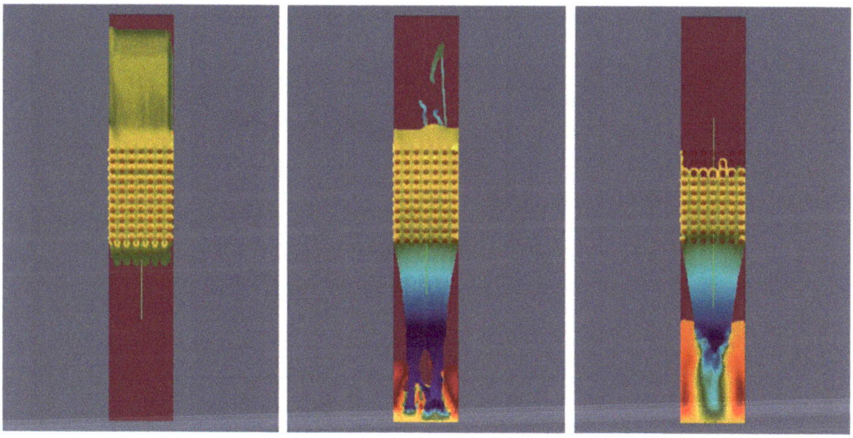

Fig. 3 Discharge velocity field at different times during the evolution of a saturated medium with a square array of solid spheres: $t = 500$ (*left*), $t = 2{,}000$ (*middle*), and $t = 3{,}400$ (*right*). Times and velocities are given in dimensionless units

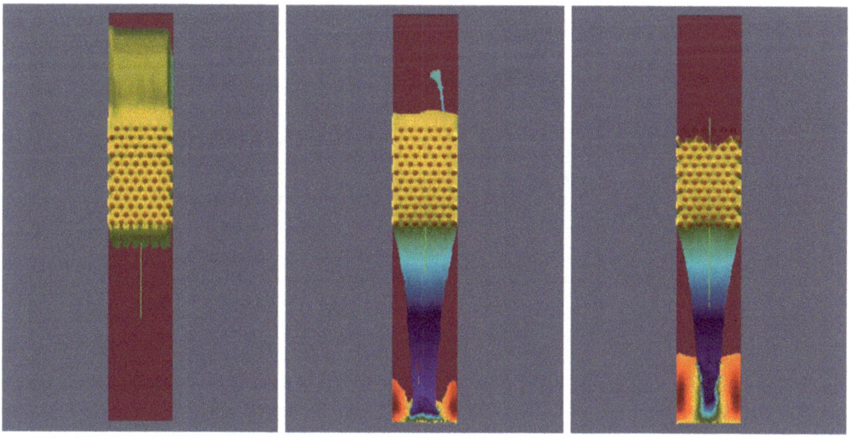

Fig. 4 Discharge velocity field at different times during the evolution of a saturated medium with an hexagonal array of solid spheres: $t = 500$ (*left*), $t = 2{,}000$ (*middle*), and $t = 3{,}400$ (*right*). Times and velocities are given in dimensionless units

during the evolution is shown in Figs. 3 and 4 for the square and hexagonal systems, respectively.

The evolution of the discharge velocities v_z at an arbitrary point in the porous matrix is plotted in Fig. 5 for all saturated models. We see that an approximate steady state is reached at different times depending on the magnitude of $\mathbf{i}$. For both types of arrays, as the magnitude of i_z is decreased, the system takes longer to achieve a steady-state regime. In addition, Fig. 6 depicts the x- and z-components of the discharge velocity as a function of time for a saturated matrix composed of a square

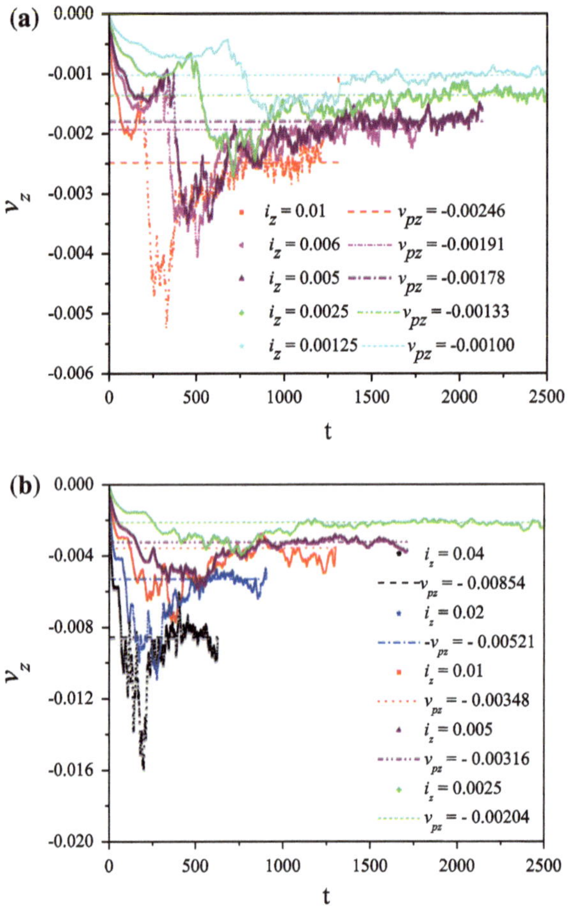

Fig. 5 Dimensionless discharge velocity as a function of time at an arbitrary point inside the porous matrix for varying magnitude of i_z. *Top*: square array; *bottom*: hexagonal array. The *straight lines* mark the mean steady-state velocities for each run

lattice of solid spheres for $i_z = 0.005$. The mean steady-state values of the velocity components are shown by the horizontal dashed lines. We see that both components reach a steady-state regime approximately at the same time as shown by the inset boxes in the first and third plots. The mean value of the steady-state velocity in the z-direction is larger than the corresponding value in the x-direction. This is expected because the main flow occurs along the z-axis where it is induced by gravity.

The mean steady-state velocity component in the z-direction, v_{pz}, increases linearly with the hydraulic z-gradient, i_z, as shown in Fig. 7. While this is true for both arrays of solid particles, the linear behaviour implies that the Darcy's law is well reproduced by the calculations. From the slopes of the linear variation, we may then calculate the hydraulic conductivity k_z. We find that $k_z = 0.13621$ for the square

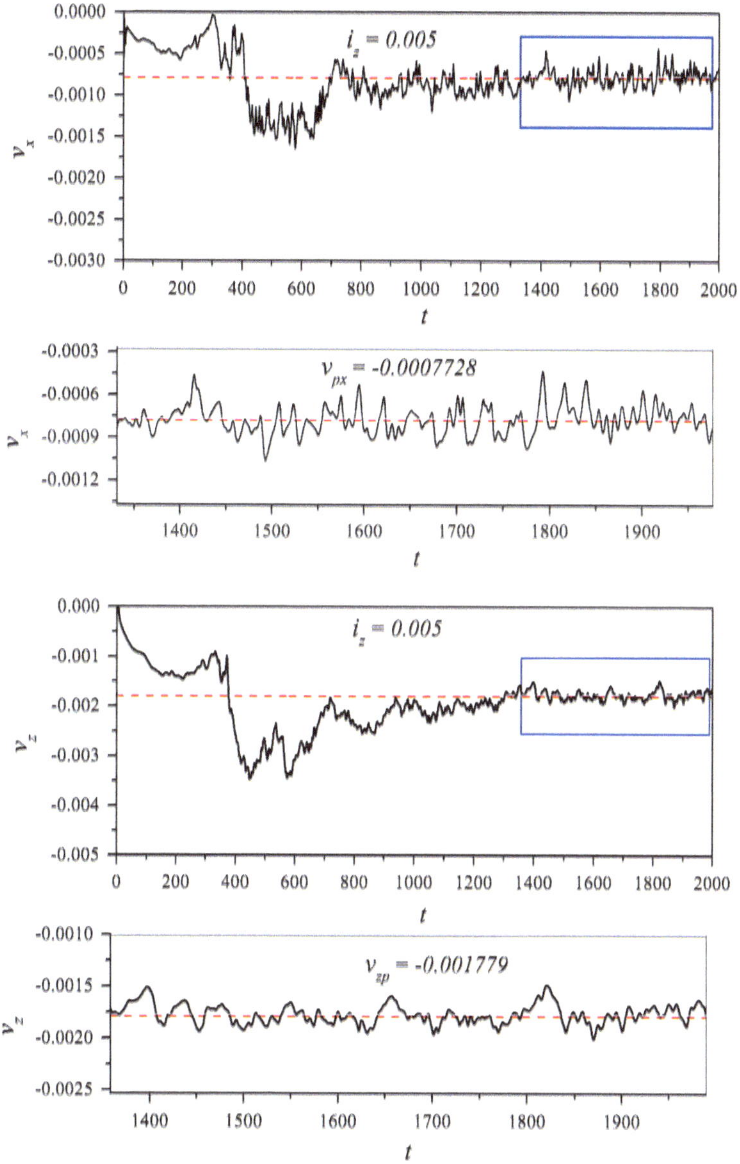

Fig. 6 Discharge velocity in the x- and z-direction for a saturated porous matrix represented by a square lattice of spherical solid particles when $i_z = 0.005$. The *inset boxes* in the *first* and *third* plots enclose the temporal region where the velocity components become steady state. The *second* and *fourth* plots are amplifications of the *inset boxes* in the *first* and *third* plots, respectively. In all boxes the *straight lines* mark the mean steady-state velocity components v_{px} and v_{pz}. All quantities are dimensionless

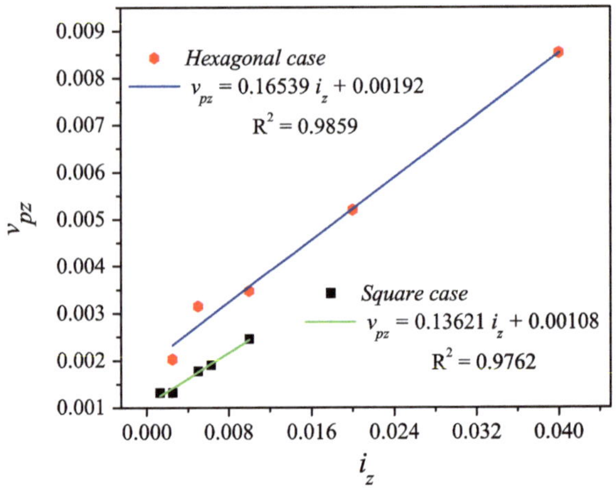

Fig. 7 Linear dependence of the z-component of the mean discharge velocity in steady state on the hydraulic z-gradient showing Darcy's law. The hydraulic conductivity obtained for the square array is $k_z = 0.13621$ and for the hexagonal array is $k_z = 0.16539$ in dimensionless units

array, while $k_z = 0.16539$ for the hexagonal case. As expected, the value of the hydraulic conductivity for the square geometry is smaller than for the hexagonal array. These values can be compared to previous 2D SPH simulations reported by Zhu et al. (1999). In order to do so we must convert our dimensionless results to dimensional form. Since for our models $\{\rho_0\} = 1{,}000$, $\{r_0\} = 0.399$, and $\{c_0\} = 10$, we have that $k_z = 0.01362$ m/s for the square case and $k_z = 0.016539$ m/s for the hexagonal case, which are comparable to the values reported by Zhu et al. (1999).

4.2 Non-Saturated Case

Similar calculation models were also carried out for the non-saturated case by varying the magnitude of i_z. The discharge velocity field at three different times during the evolution is shown in Figs. 8 and 9 for the square and hexagonal arrays, respectively. In contrast to the saturated case, no steady-state regime is observed for the flow velocity in the non-saturated media. As opposed to the square array, the hexagonal geometry produces more obstruction to the flow of fluid at similar conditions.

The discharge velocity v_z in the centre of the porous matrix is plotted in Fig. 10 as a function of time for the hexagonal array and varying magnitude of the hydraulic gradient. As the magnitude of i_z is increased higher values of v_z are achieved at the beginning of the simulation. Owing to the obstructing effects of the porous matrix, the flow velocity decreases with time and relaxes to a value close to zero. The relaxation time for the transient is shown in Fig. 11 for all model calculations, where the natural logarithm of v_z is plotted as a function of time. We see that ln v_z varies almost linearly

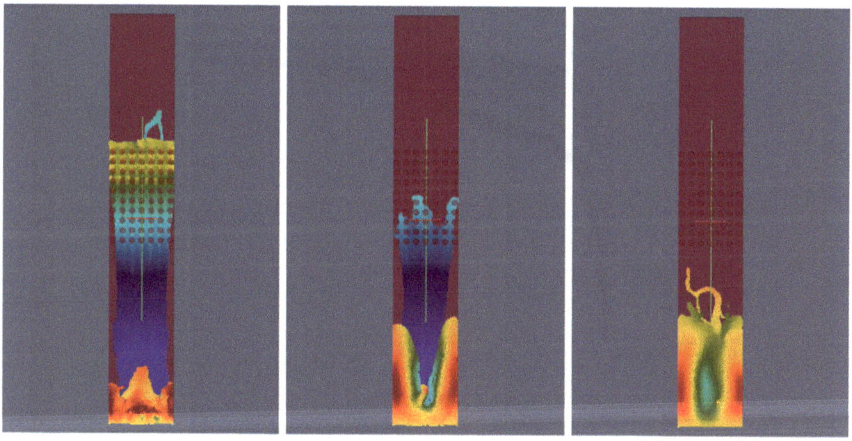

Fig. 8 Discharge velocity field at t = 1,000 (*left*), t = 1,500 (*middle*), and t = 2,000 (*right*) for the non-saturated square array. Times and velocities are given in dimensionless units

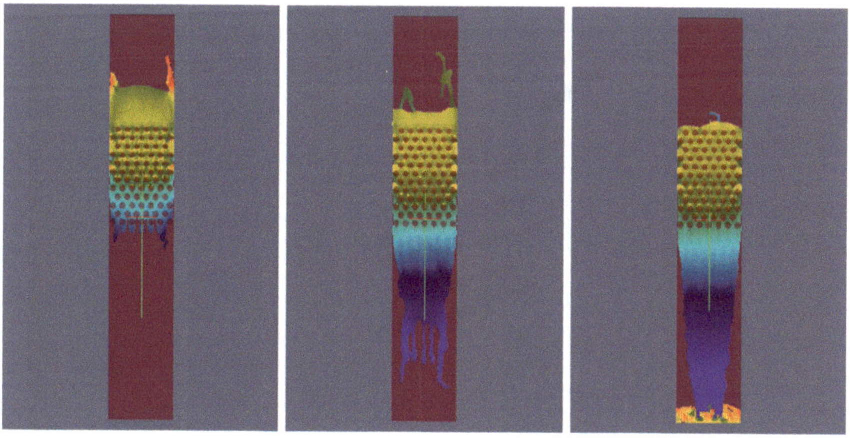

Fig. 9 Discharge velocity field at t = 1,000 (*left*), t = 1,500 (*middle*), and t = 2,000 (*right*) for the non-saturated hexagonal array. Times and velocities are given in dimensionless units

with time during relaxation. The relaxation time, τ, can then be obtained from the slope of this linear behaviour. As shown in Fig. 11, the magnitude of the relaxation time achieves a minimum for $i_z = 0.01$ and increases for higher and lower values of the hydraulic gradient.

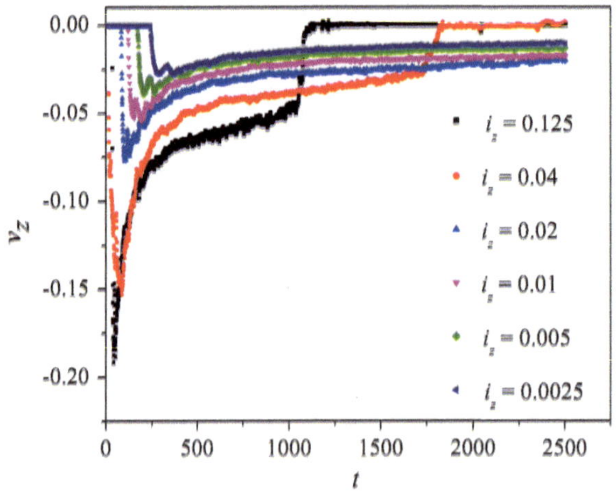

Fig. 10 Discharge velocity as a function of time for a non-saturated hexagonal array and varying values of the hydraulic gradient i_z

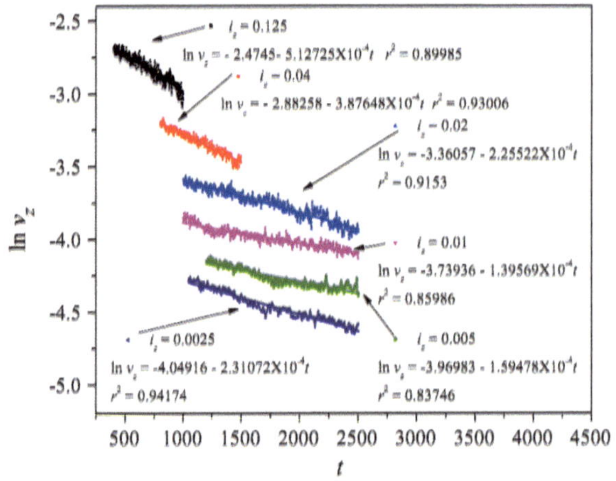

Fig. 11 Natural logarithm of v_{zp} as a function of time t for a non-saturated hexagonal array. The *plots* depict the approximate temporal interval when the system is relaxing for each value of i_z. The relaxation times are given in dimensionless units as calculated from the slopes of the linear dependences

5 Conclusions

In this chapter, we have presented exploratory three-dimensional (3D) simulations of fluid flow through saturated and non-saturated homogeneous porous media, using an improved Smoothed Particle Hydrodynamics (SPH) algorithm that has been implemented in the Dual-SPHysics code (Gesteira et al. 2010). The simulations were carried out on GPU processors at high resolution, allowing for a large number of SPH particles ($\sim 10^6$) at a low computational cost.

Two different initial configurations were tested, where the structure of the porous media was simulated using a regular 3D monolayer of solid spheres arranged in either a square or an hexagonal lattice. We find that for both geometries the fluid flow through a saturated homogeneous porous medium shows discharge velocities that are proportional to the hydraulic gradient, reproducing the Darcy's law under small body forces. Results for the derived hydraulic conductivity were shown to compare well with previous two-dimensional (2D) simulations of flow through periodic porous media (Zhu et al. 1999). In all saturated models, an approximate steady-state regime is achieved by the flow. In general, the system takes longer to achieve a steady-state flow when the magnitude of the hydraulic gradient is decreased. In contrast, the flow in non-saturated porous media never achieves a steady-state regime. In this case, the flow is more efficiently obstructed in an hexagonal array than in a square lattice. For an hexagonal lattice, the flow relaxes to values of the velocity close to zero with relaxation times $\tau \sim 10^{-4}$ in dimensionless units. More detailed simulations of these complex systems are under way for two-phase flows with different lattice arrays and geometries in order to reproduce the flow dynamics in more realistic porous media.

Acknowledgments This work has been partially supported by ABACUS and the Consejo Nacional de Ciencia y Tecnología of Mexico (CONACyT) under the project CONACyT-EDOMEX-2011-C01-165873.

References

Allen MB (1985) Numerical modelling of multiphase flow in porous media. Adv Water Resour 8:162–187

Batchelor GK (2000) An introduction to fluid dynamics. Cambridge University Press, Cambridge

Benz W (1990) Smoothed particle hydrodynamics—a review. In: Buchler JR (ed) Proceedings of the NATO advanced research workshop on the numerical modelling of nonlinear stellar pulsations problems and prospects. Kluwer Academic, Dordrecht, pp 269–288

Gesteira MG, Rogers BD, Dalrymple RA, Crespo AJC, Narayanaswamy M (2010) User Guide for The SPHysics Code.

Liu GR, Liu MB (2003) Smoothed particle hydrodynamics: a meshfree particle method. World Scientific Publishing, New Jersey

Monaghan JJ (1982) Why particle methods work. Siam J Sci Stat Comput 3:422–433

Monaghan JJ (1989) On the problem of penetration in particle methods. J Comput Phys 82:1–15

Monaghan JJ (1992) Smoothed particle hydrodynamics. Annu Rev Astron Astrophys 30:543–574

Monaghan JJ (1994) Simulating free surface flows with SPH. J Comput Phys 110:399–406

Monaghan JJ, Kos A (1999) Solitary waves on a Cretan beach. J Waterway Port Coast Ocean Eng 125:145–154

Monaghan JJ (2000) SPH without a tensile instability. J Comput Phys 159:290–311

Monaghan JJ (2005) Smoothed particle hydrodynamics. Rep Prog Phys 68:1703–1759

Tartakovsky AM, Meakin P (2006) Pore-scale modeling of immiscible and miscible fluid flows using smoothed particle hydrodynamics. Adv Water Resour 29:1464–1478

Tartakovsky AM, Meakin P, Scheibe TD, Eichler RM (2007) Simulations of reactive transport and precipitation with smoothed particle hydrodynamics. J Comput Phys 222:654–672

Tartakovsky AM (2010) Langevin model for reactive transport in porous media. Phys Rev E 82:026302

Zhu Y, Fox PJ, Morris JP (1999) A pore-scale numerical model for flow through porous media. Int J Numer Anal Methods Geomech 23:881–904

Part V
Astrophysical and Relativistic Flow

Propagation of Longitudinal Waves in Super-Radially Expanding Solar Plumes

Leonardo Di G. Sigalotti, Jordan A. Guerra and Hailleen Varela

Abstract Recent observations indicate that coronal plumes are the preferred channels for the propagation of slow magnetosonic waves from the Sun's poles to the corona. This problem is of relevance in solar physics because polar plumes are well observed exactly at the heights of the solar wind acceleration. In this chapter, we study the effects of the basal geometric spreading of polar plumes on the propagation of slow-mode waves up to $r = 5R_\odot$ by means of a non-linear analysis of the equations of hydrodynamics. We find that super-radial expansion at the base of the flux tube induces a strong dilution of the wave energy flux close to the solar surface, implying a steep decrease of the wave amplitude from the very beginning. Slow waves with periods of 7–25 min diffuse out at heights between ≈ 1.6 and $2.4R_\odot$ owing to dissipation. This result is in good agreement with recent observations.

1 Introduction

Coronal plumes are ray-like density enhancements that project outwards from the Sun's north and south poles. These structures, which are associated with the open magnetic field lines in polar coronal holes, can be seen at the Sun's limb during total solar eclipses and in visible and ultraviolet emission lines (formed at temperatures

L. Di G. Sigalotti (✉) · H. Varela
Centro de Física, Instituto Venezolano de Investigaciones Científicas, IVIC, Apartado Postal 20632, Caracas 1020-A, Venezuela
e-mail: leonardo.sigalotti@gmail.com

H. Varela
e-mail: hailleeng3@gmail.com

J. A. Guerra
Physics Department, School of Arts and Science, The Catholic University of America, 620 Michigan Avenue, Northeast, Washington, DC 20064 , USA
e-mail: Jordan.guerra@gmail.com

L. Di G. Sigalotti et al. (eds.), *Computational and Experimental Fluid Mechanics with Applications to Physics, Engineering and the Environment*, Environmental Science and Engineering, DOI:10.1007/978-3-319-00191-3_35, © Springer International Publishing Switzerland 2014

around 10^6 K) with the aid of space-borne telescopes and spectrometers. Plumes originate in photospheric unipolar magnetic flux concentrations (DeForest et al. 1997), and from SOHO observations they appear to expand super-radially in their lowest 20–30 Mm (where 1Mm $= 10^6$ m), and more slowly above that, with linear expansion ratios in the plane of the sky of about 1 at $1.05R_\odot$, 3 at 5–$6R_\odot$, and 6 at $15R_\odot$ (DeForest et al. 1997), where $R_\odot$ is the Sun's radius. Here, all radial distances and heights are measured with respect to the Sun's centre ($r = 0$), unless stated otherwise. Observations during four eclipse periods have shown an average plume width of 31 Mm at $1.05R_\odot$ as a result of super-radial expansion below these altitudes (Hiei and Takahashi 2000), suggesting that plume models based solely on a radial expansion are inconsistent with observations of polar coronal holes.

For coronal wave studies, polar plumes are interesting because they are well observed exactly at the heights of the wind acceleration, i.e., below $\sim 5\ R_\odot$, where the acceleration can be connected with the deposition of momentum carried by low-frequency waves propagating outwards. Estimation of the total mass outflow indicates that about half of the fast solar wind at $1.1R_\odot$ arises from plumes (Gabriel et al. 2003). In addition, they are natural fast magnetosonic wave guides. Moreover, since the Alfvén and the slow magnetosonic waves (i.e., longitudinal flux tube waves) are confined to the magnetic field, polar plumes act also as Alfvén and slow wave guides. Therefore, the study of magnetohydrodynamic (MHD) wave activity in coronal polar plumes is an important branch of solar coronal physics. On the other hand, their simple geometry permits the use of simpler theoretical models.

Alfvén waves are a very promising mechanism for transporting the energy from the solar surface into the corona, where they are partially reflected back down towards the Sun and dissipated by turbulent processes (Matthaeus et al. 1999; Cranmer et al. 2007; Cranmer 2009; Banerjee et al. 2009). The detection of waves in the outer solar atmosphere is made possible by analyzing the effects that these waves have on the plasma. For instance, compressional slow waves are detected as oscillations in line radiance, due to changes in plasma density, and also in the line-of-sight velocities, due to plasma motions, while transverse (Alfvén) waves give rise to only line-of-sight effects when they propagate over the plane of the sky. In the limit of incompressible Alfvén waves, they give no radiance signature and result in broadening of the spectral lines.

Polar plumes represent one structure in which longitudinal (slow compressional) waves propagate in solar winds. Such waves have been identified with periods of 7–25 min in the height range from 1.01 to $2.2R_\odot$ (DeForest and Gurman 1998; Ofman et al. 1999; Banerjee et al. 2000; Morgan et al. 2004) and the energy carried by them was estimated to be between 0.02 and 0.30 times the total energy required to heat and accelerate the fast solar wind (Ofman et al. 1999). Recent EIS/Hinode and SUMER/SOHO observations detected propagating quasi-periodic disturbances in plume and interplume regions of 15–20 min, suggesting that in the interplume region the waves are likely to be either Alfvénic or fast magnetosonic, while in plumes they are slow-mode waves not observable far off-limb because of wave dissipation (Gupta et al. 2010). Similar plume oscillations with periods of 10–30 min and propagation speeds of 100–170 km s^{-1} were also reported with AIA/SDO (Prasad et al. 2011).

A recent review on the morphology and dynamics of plumes and inter-plume regions is given by Wilhelm et al. (2011).

A theory of longitudinal flux tube waves in polar plumes was developed by Ofman et al. (1999, 2000), Cuntz and Suess (2001, 2004), and Nakariakov (2005). In radially divergent plume models, the relative amplitude of the slow-mode waves was shown to increase with height up to $\sim 1.2 R_\odot$, while at greater altitudes it decreases owing to viscous damping until complete dissipation at heights less than about $1.9 R_\odot$ (Ofman et al. 1999). Models invoking a super-radial expansion suggest that non-linear steepening of the outwardly propagating slow waves may well lead to shock formation at low coronal heights, i.e., within $1.3 R_\odot$ (Cuntz and Suess 2001) and that the influence of the solar wind flows may offset the dilution of the wave energy by the increase of the plume spreading with height (Cuntz and Suess 2004). While these models assume that the waves are generated at a height of $1.01 R_\odot$ above the solar surface, we investigate in the weak non-linear limit the effects of plume spreading on the dynamics of outwardly propagating slow-mode waves that are generated at the solar surface, i.e., at $1 R_\odot$.

2 Plume Geometry and Model

The cross-sectional area, $A(r)$, of a spreading flux tube can be parameterized as

$$A(r) = A(R_\odot)\left(\frac{r}{R_\odot}\right)^2 f(r), \tag{1}$$

where the coordinate r is measured with respect to the centre of the Sun, $R_\odot \approx 6.96 \times 10^{10}$ cm is the Sun's radius, $A(R_\odot) f(R_\odot)$ is the area of the plume at its base, and $f(r)$ is the expansion factor. Following Suess et al. (1998), the expansion factor is given by the product of two separate parts, i.e., $f(r) = f_l(r) f_g(r)$, where $f_l(r)$ is the local spreading below 35 Mm (i.e., below a height of $\approx 0.05 R_\odot$ from the Sun's surface) and $f_g(r)$ is the global spreading, which varies much slower than $f_l(r)$ below 35 Mm and is determined by the coronal hole geometry at larger distances (Wang et al. 1998). Here we use the following analytic expressions for $f_l(r)$ and $f_g(r)$ (Cuntz and Suess 2001):

$$f_l(r) = 1 + 13.31\left[1 - \exp\left(\frac{1 - r/R_\odot}{0.011}\right)\right], \tag{2}$$

$$f_g(r) = a_0 + a_1\frac{r}{R_\odot} + a_2\left(\frac{r}{R_\odot}\right)^2 + a_3\left(\frac{r}{R_\odot}\right)^3 + a_4\left(\frac{r}{R_\odot}\right)^4, \tag{3}$$

with $a_0 = 0.24974$, $a_1 = 0.76714$, $a_2 = -0.085164$, $a_3 = 0.0093196$, and $a_4 = -0.0004403$. These numbers are for a typical plume base field strength of 20 times the interplume field and a large coronal hole. From conservation of the magnetic flux density, the field strength can be written as

$$B(r) = B_0 \left(\frac{R_\odot}{r}\right)^2 \frac{f(R_\odot)}{f(r)}, \tag{4}$$

where B_0 is the field strength at the base of the plume and $f(R_\odot) = 1$. Note that when $f(r) = 1$ the field is radially divergent and we recover spherical geometry.

The plasma-β in solar plumes has values $\ll 1$ at least up to $\sim 5R_\odot$ and so the slow magnetosonic waves behave similarly to pure sound waves. In this way, the dynamics of slow-mode waves travelling along a flux tube of magnetic field strength given by Eq. (4) can be described by the standard equations of hydrodynamics. We assume that the plume is a static, isothermal, spherically stratified medium with an equilibrum density given by

$$\rho_0 = \rho_b \exp\left[-\frac{R_\odot}{H}\left(1 - \frac{R_\odot}{r}\right)\right], \tag{5}$$

where $\rho_b = 5 \times 10^{-16}$ g cm^{-3} is the density at the base of the plume and $H = 2k_B T R_\odot^2/(GM_\odot m_H)$ is the scale height. Here T is the temperature, k_B is the Boltzmann's constant, G is the gravitational constant, $M_\odot$ is the Sun's mass, and m_H is the mass of hydrogen. In addition, if we assume that the density perturbation is much smaller than ρ_0 and retain non-linear terms up to second order, the propagation of spherically-symmetric slow-mode waves in the density (ρ) and radial velocity (v) along a flux tube of expanding cross-section is given by the quasi one-dimensional flow equations

$$\frac{\partial \rho}{\partial t} + \frac{1}{Ar^2}\frac{\partial}{\partial r}\left(r^2 \rho_0 A v\right) = N_1, \tag{6}$$

$$\rho_0 \frac{\partial v}{\partial t} + c_s^2 \frac{\partial \rho}{\partial r} - \rho g = N_2, \tag{7}$$

where

$$N_1 = -\frac{1}{Ar^2}\frac{\partial}{\partial r}\left(r^2 \rho A v\right), \tag{8}$$

$$N_2 = \frac{\partial(\rho v)}{\partial t} - \frac{1}{Ar^2}\frac{\partial}{\partial r}\left(r^2 \rho_0 A v^2\right) + \frac{4\eta}{3}\frac{\partial}{\partial r}\left[\frac{1}{r^2}\frac{\partial}{\partial r}\left(r^2 v\right)\right] + \frac{4r\eta}{3A}\frac{dA}{dr}\frac{\partial}{\partial r}\left(\frac{v}{r}\right), \tag{9}$$

contain quadratic and dissipative terms, $c_s \approx 1.28 \times 10^4 \sqrt{T(K)}$ cm s^{-1} is the isothermal sound speed, $g = -GM_\odot/r^2$ is the gravitational acceleration, and

$$\eta = \eta_0 T^{5/2} = 0.72\left(\frac{m_p k_B^5}{\pi}\right)^{1/2} \frac{T^{5/2}}{e^4 \ln\lambda}, \tag{10}$$

is the compressive viscosity coefficient (Braginskii 1965), where m_p is the proton mass, e is the electron charge, and $\ln\lambda \approx 23$ is the Coulomb logarithm.

3 Results and Discussion

Equations (6) and (7) can be combined into a single wave equation for the density perturbation, which can be solved in the WKB approximation by assuming that the wavelength $\lambda \ll H$ and $\lambda \ll R_\odot$ and by introducing the small parameter $\varepsilon = \lambda/H \ll 1$. If, on the other hand, $\eta \ll c_s\rho_b\lambda$, the dissipative terms will be of the same order of ε. Now, transforming to the co-moving reference frame $\xi = r - c_s t$ and $R = \varepsilon r$, the resulting wave equation in dimensionless form will read as

$$\frac{\partial\tilde{\rho}}{\partial\tilde{r}} + \left(\frac{1}{\tilde{r}} + \frac{1}{2\tilde{H}\tilde{r}^2} + \frac{1}{2A}\frac{dA}{d\tilde{r}}\right)\tilde{\rho} + \frac{\tilde{\rho}}{\tilde{\rho}_0}\frac{\partial\tilde{\rho}}{\partial\tilde{\xi}} - \frac{\tilde{\eta}}{\tilde{\rho}_0}\left(\frac{\partial^2\tilde{\rho}}{\partial\tilde{\xi}^2} + \frac{1}{A}\frac{dA}{d\tilde{r}}\frac{\partial\tilde{\rho}}{\partial\tilde{\xi}}\right) = 0, \tag{11}$$

where we have introduced the dimensionless variables $\tilde{r} = r/R_\odot$, $\tilde{\xi} = \xi/R_\odot$, $\tilde{H} = H/R_\odot$, $\tilde{\rho} = \rho/\rho_b$, $\tilde{\rho}_0 = \rho_0/\rho_b$, and $\tilde{\eta} = 2\eta/(3\rho_b c_s R_\odot)$.

When the non-linear and dissipative terms (third and fourth terms) on the left-hand side of Eq. (11) are neglected, this equation can be easily integrated to give the linear solution

$$\tilde{\rho}(\tilde{r}) = \tilde{\rho}(1)\frac{1}{\tilde{r}}\sqrt{\frac{f(1)}{f(\tilde{r})}}\exp\left[-\frac{1}{2\tilde{H}}\left(1 - \frac{1}{\tilde{r}}\right)\right], \tag{12}$$

where $\tilde{\rho}(1)$ is the normalized perturbed density at the plume base. In this simple case, the wave amplitude changes with distance from the Sun owing to stratification and expansion of the flux tube in the direction perpendicular to the flow. If only the non-linear term is neglected, Eq. (11) can still be solved analytically for harmonic waves of the form $\tilde{\rho}(\tilde{r}, \tilde{\xi}) = \bar{\rho}(\tilde{r})\exp(i\bar{k}\tilde{\xi})$, where $\bar{k} = kR_\odot$ is a dimensionless wavenumber. However, for non-linear waves of arbitrary initial shape there is no analytical solution and Eq. (11) must be solved numerically.

The variation of the relative wave amplitude with height for typical coronal conditions is depicted in Fig. 1. The curves on each panel correspond to linear solutions of Eq. (11) when either the third and fourth terms on the left-hand side are dropped (see Eq. (12) for $\eta = 0$, left panels) or only the third term is neglected ($\eta \neq 0$, right panels). The top panels describe the dependence of the relative amplitude on height in plume models with spherical geometry, i.e., with $f(r) = 1$ so that the field lines expand radially and the flux tube resembles a truncated cone at its base. In the absence of dissipation ($\eta = 0$, top left panel), the relative amplitude grows with height

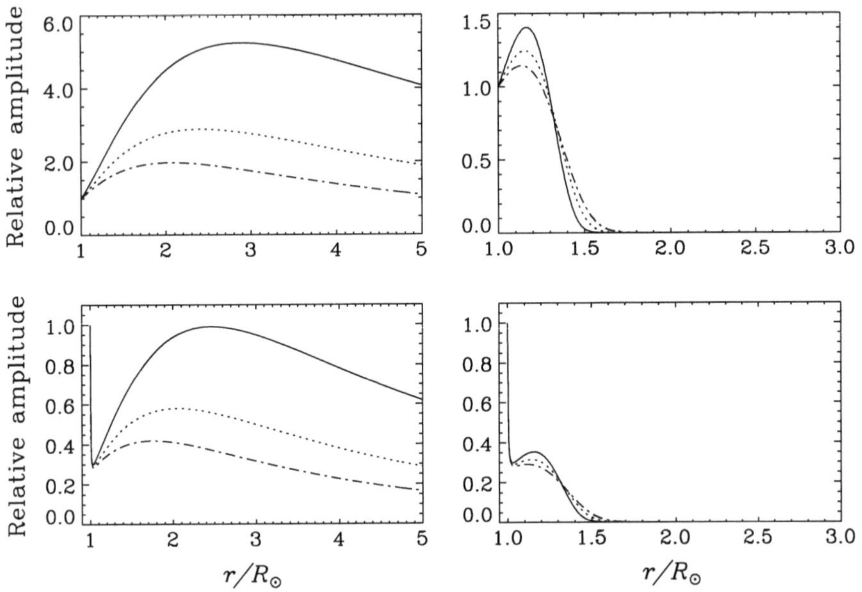

Fig. 1 Relative wave amplitude, $\tilde{\rho}(\tilde{r})[\tilde{\rho}_0(\tilde{r})\tilde{\rho}(1)]^{-1}$, as a function of distance from the solar surface, $\tilde{r} = r/R_\odot$, for three different temperatures: 1.0 MK (*solid lines*), 1.2 MK (*dotted lines*), and 1.4 MK (*dot-dashed lines*). The *top* panels correspond to plume models with spherical geometry ($f(r) = 1$, i.e., the field lines diverge radially) for $\eta = 0$ (*left*) and $\eta \neq 0$ (*right*), while the bottom panels correspond to plumes with a super-radial expansion ($f(r) = f_l(r) f_g(r)$) for $\eta = 0$ (*left*) and $\eta \neq 0$ (*right*). In all cases, the curves correspond to analytical solutions of Eq. (11) when the non-linear term is dropped

from $r = 1R_\odot$, reaches a maximum, and then decays slowly at large distances due to the expansion of the flux tube cross-section with radius. The position and strength of the maximum depends on the coronal temperature. In cooler plumes, the amplitude reaches larger values and the maximum occurs at larger radii. When viscous damping is switched on ($\eta \approx 0.096$–0.223 g cm^{-1} s^{-1} for $T = 1.0$–1.4 MK; top right panel), the waves amplify again from $r = 1R_\odot$, reaching considerably lower maximum amplitudes before decaying rapidly and dissipating completely. Waves of shorter periods ($P = 7$ min) damp out completely at heights of $\sim 1.6R_\odot$, while for longer wave periods ($P = 25$ min) full dissipation occurs at $\sim 2.2R_\odot$.

When the plume is allowed to expand super-radially from its base (bottom panels of Fig. 1), the relative wave amplitude drops sharply with height from $1R_\odot$ to about 1.03–$1.04R_\odot$. Within this small range the amplitude decays to $\sim 30\%$ of its initial value. However, most of the steep decay is completed below $\approx 1.01R_\odot$ independently of the wave parameters and properties of the medium. At heights greater than $\approx 1.05R_\odot$, where the tube expansion changes from super-radial to quasi-radial, the amplitudes grow and reach a maximum which never exceeds their initial values (bottom left panel). The sharp decay of the wave amplitude close to the solar surface is a pure geometrical effect due to a very rapid dilution of the wave energy

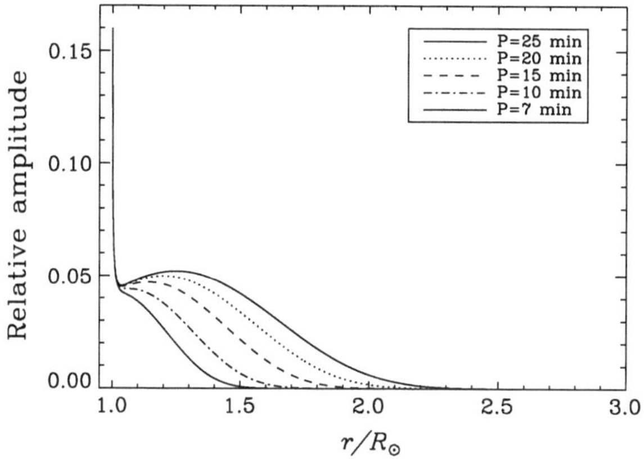

Fig. 2 Relative amplitude of non-linear waves as a function of height for different wave periods. In all runs the initial amplitude is 0.16, the plasma temperature is 1.4 MK, and the normalized viscosity is $\tilde{\eta} \approx 2.8 \times 10^{-4}$. This time the curves are obtained from numerical solution of Eq. (11) when the non-linear (quadratic) term is included

flux driven by the super-radial expansion of the flux tube at its base. In a viscous atmosphere (bottom right panel), the amplitude hardly grows after its sharp decrease owing to dissipation and complete damping occurs at heights of $\sim 1.8 R_\odot$. This result agrees with recent EIS/Hinode and SUMER/SOHO observations, which indicate that slow-mode waves in plumes become diffuse relatively close to the Sun, and not observable far off-limb, owing to dissipation, while in the inter-plume region Alfvén and fast-mode waves propagate farther in the corona with high acceleration (Gupta et al. 2010).

The relative amplitude of non-linear waves against height is depicted in Fig. 2 for different periods in the observed range. An initial amplitude of $\tilde{\rho}(1)/\tilde{\rho}_0 = 0.16$ and a coronal temperature of 1.4 MK were chosen for all runs. Similarly to the linear case, the relative amplitude drops sharply from $\tilde{r} = 1$ to $\tilde{r} \approx 1.03$ independently of the period. A similar plot for varying initial amplitudes has shown little dependence of the relative amplitude on the initial value. We see that waves with periods ≤ 10 min never grow at heights greater than $\tilde{r} \approx 1.03$ and diffuse out completely at heights less than $1.8 R_\odot$. For longer periods, the wave amplitudes hardly grow and reach a maximum between 1.2 and $1.3 R_\odot$, undergoing complete dissipation at distances not greater than $\sim 2.4 R_\odot$. Cuntz and Suess (2001) found numerically that slow-mode waves in spreading plumes may well undergo non-linear steepening and lead to shock formation at $\sim 1.3 R_\odot$. However, in their models the waves were all generated at a height of $1.01 R_\odot$, thus skipping the altitude range where a sharp reduction of the wave amplitude is predicted by our model.

4 Conclusions

We have proposed a simple model to investigate the effects of plume spreading on the dynamics of outwardly propagating slow magnetosonic waves, including the effects of stratification, compressive viscosity, and expansion of the flux tube cross-section. We find that the amplitude of the waves is strongly affected by the super-radial expansion at the base of plumes. Our model predicts a steep decay of the relative wave amplitude close to the solar surface in the height range between $1R_{\odot}$ and $\approx 1.03R_{\odot}$. However, most of the sharp decrease is seen to occur below $1.01R_{\odot}$ independently of the wave parameters and properties of the medium. Complete damping of the slow-mode waves always occurs at heights from ~ 1.6 to $2.4R_{\odot}$, which is in good agreement with recent observations.

Our models predict maximum amplitudes of the radial velocity perturbation of $\approx$3–10 km s^{-1}, which are too low compared to the estimated value of $\sim$40 km s^{-1} required to produce sufficient energy flux ($\sim 10^5$ ergs cm^{-2} s^{-1}) to heat the corona. As a next step, we plan to include the effects of heat conduction and Parker's wind flow to assess the dependence of the wave amplitudes on the wave-generation height and the level of viscous dissipation.

References

Banerjee D, O'Shea E, Doyle JG (2000) Long-period oscillations in polar plumes as observed by CDS on SOHO. Sol Phys 196:63–78

Banerjee D, Teriaca L, Gupta GR, Imada S, Stenborg G, Solanski SK (2009) Propagating waves in polar coronal holes as seen by SUMER and EIS. Astron Astrophys 499:L29–L32

Braginskii SI (1965) Transport processes in a plasma. Rev Plasma Phys 1:205–311

Cranmer SR, van Ballegooijen AA, Edgar RJ (2007) Self-consistent coronal heating and solar wind acceleration from anisotropic magnetohydrodynamic turbulence. Astrophys J Suppl Ser 171:520–551

Cranmer SR (2009) Coronal holes. Living Rev Sol Phys 6:3–63

Cuntz M, Suess ST (2001) Shock formation of slow magnetosonic waves in coronal plumes. Astrophys J 549:L143–L146

Cuntz M, Suess ST (2004) Properties of longitudinal flux tube waves III. Wave propagation in solar and stellar wind flows. Astron Astrophys 424:1003–1010

DeForest CE, Hoeksema JT, Gurman JB, Thompson BJ, Plunkett SP, Howard R, Harrison RC, Hassler DM (1997) Polar plume anatomy: results of a coordinated observation. Sol Phys 175:393–410

DeForest CE, Gurman JB (1998) Observations of quasi-periodic compressive waves in solar polar plumes. Astrophys J 501:L217–L220

Gabriel AH, Bely-Dubau F, Lemaire P (2003) The contribution of polar plumes to the fast solar wind. Astrophys J 589:623–634

Gupta GR, Banerjee D, Teriaca L, Imada S, Solanski S (2010) Accelerating waves in polar coronal holes as seen by EIS and SUMER. Astrophys J 718:11–22

Hiei E, Takahashi N (2000) Ground-based and SOHO observations of polar plumes during eclipse. Adv Space Res 25:1887–1891

Matthaeus WH, Zank GP, Oughton S, Mullan DJ, Dmitruk P (1999) Coronal heating by magnetohydrodynamic turbulence driven by reflected low-frequency waves. Astrophys J 523:L93–L96

Morgan H, Habbal SR, Li X (2004) Hydrogen Lyα intensity oscillations observed by the solar and heliospheric observatory ultraviolet coronagraph spectrometer. Astrophys J 605:521–527

Nakariakov VM (2005) Magnetohydrodynamic waves in coronal polar plumes. R Soc Lond Trans Ser A 364:473–483

Ofman L, Nakariakov VM, DeForest CE (1999) Slow magnetosonic waves in coronal plumes. Astrophys J 514:441–447

Ofman L, Nakariakov VM, Sehgal N (2000) Dissipation of slow magnetosonic waves in coronal plumes. Astrophys J 533:1071–1083

Prasad SK, Banerjee D, Gupta GR (2011) Propagating intensity disturbances in polar corona as seen from AIA/SDO. Astron Astrophys 528:L4–L8

Suess ST, Poletto G, Wang AH, Wu ST, Cuseri I (1998) The geometric spreading of coronal plumes and coronal holes. Sol Phys 180:231–246

Wang AH, Wu ST, Suess ST, Poletto G (1998) Global model of the corona with heat and momentum addition. J Geophys Res 103:1913–1922

Wilhelm K, Abbo L, Auchère F et al (2011) Morphology, dynamics and plasma parameters of plumes and inter-plume regions in solar coronal holes. Astron Astrophys Rev 19:35–104

Comparing Accretion Centres Between Rotating and Turbulent Cloud Cores

Guillermo Arreaga-García and Jaime Klapp

Abstract In this chapter we use the method of Smoothed Particle Hydrodynamics (SPH) to study the number and properties of accretion centres formed when a molecular gas cloud collapses, starting with initial conditions corresponding either to a turbulent or a rigidly rotating sphere. To do so we use a modified version of the SPH code GADGET-2, which is capable to detect when a gas particle becomes an accretion centre, inheriting the mass and momentum of all its closest neighbours. For both types of models (turbulent and uniformly rotating), we also study the effects of considering two different initial mass distributions: a uniform-density and a centrally condensed Plummer profile. We find that the turbulent models are more propense to fragment into a larger number of protostellar objects than the purely rotating clouds. However, in both types of models the average protostellar mass increases with increasing size of the kinetic energy content of the cloud.

1 Introduction

The L1544 dense core in Taurus is one that has been well-observed. In particular, Tafalla et al. (2004) have presented a multi-line and continuum study of L1544. According to these observations the core appears to have extended inward motions,

G. Arreaga-García (✉)
Departamento de Investigación en Física de la Universidad de Sonora, Apartado Postal 14740, C.P.83000 Hermosillo, Sonora, Mexico
e-mail: garreaga@cifus.uson.mx

J. Klapp
Departamento de Física, Instituto Nacional de Investigaciones Nucleares, ININ, Km. 36.5, Carretera México-Toluca, 52750 La Marquesa, Estado de México, Mexico
e-mail: jaime.klapp@inin.gob.mx

J. Klapp
Departamento de Matemáticas, Cinvestav del Instituto Politécnico Nacional (I.P.N.), 07360 México, D.F., Mexico
e-mail: jaime.klapp@hotmail.com

L. Di G. Sigalotti et al. (eds.), *Computational and Experimental Fluid Mechanics with Applications to Physics, Engineering and the Environment*, Environmental Science and Engineering, DOI:10.1007/978-3-319-00191-3_36,

thereby suggesting that it may be collapsing towards the formation of one or two low-mass stars.

Multiple fragmentation during the gravitational collapse of molecular cloud clumps and cores has been described by means of three-dimensional numerical simulations by Goodwin et al. (2004), who considered a turbulent model for L1544. In this model, fragmentation begins after a central gas clump is formed (the primary clump) and accretes mass from the surrounding gas envelope very rapidly and non-isotropically. The bulk of the rotating gas evolves towards a flattened configuration around the primary, which then develops two spiral arms (the circumstellar accretion region, hereafter CAR). When a CAR has accreted sufficient material, it can detach from the rotating central clump. The CAR becomes unstable and eventually fragments to produce multiple secondary objects.

In Arreaga-García et al. (2009), we have studied the effects of varying the extension of the cloud envelope on the collapse of an initially rotating sphere, and found that for fragmentation of the core to occur a sufficient amount of initial rotation would actually be necessary.

In order to compare the gravoturbulent models of Goodwin et al. (2004) with the initially uniformly-rotating of Arreaga-García et al. (2009), we have proposed a new set of collapse models, which may be either turbulent or uniformly rotating with initial conditions chosen to make that the central core resembles the dense core L1544. Furthermore, non-adiabatic effects during core contraction are here simulated using a barotropic equation of state as the one proposed by Boss et al. (2000). The chapter is organized as follows. In Sect. 2 we describe the initial conditions and model parameters for both the uniformly-rotating and the turbulent cores, while the numerical method is briefly outlined in Sect. 3. Section 4 presents the results and Sect. 5 contains the conclusions.

2 The Models

All models start with a gas sphere of radius $R_0 = 3.0 \times 10^{17}$ cm (≈ 0.097 pc) and central density $\rho_c = 3.0 \times 10^{-18}$ g cm^{-3}. A central condensation is modelled using a Plummer radial density profile, with the parameters ρ_c, R_c, and η entering in the definition of the Plummer's profile and having the same values as in Whithworth and Ward-Thompson (2001). The inner radius that controls the extension of the flat part of the radial distribution is $R_c = 8.06 \times 10^{16}$ cm (≈ 0.026 pc) so that the total mass contained by the gas sphere is $M_T = 5.24 M_{\odot}$. The transition from isothermal to adiabatic collapse is simulated using the barotropic equation of state (Boss et al. 2000):

$$p = c_0^2 \rho \left[1 + \left(\frac{\rho}{\rho_{crit}} \right)^{\gamma - 1} \right], \tag{1}$$

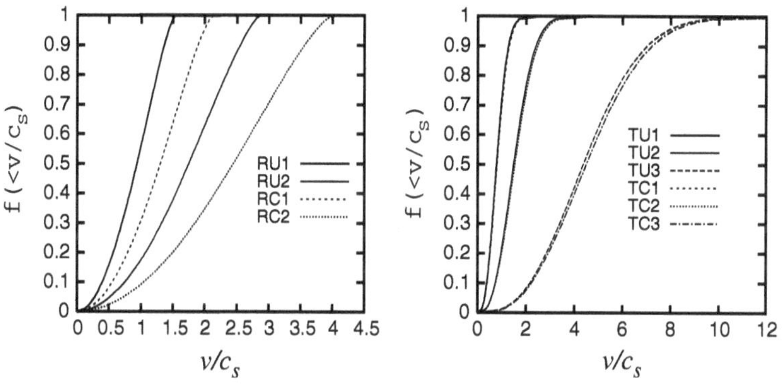

Fig. 1 Initial velocity distributions for the rotating (*left*) and turbulent (*right*) models

where $\gamma \equiv 5/3$, c_0 is the isothermal sound speed, and $\rho_{crit} = 5.0 \times 10^{-14}$ g cm^{-3} is the critical density at which the gas switches from being approximately isothermal to being adiabatic.

Table 1 lists the main parameters and results for all models. The entries in the second column identify the models, where the labels (RU, RC) and (TU, TC) correspond to initial uniform rotation with a uniform density or a central condensation and to turbulent cores with a uniform density or a central condensation, respectively. The models can be parameterized by the ratios of their thermal and kinetic energy to the gravitational energy

$$\alpha \equiv \frac{E_{therm}}{|E_{grav}|}, \quad \beta \equiv \frac{E_{kin}}{|E_{grav}|}, \tag{2}$$

respectively, as listed in the third and fourth columns of Table 1. This values are used to specify the isothermal sound speed c_0 and the initial particle velocities for the models. Starting from the fifth column, the table lists for each model the number N_{ac} of accretion centres that formed, the accretion centre average mass $M_{av}/M_{\odot}$, the total accreted mass $M_{ac}/M_{\odot}$, the evolution time t_{max}, and the maximum peak density ρ_{max}.

2.1 Initial Conditions

For all models we have used $N = 1{,}000{,}000$ SPH particles to represent the initial cloud configuration. A cubic rectangular mesh is employed as the simulation volume, which is then partitioned into small elements (cells), each of volume $\Delta x\, \Delta y\, \Delta z$. At the centre of each cell we place a particle—say, the ith particle—having a mass determined by its location according to the density profile being considered, that is: $m_i = \rho(x_i, y_i, z_i) * \Delta x\, \Delta y\, \Delta z$ with $i = 1, \ldots, N$. Next, each particle is randomly

Table 1 Models and results

Model	Label	α	β	N_{ac}	M_{av}/M_{sun}	M_{ac}/M_{sun}	t_{max} (sec)	$\rho_{max}(gr/cm^3)$
1	RC1	0.27	0.07	3	7.1892383e − 03	2.1567715e − 02	2.3197934e + 12	4.6851071e − 09
2	RU1	0.27	0.07	6	1.0574708e − 02	6.3448247e − 02	1.7146299e + 12	4.1523220e − 10
3	RC2	0.26	0.24	4	6.3262959e − 03	2.5305183e − 02	2.3702237e + 12	4.4358462e − 10
4	RU2	0.26	0.24	3	1.1299947e − 02	3.3899842e − 02	2.2441480e + 12	4.2527240e − 10
7	TC1	0.27	0.07	2	3.7652893e − 04	7.5305786e − 04	2.2189328e + 12	7.2606337e − 09
8	TU1	0.27	0.07	6	5.8568452e − 03	3.5141071e − 02	1.5381239e + 12	4.1470335e − 11
9	TC2	0.26	0.24	3	5.4242192e − 03	1.6272658e − 02	2.4206540e + 12	2.5965119e − 09
10	TU2	0.26	0.24	10	5.2657444e − 03	5.2657444e − 02	1.6137693e + 12	2.2618071e − 09
11	TC3	0.26	2.0	5	5.3688892e − 04	2.6844446e − 03	4.8866952e + 12	2.0023637e − 09
12	TU3	0.26	2.0	4	9.0702862e − 03	3.6281145e − 02	1.9894750e + 12	6.5217848e − 10

displaced from its location at the cell centre by a distance of the order $\Delta x/4.0$. We note that the SPH particles do not necessarily have the same mass.

For the rotating models, the initial velocity of particle i is given by

$$\mathbf{v}_i = \Omega_0 \times \mathbf{r}_i = (-\Omega_0, y_i, \Omega_0 x_i, 0), \tag{3}$$

where Ω_0 is the magnitude of the angular velocity.

As was done in most previous work (Bodenheimer et al. 2000; Arreaga-García et al. 2007, 2008; Arreaga-Garcia et al. 2012), where the main focus was to study the formation of binary protostellar cores, we perturb the mass of particle i according to

$$m_i \rightarrow m_i \left[1 + a \cos\left(m\, \phi_i\right)\right], \tag{4}$$

where the perturbation amplitude is set to $a = 0.1$ and the mode is fixed to $m = 2$. In the left panel of Fig. 1, we display the particle velocity distributions for all rotating models.

2.2 The Turbulent Velocity Spectrum

For the turbulent models we define a second mesh N_x, N_y, N_z, where the size of each element is $\delta x = R_0/N_x$, $\delta y = R_0/N_y$, and $\delta z = R_0/N_z$. In Fourier space the partition is therefore $\delta K_x = 1.0/\left(N_x \times \delta x\right)$, $\delta K_y = 1.0/\left(N_y \times \delta y\right)$, and $\delta K_z = 1.0/\left(N_z \times \delta z\right)$. Each Fourier mode has the components $K_x = i_{K_x} \delta K_x$, $K_y = i_{K_y}\, \delta K_y$, and $K_z = i_{K_z}\, \delta K_z$, where the indices i_{K_x}, i_{K_y}, and i_{K_z} take values on the intervals $[-N_x/2, N_x/2]$, $[-N_y/2, N_y/2]$ and $[-N_z/2, N_z/2]$, respectively. The magnitude of the wavenumber is $K = \left(K_x^2 + K_y^2 + K_z^2\right)^{1/2}$, with $K_{min} = 0$ and $K_{max} = \sqrt{3}\, N_x/(2\, R_0)$. Note that the Fourier wave can equally be described by a wave length $\lambda = 2\pi/K$ so that

$$K \approx \frac{1}{R_0}, \quad \lambda \approx R_0. \tag{5}$$

Following Dobbs et al. (2005), the component x of the particle velocity is

$$v_x = \Sigma_{-K_{max}}^{K_{max}} \left|K^{(-n-2)/2}\right| \left[K_z\, C_{K_y} \sin\left(\mathbf{K} \cdot \mathbf{r} + \Phi_{K_y}\right) - K_y\, C_{K_z} \sin\left(\mathbf{K} \cdot \mathbf{r} + \Phi_{K_z}\right)\right], \tag{6}$$

where the spectral index n was fixed to $n = -1$ and thus we have $v^2 \approx K^{-3}$. The vector $\mathbf{C}$, whose components are denoted by $\left(C_{K_x}, C_{K_y}, C_{K_z}\right)$, take values on a Rayleigh distribution. In addition, the wave phase vector Φ, given by $\left(\Phi_{K_x}, \Phi_{K_y}, \Phi_{K_z}\right)$, take random values on the interval $[0, 2\pi]$. The components of $\mathbf{C}$ are calculated using the

expression
$C = \sigma \times \sqrt{-2.0 \times \log(1.0 - u)}$, where u is a random number in the interval (0, 1) and σ is a fixed parameter set to $\sigma = 1.0$. The y and z components of the particle velocity are calculated using expressions analogous to Eq. (6). We further note that the particle velocity scales with the gas sphere size as

$$v \approx R_0^{\frac{3}{2}}. \tag{7}$$

The particle velocity distributions for the turbulent models are depicted in the right panel of Fig. 1.

3 The Computational Method

Let us now describe the modifications implemented into the GADGET-2 code for the detection of accretion centres. Any gas particle with a density higher than some specified value, say ρ_s, is a candidate to become an accretion centre. We first monitor and identify all candidate particles having densities higher than ρ_s. The spatial separation between candidate particles is next calculated: if there is one candidate at separation distances greater than $10r_{acc}$ from all other detected candidates, then this particle is chosen as an accretion centre. We define r_{acc} as the neighbour spherical radius of an accretion centre, given by $r_{acc} = 1.5(h_{min})$. In this way, the accretion centre will encompass all SPH particles that are within the sphere of radius r_{acc}. The mass and momentum of all these SPH particles will automatically be added to the accretion centre. In GADGET-2, once a particle is identified as an accretion centre its type is changed from 0 to -1 and thereafter it will no longer be advanced in time.

In order to determine the appropriate value of ρ_s we have performed a set of test model simulations for an initially uniform-density, uniformly rotating cloud model. We see that when we choose a small value for ρ_s, i.e., $\rho_s = 5.0 \times 10^{-16}$ g cm^3, then too many accretion centres are formed early in the collapse, as shown in the left panel of Fig. 2. After several test models, we found that the best value is $\rho_s = 5.0 \times 10^{-12}$ g cm^3, which produced an evolution similar to an identical model with no sink particles. This can be clearly seen by comparing the middle (with sinks) and right (without sinks) panels of Fig. 2, where the final inner structure of the core is shown for the two calculations at the same final time. Thus, all models were calculated using the parallel code GADGET-2 modified by our sink implementation with a threshold value of $\rho_s = 5.0 \times 10^{-12}$ g cm^3 for sink generation.

A full description of the code is given in Springel (2005) and so here we shall only mention some relevant aspects of it. In GADGET-2, each particle is allowed to have (i) its own smoothing length h_i and (ii) its own gravitational softening length ε_i, whose value is adjusted such that at each time step $\varepsilon_i\, h_i$ is of order unity. However, other SPH implementations use empirical formulas for ε_i in order to minimize errors in the calculation of the gravitational force (Gabbasov et al. 2006). In contrast,

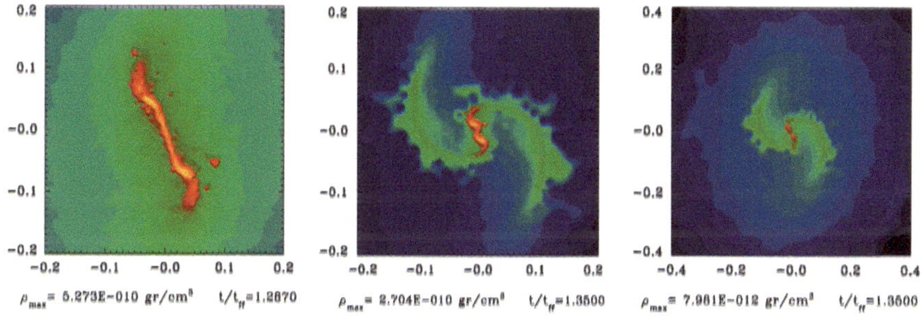

Fig. 2 Column-density distribution for three test model calculations used to choice the best value of ρ_s. All models started from initial uniform density and uniform velocity profiles. The left panel corresponds to $\rho_s = 5.0 \times 10^{-16}$ g cm^{-3} at 1.287 t_{ff}, where t_{ff} is the initial core free-fall time, the middle panel depict the central core regions at 1.35 t_{ff} for $\rho_s = 5.0 \times 10^{-12}$ g cm^{-3}. The right panel has been evovled with the Gadget 2 without sinks technique

GADGET-2 fixes the value of ε_i at each time-step by simply using the minimum value of the smoothing length for all particles, that is, $\varepsilon_i = h_{min}$ where $h_{min} = min(h_i)$ for $i = 1, 2...N$. Furthermore, GADGET-2 uses a Monaghan-Balsara form for the artificial viscosity to maintain hydrodynamical stability (Monaghan et al. 1983; Balsara 1995). For the simulations of this chapter the Courant factor was set to 0.1 and the strength of the artificial viscosity is regulated by setting $\alpha_\nu = 0.75$ and $\beta_\nu = 0.5\alpha_\nu$ (see Eq. (14) of Springel (2005)).

4 Results

We have prepared carefully the initial conditions for the gas particles in order to have a cloud under gravitational collapse. For the turbulent models, we set a bounded value of β around 2 because beyond this value, the cloud get dispersed without collapsing. For the rotating case, however, it is well-known that for values of β higher than about 0.3, the cloud is rotationally disrupted and does not collapse. Figure 3 depicts the evolution of the maximum density for all models.

4.1 Models with Initial Uniform Rotation

All the rotating models experience inside-out collapse independently of the initial density distribution. That is, the central cloud regions collapse first, while most of the gas in the envelope is left behind because of its lower rates of infall. This leads to the formation of a central dense clump within which an accretion centre forms (see Fig. 5). In the subsequent evolution, well-defined spiral arms develop around the central protostar, which then condense upon themselves allowing for fragmentation into secondary accretion centres. Fragmentation of the spiral arms occurs because

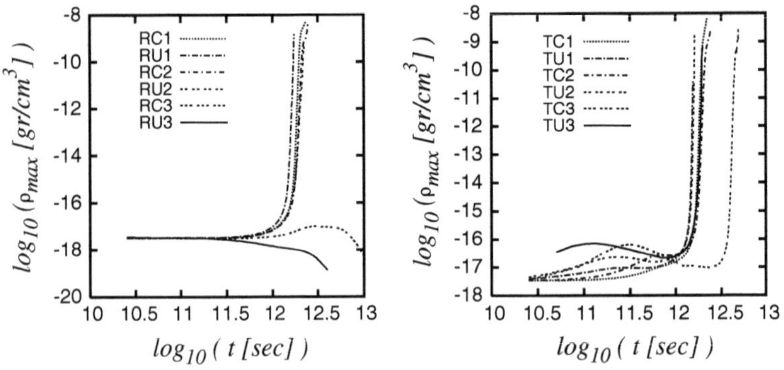

Fig. 3 Time evolution of the peak density for the uniformly-rotating (*left*) and the turbulent (*right*) models

they become unstable against growing spiral modes, as we may see in Fig. 5 for model RC1. It is interesting to mention that these secondary accretion centres follow the gas trend of the spiral arms.

4.2 Turbulent Core Models

Just in the initial stages of collapse, the turbulent gas undergoes collisions which take place simultaneously throughout the entire cloud. As a result, the gas compresses simultaneously in many places, giving rise to a swarm of interconnected pockets of gas, as we may see in the top left panel of Fig. 6 for model TU3. The artificial viscosity transforms the kinetic energy into heat, and this heat should be later radiated away in a very short timescale. Therefore, we can assume that the shock and the cloud remain isothermal at this stage.

This stage of the collapse is characterized by a smooth increase of the peak density, as we may see from Fig. 3 (right panel), followed by an equally smooth expansion where the peak density decreases. The magnitude of this expansion is seen to depend on the kinetic energy content (β) of the cloud. Note that while Fig. 3 refer to the maximum density, several other density peaks may well be evolving this way simultaneously throughout the cloud during this phase.

It is only after this expansion that the inner cloud regions truly begin to collapse, as evidenced by the rapid rise of the central density (see Fig. 3). In contrast to other turbulent collapse simulations, where turbulence is continuously replenished to prevent global collapse (Dubinski et al. 1995; Offner et al. 2008), here we have decided to include gas turbulence only at the initial time and to leave the cloud to collapse freely in order to provide a more systematic comparison between the purely rotating and the turbulent clouds.

As a result of the global cloud collapse, a central dense filamentary region soon forms, which then evolves towards much higher central densities, forming a central

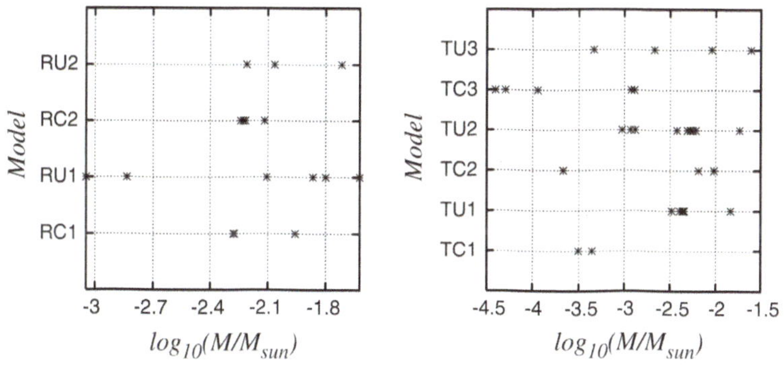

Fig. 4 Comparison of the mass of the accretion centres formed in the rotating (*left*) and turbulent (*right*) models

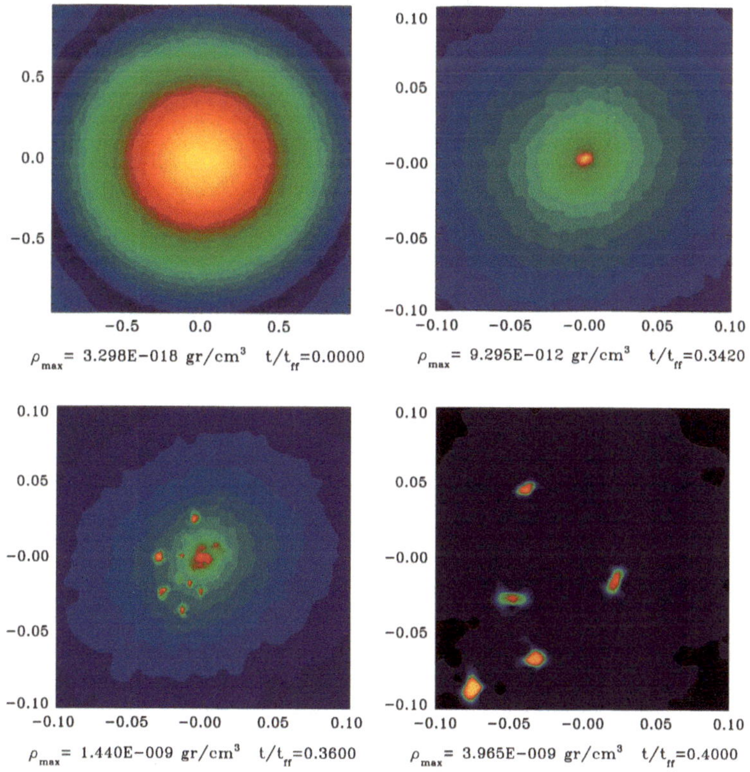

Fig. 5 Iso-density plots showing the cloud evolution for model RC1

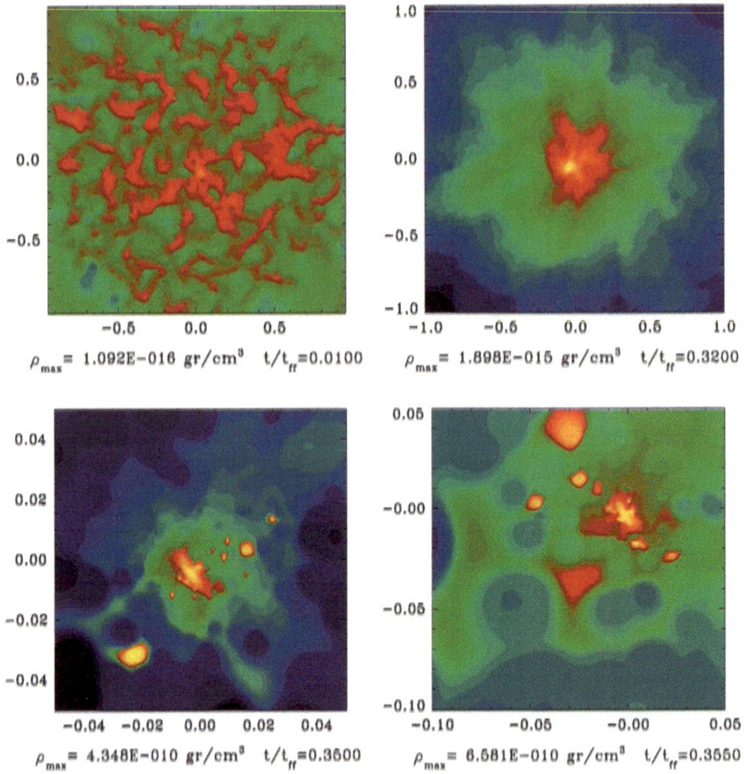

Fig. 6 Iso-density plots showing the cloud evolution for model TU3

protostar (the primary) surrounded by several other accretion centres (secondary protostars), as shown in the remaining panels of Fig. 6 for model TU3. In these models, however, the secondary protostars are formed through filament fragmentation rather than disk fragmentation. Average and integral properties for these accretion centres are reported in Table 1. The collapse of the initially centrally condensed model TC3 is displayed in Fig. 7. This time, global collapse produces a central bar-like region of high density, which condenses at its centre forming a primary protostar. As the collapse of the central regions proceeds other secondary protostars form around the primary through what resembles to be disk rather than filament fragmentation in contrast with the initially uniform-density clouds.

5 Conclusions

In this chapter, we have presented a set of calculations for the gravitational collapse of a core model whose initial structure has been chosen to fit the observational properties of the prestellar core L1544. Two sets of cloud models were considered:

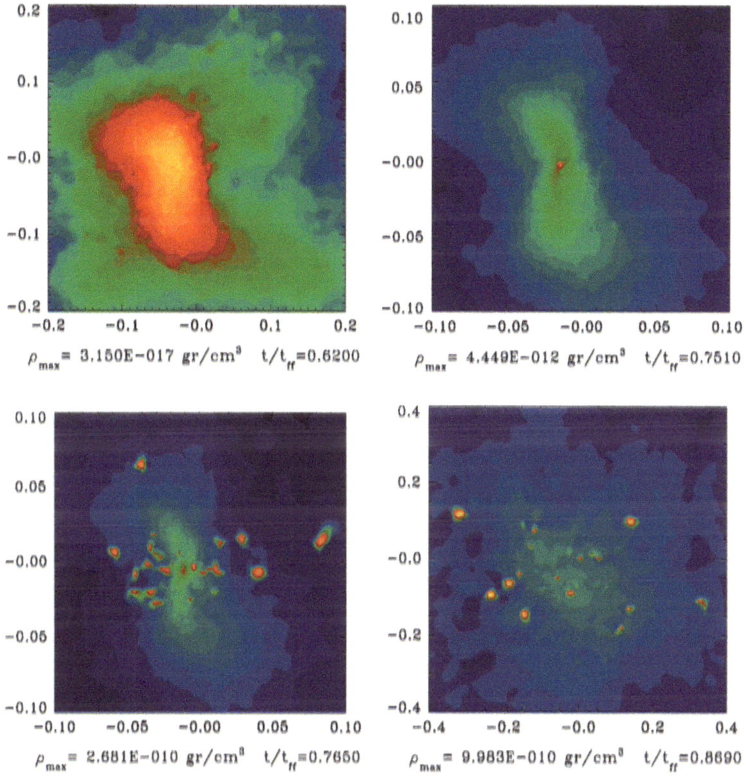

Fig. 7 Iso-density plots for the evolution of model TC3, showing the central core regions

initially uniformly-rotating and turbulent clouds with either initial uniform-density or central condensations.

As expected, the turbulent models fragmented into more protostellar objects, which were also slightly less massive, than the purely rotating clouds. The amount of initial kinetic energy, either in the form of turbulent energy or rotational energy, seems to be an important factor to determine the number of protostars formed. The efficiency of star formation appears to be higher in the uniform-density clouds than in the centrally condensed ones regardless of whether the cloud is turbulent or purely rotating. Also, in both types of models the average protostellar mass increases with increasing the kinetic energy content of the cloud.

Finally, the version of the GADGET-2 code employed in the simulations of this chapter represents a first-step version towards a full implementation of the sink-particle technique (Bate et al. 1995; Federrath et al. 2010). However, we find that for suitable values of the threshold density ρ_s this version of the code produces results that converge to identical model calculations with no sink particles, thus providing reliable fragmentation simulations.

Acknowledgments We would like to thank ACARUS-UNISON for the use of their computing facilities in the making of this chapter. This work was partially supported by the Consejo Nacional de Ciencia y Tecnología of Mexico (CONACyT) under the project Abacus CONACYT-EDOMEX-2011-C01-165873.

References

Arreaga-García G, Saucedo-Morales J, Carmona-Lemus J, Duarte-Pérez R (2008) Hydrodynamical simulations of the non-ideal gravitational collapse of a molecular gas cloud. Revista Mexicana de Astronomía y Astrofísica. 44:259–284

Arreaga-García G, Klapp-Escribano J, Gómez-Ramírez F (2009) The gravitational collapse of Plummer protostellar clouds. Astron Astrophys 509:A96

Arreaga-Garcia G, Saucedo-Morales J (2012) Physical effects of gas envelopes with different extension on the collapse of a gas core. Revista Mexicana de Astronomía y Astrofísica 48:61–84

Arreaga-García G, Klapp J, Sigalotti L. Di G (2007) Gravitational collapse and fragmentation of molecular cloud cores with GADGET-2. Astrophys J 666:290–308

Balsara DS (1995) von Neumann stability analysis of smooth particle hydrodynamics—suggestions for optimal algorithms. J Comput Phys 121:357–372

Bate MR, Bonnell IA, Price NM (1995) Modelling accretion in protobinary systems. Mon Not R Astron Soc 277:362–376

Bodenheimer P, Burkert A, Klein RI, Boss AP (2000) Multiple fragmentation of protostars. In: Mannings VG, Boss AP, Russell SS (eds) Protostars and planets IV. University of Arizona Press, Tucson, pp 675–701

Boss AP, Fisher RT, Klein RI, McKee CF (2000) The Jeans condition and collapsing molecular cloud cores: filaments or binaries? Astrophys J 528:325–335

Dobbs CL, Bonnell IA, Clark PC (2005) Centrally condensed turbulent cores: massive stars or fragmentation? Mon Not R Astron Soc 360:2–8

Dubinski J, Narayan R, Phillips TG (1995) Turbulence in molecular clouds. Astrophysical J 448:226–290

Federrath C, Banerjee R, Clark PC, Klessen RS (2010) Modeling collapse and accretion in turbulent gas clouds: implementation and comparison of sink particles in AMR and SPH. Astrophysical J 713:269–290

Gabbasov RF, Rodríguez-Meza MA, Klapp J, Cervantes-Cota JL (2006) The influence of numerical parameters on tidally triggered bar formation. Astron Astrophys 449:1043–1059

Goodwin SP, Whithworth AP, Ward-Thompson D (2004) Simulating star formation in molecular cloud cores. I. The influence of low levels of turbulence on fragmentation and multiplicity. Astron Astrophys 414:633–650

Monaghan JJ, Gingold RA (1983) Shock simulation by the particle method SPH. J Comput Phys 52:374–389

Offner SSR, Klein RI, McKee CF (2008) Driven and decaying turbulence simulations of low-mass star formation: from clumps to cores to protostars. Astrophysical J 686:1174–1194

Springel V (2005) The cosmological simulation code GADGET-2. Mon Not R Astron Soc 364:1105–1134

Tafalla M, Myers PC, Caselli P, Walmsley CM (2004) On the internal structure of starless cores. I. Physical conditions and the distribution of CO, CS, N_2H^+, and NH_3 in L1498 and L1517B. Astron Astrophys 416:191–212

Whithworth AP, Ward-Thompson D (2001) An empirical model for protostellar collapse. Astrophys J 547:317–322

Statistical Methods for the Detection of Flows in Active Galactic Nuclei Using X-Ray Spectral Lines

Luis F. Pérez and José M. Ramírez

Abstract Using robust statistical methods, we are able to detect and identify absorption lines in the X-ray spectra of quasars and active galactic nuclei taking as reference the Seyfert 1 galaxy NGC 3783. The high resolution spectrum of this object shows evidence of partially ionized gas outflowing from the centre of the system at velocities of $\approx 625 \pm 35$ km s^{-1}. This velocity differs from a previously reported value by ≈ 6 %. The understanding of these flows is important to draw a general picture of the feedback observed between the analyzed objects and the host galaxy.

1 Introduction

Active Galactic Nuclei (AGNs) are galaxies with a huge amount of energy ($\sim 10^{(42-48)}$ erg s^{-1}) released from very compact regions at the centre of the system. Astronomers believe that this energy is radiated away by a flowing accretion disk formed around a rotating, supermassive black hole (SMBH) with $\sim 10^{(6-10)} M_{\odot}$, covering the whole electromagnetic spectrum from the radio to the X-ray and gamma ray wavebands. AGNs are classified in blazars, quasars, Seyfert 1 and 2, according to the unification model of AGN species (Osterbrock and Pogge 1985). This model establishes that different observational classes of AGN are really the same object seen from different orientation angles.

In this chapter, we present an analysis of the X-ray spectrum of the Seyfert 1 NGC 3783, taken with the spatial observatory *Chandra*. In the Seyfert galaxy NGC

L. F. Pérez · J. M. Ramírez (✉)
Centro de Física,Instituto Venezolano de Investigaciones Científicas (IVIC),
Apartado Postal 20632, Caracas 1020-A, Venezuela
e-mail: josem@ivic.gob.ve

L. F. Pérez
e-mail: alfa_1100@hotmail.com

L. Di G. Sigalotti et al. (eds.), *Computational and Experimental Fluid Mechanics with Applications to Physics, Engineering and the Environment*, Environmental Science and Engineering,
DOI:10.1007/978-3-319-00191-3_37,

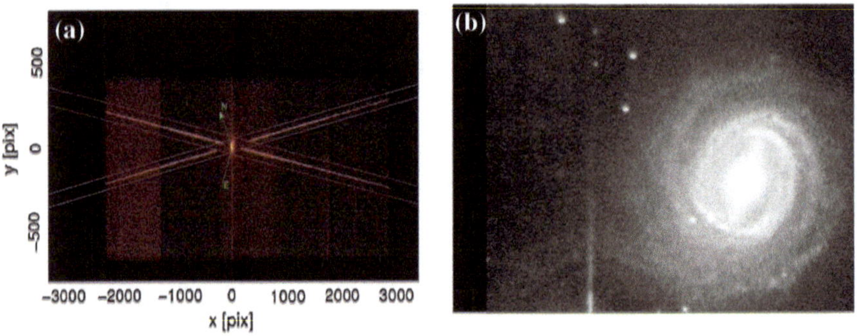

Fig. 1 **a** Image of NGC 3783 taken with the *Chandra* X-ray observatory **b** Image of the same AGN using the Hubble telescope

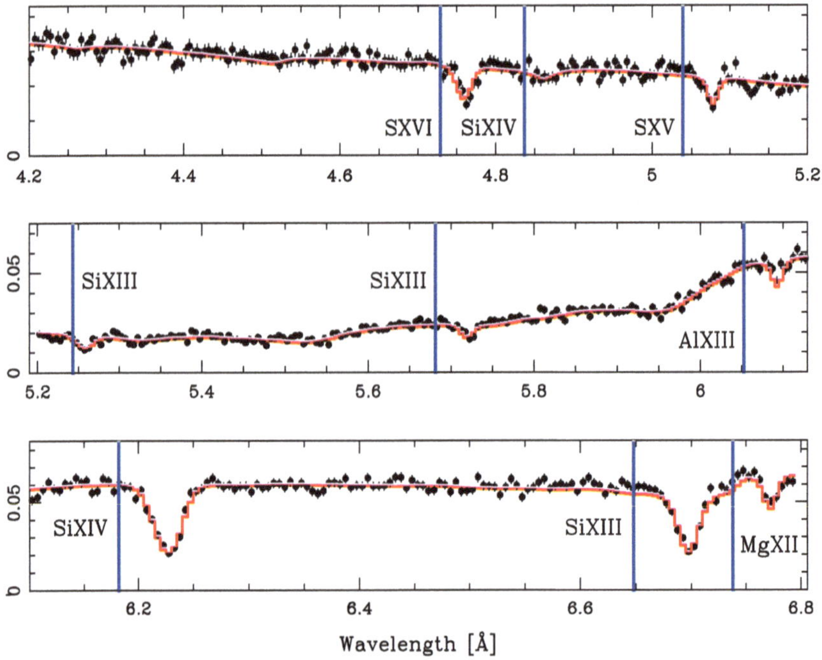

Fig. 2 X-ray spectrum of NGC 3783 in the range from 4.2 to 6.8 Å. The *vertical lines* mark the rest-frame wavelengths of the identified ions. The *solid line* shows the statistical continuum plus Gaussian model

3783, its 900 ks *Chandra* spectrum (Kaspi et al. 2002) allows for precise measurements of the radial velocities and line widths. It is seen that the shift of the spectral lines from Fe XXIII-Mg XII cover a range of velocities of $\sim$ 60–600 kms, while the lowly ionized lines Si XIII-O VII cover velocities of $\sim$ 500–1,000 kms (see Fig. 6 of Ramírez et al. (2005)). The average velocity of the warm/outflowing absorber

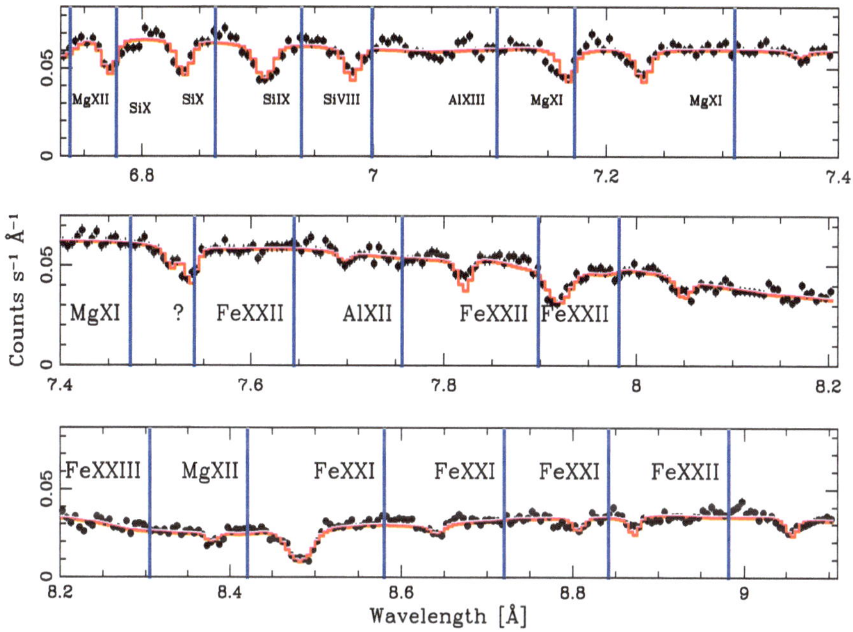

Fig. 3 X-ray spectrum of NGC 3783 in the range from 6.75 to 9.1 Å. The *vertical lines* mark the rest-frame wavelengths of the identified ions. The *solid line* shows the statistical continuum plus Gaussian model

of NGC 3783 is around ∼500 km/s. The spectrum also reveals asymmetric line profiles (Kaspi et al. 2002), showing approximately 90 % of the lines with extended blue wings. Such asymmetries were quantified by Ramírez et al. (2005). In terms of ionization, most high ionization species are seen in the short-wavelength portion of the spectrum ∼ 4–12 Å. Here, the resonant lines from Fe XXIII, Fe XXII, Fe XXI, S XVI, S XV, Si XIV, and Mg XII cover ionization parameters ξ in the range from ∼630 to ∼ 150 ergs cm s^{-1}. Figure 1 shows images of the galaxy as observed with the *Chandra* and Hubble telescopes.

2 X-Ray Spectral Lines of NGC 3783

Using specialized softwares (XSPEC 12.7.1 and CIAO 4.4), we have reduced and analyzed the data, consisting of six consecutive observations between 2000 and 2001, and then merged them to obtain a single high-resolution spectrum of 88,958 ks of exposure time. The work presented here is based on the MEG ± 1 (Medium Energy Grating) arms of the HETGS (High Energy Transmission Grating Spectrometer) on board *Chandra*. It results in 8,192 wavelength channels with 0.023 Å of instrumental resolution in the range from ∼ 4.2 to 19.2 Å. It is interesting to see the several spectral

Table 1 Line measurements

Ion (Rest λ* -Å-)	Observed wavelength (Å)	Error Error (Å)	Doppler velocity (kms^{-1})	Error (kms^{-1})	Relative velocity (kms^{-1})	Error (kms^{-1})
SXVI 4.729	4.760	0.007	1,989	444	−928	461
SXV 5.039	5.080	0.004	2,458	238	−459	269
SiXIII 5.234	5.259	0.010	2,080	573	−837	586
SiXIII 5.681	5.721	0.015	2,048	792	−869	801
AlXIII 6.053	6.094	0.003	2,245	149	−672	194
SiXIV 6.182	6.228	0.003	2,265	145	−652	191
SiX 6.778	6.841	0.017	2,759	752	−158	762
SiX 6.864	6.909	0.019	1,939	830	−978	839
SiIX 6.939	6.984	0.011	1,924	475	−993	491
SiVIII 6.999	7.061	0.023	2,631	985	−286	993
MgXII 7.106	7.170	0.010	2,702	422	−215	440
MgXI 7.310	7.371	0.079	2,553	na	−364	na
MgXI 7.473	7.511	0.022	2,481	883	−436	891
AlXII 7.757	7.826	0.023	2,639	889	−278	898
FeXXII 7.982	8.055	0.023	2,742	864	−175	873
FeXXIII 8.305	8.381	0.032	2,724	na	−193	na
MgXII 8.421	8.484	0.006	2,250	214	−667	247
FeXXI 8.580	8.643	0.027	2,201	943	−716	952
FeXXI 8.720	8.810	0.022	3,068	756	151	767
FeXXII 8.982	9.058	0.027	2,510	901	−407	910
FeXX 9.080	9.151	0.040	2,336	1,321	−581	1,327
MgXI 9.169	9.245	0.017	2,479	556	−438	570
MgIX 9.378	9.451	0.628	2,323	na	−594	na
FeXXI 9.483	9.553	0.009	2,201	285	−716	311
NeX 9.708	9.781	0.006	2,257	185	−660	223
NeX 10.239	10.321	0.030	2,391	878	−526	887
FeXVIII 10.365	10.441	0.026	2,188	752	−729	762
FeXVII 11.026	11.084	0.016	2,322	435	−595	452
NeIX 11.547	11.628	0.019	2,102	493	−815	509
FeXXII 11.780	11.861	0.051	2,058	1,298	−859	1,304
FeXXI 11.952	12.031	0.065	1,981	1,630	−936	1,635
FeXXI 11.973	12.074	0.032	2,510	801	−407	811
FeXXI 12.576	12.651	0.038	2,121	906	−796	914
FeXX 12.588	12.681	0.059	2,212	1,405	−705	1,411
FeXX 12.846	12.934	0.015	2,055	350	−862	372
FeXIX 12.946	13.035	0.020	2,054	463	−863	480
FeXVII 13.825	13.923	0.033	2,108	716	−809	726
FeXVIII 14.158	14.271	0.053	2,391	1,122	−526	1,129
FeXVIII 14.208	14.312	0.045	2,177	950	−740	958
FeXVIII 14.373	14.381	0.045	1,830	939	−1,087	947
FeXVIII 14.534	14.651	0.036	2,413	743	−504	753
OVIII 14.634	14.751	0.062	2,396	1,270	−521	1,276

(continued)

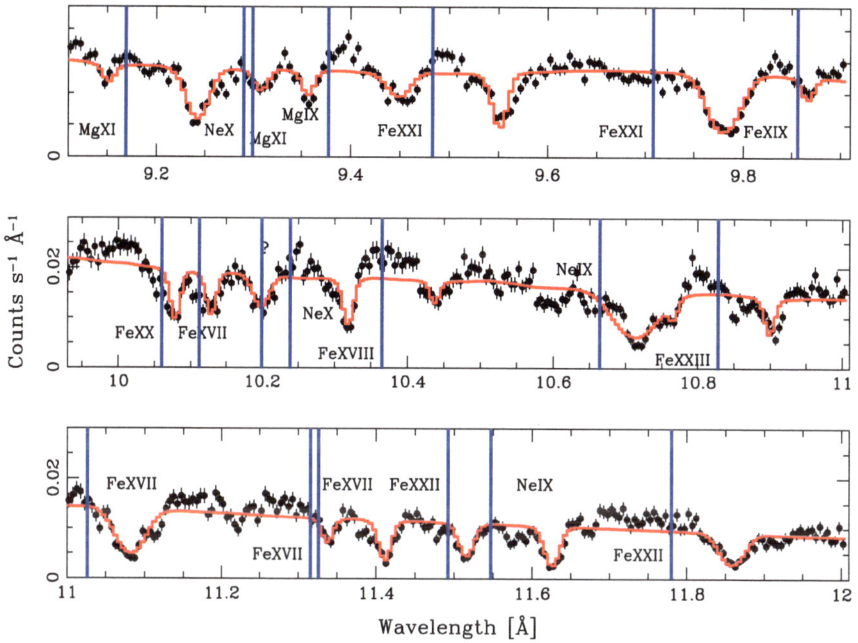

Fig. 4 X-ray spectrum of NGC 3783 in the range from 9.1 to 12.0 Å. The *vertical lines* mark the rest-frame wavelengths of the identified ions. The *solid line* shows the statistical continuum plus Gaussian model

Table 1 (continued)

Ion (Rest λ* -Å-)	Observed wavelength (Å)	Error Error (Å)	Doppler velocity (kms^{-1})	Error (kms^{-1})	Relative velocity (kms^{-1})	Error (kms^{-1})
OVIII 14.821	14.945	0.036	2,500	728	−417	739
FeXVII 15.014	15.121	0.037	2,138	739	−779	749
OVIII 15.176	15.295	0.030	2,336	593	−581	606
FeXVII 15.261	15.381	0.035	2,354	688	−563	699
OVIII 16.006	16.132	0.096	2,343	na	−574	na
OVII 17.396	17.528	0.090	2,258	1,551	−659	1,556
OVII 17.768	17.862	0.089	2,248	1,502	−669	1,507
OVIII 18.969	19.102	0.055	2,093	869	−824	878

features produced by H- and He-like ions in this band. Figures 2, 3 and 4 show the X-ray spectrum of NGC 3783 along with the theoretical global model we have used to detect and measure the line centroids.

The quality of the fit and the extraction of the errors were quantitatively measured using a χ^2-statistics (which gives a measure of the quadratic deviation between the data and the model). Furthermore, a single absorbed power-law was used for the global continuum and a Gaussian model for the characterization of the narrow-line features.

3 Results

We have measured a total number of 78 transitions of Si, S, Al, Mg, Fe, Ne and O. However, we were not able to identify about seven features associated to the measured lines in our database, thus giving the method an efficiency of $\sim 91.76\,\%$.

Once the line identification is completed, we compare the measured line centroids to the rest-frame wavelengths to obtain the radial velocities of the outflows using Doppler shift analysis. It is worth mentioning that for the velocity of the galaxy we have used a corrected value of $v_{\rm gal} = 2,917 \pm 2$ km s^{-1}.

The results are summarized in Table 1, which lists all ions identified, the measured wavelengths, and the Doppler and relative velocities observed in the flow. The errors in the wavelengths and velocities are also given.

4 Conclusions

We have detected and identified absorption lines in the X-ray spectrum of the Seyfert 1 NGC 3783 galaxy in the range from ~ 4.2 to 19.2 Å, with the purpose of measuring the average velocity of the partially ionized gas outflowing from its centre.

We find an average outflow velocity of -625 ± 35 km s^{-1}, which is of the same order of the value measured by Kaspi et al. (2002), with a difference of only 6 %. These flows have their likely origin in the accretion disk surrounding a supermassive black hole in the centre of the galaxy along with violent supernova explosions taking place in sites of star formation, and/or in strong magnetic forces that accelerate the flows to the observed velocities. Further work on other objects and more complex models would be required to improve our understanding of these systems.

Acknowledgments Some of the measurements were made during the stay of one of us (J. M. R.) at NASA Goddard Space Flight Center sponsored by the US National Science Foundation program. J. M. R. is indebted to Dr. Timothy Kallman. The great effort made by all members of the Chandra team is also kindly acknowledged.

References

Kaspi S, Brandt WN, George IM, Netzer H, Crenshaw DM, et al. (2002) The ionized gas and nuclear environment in NGC 3783. I. Time-averaged 900 kilosecond Chandra grating spectroscopy. Astrophys J 574:643–662

Osterbrock DE, Pogge RW (1985) The spectra of narrow-line Seyfert 1 galaxies. Astrophys J 297:166–176

Ramírez JM, Bautista M, Kallman T (2005) Line asymmetry in the Seyfert galaxy NGC 3783. Astrophys J 627:166–176

Reproducing the X-Ray Soft Step @ 0.9 keV Observed in the Spectrum of Ark 564 Using Reflection Models

José M. Ramírez and Snell Rojas

Abstract Using reliable atomic data, we attempt to reproduce the global 100 ks X-ray spectrum of the narrow-line Seyfert 1 Galaxy Ark 564, observed with the Low Energy Transmission Grating Spectrometer (LETGS) on board *Chandra*. In order to do this, we use accretion disk and reflection flow models.

1 Introduction

Narrow-line Seyfert 1 galaxies (NLS1s) are a subclass of the "normal" Seyfert 1 galaxies, which exhibit in their optical spectra H_β lines with full width at half maximum FWHM $<$ 2000 km $s^{-1}$and strong Fe II emission (Osterbrock and Pogge 1985; Boroson and Green 1992). X-ray observations of these objects reveal extreme spectral and temporal properties, with Ark 564 being perhaps the more representative and the most observed Seyfert of this category. NLS1s show a "soft X-ray excess emission", which has been the subject of an intense debate for at least two decades. In particular, Pounds et al. (1995) proposed that this soft excess might be produced by an optically thick emission from the accretion disk. However, the temperature of this optically thick region ($kT \sim$ 0.1–0.15 keV) is to high to support a disk around a supermassive black hole (SMBH).

On the other hand, the soft excess is well fitted with a blackbody, which has a roughly constant temperature of $\sim$ 0.1–0.2 keV over a wide range of SBH masses (Gierliński and Done 2004). However, if it is thermal, this temperature could be explained by a slim disk in which the temperature is raised by photon trapping

J. M. Ramírez (✉) · S. Rojas
Centro de Física, Instituto Venezolano de Investigaciones Científicas IVIC, Apartado Postal 20632, Caracas1020-A, Venezuela
e-mail: josem@ivic.gob.ve

S. Rojas
e-mail: snell0808@hotmail.com

L. Di G. Sigalotti et al. (eds.), *Computational and Experimental Fluid Mechanics with Applications to Physics, Engineering and the Environment*, Environmental Science and Engineering, DOI:10.1007/978-3-319-00191-3_38, © Springer International Publishing Switzerland 2014

(Czerny et al. 2003; Gierliński and Done 2004; Crummy et al. 2006). A further explanation for the soft X-ray step is related to the atomic physics and considers that the detected X-ray emission could be dominated by reflection off the walls of ring-like structures formed by clumping of cold and dense material owing to instabilities of the accretion disk (Fabian et al. 2002).

2 Observation of Ark 564

After having properly extracted both the source and the background spectra from each arm of the dispersed spectrum as received by the arrange HRC + LETG (i.e., High Resolution Camera plus Low Energy Transmission Grating on board Chandra), we have merged them to increase the signal-to-noise ratio (S/N) of the final spectrum. The analysis presented here is entirely based on this merged spectrum. The effective areas (EA) for the dispersion orders 2–10 employed in the fitting procedures were built up using the standard CIAO task `mkgarf` and `fullgarf`. For the first order we used the corrected EA of Beuermann et al. (2006), while EA order 1–10 (positive and negative orders) were added together for use with the corresponding merged spectrum. A log of the observation is shown in Table 1.

Figure 1 shows the best-fit single power-law model to the data, where an emission excess is clearly seen. We note that this is qualitatively similar to the soft excess reported in Comastri et al. (2001); Vignali et al. (2004) and Papadakis et al. (2007). In order to investigate the nature of this soft excess (or soft step), which is still subject to debate (Done and Nayakshin 2007; Ramírez 2013), we also fit several other components to describe the soft band. The extra components are: the single blackbody model [`TBabs*(zpowerlw + bb)`] and the single accretion disk reflection model model [`TBabs*(zpowerlw + xillver)`] (García and Kallman 2010). All models account for the galactic absorption in the line-of-sight towards Ark 564, of $N_H = 6.4 \times 10^{20}\mathrm{cm}^{-2}$, using the TBabs model included in xspec (Wilms et al. 2000).

Table 1 We merge the dispersion orders +1 and −1 to increase the signal-to-noise (S/N)

Log of the LETGS observation of Ark 564	
Telescope	Chandra
Instrument	HRC + LETG
Channel type	PHA
Source exposure time	199 ks

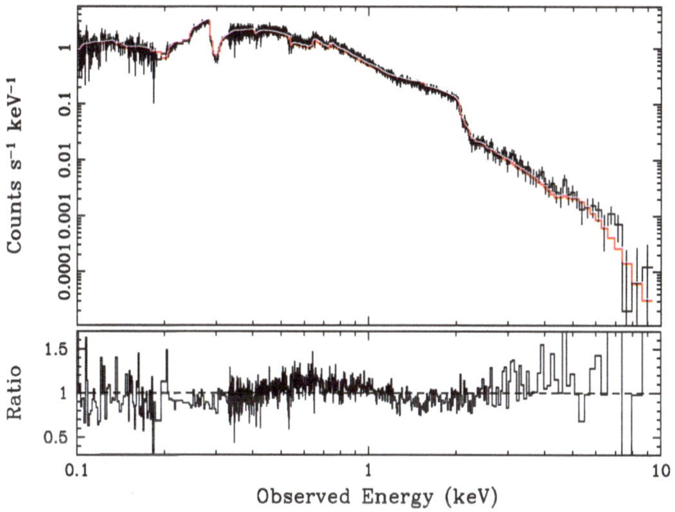

Fig. 1 Single absorbed power-law best-fit to the LETGS X-ray spectrum of Ark 564 (*points with error bars*). With this model we are able to observe significant residuals around 0.9 keV and at high energies (5–8 keV), which are believed to be an "excess" in emission

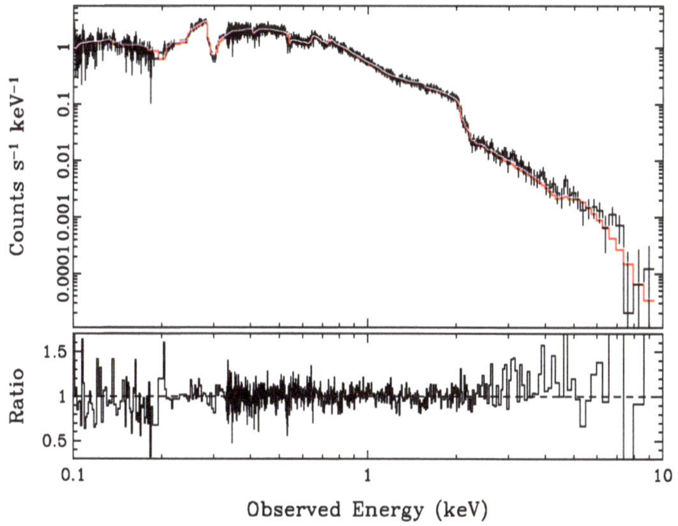

Fig. 2 The solid line is an absorbed power-law plus a black body model plotted over the LETGS X-ray spectrum of Ark 564 (*points with error bars*). There is only a slight improvement over the power-law. See the text for details

3 Results

We have measured in quantitative terms the quality of the fit using χ^2-statistics. This gives a measure of the quadratic deviation between the data and the model. In the case of the single power-law, we obtain $\chi^2/dof = 1.89$, where dof is the degree of freedom, defined as the number of data channels minus the number of parameters in the model. When a blackbody model is added to the absorbed power-law, in an attempt to characterize the residual at soft energies (~0.9 keV), the power-law fit improves only slightly with $\chi^2/dof = 1.87$. The results are shown in Fig. 2.

Given that the temperature of the optically thick region ($kT \sim$ 0.1–0.15 keV) is to high for supporting a disk around a SMBH, it is necessary to use a more sophisticated model. Accretion disks around SMBHs reflect the emission from the central region, re-processing high-energy photons, and producing emission and absorption features. These features form complexes, which actually improve the description of the data around 0.9 keV, with $\chi^2/d_{of} = 1.86$. Tables 2 and 3 show a comparison between `TBabs*zpowerlw`, `TBabs*(zpowerlw + bb)`, and `TBabs*(zpowerlw + xillver)` with their respective parameter values. Figure 3 shows the best-fit physical model, where the solid line was obtained using the partially ionized accretion disk model `TBabs*(zpowerlw + xillver)`.

Table 2 Parameter values of the model TBabs*zpowerlw

Model: TBabs*zpowerlw			
Component	Parameter	Unit	Value
TBabs	nH	10^{22}	0.064
powerlaw	PhoIndex		2.886
powerlaw	norm		0.012

Table 3 Parameter values of the models `TBabs*(zpowerlw + bb)` and `TBabs*(zpowerlw + xillver)`

Model: TBabs*(zpowerlw + bb)				TBabs*(zpowerlw + xillver)			
Component	Parameter	Unit	Value	Component	Parameter	Unit	Value
TBabs	nH	10^{22}	0.064	TBabs	nH	10^{22}	0.064
powerlaw	PhoIndex		2.894	zpowerlw	PhoIndex		2.861
powerlaw	norm		0.0121	zpowerlw	Redshift		0.024
bbody	kT	keV	0.137	zpowerlw	norm		0.088
bbody	norm		2.804×10^{-4}				
				xillver	log(xi)		1.931
				xillver	Fe/solar		1.000
				xillver	Redshift		0.024

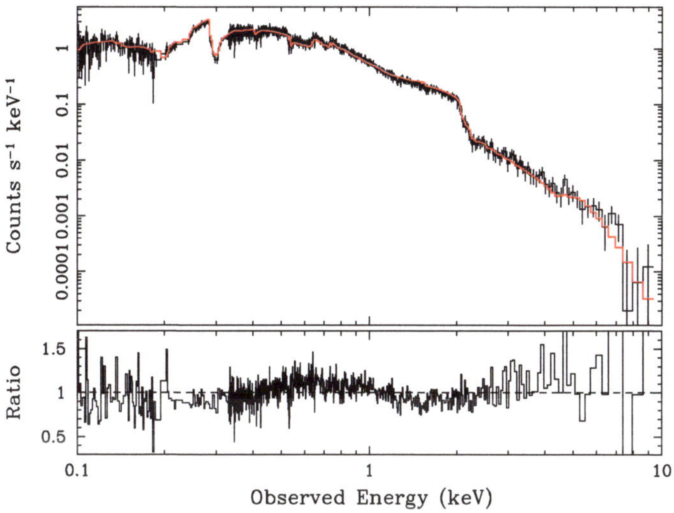

Fig. 3 The solid line is an absorbed `TBabs*(zpowerlw + xillver)` model (García and Kallman 2010), plotted over the LETGS X-ray spectrum of Ark 564 (*points with error bars*). This is the best-fit physical model. See the text for details

4 Conclusions

The main conclusions from this work are:

- We have studied in detail the global (0.1–10 keV) LETGS spectrum of Ark 564. Reproducing this spectrum will certainly require more sophisticated and complex models than just semi-empirical models, like the power-law plus blackbody model, or put-by-hands absorption edges (Matsumoto et al. 2004; Ramírez et al. 2008; Ramírez 2013).
- The best-global-fit in this work corresponded to the partially ionized accretion disk model `Tbabs*(zpowerlw + xillver)` (García and Kallman 2010), which uses a self-consistent photoionization model with reliable atomic data. This fit includes a more comprehensive set of absorption and emission lines than it has previously been tried for this kind of object.
- Future work must focus on the use of more complex models, which combined with high-resolution/quality observations are expected to improve the statistics.

Acknowledgments The data used in this work were taken during the stay of one of us (J. M. R.) at MPE sponsored by the *Chandra* GTO program. We are indebted to P. Predehl, V. Burwitz, and S. Komossa for kindly supporting the proposal of this observation. The efforts made by all members of the *Chandra* team are also kindly acknowledged.

References

Beuermann K, Burwitz V, Rauch T (2006) Establishing HZ43 A, Sirius B, and RXJ185635-3754 as soft X-ray standards: a cross-calibration between Chandra LETG + HRC-S, the EUVE spectrometer, and the ROSAT PSPC. Astron Astrophys 458:541–552

Boroson TA, Green RF (1992) The emission-line properties of low-redshift quasi-stellar objects. Astrophys J Suppl Ser 80:109–135

Comastri A, Stirpe GM, Vignali C, Brandt WN, Leighly KM, et al. (2001) BeppoSAX observations of narrow-line Seyfert 1 galaxies. II. Ionized iron features in Arakelian 564. Astron Astrophys 365:400–408

Crummy J, Fabian AC, Gallo L, Ross RR (2006) An explanation for the soft X-ray excess in active galactic nuclei. Mon Not R Astron Soc 365:1067–1081

Czerny B, Nikołajuk M, Różańska A, Dumont A-M, Loska Z, Zycki PT (2003) Universal spectral shape of high accretion rate AGN. Astron Astrophys 412:317–329

Done C, Nayakshin S (2007) Can the soft excess in AGN originate from disc reflection? Mon Not R Astron Soc 377:L59–L63

Fabian AC, Ballantyne DR, Merloni A, Vaughan S, Iwasawa K, Boller T (2002) How the X-ray spectrum of a narrow-line Seyfert 1 galaxy may be reflection-dominated. Mon Not R Astron Soc 331:L35–L39

García J, Kallman TR (2010) X-ray reflected spectra from accretion disk models. I. Constant density atmospheres. Astrophys J 718:695–706

Gierliński M, Done C (2004) Is the soft excess in active galactic nuclei real? Mon Not R Astron Soc 349:L7–L11

Matsumoto C, Leighly KM, Marshall HL (2004) A Chandra HETGS observation of the narrow-line Seyfert 1 galaxy Arakelian 564. Astrophys J 603:456–462

Osterbrock DE, Pogge RW (1985) The spectrum of narrow-line Seyfert 1 galaxies. Astrophys J 297:166–176

Papadakis IE, Brinkmann W, Page MJ, McHardy I, Uttley P (2007) XMM-Newton observation of the NLS1 galaxy Ark 564. I. Spectral analysis of the time-average spectrum. Astron Astrophys 461:931–942

Pounds KA, Done C, Osborne JP (1995) RE 1034 + 39: a high-state Seyfert galaxy? Mon Not R Astron Soc 277:L5–L10

Ramírez JM, Komossa S, Burwitz V, Mathur S (2008) Chandra LETGS spectroscopy of the quasar MR 2251–178 and its warm absorber. Astrophys J 681:965–981

Ramírez JM (2013) Chandra LETGS observation of the variable NLS1 galaxy Ark 564. Astron Astrophys 551:A95

Vignali C, Brandt WN, Boller T, Fabian AC, Aughan VS (2004) Arakelian 564: an XMM-Newton view. Mon Not R Astron Soc 347:854–860

Wilms J, Allen A, McCray R (2000) On the absorption of X-rays in the interstellar medium. Astrophys J 542:914–924

Dynamics of Relativistic, Dissipative and Anisotropic Self-Gravitating Fluids

Orlenys Troconis

Abstract This chapter deals with the study of dissipative, locally anisotropic, and spherically symmetric self-gravitating fluids. The analysis is based on a full causal approach, where the dynamical equations are coupled to causal transport equations for the heat flux, shear, and bulk viscosity in the context of the Müller-Israel-Stewart theory by including the thermodynamic viscous/heat coupling coefficients.

1 Introduction

Since the seminal paper by Oppenheimer and Snyder (1939), most of the work devoted to study the gravitational collapse from the relativistic point of view has dealt with spherically symmetric fluid distributions, because this symmetry describes with a good approximation many physical real situations.

Many existing models have considered the fluid to be locally isotropic. However, small deviations from this condition may affect the system evolution, eventually leading to an explosive collapse (Herrera 1996). Here we assume that the fluid is locally anisotropic and dissipative due to the effects of viscosity (shear and bulk viscosity) (Herrera and Santos 1997; Herrera et al. 2009). This latter assumption is justified because the gravitational collapse is by itself a naturally dissipative process (Herrera and Santos 2004), where the collapse of a massive star represents a good example. However, it seems that the only plausible mechanism to carry away the bulk of the binding energy of the collapsing star, leading to a neutron star or black hole, is the neutrino emission (Kazanas and Schramm 1979). In fact, a characteristic feature of the evolution of a massive star is the emission of massless particles (photons and/or neutrinos).

O. Troconis (✉)
Escuela de Física, Universidad Central de Venezuela, Avenida Los Illustres, Los Changuaramos, Apartado Postal 20513, Caracas 1020-A, Venezuela
e-mail: otroconis@fisica.ciens.ucv.ve

L. Di G. Sigalotti et al. (eds.), *Computational and Experimental Fluid Mechanics with Applications to Physics, Engineering and the Environment*, Environmental Science and Engineering, DOI:10.1007/978-3-319-00191-3_39,

If the particles that transport energy have a mean free path much smaller than the typical length of the object, then the diffusion approximation can be used to describe the transport of energy. For instance, for a main sequence star like the sun, the mean free path of photons at the centre is of the order of 2 cm, which is much smaller than the sun's radius, and hence radiation transfer is very well described by a diffusion process (Kippenhann and Weigert 1990). On the other hand, observational data collected from the supernova 1987A indicates that the regime of radiation transport prevailing during the emission process is the diffusion approximation (Lattimer 1988). Conversely, if the mean free path of particles is larger than the typical length of the object, then the free streaming approximation may be used. In this work, we shall include simultaneously both limiting cases.

The dynamical equations governing the relativistic collapse in the adiabatic and streaming out approximation were developed by Misner and Sharp (1964) and Misner (1965), respectively. An extension of these equations was presented by Herrera and Santos (2004), who included the effects of dissipation in the form of a radial heat flux, satisfying a causal transport equation. In their case, however, the fluid was assumed to be shear-free and therefore viscous dissipation as well as the thermodynamic viscous/heat coupling coefficients were not taken into account. The importance of viscosity in the evolution of self-gravitating systems has been outlined by Collins and Wainwright (1983) and references therein. In a more recent work, Di Prisco et al. (2007) introduced the effects of shear viscosity. However, in their model the transport equation corresponds to the standard Eckart theory of relativistic irreversible thermodynamics (Eckart 1940; Landau and Lifshitz 1968), which predicts that the disturbances propagate at infinite velocity. In addition, their model also neglects the viscous/heat coupling coefficients.

The aim of this chapter is to present a dynamical description of the gravitational collapse within the framework of the Misner approach, for a dissipative, spherically symmetric, relativistic fluid, taking into account both limiting cases of the radiative transport (i.e., diffusion and streaming out) as well as the shear and bulk viscosity and including the thermodynamic viscous/heat coupling coefficients. In this way, the transport equations will include all relevant dissipative variables from a causal point of view.

2 Dynamics of the Viscous and Dissipative Gravitational Collapse

In this section we describe a viscous, dissipative, and self-gravitating fluid bounded by a spherical surface Σ. The description is made from a causal point of view by coupling the dynamical equations to the causal transport equations for the heat flux, the shear, and bulk viscosity in the context of the Müller-Israel-Stewart theory (Müller 1967; Israel 1976; Israel and Stewart 1976) and by including the viscous/heat coupling coefficients.

We start by noting that the line element in co-moving coordinates inside Σ, can be written as:

$$ds_-^2 = -A^2dt^2 + B^2dr^2 + (Cr)^2(d\theta^2 + \sin^2\theta d\phi^2), \tag{1}$$

where A, B, and C are functions of the radius r and time t and are assumed to be positive. The coordinates are identified as $x^0 = t$, $x^1 = r$, $x^2 = \theta$ and $x^3 = \phi$.

The system consists of a fluid with energy density ρ, pressure P, radial heat flux q^α, radiation density ε, shear viscosity $\pi_{\alpha\beta}$, and bulk viscosity Π. Therefore, the energy momentum tensor inside Σ has the form

$$T^-_{\alpha\beta} = (\rho + P + \Pi)\, u_\alpha u_\beta + (P + \Pi)g_{\alpha\beta} + q_\alpha u_\beta + q_\beta u_\alpha + \varepsilon l_\alpha l_\beta + \pi_{\alpha\beta}, \tag{2}$$

where u^α is the four-velocity of the fluid and l^α is the radial null four-vector. These satisfy the relations:

$$\begin{aligned} &u^\alpha u_\alpha = -1, u^\alpha q_\alpha = 0, l^\alpha u_\alpha = -1, l^\alpha l_\alpha = 0, \\ &\pi_{\mu\nu}u^\nu = 0, \pi_{[\mu\nu]} = 0, \pi^\alpha_\alpha = 0. \end{aligned} \tag{3}$$

Furthermore, since we have assumed comoving coordinates then

$$u^\alpha = A^{-1}\delta^\alpha_0, \quad q^\alpha = qB^{-1}\delta^\alpha_1, \quad l^\alpha = A^{-1}\delta^\alpha_0 + B^{-1}\delta^\alpha_1, \tag{4}$$

where q is a function of r and t. The shear viscosity components are:

$$\pi_{0\alpha} = 0 \;\; , \pi^1_1 = -2\pi^2_2 = -2\pi^3_3 \; , \tag{5}$$

which can be written in compact form as

$$\pi_{\alpha\beta} = \Omega(\chi_\alpha\chi_\beta - \frac{1}{3}h_{\alpha\beta}), \tag{6}$$

where $\Omega = \frac{3}{2}\pi^1_1$, χ^α is a unit four-vector along the radial direction and $h_{\alpha\beta}$ is the projector onto the hypersurface orthogonal to the four-velocity, satisfying the relations

$$\chi^\alpha\chi_\alpha = 1, \quad \chi^\alpha u_\alpha = 0, \quad \chi^\alpha = B^{-1}\delta^\alpha_1, \tag{7}$$

$$h_{\alpha\beta} = g_{\alpha\beta} + u_\alpha u_\beta. \tag{8}$$

In the standard irreversible thermodynamics (Maartens 1996; Chan 2000), we have that

$$\pi_{\alpha\beta} = -2\eta\sigma_{\alpha\beta}, \quad \Pi = -\zeta\Theta, \tag{9}$$

where η and ζ denote the coefficient of shear and bulk viscosity, respectively, $\sigma_{\alpha\beta}$ is the shear tensor, and Θ is the expansion. However, the irreversible thermodynamics is not a causal theory, because it predicts an infinite velocity for the propagation of information. Since we are interested in a full causal picture of the dissipative variables, we shall then use the transport equations as derived in the context of the Müller-Israel-Stewart theory.

3 Dynamical Equations

The non-trivial components of the Bianchi identities $(T^{-\alpha\beta})_{;\beta} = 0$ lead to

$$\begin{aligned} T^{-\mu\nu}_{\ \ ;\nu} u_\mu = & -\frac{1}{A}(\dot{\rho} + \dot{\varepsilon}) - \frac{1}{B}(q' + \varepsilon') - 2(q+\varepsilon)\frac{(ACr)'}{ABCr} + \\ & -\frac{2}{A}\frac{\dot{C}}{C}\left(\rho + P + \Pi + \varepsilon - \frac{\Omega}{3}\right) - \frac{1}{A}\frac{\dot{B}}{B}\left(\rho + P + \Pi + 2\varepsilon + \frac{2}{3}\Omega\right) \\ = & \, 0 \end{aligned} \tag{10}$$

and

$$\begin{aligned} T^{-\mu\nu}_{\ \ ;\nu} \chi_\mu = & \frac{1}{A}(\dot{q} + \dot{\varepsilon}) + \frac{2}{A}\frac{(BC)^{\cdot}}{BC}(q+\varepsilon) \\ & + \frac{1}{B}\left(P' + \Pi' + \varepsilon' + \frac{2}{3}\Omega'\right) + \frac{1}{B}\frac{A'}{A}\left(\rho + P + \Pi + 2\varepsilon + \frac{2}{3}\Omega\right) \\ & + \frac{2}{B}\frac{(Cr)'}{Cr}(\varepsilon + \Omega) = 0. \end{aligned} \tag{11}$$

where the primes mean derivation with respect to r and the dots mean derivation with respect to time.

In order to study the dynamical properties of the system, let us first introduce the proper time derivative, D_T, and the proper radial derivative, D_R, given by

$$D_T = \frac{1}{A}\frac{\partial}{\partial t}, \tag{12}$$

and

$$D_R = \frac{1}{R'}\frac{\partial}{\partial r}, \tag{13}$$

respectively, where $R = Cr$ defines the proper radius of a spherical surface inside Σ. We then calculate the velocity U of the collapsing fluid as the variation of the proper radius with respect to the proper time, i.e.,

$$U = r D_T C < 0 \ \text{(in the case of collapse)}. \tag{14}$$

Recalling that the mass function introduced by Misner and Sharp (1964) is defined by

$$m = \frac{(Cr)^3}{2} R_{23}{}^{23} = \frac{Cr}{2}\left\{\left(\frac{r\dot{C}}{A}\right)^2 - \left[\frac{(Cr)'}{B}\right]^2 + 1\right\}, \tag{15}$$

and using Eqs. (10) and (11) in combination with the definitions (12–14), we find, after some calculations, the following equation (Herrera et al. 2009):

$$\begin{aligned}
&\left(\rho + P + \Pi + 2\varepsilon + \frac{2}{3}\Omega\right) D_T U = \\
&-\left(\rho + P + \Pi + 2\varepsilon + \frac{2}{3}\Omega\right)\left[\frac{m}{R^2} + 4\pi R\left(P + \Pi + \varepsilon + \frac{2}{3}\Omega\right)\right] \\
&- E^2\left[D_R\left(P + \Pi + \varepsilon + \frac{2}{3}\Omega\right) + \frac{2}{R}(\varepsilon + \Omega)\right] \\
&- E\left[D_T q + D_T \varepsilon + 4(q + \varepsilon)\frac{U}{R} + 2(q + \varepsilon)\sigma\right].
\end{aligned} \tag{16}$$

where the quantity E is

$$E \equiv \frac{(Cr)'}{B} = \left[1 + U^2 - \frac{2m(t,r)}{Cr}\right]^{1/2}. \tag{17}$$

Equation (16) is the dynamical equation. It shows the equivalence principle because the expression inside the round brackets on the left-hand side is identical to the one inside the first round brackets on the right-hand side, and it represents the inertial (or passive gravitational) mass. The first term on the right-hand side of the equation is the gravitational force. Note that the expression inside the square brackets in the first term on the left side, which represents the active gravitational mass, is affected by dissipation. Furthermore, inside the second square brackets there are two different contributions. The first one is the gradient of the total effective pressure (which includes the radiation pressure and the influence of the shear and bulk viscosity). The second contribution comes from the local anisotropy in the pressure induced by the radiation pressure and the shear viscosity. Finally, the last square brackets contain different contributions due to dissipative processes. In particular, the third contribution term is positive in the case of collapse (i.e., $U < 0$), showing that the heat flux $q > 0$ and the radiation $\varepsilon > 0$ both reduce the rate of collapse. The last term describes the coupling of the dissipative flux with the shear of the fluid.

4 Transport Equation

The main purpose of this work is to provide a full causal description of the viscous dissipative gravitational collapse. This implies that all dissipative variables must satisfy the corresponding transport equations derived from the causal thermodynamics, using the Müller-Israel-Stewart second-order phenomenological theory for dissipative fluids (Müller 1967; Israel 1976; Israel and Stewart 1976).

The general expression for the entropy four-current in the context of the Müller-Israel-Stewart theory reads as follows

$$S^{\mu} = Snu^{\mu} + \frac{q^{\mu}}{T} - (\beta_0 \Pi^2 + \beta_1 q_{\nu} q^{\nu} + \beta_2 \pi_{\nu\kappa}\pi^{\nu\kappa})\frac{u^{\mu}}{2T} + \frac{\alpha_0 \Pi q^{\mu}}{T} + \frac{\alpha_1 \pi^{\mu\nu} q_{\nu}}{T}, \quad (18)$$

where n is the particle number density, T is the temperature, $\beta_A(\rho, n)$ are thermodynamic coefficients for different contributions to the entropy density, and $\alpha_A(\rho, n)$ are the thermodynamic viscous/heat coupling coefficients.

To derive the transport equation for each dissipative variable, we follow the standard procedure starting from the requirement that $S^{\alpha}_{\alpha} \geq 0$ (Herrera et al. 2009). After some calculations, we arrive at the transport equations:

$$\begin{aligned} \tau_0 \dot{\Pi} = & -\left(\zeta + \frac{\tau_0}{2}\Pi\right) A\Theta + \frac{A}{B}\alpha_0 \zeta \left[q' + q\left(\frac{A'}{A} + \frac{2(rC)'}{rC}\right)\right] \\ & - \Pi \left[\frac{\zeta T}{2}\left(\frac{\tau_0}{\zeta T}\right)^{\cdot} + A\right], \end{aligned} \quad (19)$$

$$\begin{aligned} \tau_1 \dot{q} = & -\frac{A}{B}\kappa T' \left(1 + \alpha_0 \Pi + \frac{2}{3}\alpha_1 \Omega\right) \\ & - T\frac{A}{B}\kappa \left[\frac{A'}{A} - \alpha_0 \Pi' - \frac{2}{3}\alpha_1 \left(\Omega' + \left(\frac{A'}{A} + 3\frac{(rC)'}{rC}\right)\Omega\right)\right] \\ & - q\left[\frac{\kappa T^2}{2}\left(\frac{\tau_1}{\kappa T^2}\right)^{\cdot} + \frac{\tau_1}{2}A\Theta + A\right], \end{aligned} \quad (20)$$

and

$$\tau_2 \dot{\Omega} = -2\eta A\sigma + 2\eta \alpha_1 \frac{A}{B}\left(q' - q\frac{(rC)'}{rC}\right) - \Omega\left[\eta T\left(\frac{\tau_2}{2\eta T}\right)^{\cdot} + \frac{\tau_2}{2}A\Theta + A\right]. \quad (21)$$

where τ_0, τ_1, and τ_2 are the relaxation times for the bulk viscosity, the heat flux, and the shear viscosity, respectively, given by

$$\tau_0 = \zeta\beta_0, \ \tau_1 = \kappa T \beta_1, \ \tau_2 = 2\eta\beta_2. \quad (22)$$

Now coupling the transport equations to the dynamical Eq. (16), we find that

$$\left(\rho + P + \Pi + 2\varepsilon + \frac{2}{3}\Omega\right)(1-\Lambda)D_T U = (1-\Lambda)F_{grav} + F_{hyd}$$
$$+ \frac{\kappa E^2}{\tau_1}\left\{D_R T\left(1 + \alpha_0 \Pi + \frac{2}{3}\alpha_1 \Omega\right) - T\left[\alpha_0 D_R \Pi + \frac{2}{3}\alpha_1\left(D_R\Omega + \frac{3}{R}\Omega\right)\right]\right\}$$
$$+ E\left[\frac{\kappa T^2 q}{2\tau_1}D_T\left(\frac{\tau_1}{\kappa T^2}\right) - D_T\varepsilon\right] - E\left[\left(\frac{3q}{2} + 2\varepsilon\right)\Theta - \frac{q}{\tau_1} - 2(q+\varepsilon)\frac{U}{R}\right], \tag{23}$$

where F_{grav} and F_{hyd} represent the gravitational force and the hydrodynamical force, respectively, with

$$F_{grav} = -\left(\rho + P + \Pi + 2\varepsilon + \frac{2}{3}\Omega\right) \times \left[m + 4\pi\left(P + \Pi + \varepsilon + \frac{2}{3}\Omega\right)R^3\right]\frac{1}{R^2}, \tag{24}$$

$$F_{hyd} = -E^2\left[D_R\left(P + \Pi + \varepsilon + \frac{2}{3}\Omega\right) + 2(\varepsilon + \Omega)\frac{1}{R}\right], \tag{25}$$

and

$$\Lambda = \frac{\kappa T}{\tau_1}\left(\rho + P + \Pi + 2\varepsilon + \frac{2}{3}\Omega\right)^{-1}\left(1 - \frac{2}{3}\alpha_1\Omega\right). \tag{26}$$

At this point, we can express Θ in terms of the dissipative variables using Eq. (19) and inserting the result back into Eq. (23), to obtain

$$\left(\rho + P + \Pi + 2\varepsilon + \frac{2}{3}\Omega\right)(1-\Lambda+\Delta)D_T U = (1-\Lambda+\Delta)F_{grav} + F_{hyd}$$
$$+ \frac{\kappa E^2}{\tau_1}\left\{D_R T\left(1 + \alpha_0 \Pi + \frac{2}{3}\alpha_1 \Omega\right) - T\left[\alpha_0 D_R \Pi + \frac{2}{3}\alpha_1\left(D_R\Omega + \frac{3}{R}\Omega\right)\right]\right\}$$
$$- E^2\left(\rho + P + \Pi + 2\varepsilon + \frac{2}{3}\Omega\right)\Delta\left(\frac{D_R q}{q} + \frac{2q}{R}\right)$$
$$+ E\left[\frac{\kappa T^2 q}{2\tau_1}D_T\left(\frac{\tau_1}{\kappa T^2}\right) - D_T\varepsilon\right] + E\left[\frac{q}{\tau_1} + 2(q+\varepsilon)\frac{U}{R}\right]$$
$$+ E\frac{\Delta}{\alpha_0\zeta q}\left(\rho + P + \Pi + 2\varepsilon + \frac{2}{3}\Omega\right)\left\{\left[1 + \frac{\zeta T}{2}D_T\left(\frac{\tau_0}{\zeta T}\right)\right]\Pi + \tau_0 D_T \Pi\right\}, \tag{27}$$

where

$$\Delta = \alpha_0\zeta q\left(\rho + P + \Pi + 2\varepsilon + \frac{2}{3}\Omega\right)^{-1}\left(\frac{3q + 4\varepsilon}{2\zeta + \tau_0\Pi}\right). \tag{28}$$

Equation (27) allows us to describe the dynamics of gravitational collapse in terms of all the dissipative variables, including the viscous/heat coupling coefficients for a full causal approach.

5 Conclusions

In this chapter, we have studied the dynamics of relativistic, dissipative, and anisotropic self-gravitating fluids under the assumption of spherical symmetry. The analysis was based on a full causal approach, where the dynamical equations are coupled to causal transport equations for the heat flux, shear, and bulk viscosity in the context of the Müller-Israel-Stewart theory, including the thermodynamic viscous/heat coupling coefficients.

We have derived the equation that governs the gravitational collapse of a spherically symmetric object by including all the dissipative variables. The equation shows how the active gravitational mass is affected by dissipation, while local anisotropy in the pressure is induced by the radiation pressure and the shear viscosity. In addition, we find that the rate of collapse is reduced by radiation and the flux of heat and that the viscous/heat coupling coefficients affect the inertial (or passive gravitational) mass.

Acknowledgments I thank my advisors Luis Herrera Cometta and Alicia Di Prisco for their kind support. I also extend my acknowledgements to the organizing committee of the first workshop of the Venezuelan Society of Fluid Mechanics (FLUIDOS2012) for financial support and hospitality during the event.

References

Chan R (2000) Radiating gravitational collapse with shear viscosity. Mon Not R Astron Soc 316:588–604

Collins CB, Wainwright J (1983) Role of shear in general-relativistic cosmological and stellar models. Phys Rev D 27:1209–1218

Eckart C (1940) The thermodynamics of irreversible processes. III. Relativistic theory of the simple fluid. Phys Rev 58:919–924

Herrera L (1996) II Escuela Venezolana de Relatividad. Universidad de Los Andes, Mérida, Venezuela, Campos y Astrofísica

Herrera L, Santos NO (1997) Local anisotropy in self-gravitating systems. Phys Rep 286:53–130

Herrera L, Santos NO (2004) Dynamics of dissipative gravitational collapse. Phys Rev D 70:084004

Herrera L, Di Prisco A, Fuenmayor E, Troconis O (2009) Dynamics of viscous dissipative gravitational collapse: a full causal approach. Int J Mod Phy D 18:129–145

Israel W (1976) Nonstationary irreversible thermodynamics: a causal relativistic theory. Annals Phys 100:310–331

Israel W, Stewart J (1976) Thermodynamics of nonstationary and transient effects in a relativistic gas. Phys Lett A 58:213–215

Kazanas D, Schramm DN (1979) Neutrino competition with gravitational radiation during collapse. In: Smarr L (ed) Sources of gravitational radiation. Cambridge University Press, Cambridge
Kippenhann R, Weigert A (1990) Stellar structure and evolution. Springer, Berlin
Landau LD, Lifshitz EM (1968) Fluid mechanics. Pergamon Press, Oxford
Lattimer JM (1988) Supernova theory and the neutrinos from SN1987a. Nucl Phys A 478:199–217
Maartens R (1996) Causal thermodynamics in relativity. arXiv:astro-ph/9609119.
Misner CW, Sharp DH (1964) Relativistic equations for adiabatic, spherically symmetric gravitational collapse. Phys Rev 136:571–576
Misner CW (1965) Relativistic equations for soherical gravitational collapse with escaping neutrinos. Phys Rev 137:1360–1364
Müller I (1967) Zum Paradoxon der Wärmeleitungstheorie. Zeitschrift für Physik 198:329–344
Oppenheimer JR, Snyder H (1939) On continued gravitational contraction. Phys Rev 56:455–459
Di Prisco A, Herrrera L, Le Denmat G, MacCallum MAH, Santos NO (2007) Nonadiabatic charged spherical gravitational collapse. Phys Rev D 76:064017

Hydrodynamic Version of the Equation of Motion of a Charged Complex Scalar Field

Mario A. Rodríguez-Meza and Tonatiuh Matos

Abstract In this chapter we derive the equation for a charged complex scalar field in its hydrodynamic form. This is done by re-writing the Klein-Gordon (KG) equation for the complex scalar field as a new Gross-Pitaevskii (GP)-like equation. In particular, we use as the potential of the scalar field the Mexican-hat potential, and assume that the field is in a thermal bath with a one loop contribution. We interpret the new GP equation as a finite temperature generalization of the GP equation for a charged field. From its hydrodynamic form, we derive the corresponding thermodynamics and obtain a generalized first law for a charged Bose-Einstein Condensate (BEC).

1 Introduction

Nowadays, several astronomical observations are giving evidence for the existence of the so-called dark matter, which occupies approximately 26 % of the Universe; the remaining part being occupied by another dark component, which is usually associated to a fundamental constant. The standard cosmological model, which has been developed for describing the Universe and its evolution, is based on assuming that dark matter is a cold component with a dust-type equation of state and a cosmological constant (Λ) for the other dark component. This model is usually referred to as the ΛCDM model. However, despite the successes of the ΛCDM cosmology

M. A. Rodríguez-Meza (✉)
Departamento de Física, Instituto Nacional de Investigaciones Nucleares, ININ, Km. 36.5, Carretera México-Toluca, 52750 La Marquesa, Estado de México, Mexico
e-mail: marioalberto.rodriguez@inin.gob.mx

T. Matos
Departamento de Física, Centro de Investigación y de Estudios Avanzados (Cinvestav) del Instituto Politécnico Nacional (I.P.N.),
Apartado Postal 14-740, 07000 México, D. F., Mexico
e-mail: tmatos@fis.cinvestav.mx

L. Di G. Sigalotti et al. (eds.), *Computational and Experimental Fluid Mechanics with Applications to Physics, Engineering and the Environment*, Environmental Science and Engineering, DOI:10.1007/978-3-319-00191-3_40,

at describing the large-scale structure of the Universe, the model remains far from perfect. A detail of significant difference comes in the density profiles of dark matter halos, which are predicted by the model to have central cusps, while observations of late-type galaxies tend to favour a constant density core. Alternatively, a scalar field has been proposed as a dark matter model (see Suárez et al. (2013) for a review). In this model, the nature of dark matter is a fundamental real scalar field. In short, the scalar field Lagrangian is comprised of two terms: a kinetic term and a potential term. We then fix the form of the potential, and with the resultant equations we try to explain some observed astronomical data as, for example, the rotation curves of spiral galaxies. In particular, this scalar field model predicts that the galactic dark halos are core-type. We refer the reader to Suárez et al. (2013) for more details as well as for in-depth comparisons between observations and the ΛCDM model.

In this chapter, we extend the model by considering that the scalar field is a complex field and by adding an electromagnetic term in the Lagrangian. This allows us to derive a hydrodynamic version of the scalar field, where the potential exhibits a symmetry breaking and a possible thermodynamics (see Matos and Suárez (2011) for the real case). Symmetry breaking is normally associated to phase transitions in other areas of physics, and therefore its importance.

The chapter is organized as follows. In Sect. 2 we present a general Lagrangian for the complex scalar field, which includes a term with electromagnetic fields, and show how the symmetry appears and is broken. In Sect. 3 we show how the Klein-Gordon (KG) equation transforms into a generalized Gross-Pitaevskii (GP) equation, while in Sects. 4–6 we derive the hydrodynamic version of the generalized GP equation, analyze its Newtonian limit, and derive the corresponding thermodynamics, respectively. Finally, Sect. 7 summarizes the relevant conclusions.

2 Gauge Symmetry Breaking

The Lagrangian model having a local $U(1)$ symmetry is

$$\mathscr{L} = \left(\nabla_\mu \Phi + ieA_\mu \Phi\right)\left(\nabla^\mu \Phi^* - ieA^\mu \Phi^*\right) + V(|\Phi|) - \frac{1}{4}F^{\mu\nu}F_{\mu\nu}, \tag{1}$$

where V is the scalar field potential—a double-well interacting Mexican-hat potential for a complex scalar field (hereafter SF) $\Phi(\mathbf{x}, t)$. The scalar field is in thermal equilibrium with a reservoir at temperature T. The thermal interaction between the scalar field and the thermal bath is computed up to one loop of correction, and is given by Kolb and Turner (1990), Matos and Suárez (2011)

$$V(\Phi) = -m^2\Phi\Phi^* + \frac{\lambda}{2}(\Phi\Phi^*)^2 + \frac{\lambda}{4}T^2\Phi\Phi^* - \frac{\pi^2}{90}T^4. \tag{2}$$

This result includes both quantum and thermal contributions. The parameter m is the mass of the scalar field and λ gives the scalar field self-interaction, which is related to a "pressure" obeying an equation of state of the form $p \propto \rho$. Here we use reduced units so that $\hbar = 1, c = 1$, and $k_B = 1$. We can plot this potential as a function of the complex SF (with $m = 1$ and $\lambda = 1$), and observe that for high enough values of the temperature (say, $T = 5$) the form of the potential close to the minimum resembles a paraboloid of revolution and that when the temperature is lowered the symmetry is broken (around $T = 2$), the minimum at the centre of the complex SF plane vanishes, and there appears the Mexican-hat behaviour ($T = 1$).

The dynamics of a SF is governed by the KG equation, which comes by computing the extremum of the action corresponding to the Lagrangian in Eq. (1). This is the equation of motion of a field composed of spinless particles

$$\Box_E^2 \Phi + \frac{dV}{d\Phi^*} - 2m^2 \varphi \Phi = 0, \tag{3}$$

where we have added a first-order, self-interaction potential φ to the SF, which we will justify below. For a charged field, the D'Alambertian operator is given by

$$\Box_E^2 \equiv \left(\nabla_\mu + ieA_\mu\right)\left(\nabla^\mu + ieA^\mu\right), \tag{4}$$

where $A_\mu = (\boldsymbol{A}, \varphi)$ is the electromagnetic four potential. Notice that we can re-write the D'Alambertian as

$$\Box_E^2 = (\nabla + ie\boldsymbol{A}) \cdot \nabla - (\frac{\partial}{\partial t} + ie\varphi)\frac{\partial}{\partial t} + ie\nabla_\mu A^\mu - e^2 A_\mu A^\mu \,. \tag{5}$$

In what follows, we will use the Lorentz gauge $\nabla_\mu A^\mu = 0$. Moreover, it is convenient to consider the total potential V_T by adding to the potential V the self-interaction contribution and the term $e^2 A_\mu A^\mu = e^2 A^2$ so that

$$\begin{aligned} V_T(\Phi) = & -m^2 \Phi\Phi^* + \frac{\lambda}{4} T^2 \Phi\Phi^* - e^2 A^2 \Phi\Phi^* \\ & + \frac{\lambda}{2}(\Phi\Phi^*)^2 - \frac{\pi^2}{90} T^4 - 2m^2 \varphi\, \Phi\Phi^* \,. \end{aligned} \tag{6}$$

Therefore, the KG equation can be written as

$$\Box^2 \Phi + \frac{dV_T}{d\Phi^*} = 0, \tag{7}$$

where now $\Box^2 = \nabla_\mu \nabla^\mu$. Thus, we can define an effective mass by

$$m_{eff} = \sqrt{m^2 + e^2 A^2}\,, \tag{8}$$

where the electromagnetic potential term is the Proca mass.

The total potential in Eq. (6) has a minimum in $\Phi = 0$ when the temperature $T > T_c$. If $T < T_c$, the point $\Phi = 0$ becomes a maximum and the potential in Eq. (2) has a circular zone of minima with radius

$$\Phi_{min} = \frac{1}{2}\sqrt{T_c^2 - T^2}. \tag{9}$$

The critical temperature where the minimum of the potential $\Phi = 0$ becomes a maximum and the symmetry is broken is given by

$$T_c = \frac{2}{\sqrt{\lambda}}\sqrt{m_{eff}^2 + 2m^2\varphi}. \tag{10}$$

We refer the reader to Castellanos and Matos (2012) for details on BEC and this critical temperature.

3 The Generalized Gross-Pitaevskii Equation

For the SF we now perform the transformation

$$\Phi = \Psi\, \mathrm{e}^{-imc^2t/\hbar}$$

where we have returned to normal units, except for the temperature which is given in energy units ($k_B = 1$).

In terms of the function Ψ and temperature T, the KG equation (3) reads

$$\begin{aligned} \mathrm{i}\hbar\dot{\Psi} + \frac{\hbar^2}{2m}\Box^2\Psi + M\frac{\lambda}{2mc^2}|\Psi|^2\Psi - mc^2\varphi\Psi + ec\varphi\Psi \\ + \frac{\lambda T^2}{8mc^2}\Psi = 0, \end{aligned} \tag{11}$$

where $\dot{\Psi} = \partial\Psi/\partial t$. Note that the complex conjugate equation can be described in the same way.

Equation (11) is an exact equation for the field $\Psi(\mathbf{x}, t)$, where φ defines the external potential acting on the system and the terms containing λ represent the interaction potential within the system. When $T \to 0$ and in the non-relativistic limit, $\Box^2 \to \nabla^2$ and Eq. (11) becomes the GP equation for the Bose-Einstein Condensates (BEC). We also note that the static limit of Eq. (11) corresponds to the well-known Ginzburg-Landau equation. Therefore, we shall consider Eq. (11) as a generalization to the GP equation for the description of a complex charged SF in a thermal bath at finite temperature (Castellanos and Matos 2012).

4 The Hydrodynamic Version

In this section, we transform the generalized GP equation (11) into its analog hydrodynamical version. To do so the ensemble wave function Ψ will be expressed in terms of a modulus n and a phase S as

$$\Psi = \sqrt{n}\,e^{iS}, \tag{12}$$

where the phase $S(\boldsymbol{x}, t)$ is assumed to be a real function. As usual this phase will define the velocity. We can interpret $n(\boldsymbol{x}, t) = \rho/M_T$ as the ratio of the number density of particles in the condensed state, i.e., $\rho = mn_0 = mN_0/L^3$, (with N_0 being the number of particles in condensed state) over the total mass of the particles in the system, M_T. Both S and n are functions of time and position. In passing, we recall that the concept of symmetry breaking is often used as a sufficient condition for BEC (Castellanos and Matos 2012).

We have that the SF can oscillate around the $\Phi = 0$ minimum, while below the critical temperature T_c, SF oscillates close to the "ring" minimum zone. We see that for a real scalar field the oscillation occurs around the minimum values $\Phi^2_{min} = (T_c^2 - T^2)/4$, while the density oscillates around $n = \kappa^2(T_c^2 - T^2)/4$ as we may see from Eq. (9).

In order to derive the hydrodynamic equations, we perform the Madelung transformation (12) in the generalized GP equation (11). After some algebraic steps, we obtain the following equations:

$$\dot{n} + \nabla\cdot(n\mathbf{v}) - \frac{h}{2me}\frac{\partial j}{\partial t} = 0\,, \tag{13a}$$

$$\begin{aligned}\dot{\mathbf{v}} + (\mathbf{v}\cdot\nabla)\mathbf{v} - \frac{ce}{m}(\mathbf{E} + \mathbf{v}\times\mathbf{B}) = \\ -c^2\nabla\varphi + \frac{\lambda}{m^2c^2\kappa^2}\nabla n + \frac{h^2}{m^2}\nabla\left(\frac{\nabla^2\sqrt{n}}{\sqrt{n}}\right) \\ + \frac{h}{2emc^2n}\frac{\partial \mathbf{v} j}{\partial t} - \frac{\hbar^2}{m^2}\nabla\left(\frac{\partial_t^2\sqrt{n}}{\sqrt{n}}\right) + \frac{\lambda}{4m^2}T\nabla T,\end{aligned} \tag{13b}$$

where $\mathbf{E}$ and $\mathbf{B} = \nabla\times\mathbf{A}$ are the electric and the magnetic field vectors, respectively. Here we have also made the following definitions for the fluxes

$$\mathbf{j} = 2en(\nabla S + \frac{e}{\hbar}\mathbf{A}), \quad j = 2en(\dot{S} + \frac{e}{\hbar}\varphi), \quad j_\mu = (\mathbf{j}, j), \tag{14a}$$

and the velocity

$$\mathbf{v} \equiv \frac{\hbar}{m}(\nabla S + e\mathbf{A})\,. \tag{15}$$

Notice that in Eq. (13b) h enters on the right-hand side through the term containing the gradient of $\nabla^2\sqrt{n}/\sqrt{n}$. This term is usually called the "quantum pressure" and is a direct consequence of the Heisenberg uncertainty principle. It reveals the importance of quantum effects in interacting gases. Now, multiplying Eq. (13b) by n, we obtain the form

$$n\dot{\mathbf{v}} + n(\mathbf{v}\cdot\nabla)\mathbf{v} = n\mathbf{F}_E + n\mathbf{F}_\varphi - \nabla p + n\mathbf{F}_Q + \nabla\sigma, \tag{16}$$

where $\mathbf{F}_E = \frac{e}{m}(\mathbf{E} + \mathbf{v}\times\mathbf{B})$ is the electromagnetic force, $\mathbf{F}_\varphi = -\nabla\varphi$ is the force associated to the external potential φ, p is interpreted as the pressure of the SF gas that satisfies the equation of state $p = wn^2$ (where $w = -\lambda/4m^2$ is an interaction parameter), ∇p is the pressure gradient force, $\mathbf{F}_Q = -\nabla U_Q$ is the quantum force associated to the quantum potential

$$U_Q = -\frac{\hbar^2}{2m^2}\left(\frac{\nabla^2\sqrt{n}}{\sqrt{n}}\right), \tag{17}$$

and $\nabla\sigma$ is defined as

$$\begin{aligned}\nabla\sigma = \frac{\hbar}{2me}\dot{\mathbf{v}}j + \frac{1}{4}\frac{\lambda}{m^2}nT\nabla T \\ + \zeta\nabla(\ln n)^{\cdot} - \frac{\hbar^2 n}{4m^2}\nabla\left(\frac{\ddot{n}}{n}\right),\end{aligned} \tag{18}$$

where the coefficient ζ is given by

$$\zeta = \frac{\hbar^2}{4m^2}\left[-\nabla\cdot(n\mathbf{v}) + \frac{\hbar}{2me}j\right],$$

and the term $\nabla(\ln n)^{\cdot}$ can be written as

$$\nabla(\ln n)^{\cdot} = -\nabla(\nabla\cdot\mathbf{v}) - \nabla[\nabla(\ln n)\cdot\mathbf{v}] + \frac{1}{m}\nabla\left[\frac{1}{n}\frac{\partial j}{\partial t}\right].$$

Equations (13) are effectively the hydrodynamical representation of Eq. (11) and they are completely equivalent to it.

5 The Newtonian Limit

The system of equations (13) can be simplified if we neglect second-order time derivatives as well as the products of time derivatives. In this limit we arrive to the Newtonian system of equations

$$\dot{n} + \nabla \cdot (n\mathbf{v}) = 0, \tag{19a}$$

$$n\dot{\mathbf{v}} + n(\mathbf{v} \cdot \nabla)\mathbf{v} = n\mathbf{F}_E + n\mathbf{F}_\varphi - \nabla p + n\mathbf{F}_Q + \nabla\sigma. \tag{19b}$$

Equations (19a) and (19b) are, respectively, the continuity and the momentum equations of fluid mechanics. However, Eq. (19b) contains forces due to the external potential, the pressure gradient, the viscous friction due to the interactions of the condensate, and to the quantum nature of the equations. In particular, the quantity $\nabla(\ln n)^{\cdot}$ plays a very important role. In this limit it reads

$$\nabla(\ln n)^{\cdot} = -\nabla(\nabla \cdot \mathbf{v}) - \nabla[\nabla(\ln n) \cdot \mathbf{v}],$$

so that

$$\nabla\sigma = \frac{1}{4}\frac{\lambda}{m} nT\nabla T - \zeta\,[\nabla(\nabla \cdot \mathbf{v}) + \nabla[\nabla(\ln n) \cdot \mathbf{v}]], \tag{20}$$

with

$$\zeta = -\frac{\hbar^2}{4m^2}\nabla \cdot (n\mathbf{v}).$$

We interpret the term $\nabla\sigma$ as the viscous stress force. It contains terms involving the gradients of the temperature and the divergence of the velocity and density (dissipative contributions). The measurement of the temperature dependence in this thermodynamic quantity at the phase transitions might reveal important information on the behaviour of the gas due to particle interactions.

6 The Thermodynamics

Here we derive the equations of thermodynamics from the hydrodynamic representation. We start by deriving a conservation equation for a function α, using the following relation (Matos and Suárez 2011):

$$(n\alpha)^{\cdot} = n\dot{\alpha} + \alpha\dot{n}, \tag{21}$$

where α can be either φ or U_Q. If we next combine Eq. (21) with the continuity equation

$$\dot{n} + \nabla \cdot (n\mathbf{v}) = 0 \tag{22}$$

we obtain that

$$(n\alpha)^{\cdot} + \nabla \cdot (n\alpha\mathbf{v}) = -n\mathbf{v} \cdot \mathbf{F}_\alpha + n\dot{\alpha}.$$

The treatment of σ using the above procedure is more difficult because in general we do not know it explicitly.

In general, we know that for a non-relativistic system, its total energy density ε is the sum of the kinetic, potential, and internal energy contributions. However, in this case a fourth term, U_Q, should be added due to the quantum potential so that

$$\varepsilon = ne = \frac{1}{2}nv^2 + n\varphi + nu + nU_Q + \psi_E, \tag{23}$$

where u is the internal energy of the system and

$$\Psi_E = \frac{e}{m}(\varphi - \mathbf{v}\cdot\mathbf{A}), \tag{24}$$

is the electromagnetic energy potential, defined in terms of the vector potential $\mathbf{A}$ and the electric potential φ.

Thus, from relation (23) we have that u must satisfy the equation

$$(nu)^{\dot{}} + \nabla\cdot\mathbf{J}_u - \nabla\cdot\mathbf{J}_\rho + n\dot{\varphi} = -p\nabla\cdot\mathbf{v}, \tag{25}$$

where $\mathbf{J}_u$ is the energy current (produced by an energy flux and a heat flux $\mathbf{J}_q$) given by

$$\mathbf{J}_u = nu\mathbf{v} + \mathbf{J}_q + \mathbf{J}_B - p\mathbf{v}$$

where $\nabla\cdot\mathbf{J}_q = \mathbf{v}\cdot(\nabla\sigma)$ and $\nabla\cdot\mathbf{J}_B = \mathbf{v}\cdot(n\mathbf{j}_B)$. Here $\mathbf{j}_B$ is given by the continuity-like equation for the vector potential $\mathbf{A}$

$$\frac{\partial\mathbf{A}}{\partial t} + (\mathbf{v}\cdot\nabla)\mathbf{A} = -(\mathbf{A}\cdot\nabla)\mathbf{v} + \frac{m}{e}\mathbf{j}_B. \tag{26}$$

The above expression for the energy current is related to the velocity and temperature gradient in the condensate. It shows explicitly the temperature dependence of the thermodynamic equations. With these definitions at hand we can re-write Eq. (25) as

$$(nu)^{\dot{}} + \nabla\cdot(nu\mathbf{v} + \mathbf{J}_q + \mathbf{J}_B - p\mathbf{v} - \mathbf{J}_\rho) + n\dot{\varphi} = -p\nabla\cdot\mathbf{v}. \tag{27}$$

To find the thermodynamical quantities of the system in equilibrium (by taking p as a constant on a volume L), we restrict the system to the regime where the auto-interacting potential is constant in time. With this provision Eq. (27) becomes

$$(nu)^{\dot{}} + \nabla\cdot(nu\mathbf{v} + \mathbf{J}_q + \mathbf{J}_B - p\mathbf{v} - \mathbf{J}_\rho) = -p\nabla\cdot\mathbf{v}, \tag{28}$$

where the terms on the left-hand side represent the change in the internal energy of the system, $-p\nabla\cdot\mathbf{v}$ is the work done by the pressure, and $\nabla\cdot\mathbf{v}$ is related to changes in the volume. The flux $\mathbf{J}_q$ contains terms related to the heat generated by the temperature gradients ∇T and to dissipative (viscous) forces $\sim\nabla(\nabla\cdot\mathbf{v})$. Finally, but not less important, we have the flux term $\mathbf{J}_\rho$, which owes its existence to the gradients of the quantum potential defined by relation (17).

Integration of Eq. (28) over a closed volume V yields

$$\frac{\mathrm{d}}{\mathrm{d}t}\int nu\,\mathrm{d}V + \oint(\mathbf{J}_q + \mathbf{J}_B + p\mathbf{v})\cdot\mathbf{n}\,\mathrm{d}S - \oint \mathbf{J}_\rho\cdot\mathbf{n}\,\mathrm{d}S = -p\frac{\mathrm{d}}{\mathrm{d}t}\int \mathrm{d}V,$$

where S is the surface enclosing the volume region and $\mathbf{n}$ is its unit normal vector. This equation, which is a continuity-like equation for the internal energy, describes the thermodynamics of the system in a way analogous to the first law of thermodynamics (for the KG equation or a BEC), which reads

$$\mathrm{d}U = \hat{\mathrm{d}}Q + \hat{\mathrm{d}}Q_B + \hat{\mathrm{d}}A_Q - p\mathrm{d}V, \tag{29}$$

where the internal energy is $U = \int nu\,\mathrm{d}V$. Its change is due to a combination of heat $Q + Q_B + A_Q$ added to the system and work done on the system (pressure dependent). In particular,

$$\frac{\hat{\mathrm{d}}A_Q}{\mathrm{d}t} = \frac{\hbar^2}{4m^2}\oint n(\nabla\ln n)^{\cdot}\cdot\mathbf{n}\mathrm{d}S = \oint n\mathbf{v}_\rho\cdot\mathbf{n}\,\mathrm{d}S,$$

is the quantum heat flux due to the quantum nature of the KG equation.

The first and third terms on the right-hand side of Eq. (29) make the crucial difference between a classical and a quantum first law of thermodynamics. Furthermore, for the magnetic heat we have

$$\begin{aligned}\frac{\hat{\mathrm{d}}Q_B}{\mathrm{d}t} &= \int \nabla\cdot\mathbf{J}_B\,dV = \int \mathbf{v}\cdot(n\mathbf{j}_B)\,dV \\ &= \frac{m}{e}\int n\left(\frac{\partial\mathbf{A}}{\partial t} + (\mathbf{v}\cdot\nabla)\mathbf{A} + (\mathbf{A}\cdot\nabla)\mathbf{v}\right)\cdot\mathbf{v}\,dV,\end{aligned} \tag{30}$$

where the vector potential $\mathbf{A}$ fulfils the Maxwell equations, which in terms of the fluxes (14) reads

$$F^{\mu\nu}{}_{,\nu} = -j_\nu, \tag{31}$$

where $F_{\mu\nu} = \partial_\mu A_\nu - \partial_\nu A_\mu$.

To complete the description we write the Maxwell equations in terms of the vector and the electric potentials

$$\Box \mathbf{A} = -\frac{1}{\kappa^2}\mathbf{j}, \tag{32a}$$

$$\Box \varphi = -\frac{1}{\kappa^2} j, \tag{32b}$$

where we have used the Lorentz gauge. Note that the fluxes contain information on the fluid velocity and the electromagnetic field as well.

7 Conclusions

We have studied a $U(1)$ symmetry breaking term and the temperature contribution to the effective Mexican-hat potential of a system of weakly interacting bosons, and derived the hydrodynamic version of the complex scalar field equation. In addition, from the hydrodynamic form of the equations we have also derived the corresponding thermodynamics for the Gross-Pitaevskii equation.

References

Castellanos E, Matos T (2012) Critical temperature associated to symmetry breaking of Klein-Gordon fields versus condensation temperature in a weakly interacting Bose-Einstein gas. arXiv:1202.3793v2, 6 pp

Kolb EW, Turner MS (1990) The early universe. Addison-Wesley Publishing, New York

Matos T, Suárez A (2011) Finite temperature and dissipative corrections to the Gross-Pitaevskii equation from $\lambda\phi^4$ one-loop contributions. Europhys Lett 96:56005

Suárez A, Robles VH, Matos T (2013) A review of the scalar field/Bose-Einstein condensate dark matter model. arXiv:1302.0903v1, 32 pp.

MIX
Papier aus verantwortungsvollen Quellen
Paper from responsible sources
FSC® C105338

If you have any concerns about our products,
you can contact us on
ProductSafety@springernature.com

In case Publisher is established outside the EU,
the EU authorized representative is:
Springer Nature Customer Service Center GmbH
Europaplatz 3, 69115 Heidelberg, Germany

Printed by Libri Plureos GmbH
in Hamburg, Germany